$Q_1, Q_2,$ and Q_3	Quartiles	55
r	Coefficient of correlation	425
R_1, R_2, R_3	Rank sums	448
$r \times c$ table	Contingency table	377
r_s	Rank-correlation coefficient	455
ρ (rho)	Population correlation coefficient	428
s	Sample standard deviation	69
s^2	Sample variance	69
σ (sigma)	Standard deviation of population, probability distribution, or continuous distribution	68, 221, 237
σ^2	Variance of population, probability distribution, or continuous distribution	69, 221, 237
Σ (sigma)	Summation	45
$\sigma_{\bar{x}}$	Standard deviation of sampling distribution of $\bar{x}$	272
s_e	Standard error of estimate	418
SK	Pearsonian coefficient of skewness	86
SSE	Error sum of squares	350
SST	Total sum of squares	350
SS(Tr)	Treatment sum of squares	350
S_{xx}	Statistic for the computing formula of s and for least-squares calculations	70, 41
S_{yy}, S_{xy}	Statistics for computing formulas of least-squares calculations	411
t	t statistic	297
t_α	Critical value of t	298
$T_{i.}$	Total of values in ith sample	353
$T_{..}$	Grand total for all the samples	353
u	Total number of runs	450
$u_{\alpha/2}$ and $u'_{\alpha/2}$	Critical values of u	450
$U_1, U_2,$ or U	Statistics for U test	443
U'_α	Critical value of U	443
$\cup$	Union, commonly read as *or*	146
V	Coefficient of variation	76
$\dfrac{x}{n}$	Sample proportion	304
$\bar{x}$	Sample mean	45
$\bar{\bar{x}}$	Grand mean of combined data	48
$\tilde{x}$	Sample median	52
$\bar{x}_w$	Weighted mean	42
$\hat{y}$	Value of y on regression line	408
z	Standard unit	76
z	z statistic	237
z_α	Critical value of z	245

Statistics
A First Course
Seventh Edition

John E. Freund
Arizona State University

Benjamin M. Perles
Suffolk University

Prentice Hall
Upper Saddle River, New Jersey 07458

Library of Congress Cataloging-in-Publication Data

Freund, John E.
   Statistics: a first course.—7th ed. / John E. Freund, Benjamin
Perles.
     p.   cm.
   Includes bibliographical references and index.
   ISBN 0-13-959909-6
   1. Statistics.   I. Perles, Benjamin M.   II. Title.
QA276.12.F74   1999
519.5—dc21   98-35505
                                  CIP

Executive Editor: **Ann Heath**
Editorial Assistant/Supplements: **Mindy McClard**
Editorial Director: **Tim Bozik**
Editor-in-Chief: **Jerome Grant**
AVP, Production and Manufacturing: **David W. Riccardi**
Production Editor: **Elaine W. Wetterau**
Senior Managing Editor: **Linda Mihatov Behrens**
Executive Managing Editor: **Kathleen Schiaparelli**
Marketing Manager: **Melody Marcus**
Marketing Assistant: **Amy Lysik**
Manufacturing Buyer: **Alan Fischer**
Manufacturing Manager: **Trudy Pisciotti**
Art Director: **Maureen Eide**
Assistant to Art Director: **John Christiana**
Associate Creative Director: **Amy Rosen**
Director of Creative Services: **Paula Maylahn**
Art Editor: **Michele Giusti**
Art Manager: **Gus Vibal**
Interior Designer: **Rosemarie Votta**
Compositor: **Preparé/Emilcomp**
Cover Credit: **Telegraph Colour Library/FPG International LLC**

ISBN 0-13-959909-6

Prentice-Hall International (UK) Limited, *London*
Prentice-Hall of Australia Pty. Limited, *Sydney*
Prentice-Hall Canada, Inc., *Toronto*
Prentice-Hall Hispanoamericana, S.A., *Mexico*
Prentice-Hall of India Private Limited, *New Delhi*
Prentice-Hall of Japan, Inc., *Tokyo*
Simon & Schuster Asia Pte. Ltd., *Singapore*
Editora Prentice-Hall do Brasil, Ltda., *Rio de Janeiro*

Contents

Preface

Since the first edition of this book was published, there have been great changes in the teaching of statistics and in textbooks designed to support these efforts. Not only has there been a pronounced shift in emphasis from descriptive methods toward inferential statistics, but there has been a change in the level at which first courses are being taught.

Formerly directed to college juniors and seniors, statistics is now also taught to college freshmen and sophomores, and in some advanced high-school programs. Our aim, as in the preceding editions, is to reach the student earlier in his or her education. We meet this objective through the organization of the subject matter, simplified language, the format and notation, and, above all, through a wealth of examples and exercises. Attention to mathematical precision has been enhanced without complicating the presentation. This well-integrated, carefully crafted textbook provides a strong underpinning for subsequent academic and professional activities.

Exercises are placed at the ends of sections to make it easier for both students and instructors to locate exercises relevant to given topics. There are five review segments intended to reinforce the concepts of immediately preceding chapters.

Important formulas, definitions, and rules are highlighted within colored boxes. The frequently used tables for normal and t distributions are repeated on the inside back cover; and the inside front cover contains an indexed glossary of statistical symbols.

As in the previous editions, controversy has not been avoided. The reader is exposed to the weaknesses of statistical techniques, as well as their strengths. It is hoped that this honest approach will provide a stimulus as well as a challenge.

New to This Edition

The Seventh Edition has been thoroughly revised and includes many incremental improvements in the organization and content. Major changes include:

- Three new sections have been added to Chapter 1 to provide a more complete introduction to the types of data and data collection.
- A new section on Dot Diagrams has been added to Chapter 2.
- The coverage of Estimation and Hypothesis Testing has been reorganized. Confidence Intervals are now presented in a separate chapter, followed by several chapters devoted to hypothesis testing.

The problem sets have been extensively revised.

- More than 30% of the problems are new or revised.
- Many additional real data based problems have been added.
- Internal referencing within the problem sets has been minimized. Where

necessary, pertinent data has been repeated to free students from having to flip back to previous problems.

Increased use of Technology has been added.

- The text is illustrated throughout with computer output. This edition uses, but does not require MINITAB. Minitab printouts, showing both output and session commands, are set off in boxes and are referenced in the examples. MINITAB users will be pleased by the new opportunity for students to purchase this textbook packaged with a CD-Rom containing MINITAB Release 12.0 Student Edition and the data files for this book.

- Although it is not required, this edition introduces the graphing calculator and acquaints the reader with its numerous benefits. **Using Technology: The Graphing Calculator** boxes appear throughout the book. These boxes provide keystroke-level instruction for using the TI-83 to find solutions to selected examples in the book.

- A new **Appendix A: TI-83 Tips**, written by Wendy Metzger of Palomar College, provides additional help in using the graphing calculator.

- Many new problems are designed to be solved with the aid of a computer or a graphing calculator and are identified with the technology 🖫 icon.

HALLMARK FEATURES

The entire text is built around the authors' goal of presenting statistics at a level that is appropriate and interesting for beginning students. *In spite of their obvious value, computers and graphing calculators can safely be eliminated by readers of this textbook without loss of continuity.*

Review Sections are presented at two-or-three-chapter intervals throughout the book. Five review sections reinforce connections outside the context of a particular section or chapter. Identified with a pale blue screen, they include:

- a list of Achievements that students should have mastered after studying the group of chapters;

- a checklist of key terms annotated with specific page references;

- a cumulative section of Review Exercises that tests concepts from the immediately preceding chapters.

Representative problems within each exercise set are highlighted with a ✓ to identify them as **Practice Exercises**. Fully worked out solutions to these exercises are presented at the end of each chapter in the **Solutions to Practice Exercises** section.

Supplements for the Instructor

The supplements for the seventh edition have been revised to reflect the revisions of the text.

Instructor's Solutions Manual (by Benjamin M. Perles and John E. Freund) (ISBN: 0-13-020147-2)
Solutions to all of the exercises are given in this manual. Careful attention has been paid to ensure that all methods of solution and notation are consistent with those used in the core text.

Test Item File (ISBN: 0-13-020145-6) (by Daniel Mihalko, Western Michigan University)
Thoroughly revised, the Test Bank includes Multiple Choice, Fill-in, and True/False questions that correlate to problems presented in the text.

Windows PH Custom Test (ISBN: 0-13-020634-2)
Incorporates three levels of test creation: (1) selection of questions from a test bank, (2) addition of new questions with the ability to import test and graphics files from WordPerfect, Microsoft Word, and Wordstar, and (3) algorithmic generation of multiple questions from a single-question template. The PH Custom Test has a full-featured graphics editor supporting the complex formulas and graphics required by the statistics discipline.

Data Disk (ISBN: 0-13-020144-8)
The larger data sets used in problems and exercises in the book are available on a data disk. The Data files may also be downloaded from the Prentice Hall ftp site://www.prenhall.com/pub/esm/statistics.s027/freund/statistics7e/data.

New York Times Supplement (ISBN: 0-13-973520-8)
Copies of this supplement may be requested from Prentice Hall by instructors for distribution in their classes. This supplement contains high-interest articles published recently in the *New York Times* that relate to topics covered in the text.

Supplements Available for Purchase by Students

Student's Solution Manual (by Benjamin M. Perles and John E. Freund) (ISBN: 0-13-020146-4)
Fully worked out solutions to all of the odd-numbered exercises are provided in this manual. Careful attention has been paid to ensure that all methods of solution and notation are consistent with those used in the core text.

Text and MINITAB 12.0 Student Edition Integrated Package (ISBN: 0-13-021376-4)
A CD-Rom containing the MINITAB Release 12.0 Student Edition and the data files from the text may be purchased as a package with the textbook for a small additional charge.

A MINITAB Guide to Statistics (by Ruth Meyer and David Krueger) (ISBN: 0-13-784232-5)
This manual assumes no prior knowledge of MINITAB. Organized to correspond to the table of contents of most statistics texts, this manual provides step-by-step instruction for using MINITAB in statistical analysis.

TI-83 Graphing Calculator Manual for Statistics (by Stephen Kelly) (ISBN: 0-13-020911-2)
This brief, spiral bound manual provides simple, keystroke instructions for using the TI-83 graphing calculator in statistics. Class tested for many years, this manual is the perfect answer for frustrated students and professors.

Student Versions of SPSS and SYSTAT
Student versions of SPSS, the award-winning and market-leading commercial data analysis package, and SYSTAT are available for student purchase. Details on all current products, including integrated packages, are available from the publisher or via the SPSS website at http://www.spss.com.

Text and SPSS 8.0 for Windows Student Version Integrated Package (ISBN: 0-13-021378-0)

ConStatS (by Tufts University) (ISBN: 0-13-502600-8)
ConStatS is a set of Microsoft-Windows-based programs designed to help college students understand concepts taught in a first-semester course on probability and statistics. ConStatS helps to improve students' conceptual understanding of statistics by engaging them in an active, experimental style of learning. A companion ConStatS workbook (ISBN: 0-13-522848-4) that guides students through the laboratories and ensures that they gain the maximum benefit is also available.

For additional information about texts and other materials available from Prentice Hall, visit us on-line at http//.www@prenhall.com.

ACKNOWLEDGMENTS

The authors are indebted to Professor E. S. Pearson and the Biometrika trustees for permission to reproduce parts of Tables 8 and 18 from their *Biometrika Tables for Statisticians*; to Prentice-Hall, Inc., to reproduce part of Table 2 of R. A. Johnson and W. W. Wichern's *Applied Multivariate Statistical Analysis*; to the Addison-Wesley Publishing Company to base Table VII on Table 11.4 of D. B. Owen's *Handbook of Statistical Tables*; and to the editor of the *Annals of Mathematical Statistics* to reproduce the material in Table VIII. The changes in this edition are the authors' responsibility.

We express appreciation to Wendy Metzger for writing Appendix A: TI-83 Tips and to Sarah Street for accuracy checking the text and solutions manuals. The Prentice Hall staff of Ann Heath, Mindy McClard, Elaine Wetterau, Linda Behrens, Alan Fischer, Maureen Eide, Melody Marcus, and Amy Lysik helped greatly with all phases of the text development, production, and marketing effort.

Our gratitude is also extended to the many students and colleagues whose suggestions have contributed greatly to this and to previous editions; in particular to E. Ray Bobo, Georgetown University; Tom Brieske, Georgia State University; Neil S. Dickson, Weber State University; Dale O. Everson, University of Idaho; Elizabeth Farber, Bucks County Community College; David A. Ford, Emory University; Ronald Friesen, Bluffton College; Frank

Gunnip, Oakland Community College; Steven L. Harvey, Seminole Junior College; John Heinen, Regis College; Tom Hogenkamp, Erie Community College; John Van Iwaarden, Hope College; Elizabeth S. Low, University of Vermont; Raymond Mueller, DeVry Institute of Technology; Iosif Pinelis, Michigan Technological University; Mark Sherrin, Flagler College; James Siefert, ICM School of Business; Lloyd B. Smith, Jr., Lenoir-Rhyne College; Jake Uhrich, South Puget Sound Community College; Ian Walton, Mission College; and Arthur Larry Wright, University of Arizona.

JOHN E. FREUND
BENJAMIN M. PERLES

1 Introduction

Statistics deals with all aspects of the collection, processing, presentation, and interpretation of measurements or observations, that is, with all aspects of the handling of **data**. Thus, data constitutes the raw material we deal with in statistics, and its collection is of major concern in any statistical investigation. Data is obtained by taking measurements, by counting, by asking questions, or by referring to data made available in published form. Note that we said "data constitutes" and "data is," even though *data* is the plural form of *datum*, a term that in actual practice is rarely used. We do this because all the data is usually regarded as one unit of information and, hence, used with a singular verb.[†]

In view of the importance of data as the raw material that we deal with in statistics, the first four sections of this chapter concern various aspects of data—whether data is numerical or, originally at least, not numerical; to what extent numerical data can be treated by arithmetical techniques; whether data constitutes a sample; and whether data may be regarded as being good (usable for what it is intended) or bad (perhaps unintentionally biased or intentionally misleading).

[†] In more formal writing the word *data* is plural, and we would write "Thus, these data constitute the raw material...."

In Sections 1.5, 1.6, and 1.7 we shall discuss some general questions about statistics, its past and present, its study, and its future development.

1.1 Numerical Data and Categorical Data

Fundamentally, there are two types of data: numerical and categorical. **Numerical data** contains numbers that can be treated by ordinary arithmetical methods. For instance, if we counted the number of passengers on three buses, we might get 32, 41, and 28 passengers. To obtain the total number of passengers, we simply add the values 32, 41, and 28 and obtain 101. If necessary, we can multiply, divide, subtract, raise to powers, and extract roots of these values.

Categorical data results from data being sorted into nonnumerical categories. For instance, if an interviewer determines whether a person is single (never married), married, a widow or widower, or divorced, this information has been categorized. Questions that result in a choice of answers such as yes or no; agree or disagree; true or false; or poorly, acceptably; or superior tasting; result in categorical data. For ease in manipulating categorical data, it is sometimes coded. In the marital status illustration, the categories of single (never married), married, widow or widower, or divorced could be assigned code numbers of 1, 2, 3, and 4, respectively. Exercise 1.3 gives a simple example of how coding may be used.

1.2 Nominal, Ordinal, Interval, and Ratio Data

Data may also be classified as nominal, ordinal, interval, or ratio. The coded numbers of the previously described marital data are referred to as **nominal data** and are incapable of being manipulated arithmetically. Other examples of nominal data include the numbers on the jerseys of football players, and the identification of undergraduate college students as freshmen, sophomores, juniors, or seniors. Nominal data is the weakest level of measurement. For the marital data, we cannot add the coded numbers 1 and 2 to obtain 3, since this would result in the nonsensical statement that single + married = widow or widower. Nominal data, however, focuses on the frequency in a category and shows clearly the number of respondents who fall into each marital class.

Sometimes data is recorded in numerical form, but restated in categorical form. A meteorologist might record air temperatures in degrees Fahrenheit, but the resulting weather report might describe the data categorically as hot, cold, or normal for this time of year.

Ordinal data can be rank ordered. For instance, the five teams constituting a baseball league might be individually ranked as being in first place, in second place, in third place, in fourth place, and in last place. In connection with ordinal data, **inequality signs** can be used, but $<$ and $>$ have a broader meaning than the usual "less than" and "greater than." The sign $>$ can mean happier than, more painful than, has a better flavor than, louder than, or harder than, while $<$ would have the opposite meaning. In the army the rank of general $>$ colonel $>$ major $>$ captain $>$ first lieutenant $>$ second lieutenant. In finance, according to a recent issue of *Fortune* magazine, the six largest U.S. corporations ranked by revenue are General Motors $>$ Ford Motor $>$ Exxon $>$ Wal-Mart $>$ General Electric $>$ International Business Machines.

If we can form differences, but not multiply or divide, we refer to the data as **interval data**. Suppose that a meteorological map gives us temperatures of $40°$, $60°$, $80°$, and $100°F$. We can say that $40°F < 60°F$, which means that $40°F$ is colder than 60°F; or we can say that 60°F > 40°F, which means that 60°F is warmer than 40°F. Or we can say that 100°F − 80°F = 60°F − 40°F, which means that the temperature difference of 20°F is the same on both sides of the equation. But we cannot say that 80°F is twice as hot as 40°F, although $\frac{80.00}{40.00} = 2.00$. The falsity of this statement can be seen if we convert the two temperatures to the Celsius scale, where 80°F corresponds to 26.67°C, and 40°F corresponds to 4.44°C. If we divide the two temperatures as before, we get $\frac{26.67}{4.44} \approx 6.01$, which implies that 26.67 is 6.01 times hotter than 4.44, and not merely twice as hot. Actually, both statements are incorrect, and if you wish to pursue this matter further, you would need to know that absolute zero (0°A), which is the lowest theoretical temperature that a gas can reach, is − 460° on the Fahrenheit scale and − 273° on the Celsius scale. The zero values for the Fahrenheit, Celsius, and absolute scales are found at different temperatures.

For **ratio data** we can form quotients, that is, divide one quantity by another quantity of the same kind. Both the dividend and divisor must be expressed in the same units, such as inches, gallons, or dollars, and the unit must have a true zero as its origin. The resulting ratio is a pure, or abstract, number and is not expressed in units. The ratio of $2 to $5 is expressed as $\frac{2}{5}$ or 0.40. Note that the ratio omits the dollar sign.

According to U.S. law, the ratio of the length to the width of an official U.S. flag must be 1.9. If an official flag is 2 feet wide, it must therefore be 2 × 1.9 = 3.8 feet long. If it is 150 centimeters wide, it must be 150 × 1.9 = 285 centimeters long.

1.3 Sample Data and Populations

In the introduction to this chapter we said that "statistics deals with all aspects of the collection, processing, presentation, and interpretation of measurements or observations, that is, with all aspects of the handling of data." This broad definition of statistics includes both populations and samples, which we distinguish as follows:

> If a set of data consists of all conceivably possible (or hypothetically possible) observations of a certain phenomenon, we call it a "**population**"; if a set of data consists of only a part of these observations, we call it a **sample**.

We added the phrase "hypothetically possible" to take care of such clearly hypothetical situations in which we consider the outcomes (heads or tails) of 12 flips of a coin to be a sample from the population of all possible flips of the coin, or we consider the weights of ten 30-day-old calves to be a sample of the weights of all (past, present, and future) 30-day-old calves, or we consider four determinations of the copper content of an ore to be a sample of all possible determinations of the copper content of the ore. In fact, we often consider the results of an experiment to be a sample of what we might obtain if the experiment were repeated over and over again.

The term *population* comes from the origin of statistics, which attempted to describe human populations. Even though it may sound strange to refer to the heights of all the trees in a forest as a "population of heights" or to the speeds of all the cars passing a checkpoint as a "population of speeds," in statistics *population* is a technical term with a meaning of its own.

Although we are free to call any set of data a population, what we do in practice depends on the context in which the data is to be viewed. For instance, the complete figures for a recent year, giving the weekly amounts of sales tax collected by the 32 restaurants of a fast-food chain, can be viewed as either a population or a sample. If a tax collector, auditing the records of this fast-food chain, is interested only in the figures for the given year, they constitute a population; on the other hand, if the management of the chain wants to estimate future collections of sales tax or guess at the amounts of sales tax that were collected by its competitors, the original data must be viewed as a sample.

■ **EXAMPLE**   According to a recent edition of *Current Population Reports* by the U.S. Bureau of the Census, the following are the percentages of persons not covered by health insurance in the New England states.

State	Percent not covered
Massachusetts	12.5
Rhode Island	11.5
Connecticut	10.4
Maine	13.1
New Hampshire	11.9
Vermont	8.6

Give one example each of a problem in which this set of data would be viewed as

(a) a population;

(b) a sample.

Solution   (a) The data would constitute a population if a health insurance company used it to explore the possibility of marketing its health insurance policies exclusively in New England.

(b) The data would constitute a sample if the health insurance company used it to estimate the percentage of people not covered by health insurance in the United States.   ■

1.4 Biased Data

In all statistical studies that use samples, great care must be exercised to ensure that the data lends itself to valid generalizations. A key issue here is the question of **bias**. A sample is said to be biased if it is not representative of the population that it is supposed to represent. Every precaution must be taken to avoid inadvertent biases. It is, of course, unethical to introduce deliberate biases to prove particular points.

Direct information is usually gathered by personal interviews, telephone calls, mail questionnaires, or any combination of the three. There is no general answer as to which method is best, although mail questionnaires are often preferred, because they are easier to get into homes and businesses that are widely dispersed. For one reason or another, however, many people fail to return even well-constructed questionnaires unless the questions are devoted to matters in which they are interested. This may introduce a bias that makes it very difficult to arrive at valid conclusions. Whether used for personal interviews or sent through the mail, the construction of questionnaires is the subject of extensive study. Statistical investigations have often been challenged on the grounds that questions were not properly asked. The question "Why do you prefer Coca-Cola to Pepsi-Cola?" is apt to elicit a different reply than "Which do you prefer, Coca-Cola or Pepsi-Cola?"

Mail questionnaires are often cheaper and easier to distribute, but an advantage of the personal interview is that a tactful interviewer can often obtain answers to personal questions relating to age, health, and income that many persons would refuse to answer in a mail questionnaire.

Telephone calls are often used to get on-the-spot responses, for example, to television viewing and other habits. It is exposed to the criticism, however, that it does not reach homes where there are no telephones and thus eliminates a portion of persons who may be relevant to the survey. The widespread use of telephone answering machines in homes and businesses also reduces the response rate. Also, telephone calls made during normal working hours may fail to reach residents, especially in homes where there is more than one worker in the family.

Exercises

Exercises 1.1, 1.4, *and* 1.7 *are practice exercises*; *their complete solutions are given on page* 12.

1.1 Which of the following result in numerical data and which result in categorical data?
  (a) Counting the number of potatoes in a sack of potatoes.
  (b) Measuring the number of gallons of water that can flow through a pipe in a minute.
  (c) Participating in a taste test for a new brand of soft drink for which the respondent's choices are poor flavor, satisfactory flavor, and very good flavor.
  (d) Responding to a mail questionnaire in which we are asked to approve or disapprove each statement in a list of 10 statements.
  (e) Medical doctors implementing a system of priorities for treatment of victims following a major disaster.

1.2 Which of the following are nominal data and which are ordinal data?
  (a) Coding of military personnel by branch of military service for use in a computer analysis, if 1 represents Air Force, 2 represents Army, 3 represents Coast Guard, 4 represents Marine Corps, and 5 represents Navy.
  (b) The numbers on the jerseys of basketball players.
  (c) The numbers on a lottery ticket on which the buyer is permitted to select his or her own set of numbers.
  (d) Students of a graduating college class are ranked as first in class, second in class, third in class, ….
  (e) Academic ranks of faculty as professor, associate professor, assistant professor, and instructor.

1.3 Referring to the marital status categories in Section 1.1, suppose that we code *single* (never married) as 1, *married* as 2, *widow or widower* as 3, and *divorced* as 4. If an interviewer obtains marital information from 24 respondents, determine how many are *single* (never married), *married*, *widow or widower*, and *divorced* if they are coded as follows:

1	3	2	2	2	1	4	2	3	4	1	4
3	1	1	2	1	4	1	3	2	1	1	4

1.4 Suppose that we are given complete information about the number of life insurance policies sold by the Metropolitan Life Insurance Company during a recent year and the amount of each policy. Give one example each of problems in which we would consider this set of data to be
   (a) a population;   (b) a sample.

1.5 Suppose that we are given the College Board scores of all freshmen entering a certain university in a recent year. Give one example each of problems in which we would consider this set of data to be
   (a) a population;   (b) a sample.

1.6 Suppose that we are given complete information about salaries paid to secretaries in Chicago, Illinois, in a recent month. Give one example each of problems in which we would consider this set of data to be
   (a) a population;   (b) a sample.

1.7 "Bad" statistics may well result from asking questions in the wrong way or of the wrong persons. Explain why the following may lead to useless data.
   (a) To determine public sentiment about a certain foreign trade restriction, an interviewer asks voters, "Do you feel that this unfair practice should be stopped?"
   (b) To predict a municipal election, a public opinion poll telephones persons selected haphazardly from the city's telephone directory.
   (c) In a study of art appreciation, persons are asked whether they like Indian art.

1.8 "Bad" statistics may also result from asking questions in the wrong place or at the wrong time. Explain why the following may lead to useless data.
   (a) A house-to-house survey is made during weekday mornings to study consumer reaction to certain convenience foods.
   (b) To predict an election, a poll taker interviews persons coming out of a building that houses the national headquarters of a political party.
   (c) To determine what the average person spends on a vacation, a researcher interviews the passengers on a luxury cruise.

1.9 Explain why the following may lead to useless data.
   (a) To determine the proportion of improperly sealed cans of coffee, a quality-control inspector examines every fiftieth can coming off a production line.
   (b) To determine the average annual income of its graduates 10 years after graduation, a college's alumni office sent questionnaires in 1999 to all members of the class of 1989, and the estimate was based on the questionnaires returned.
   (c) To study executives' reaction to its copying machines, the Xerox corporation hires a research organization to ask executives the question, "How do you like using Xerox copies?"

1.5
Statistics, Past and Present

The origin of the material we shall study in this book may be traced to two areas of interest, which, on the surface, have very little in common: government (political science) and games of chance.

Governments have long used censuses to count persons and property. A famous example is the census reported in the *Domesday Book* of William of Normandy, completed in the year 1086, which covered most of England, listing its economic resources and including property owners and the land they owned. In the first U.S. census in 1790, government agents merely counted the population; but more recent U.S. censuses have become much wider in scope, providing a wealth of information about the population and the economy, and they are conducted every 10 years. Census data is collected in years ending in zero such as 1980, 1990, 2000, and 2010.

The problem of describing, summarizing, and analyzing census data led to the development of methods that, until recently, constituted almost all there was to the subject of statistics. These methods, which originally consisted mainly of presenting data in the form of tables and charts, constitute what we now call **descriptive statistics**. This includes anything done to data that is designed to summarize or describe it without going any further, that is, without trying to infer anything that goes beyond the data itself. For instance, if a newspaper reports net paid circulations of 172,316 in 1989 and 207,185 in 1999 and we perform the calculation to show that there was an increase of 20.2%, our work belongs to the field of descriptive statistics. This would not be the case, however, if we used the given data to predict the newspaper's circulation in the year 2008.

Although descriptive statistics is an important branch of statistics and continues to be widely used, statistical information usually arises from samples (from observations made on only part of a large set of items), and thus its analysis will require generalizations that go beyond the data. As a result, an important feature of the growth of statistics in this century has been the shift in emphasis from methods that merely describe to methods of **statistical inference** that serve to make generalizations.

Such methods are required, for instance, to predict the operating life span of a sewing machine (on the basis of the performance of several such machines); to estimate the assessed value of all privately owned property in Orange County, California in the year 2005 (on the basis of business trends, population projections, and so forth); to compare the effectiveness of two reducing diets (on the basis of the weight losses of persons who have been on the diets); to determine the optimum dose of a medication (on the basis of tests performed with volunteer patients from selected hospitals); or to predict the flow of traffic on a freeway that has not yet been built (on the basis of past traffic counts on alternate routes).

In each situation described in the preceding paragraph, there are uncertainties because only partial, incomplete, or indirect information is available, and it is with the use of the methods of statistical inference that we judge the merits of the results and, perhaps, suggest a "most profitable" choice, a "most promising" prediction, or a "most reasonable" course of action.

In view of the uncertainties, we handle problems like these with statistical methods that find their origin in games of chance. Although the mathematical study of games of chance dates back to the seventeenth century, it was not until the early part of the nineteenth century that the theory developed for "heads or tails," "red or black," or "even or odd" was also applied to real-life situations in which the outcomes were "boy or girl," "life or death," or "pass or fail." Thus, **probability theory** was applied to many problems in the behavioral, natural, and social sciences, and nowadays it provides an important tool for the analysis of any situation (in science, business, or everyday life) that in some way involves an element of uncertainty or chance. In particular, it provides the basis for the methods we use when we generalize from observed data, that is, when we use the methods of statistical inference.

1.6
The Study of Statistics

The scope of statistics and the need to study statistics have grown enormously in the last few decades for two reasons. One is the increasingly quantitative approach employed in all the sciences, as well as in business and many other activities that directly affect our lives. This includes the use of mathematical techniques in such problems as the evaluation of antipollution controls, inventory planning, the analysis of cloud formations, the study of diet and longevity, and the evaluation of teaching techniques.

The other reason is that the amount of statistical information that is collected, processed, and disseminated to the public for one reason or another has increased almost beyond comprehension, and what part of it is "good" statistics and what part is "bad" statistics is anybody's guess. To act as watchdogs, more and more persons with some knowledge of statistics are needed to take an active part in the collection and statistical analysis of the data and, what is equally important, in all the preliminary planning. Without the latter, it is frightening to think of all the things that can go wrong. The results of costly studies can be completely useless if, for example, questions are ambiguous or asked in the wrong way, instruments are poorly adjusted, or all relevant factors are not taken into account.

This book is a general statistics book that provides preparation in the basic statistics underlying the many specialties in which a student may major. Therefore, the exercises and examples utilized throughout the text are drawn from an assortment of disciplines. This is possible because the solution of statistical problems is very much the same regardless of the data source.

To clarify this point, let us examine the following example, which comes from the field of nutrition (see Section 11.2).

> Suppose that a nutritionist claims that at least 75% of the preschool children in a certain country have protein-deficient diets, and that a sample survey reveals that this is true for 206 preschool children in a sample of 300. Test the claim at the 0.05 level of significance.

This example is followed by a step-by-step solution, which the student can use as a model for similar problems.

If this book specialized in medical statistics, the same example might have been rephrased as follows:

Suppose that a medical doctor claims that at least 75% of his patients suffer from obesity, and that a sample survey conducted by a medical research team reveals that this is true for 206 similar patients in a sample of 300. Test the claim at the 0.05 level of significance.

As the reader may discern, the step-by-step solution to the nutritionist example provides a model for the solution of this medical problem, and the answer is, of course, identical.

Similarly, if this were a book which specialized in, say, business statistics, the nutritionist example might have been rephrased as follows:

Suppose that the manager of a supermarket in a chain of supermarkets asserts that 75% of her customers prefer nationally advertised brands of dishwasher detergent over the supermarket's house brand of detergent. A market research team observes that 206 of 300 purchasers of dishwasher detergent bought the nationally advertised brand. Test the manager's assertion at the 0.05 level of significance.

Again, the nutritionist example provides a model for the solution of this business problem, and the answer is identical.

In contrast to this textbook, which presents a general introduction to the subject of statistics, numerous books have been written on business statistics, educational statistics, medical statistics, psychological statistics,…, and even on statistics for historians. Although problems arising in these various disciplines sometimes require special statistical techniques, none of the basic methods described in this book is restricted to any particular field of application. In the same way in which $3 + 3 = 6$ regardless of whether we are adding dollar amounts, horses, or trees, the methods that we present provide appropriate *statistical models*, regardless of whether the data are Intelligence Quotients, tax payments, reaction times, test scores, or humidity readings.

1.7 Statistics: What Lies Ahead

Although the point is arguable, in the last few decades the emphasis in statistics has shifted from summarizing data by means of charts and tables to making inferences (that is, generalizations) on the basis of partial, incomplete, or indirect information. This is not meant to imply, however, that the subject of statistics has now become stable and inflexible and that it has ceased to grow. Aside from the fact that new statistical techniques are constantly being developed to meet particular needs, the whole philosophy of statistics continues to be in a state of change. For example, attempts have been made to treat all problems of statistical inference within the framework of a unified theory called **decision theory**, which, so to speak, covers everything "from cradle to grave." One main feature of this theory is that we must account for all the consequences that can arise when we base decisions on statistical data. This poses serious problems, because it is generally difficult, if not impossible, to put cash values on the consequences of one's actions. For instance, how can we put a cash value on the consequences of the decision of whether or not to market a new medication, especially if the wrong decision may well involve the loss of human lives?

However, some statisticians suggest that the emphasis has swung too far from descriptive statistics to statistical inference; rightly so, they feel that the solution of many problems requires only descriptive methods. To accommodate their needs, new descriptive techniques have recently been developed under the general heading of **exploratory data analysis**. These will be introduced in Sections 2.2 and 3.3.

We have mentioned all this mainly to impress upon the reader that statistics, like most other fields of learning, is not static. Indeed, it is difficult to picture what a beginning course in statistics will be like 20 years hence. Certain aspects will probably still be the same, which includes the role of probability theory in the foundations of statistics, as well as certain data-summarizing techniques that have been very useful in the past and will undoubtedly continue to be widely used in the future.

Exercises

Exercises 1.10 *and* 1.13 *are practice exercises; their complete solutions are given on page 12.*

1.10 In four successive history tests a student received grades of 45, 73, 77, and 86. Which of the following conclusions can be obtained from these figures by purely descriptive methods and which require generalizations? Explain your answers.
   (a) Only one of the grades exceeds 85.
   (b) The student's grades increased from each test to the next.
   (c) The student must have studied harder for each successive test.
   (d) The difference between the highest and lowest grades is 41.

1.11 Mary and Jean are real estate salespersons. In the first three months of 1999 Mary sold 3, 6, and 2 one-family homes and Jean sold 4, 0, and 5 one-family homes. Which of the following conclusions can be obtained from these figures by purely descriptive methods and which require generalizations? Explain your answers.
   (a) During the three months Mary sold more one-family homes than Jean.
   (b) Mary is a better real estate salesperson than Jean.
   (c) Mary sold at least two one-family homes during each of the three months.
   (d) Jean probably took her annual vacation during the second month.

1.12 The paid attendance of a minor league baseball team's first four home games was 5,308, 4,030, 6,386, and 5,770 in the year 1998 and 6,274, 5,883, 7,615, and 1,312 in the year 1999. Which of the following conclusions can be obtained from these figures by purely descriptive methods and which require generalizations? Explain your answers.
   (a) The fourth 1999 figure was probably recorded incorrectly and should have been 7,312 instead of 1,312.
   (b) Among the eight games, the paid attendance for any one game was highest in 1999.

(c) Among the eight games, the paid attendance in 1999 exceeded 6,000 more often than in 1998.

(d) Since the paid attendance at each of the first three home games was higher in 1999 than in 1998, the weather must have been better on those days.

 1.13 Driving the same model car, five persons averaged 22.5, 21.7, 23.0, 22.5, and 21.8 miles per gallon. Which of the following conclusions can be obtained by purely descriptive methods and which require generalizations? Explain your answers.

(a) More often than any of the other figures, the drivers averaged 22.5 miles per gallon.

(b) The second and fifth persons must have done more city driving than the others.

(c) None of the averages differs from 22.0 by more than 1.0.

(d) If the whole experiment were repeated, none of the drivers would average less than 21.0 or more than 24.0 miles per gallon.

1.14 Three oranges that a person bought at a supermarket weighed 9, 8, and 13 ounces. Which of the following conclusions can be obtained from these data by purely descriptive methods and which require generalizations? Explain your answers.

(a) The average weight of the three oranges is 10 ounces.

(b) The average weight of oranges sold at that supermarket is 10 ounces.

1.15 According to the secretary of the Senate of the United States, the numbers of Democratic and Republican senators in the 100th to the 105th Congress were as follows:

Congress	Number of Democratic senators	Number of Republican senators
100th	54	46
101st	57	43
102nd	57	43
103rd	56	44
104th	47	53
105th	45	55

Which of the following conclusions can be obtained from these figures by purely descriptive methods and which require generalizations?

(a) In the 100th Congress the number of Democratic senators exceeded the number of Republican senators.

(b) Neither the Democrats nor the Republicans will ever attain a two-thirds majority in the Senate.

(c) The maximum number of Democratic senators in the 100th to the 105th Congress was 57, and the maximum number of Republican senators in the 100th to the 105th Congress was 55.

(d) The smallest number of senators of either party was 43.

(e) The number of Democratic senators has steadily declined in recent years and will probably continue to decline.

(f) In the near future, the Republican senators will continue to increase their majority.

1.16 A statistically minded lawyer has his office on the third floor of a very tall office building, and whenever he leaves his office he records whether the first elevator that stops at his floor is going up or coming down. Having done this for some time, he discovers that the vast majority of the time the first elevator that stops is going down. Comment on his conclusion that fewer elevators are going up than are coming down.

✓ Solutions to Practice Exercises

1.1 (a) Numerical.
 (b) Numerical.
 (c) Categorical.
 (d) Categorical.
 (e) Categorical.

1.4 (a) The data is a population if we are performing a study of the data for that particular year only.
 (b) The data is a sample if we use the information to estimate the numbers and amounts of the policies for subsequent years.

1.7 (a) This is called "begging the question," because the interviewer suggests to the voters that the practice is, in fact, unfair.
 (b) Persons selected from a telephone directory will generally not provide a satisfactory cross section of all persons eligible to vote. One reason is that some persons choose to have unlisted numbers. Others do not have a phone.
 (c) The term *Indian art* is ambiguous; some persons may respond with reference to the work of American Indians, while others may be thinking about art produced in India.

1.10 (a) The conclusion merely describes the data, because it can be seen that 86 exceeds 85, but 45, 73, and 77 do not.
 (b) The conclusion merely describes the data, since 73 exceeds 45, 77 exceeds 73, and 86 exceeds 77.
 (c) This is a generalization, because there can be many other reasons for the increases in the grades. For instance, the student may have felt better physically when he got the higher grades, or he may have been just lucky in studying the exact material asked for in the tests.
 (d) This conclusion merely describes the data; the highest grade is 86, the lowest grade is 45, and their difference is $86 - 45 = 41$.

1.13 (a) Since 22.5 is the only figure occurring more than once, the conclusion merely describes the data.
 (b) This is a generalization; perhaps the second and fifth persons are poorer drivers or drove in more traffic.
 (c) Since $22.5 - 22.0 = 0.5, 21.7 - 22.0 = -0.3, 23.0 - 22.0 = 1.0, 22.5 - 22.0 = 0.5$, and $21.8 - 22.0 = -0.2$, and all these differences are numerically (in magnitude) less than or equal to 1, the conclusion merely describes the data.
 (d) This is a generalization; under different traffic conditions or with specially good or bad luck with regard to red lights, one of the drivers might average less than 21.0 or more than 24.0 miles per gallon.

2 Summarizing Data: Listing and Grouping

The collecting of statistical data has grown enormously in recent years. Consequently, it has become virtually impossible to keep up with even a small part of the things that directly affect our lives unless this information is disseminated in predigested, or summarized, form. The whole matter of putting large masses of data into usable form has always been important, but it has multiplied greatly in the last few decades. This has been due partly to the development of computers that now make it possible to accomplish in seconds what was previously left undone because it would have taken months or years, and partly to the deluge of data generated by the increasingly quantitative approach of the sciences, especially the behavioral and social sciences, where nearly every aspect of human life is measured in one way or another.

The most common method of summarizing data is to present it in condensed form in tables or charts, and at one time this took up the better part of an elementary course in statistics. However, there is so much else to learn in statistics that very little time is now devoted to this kind of work.

In Sections 2.1 and 2.2 we present methods of **listing** (rearranging) sets of data so as to simplify their use. By *listing*, we refer to any kind of treatment in which the identity of each value (or item) remains preserved. In other words, the number 25 remains a 25 and a hospital patient remains a hospital patient, although the order in which the values (or items) are presented may well be changed.

In Sections 2.3 and 2.4 we shall see how a solid grasp of a set of data can also be obtained by **grouping** the data into a number of classes (intervals or categories) and then presenting the result in the form of a table or a pictorial chart.

2.1 Dot Diagrams

Once the data has been tallied (that is, we have counted how many times each value occurs), the results may be presented in the form of a **dot diagram**. As illustrated in Figures 2.1 and 2.2, in a dot diagram we simply indicate by means of dots how many times each value (or item) occurs.

■ **EXAMPLE**   The following data pertains to the number of times per week that the departure of an airline's 48 daily flights was delayed:

$$
\begin{array}{cccccccccccc}
2 & 1 & 5 & 0 & 1 & 3 & 2 & 0 & 7 & 1 & 3 & 4 \\
2 & 4 & 1 & 2 & 2 & 5 & 1 & 3 & 4 & 3 & 1 & 1 \\
3 & 2 & 6 & 4 & 1 & 0 & 2 & 2 & 3 & 5 & 2 & 3 \\
0 & 2 & 4 & 1 & 1 & 3 & 2 & 3 & 5 & 2 & 4 & 4
\end{array}
$$

Construct a dot diagram.

Solution   Counting how many times each value occurs, we get the dot diagram of Figure 2.1. It gives a clearer picture of the situation than the original numbers shown in the four rows of 12 numbers each.   ■

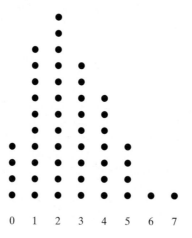

FIGURE 2.1 Dot diagram of weekly departure delays.

There are various ways in which dot diagrams can be modified. For instance, instead of dots we can use other symbols such as ×'s or ∗'s. Also, we could align the dots horizontally rather than vertically.

■ **EXAMPLE**   Thirty persons were asked to name their favorite color, and their responses were as follows:

blue	red	green	blue	red	blue
brown	blue	red	red	red	yellow
white	red	blue	green	blue	blue
orange	green	blue	blue	blue	red
green	blue	red	blue	brown	green

Solution   Counting the number of times that each color occurs, we get the dot diagram of Figure 2.2. With the colors thus ordered according to their frequencies, this kind of dot diagram is also referred to as a **Pareto diagram**.   ■

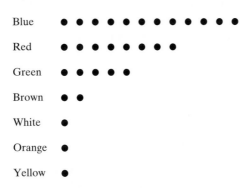

FIGURE 2.2 Dot diagram of color preferences.

2.2 Stem-and-Leaf Displays

When we deal with large sets of data, the mere gathering of the information is no small task. Usually, much more has to be done, however, to make large sets of numbers comprehensible. Consider, for example, the following values, which represent the average travel time to work of 100 employees in a large downtown office building. The original travel times were measured to the nearest minute, and each value represents an employee's average time over five consecutive workdays.

44.0	35.4	28.4	37.0	46.0	35.4	19.4	20.4	56.4	43.2
36.2	38.4	49.2	31.8	86.4	12.6	27.4	14.0	39.4	39.4
15.8	28.8	38.0	44.0	38.4	74.0	23.0	11.4	39.8	30.2
29.2	40.6	49.6	30.4	12.2	123.8	42.0	47.0	32.4	39.2
35.2	56.4	31.0	45.0	90.2	100.0	39.0	37.0	49.4	28.2
12.6	27.0	47.8	52.6	41.0	40.0	28.0	23.6	37.6	37.8
30.0	45.8	18.0	41.0	22.6	24.2	89.6	90.4	43.0	29.8
56.2	24.8	12.6	53.6	125.4	16.2	39.0	40.8	33.6	39.4
45.6	37.4	18.0	50.6	103.4	52.4	20.2	64.6	22.2	60.0
42.2	42.0	16.2	108.2	44.0	42.6	39.4	37.6	41.4	40.4

What can be done to make this mass of information more usable? Some persons find it helpful to list such data ordered from low to high or high to low, although sorting a large set of numbers according to size can be a surprisingly difficult task. For these travel times, we get

11.4	12.2	12.6	12.6	12.6	14.0	15.8	16.2	16.2	18.0
18.0	19.4	20.2	20.4	22.2	22.6	23.0	23.6	24.2	24.8
27.0	27.4	28.0	28.2	28.4	28.8	29.2	29.8	30.0	30.2
30.4	31.0	31.8	32.4	33.6	35.2	35.4	35.4	36.2	37.0
37.0	37.4	37.6	37.6	37.8	38.0	38.4	38.4	39.0	39.0
39.2	39.4	39.4	39.4	39.4	39.8	40.0	40.4	40.6	40.8
41.0	41.0	41.4	42.0	42.0	42.2	42.6	43.0	43.2	44.0
44.0	44.0	45.0	45.6	45.8	46.0	47.0	47.8	49.2	49.4
49.6	50.6	52.4	52.6	53.6	56.2	56.4	56.4	60.0	64.6
74.0	86.4	89.6	90.2	90.4	100.0	103.4	108.2	123.8	125.4

and it can be seen, for example, that the smallest value is 11.4, the largest value is 125.4, and many of the values are close to 40 minutes.

The list of numbers, even in sorted form, is still a ponderous piece of information, and it will be helpful to have other ways of dealing with these values. A method developed in the late 1960s, the **stem-and-leaf display**, will give a good overall impression of the data.

To illustrate this technique, consider the following scores on a test of physical coordination given to 20 students who had consumed an amount of alcohol equal to 0.1% of their weight.

| 69 | 84 | 52 | 93 | 61 | 74 | 79 | 65 | 88 | 63 |
| 57 | 64 | 67 | 72 | 74 | 55 | 82 | 61 | 68 | 77 |

Now break each number into its tens and units digits, tallying together values that share the tens digit. That is, we will think of the number 69 as 6 | 9. The tens digits will then be aligned vertically with the units digits displayed to the side. For the set of 20 physical coordination scores, the picture is this:

```
5 | 2 7 5
6 | 9 1 5 3 4 7 1 8
7 | 4 9 2 4 7
8 | 4 8 2
9 | 3
```

The first row of the display, 5 | 2 7 5, tells us that the list contains the values 52, 57, and 55. The second row tells us that the list contains eight values in the sixties.

This table is referred to as a stem-and-leaf display, since each row represents a **stem** position, and each digit to the right of the vertical line can be thought of as a **leaf**. To make this stem-and-leaf display, begin with the stems only:

```
5 |
6 |
7 |
8 |
9 |
```

This step need not be perfect; after all, it is easy to add more stem positions at the top or bottom. Then mark the leaves by going through the data items sequentially. After the first three values (69, 84, 52), the stem-and-leaf display will look like this:

```
5 | 2
6 | 9
7 |
8 | 4
9 |
```

After just one pass through the data, the complete stem-and-leaf display will be done.

The stem-and-leaf display conveys the same information as the original list, but it is much more compact. The stem-and-leaf display highlights the important aspects of the data. For instance, in this case it reveals immediately that most of the values are in the sixties.

The stem-and-leaf display accomplishes most of the task of sorting the values. Many persons like to complete the sorting by ordering the leaves on each branch as well. In this example, it gives the following:

```
5 | 2  5  7
6 | 1  1  3  4  5  7  8  9
7 | 2  4  4  7  9
8 | 2  4  8
9 | 3
```

Not every set of values can be placed into a stem-and-leaf display this easily. Treatment of more complicated situations is a matter of taste. Let us consider the 100 average travel times noted before. These numbers were given to *tenths* of minutes. In making the stem-and-leaf display, we recommend ignoring the tenths rather than rounding the values to the nearest minute. This introduces a half-minute bias to the display, but produces with less effort nearly the same result as proper rounding. We will use the tens digit to label the stem, and this produces the following display:

```
 1 | 122224566889
 2 | 0022334477888899
 3 | 000112355567777788899999999
 4 | 00001112222334445555677999
 5 | 0223666
 6 | 04
 7 | 4
 8 | 69
 9 | 00
10 | 038
11 |                    NOTE:   7 | 4   means   74 minutes
12 | 35                        12 | 3   means 123 minutes
13 |
```

This stem-and-leaf display requires looking through the original list just once. The leaves can then be sorted. Indeed, this picture conveys the message of the original list in a very clean pictorial form. It is helpful to attach a note, as we did, to help persons read the display. This is certainly helpful when the stem labels are not the tens digits.

Exercises

Exercises 2.1, 2.4, and 2.10 are practice exercises; their complete solutions are given on page 40.

 2.1 The number of passengers in 22 automobiles traveling on a highway is

 2 3 2 4 3 1 1 6 0 1 3 2 1 4 0 1 1 3 0 1 2 5

Construct a dot diagram that shows the number of cars carrying 0, 1, 2, 3, 4, 5, or 6 passengers.

2.2 According to the U.S. National Oceanographic and Atmospheric Administration, the average number of days per month when there was precipitation (of 0.01 inch or more) in New York City for the period of record through 1995 is as follows:

Month	Number of days	Month	Number of days
January	11	July	11
February	10	August	10
March	11	September	8
April	11	October	8
May	11	November	9
June	10	December	10

Construct a dot diagram that shows the number of months in which there were 8, 9, 10, or 11 days of precipitation.

2.3 An automobile dealer complains to a manufacturer's representative that there are numerous defects in the paint of 25 automobiles received in a shipment. The manufacturer's representative checked each car and found the following defects:

1, 3, 0, 2, 1, 2, 2, 1, 2, 1, 1, 1, 2, 2, 3,

1, 1, 2, 1, 2, 2, 1, 3, 2, and 2

Construct a dot diagram that shows the number of automobiles that have 0, 1, 2, and 3 paint defects.

 2.4 Forty-two people attending a conference were offered their choice of nonalcoholic beverages and selected the following:

BEVERAGE	NUMBER
Coffee	15
Soda pop	10
Juice	7
Tea	6
None	4
Total	42

Construct a Pareto diagram like that of Figure 2.2 with the largest value plotted on top and the remaining values plotted below in descending order.

2.5 Pareto charts are often used in quality control to point out the relative importance of different types of defects. A quality control inspector observed the following types and numbers of defects in a large production run of electrically operated toys:

TYPE OF DEFECT	NUMBER OF DEFECTS
Broken	1
Paint defects	18
Loose or missing parts	4
Faulty electrical connections	2
All other defects	3
Total	28

(a) Construct a Pareto diagram like that of Figure 2.2 with the largest number of defects on top and the remaining values plotted below in descending order.

(b) What message does the diagram of part (a) send to the inspector?

2.6 The following are the standardized reading scores of 16 high school students:

120 105 112 108 102 117 100 108 103 107 115 143 98 126 103 114

Construct a stem-and-leaf display with the stem labels 9, 10,…, and 14.

2.7 The following are the weights of 24 applicants for jobs with a city's fire department:

216 170 194 212 194 205 186 190 181 198 204 223
169 226 196 175 207 183 199 187 203 218 187 192

Construct a stem-and-leaf display with stem labels 16, 17, 18, 19, 20, 21, and 22.

2.8 We have illustrated stem-and-leaf displays with one-digit leaves, but sometimes it is more convenient to display them with, say, two-digit leaves, three-digit leaves, or more. For example, the numbers 247, 139, 223, 148, and 115 can be displayed as

```
1 | 39   48   15

2 | 47   23
```

or, ordering the leaves as

```
1 | 15   39   48

2 | 23   47
```

Construct a stem-and-leaf display with two-digit leaves and with stem labels 1, 2, 3, 4, and 5 for the following weekly earnings, in dollars, of 15 salespersons:

305 255 319 167 270 291 512 283 334 362 188 217 440 195 408

2.9 List the data that correspond to the following stems of stem-and-leaf displays:

(a) 1	0	1	1	2	5	7	8	
(b) 12	0	2	3	3	5			
(c) 0.6	0	2	3	6				
(d) 1.5	0	1	1	3	4	6		

2.10 If we want to construct a stem-and-leaf display with more stems than there would be otherwise, we can divide each stem position in two. Use the first stem position to hold leaves 0, 1, 2, 3, and 4, and use the second stem position to hold leaves 5, 6, 7, 8, and 9. For the data on page 16, which are

69 84 52 93 61 74 79 65 88 63
57 64 67 72 74 55 82 61 68 77

we would thus get the **double-stem display**

5∗	2			
5·	5	7		
6∗	1	1	3	4
6·	5	7	8	9
7∗	2	4	4	
7·	7	9		
8∗	2	4		
8·	8			
9∗	3			

where we doubled the number of stem positions by cutting in half the interval covered by each tens digit. Construct a double-stem display for the data of Exercise 2.7.

2.11 The following are the ages of 30 heads of household in a retirement community:

68 81 61 62 76 65 69 73 78 60 64 74 57 70 68
66 83 71 59 66 61 65 85 72 76 65 67 73 72 67

Construct a double-stem display (see Exercise 2.10) for these values.

2.12 The following list gives the dollar value of 30 consecutive transactions at the express checkout of a particular supermarket.

18.20	22.45	9.20	10.88	26.62
13.46	5.20	13.55	26.22	24.00
15.20	12.85	6.40	11.55	6.62
29.86	8.20	10.35	6.32	18.50
5.50	22.85	9.20	20.55	26.62
19.46	4.20	19.75	21.32	28.00

Construct a double-stem display, using the tens digit to label the stem. You will have to ignore some of the digits.

2.13 According to the U.S. Bureau of Justice, the rate of full-time sworn police officers (the number of police officers per 10,000 population) in the midwestern states during a recent year is as follows:

State	Rate	State	Rate
Illinois	31	Missouri	22
Indiana	18	Nebraska	19
Iowa	17	North Dakota	16
Kansas	22	Ohio	19
Michigan	21	South Dakota	16
Minnesota	16	Wisconsin	23

Construct a stem-and-leaf display with stem labels 1, 2, and 3.

2.14 The National Council for Education Statistics, U.S. Department of Education, gives the expenditure per pupil (in hundreds of dollars) for the year 1995 in the 50 states and in Washington, D.C. These are as follows:

```
44.0  89.6  47.8  44.6  49.9  54.4  88.2  70.3  93.3  57.2  51.9  60.8
42.1  61.4  58.3  54.8  58.2  52.2  47.6  64.2  72.4  72.9  69.8  60.0
40.8  53.8  56.9  59.4  51.6  58.6  97.7  45.9  96.2  50.8  47.8  61.6
48.4  64.4  71.1  74.7  48.0  47.8  43.9  52.2  36.6  67.5  53.3  59.0
61.0  69.3  61.6
```

(a) Construct a stem-and-leaf display with the stem labels 3, 4, 5, 6, 7, 8, and 9. Omit the values after the decimal point.

(b) Construct a double-stem display for the same data. Use the first stem position to hold leaves 0, 1, 2, 3, and 4, and use the second stem position to hold leaves 5, 6, 7, 8, and 9. Omit the values after the decimal point.

2.3 Frequency Distributions

The previous section confirms our experience that untreated raw data is difficult to deal with. The dot diagram and the stem-and-leaf display were introduced as techniques to organize numbers and make them more comprehensible. In this section we discuss the more traditional technique of using frequency distributions to classify, or group, data.

Consider the problem of a social scientist who wants to examine the ages of death-row inmates, that is, persons awaiting execution, in the United States. He will likely turn to the *Statistical Abstract of the United States*, which is a reliable source for information of this nature.

A good overall understanding of a large set of data can often be created by grouping the data into a number of classes (intervals or categories), and our social scientist finds that the desired information is presented as follows for the year 1995, from the 1997 edition of the *Statistical Abstract of the United States*.

PERSONS UNDER SENTENCE OF DEATH—1995	
AGE	NUMBER OF PERSONS
Under 20 years	20
20–24 years	264
25–34 years	1,068
35–54 years	1,583
55 years and over	119
Total	3,054

A table like this is called a **frequency distribution**, or simply a **numerical distribution**. It shows how the ages of 3,054 death-row inmates are distributed. The data are grouped according to the numeric value, age. In some instances we group information according to nonnumeric criteria, such as color, geographic region, consumer preference, or medical diagnosis.

In the preceding example, each class covered more than one possible value. For instance, the second class covered the values 20, 21, 22, 23, and 24. Each class may also cover a single value, as illustrated by the following example based on a study in which 400 persons were asked how many full-length movies they had seen on television during the preceding week.

NUMBER OF MOVIES	NUMBER OF PERSONS
0	72
1	106
2	153
3	40
4	18
5	7
6	3
7	0
8	1
Total	400

The following display illustrates a **categorical distribution**. These nonnumerical categories collectively constitute the values of all products and services sold by establishments in the luggage and personal leather goods industry for a recent year. The information was provided by the U.S. Department of Commerce and other sources.

	MILLIONS OF DOLLARS
Leather and sheep-lined clothing	201
Leather gloves and mittens	155
Luggage	960
Women's handbags and gloves	269
Personal leather goods	371
Total	1,956

Frequency distributions present data in a relatively compact form, give a good overall picture, and contain information that is adequate for many purposes; but some things can be obtained only from the original data. For instance, the distribution given previously on death-row inmates cannot tell us how many are 23 years old or how many are over 60. Similarly, the study of the personal leather goods industry might have provided data concerning imports, exports, other products, and competing merchandise. Nevertheless, frequency distributions present data in a more usable form, and the price we pay—the loss of certain information—is usually a favorable exchange.

The construction of numerical distributions consists essentially of four steps: (1) choosing the classes, (2) sorting (or tallying) the data into these classes, (3) counting the number of items in each class, and (4) displaying the results in the form of a chart or table. Since the second and third steps are purely mechanical and the fourth step is a matter of craftsmanship and taste, we shall concentrate here on the first step, choosing suitable classifications.

This first step involves choosing the number of classes and the range of values that each class should cover, that is, from where to where each class should go. Both of these choices are arbitrary to some extent, but they depend on the nature of the data and on the purpose that the distribution is to serve. The following are some rules that are generally observed.

> *We seldom use fewer than 6 or more than 15 classes; the exact number we use in a given situation depends mainly on the number of measurements or observations that we have to group.*

Clearly, we would lose more than we gain if we group 6 observations into 12 classes with many of them empty, and we would probably give away too much information if we group 10,000 measurements into 3 classes.

> *We always make sure that each item (measurement or observation) goes into one and only one class.*

To this end, we must make sure that the smallest and largest values fall within the classification, that none of the values can fall into possible gaps between successive classes, and that the classes do not overlap, that is, successive classes have no values in common.

> *Whenever possible, we use classes covering equal ranges (or intervals) of values.*

If we can, we also use numbers for these ranges that are easy to work with, such as 5, 10, or 100, for this will facilitate constructing, reading, and using the distribution.

In connection with these rules, the age distribution given for death-row inmates is not a good example, but, presumably, the government statisticians had good reasons for choosing the classes as they did. There are only five classes, none of the classes covers equal ranges of values, and the "55 and over" class is *open*; for all we know the oldest person may be 75 or 85.

In general, we refer to classes of the "less than," "or less," "more than," and "or more" type as **open classes**, and they are used to reduce the number of classes that are required when a few of the values are much smaller (or much greater) than the rest. However, they should be avoided when possible for, as we shall see, they make it difficult or even impossible to calculate certain further descriptions that may be of interest.

Insofar as the second rule is concerned, it is important to watch whether the data is given to the nearest dollar or to the nearest cent, whether it is given to the nearest inch or to the nearest tenth of an inch, whether it is given to the nearest ounce or to the nearest hundredth of an ounce, and so forth. For instance, to group the weights of certain animals, we could use the first of the following three classifications if the weights are given to the nearest kilogram, the second if the weights are given to the nearest tenth of a kilogram, and the third if the weights are given to the nearest hundredth of a kilogram.

Weight (kilograms)	Weight (kilograms)	Weight (kilograms)
10–14	10.0–14.9	10.00–14.99
15–19	15.0–19.9	15.00–19.99
20–24	20.0–24.9	20.00–24.99
25–29	25.0–29.9	25.00–29.99
30–34	30.0–34.9	30.00–34.99

To illustrate what we have been discussing in this section, let us now go through the actual steps of grouping a given set of data into a frequency distribution.

■ **EXAMPLE**   Construct a distribution of the following data on the amount of time (in hours) that 80 college students devoted to leisure activities during a typical school week:

23	24	18	14	20	24	24	26	23	21
16	15	19	20	22	14	13	20	19	27
29	22	38	28	34	32	23	19	21	31
16	28	19	18	12	27	15	21	25	16
30	17	22	29	29	18	25	20	16	11
17	12	15	24	25	21	22	17	18	15
21	20	23	18	17	15	16	26	23	22
11	16	18	20	23	19	17	15	20	10

Solution   Since the smallest value is 10 and the largest is 38, we might choose the six classes 10–14, 15–19, 20–24, 25–29, 30–34, and 35–39; or we might choose the eight classes 10–13, 14–17, 18–21, 22–25, 26–29, 30–33, 34–37, and 38–41, to mention two possibilities. Note that in each case the classes accommodate all the data, they do not overlap, and they are all of the same size.

Generally, we prefer classes that begin or end in multiples of 5 or 10; accordingly we will decide on the first of these classifications. We now tally the 80 observations and get the results shown in the following table:

HOURS	TALLY	FREQUENCY
10–14	┼┼┼ ///	8
15–19	┼┼┼ ┼┼┼ ┼┼┼ ┼┼┼ ┼┼┼ ///	28
20–24	┼┼┼ ┼┼┼ ┼┼┼ ┼┼┼ ┼┼┼ //	27
25–29	┼┼┼ ┼┼┼ //	12
30–34	////	4
35–39	/	1
	Total	80

The numbers given in the right-hand column of this table, which show how many items fall into each class, are called the **class frequencies**. Also, the smallest and largest values that can go into any given class are referred to as its **class limits**, and in our example they are 10 and 14, 15 and 19, 20 and 24,..., and 35 and 39. More specifically, 10, 15, 20,..., and 35 are called the **lower class limits**, and 14, 19, 24,..., and 39 are called the **upper class limits**.   ■

The amounts of time that we grouped in our example were all given to the nearest hour, so that the first class actually covers the interval from 9.5 hours to 14.5 hours, the second class covers the interval from 14.5 hours to 19.5 hours, and so forth. It is customary to refer to these numbers as the **class boundaries** or the **"real" class limits**. In actual practice, class limits are used much more widely than class boundaries, and we have mentioned them here mainly because they will be needed in Chapter 3 for calculating certain descriptive measures of a distribution.

Also, since the upper class boundary of any one class always equals the lower class boundary of the next class, there is some disagreement as to what one should do about values that actually equal one of the class boundaries. For instance, MINITAB always places observations that fall on class boundaries into the class to the right (except those in the class farthest to the right, which are placed in the class to the left). To avoid situations like this, which create a lack of symmetry, it is desirable always to choose classes for which none of the values that we intend to group can actually equal one of the class boundaries.

Numerical distributions also have what we call **class marks** and **class intervals**. Class marks are simply the midpoints of the classes, and they are obtained by adding the upper and lower limits of a class (or its upper and lower boundaries) and dividing by 2. A class interval is the length of a class, or the range of values that it can contain, and it is given by the difference between its class boundaries. If the classes of a distribution are all of equal length, their common class interval, which we refer to as the **class interval of the distribution**, is also given by the difference between any two successive class marks.

■ **EXAMPLE**   Find the class marks and the class interval of the distribution obtained in the preceding example.

Solution   The class marks are

$$\frac{10 + 14}{2} = 12, \quad \frac{15 + 19}{2} = 17, \ldots, \quad \text{and} \quad \frac{35 + 39}{2} = 37$$

The intervals of the classes are $14.5 - 9.5 = 5$, $19.5 - 14.5 = 5, \ldots$, and $39.5 - 34.5 = 5$; since they are all equal, we say that the class interval of the distribution is 5. Note that class intervals are *not* given by the differences between the respective class limits, which in our example all equal 4, and not 5.   ■

There are essentially two ways in which frequency distributions can be modified to suit particular needs. One way is to convert a distribution into a

percentage distribution by dividing each class frequency by the total number of items grouped and then multiplying by 100%.

■ **EXAMPLE**   Convert the distribution of the amounts of time the 80 college students devoted to leisure activities into a percentage distribution.

Solution   The first class contains

$$\frac{8}{80} \cdot 100\% = 10\% \text{ of the data,}$$

the second class contains

$$\frac{28}{80} \cdot 100\% = 35\% \text{ of the data,}$$

the third class contains

$$\frac{27}{80} \cdot 100\% = 33.75\% \text{ of the data,}$$

..., and the sixth class contains

$$\frac{1}{80} \cdot 100\% = 1.25\% \text{ of the data}$$

These results are shown in the following table:

HOURS	PERCENTAGE
10–14	10
15–19	35
20–24	33.75
25–29	15
30–34	5
35–39	1.25
Total	100

The other way of modifying a frequency distribution is to convert it into a "less than," "or less," "more than," or "or more" **cumulative distribution**. To construct a cumulative distribution, we simply add the class frequencies, starting either at the top or at the bottom of the distribution.

■ **EXAMPLE**   Convert the distribution for leisure activity times of college students into a cumulative "less than" distribution.

Solution   Since none of the values is less than 10, 8 are less than 15, 8 + 28 = 36 are less than 20, 8 + 28 + 27 = 63 are less than 25,..., the result is shown in the following table:

HOURS	CUMULATIVE FREQUENCY
Less than 10	0
Less than 15	8
Less than 20	36
Less than 25	63
Less than 30	75
Less than 35	79
Less than 40	80

Instead of "less than 10," we could also have written "less than or equal to 9"; instead of "less than 15," we could also have written "less than or equal to 14"; and so on. When cumulative distributions are generated by means of computers, they are usually given as "less than or equal to" distributions.

In the same way, we can also convert a percentage distribution into a **cumulative percentage distribution** by adding the percentages, starting either at the top or at the bottom of the distribution. For our example, we get the following:

HOURS	CUMULATIVE PERCENTAGE
Less than 10	0
Less than 15	10
Less than 20	45
Less than 25	78.75
Less than 30	93.75
Less than 35	98.75
Less than 40	100

Exercises

Exercises 2.15, 2.18, 2.21, 2.25, and 2.28 are practice exercises; their complete solutions are given on pages 40 and 41.

2.15 In New York City 120 buildings are from 27 to 110 stories tall. Indicate the limits of five classes into which the number of these buildings might be grouped.

2.16 The average speed of the winners of the annual Indianapolis 500 automobile race from 1957 to 1997, in a race that is usually 500 miles in length, varied from 133.791 to 185.981 miles per hour. Indicate limits of six classes into which these winning speeds might be grouped.

2.17 The circulations of the 100 leading U.S. newspapers in the year 1996, according to the *1997 Editor and Publisher International Yearbook*, varied from 103, 522 to 1,783,532. Indicate the limits of eight classes into which these values (numbers of copies) might be grouped.

2.18 The numbers of empty seats on flights from Dallas to New Orleans are grouped into a table with the classes 0–9, 10–19, 20–29, 30–39, and 40 or more. Will it be possible to determine exactly from this table the number of flights on which there were
  (a) at least 20 empty seats;
  (b) more than 20 empty seats;
  (c) more than 19 empty seats;
  (d) at least 19 empty seats;
  (e) exactly 19 empty seats?

2.19 The declared values of packages mailed from a foreign country are grouped into a distribution with the following classes: $0.00–$49.99, $50.00–$99.99, $100.00–$149.99, $150.00–$199.99, and $200.00 or more. Will it be possible to determine exactly from the distribution the number of packages valued at
  (a) more than $100.00;
  (b) at least $100.00;
  (c) exactly $100.00;
  (d) more than $99.99?

2.20 The following is the distribution of the weights of 125 mineral specimens collected on a field trip:

WEIGHT (GRAMS)	NUMBER OF SPECIMENS
0–19.9	19
20.0–39.9	38
40.0–59.9	35
60.0–79.9	17
80.0–99.9	11
100.0–119.9	3
120.0–139.9	2
Total	125

If possible, find how many of the specimens weigh
  (a) at most 59.9 grams;
  (b) less than 40.0 grams;
  (c) more than 100.0 grams;
  (d) 80.0 grams or less;
  (e) exactly 20.0 grams;
  (f) anywhere from 40.0 to 80.0 grams.

2.21 The numbers of students who are absent each day from a certain school are grouped into a distribution having the classes 0–14, 15–29, 30–44, and 45–59. Determine
  (a) the lower class limits;
  (b) the upper class limits;
  (c) the class marks;
  (d) the class interval of the distribution.

2.22 The numbers of suitcases lost each week by an airline on flights between Honolulu and San Francisco are grouped into a distribution having the classes 0–2, 3–5, 6–8, 9–11, 12–14, and 15–17. Find
  (a) the lower class limits;
  (b) the upper class limits;
  (c) the class marks;
  (d) the class interval of the distribution.

2.23 The number of congressmen absent each day during a session of Congress are grouped into a distribution having the classes 0–19, 20–39, 40–59, 60–79, 80–99, 100–119, 120–139, and 140–159. Determine

(a) the lower class limits;
(b) the upper class limits;
(c) the class marks;
(d) the class interval of the distribution.

2.24 The numbers of nurses on duty each day at a hospital are grouped into a distribution having the classes 20–34, 35–49, 50–64, 65–79, and 80–94. Find
(a) the class limits;
(b) the class boundaries;
(c) the class marks;
(d) the class interval of the distribution.

2.25 The class marks of a distribution of the daily number of traffic accidents reported in a county are 4, 11, 18, and 25. Find
(a) the class boundaries;   (b) the class limits.

2.26 The class marks of a distribution of the daily number of calls received by a small cab company are 22, 27, 32, 37, 42, 47, and 52. Find
(a) the class boundaries;   (b) the class limits.

2.27 The class marks of a distribution of temperature readings, given to the nearest degree Fahrenheit, are 36, 45, 54, 63, 72, 81, and 90. Find
(a) the class boundaries;   (b) the class limits.

2.28 To group sales invoices ranging from $5.00 to $30.00, a clerk uses the following classification: $5.00–$9.99, $10.00–$15.99, $15.00–$19.99, $20.00–$24.90, and $25.00–$29.99. Explain where difficulties might arise.

2.29 To group data on the number of rainy days reported by a weather station for the month of May during the last 50 years, a meteorologist uses the classes 0–5, 6–10, 12–17, 18–23, and 23–30. Explain where difficulties might arise.

2.30 The following are the body weights (in grams) of 50 immature rats used in a study of vitamin deficiencies:

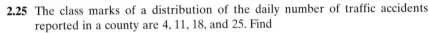

136	92	115	118	121	137	132	120	104	125
119	115	101	129	87	108	110	133	135	126
127	103	110	126	118	82	104	137	120	95
146	126	119	119	105	132	126	118	100	113
106	125	117	102	146	129	124	113	95	148

(a) Group these weights into a table having the classes 80–89, 90–99, 100–109, 110–119, 120–129, 130–139, and 140–149 grams.
(b) Convert the distribution obtained in part (a) into a cumulative "or more" distribution.

2.31 Convert the distribution obtained in part (a) of Exercise 2.30 into a
(a) percentage distribution;
(b) cumulative "less than or equal to" percentage distribution.

2.32 The following are the grades that 50 students obtained on an accounting test:

73	65	82	70	45	50	70	54	32	75
75	67	65	60	75	87	83	40	72	64
58	75	89	70	73	55	61	78	89	93
43	51	59	38	65	71	75	85	65	85
49	97	55	60	76	75	69	35	45	63

(a) Prepare a stem-and-leaf display of these values. Use the tens digits to form the stem.

(b) Group these grades into a distribution having the classes 30–39, 40–49, 50–59,..., and 90–99.

(c) Convert the distribution obtained in part (b) into a cumulative "less than" distribution, beginning with "less than 30."

2.33 The following is a distribution of the ages of the members of a video dating service for single persons:

AGE (YEARS)	FREQUENCY
20–24	129
25–29	221
30–34	310
35–39	163
40–44	105
45–49	62
50–54	10
Total	1,000

(a) Convert the distribution into a cumulative "less than or equal to" distribution.

(b) Convert the distribution into a cumulative "or more" distribution.

2.34 The following are the miles per gallon obtained with 40 tankfuls of gas:

```
24.1  25.0  24.8  24.3  24.2  25.3  24.2  23.6  24.5  24.4
24.5  23.2  24.0  23.8  23.8  25.3  24.5  24.6  24.0  25.2
25.2  24.4  24.7  24.1  24.6  24.9  24.1  25.8  24.2  24.2
24.8  24.1  25.6  24.5  25.1  24.6  24.3  25.2  24.7  23.3
```

(a) Prepare a double-stem plot of these values. Use the stem labels 23*, 23·, 24*, 24·, 25*, and 25·.

(b) Group these figures into a distribution having the classes 23.0–23.4, 23.5–23.9, 24.0–24.4, 24.5–24.9, 25.0–25.4, and 25.5–25.9.

(c) Convert the distribution obtained in part (b) into a cumulative "or more" distribution, beginning with "23.0 or more" and ending with "26.0 or more."

2.35 In the 40 lectures of a psychology class

```
2  1  5  0  0  3  2  1  1  4  3  1  0  1  2  1  3  6  2  2
1  0  2  1  1  3  4  1  0  2  1  0  1  2  4  1  3  1  2  3
```

students were absent. Construct a distribution showing how many times 0, 1, 2, 3, 4, 5, or 6 students were absent.

2.36 A survey made at a resort city showed that 50 tourists arrived by the following means of transportation:

car, train, plane, plane, plane, bus, train, car, car, car, plane, car, plane train, car, car, bus, car, plane, plane, train, train, plane, plane, car, car train, car, car, plane, car, car, plane, bus, plane, bus, car, plane, car, car, train, train, car, plane, bus, plane, car, car, train, bus

Construct a categorical distribution showing the frequencies corresponding to the different means of transportation.

2.37 Asked to rate the maneuverability of a car as excellent, very good, good, fair, poor, or very poor, 40 drivers responded as follows:

very good, good, good, fair, excellent, good, good, good, very good, poor, good, good, good, good, very good, good, fair, good, good, very poor, very good, fair, good, good, excellent, very good, good, good, good, fair, fair, very good, good, very good, excellent, very good, fair, good, good, very good

Construct a distribution showing the frequencies corresponding to the different ratings of the maneuverability of the car.

2.4 Graphical Presentations

When frequency distributions are constructed primarily to condense large sets of data and display them in an "easy to digest" form, they are often presented graphically. The most common form of graphical presentation of a frequency distribution is the **histogram**. The data in the example on page 24 is shown as a histogram in Figure 2.3.

A histogram is constructed by representing the measurements or observations that are grouped on a horizontal scale, the class frequencies on a vertical scale, and drawing rectangles whose bases equal the class intervals and whose heights are determined by the corresponding class frequencies. The markings on the horizontal scale can be the class limits as in Figure 2.3, the class boundaries, the class marks, or arbitrary key values. For easy readability it is usually preferable to indicate the class limits, although the rectangles actually go from one class boundary to the next. Histograms cannot be used

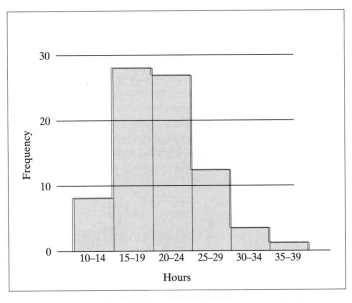

FIGURE 2.3 Histogram of the distribution of the amounts of time that 80 students devoted to leisure activities, during a typical school week.

in connection with frequency distributions having open classes, and they must be used with extreme care when the class intervals are not all equal. In that case, it is best to represent the class frequencies by the areas of the rectangles instead of their heights.

Optional Figure 2.3 presents the data of Figure 2.3 in similar fashion, except that the histogram is reproduced from a TI-83 graphing calculator.

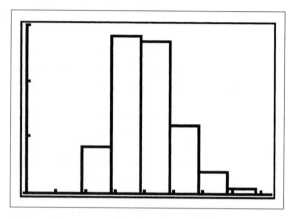

OPTIONAL FIGURE 2.3 The data here is identical to the data of Figure 2.3. The histogram is reproduced from the display screen of a TI-83 graphing calculator.

This optional figure is intended primarily for users of graphing calculators and can be omitted, without loss of continuity, by other readers. It is nevertheless of general interest to observe that we can obtain a histogram from a graphing calculator by entering the numerical data and following specific procedures. We do not actually draw the rectangles. The results of Figures 2.3 and Optional Figure 2.3 are comparable, except that the graphing calculator does not provide verbal or numerical labels.

For large sets of data, it may be convenient to construct histograms directly from raw data by using a computer package designed specially for data analysis. For instance, the MINITAB printout in Figure 2.4 shows a histogram of this distribution with the class marks 12, 17, 22, 27, 32, and 37 (as given on page 25 for the data dealing with the amounts of time that 80 students devoted to leisure activities).[†] Here, SET C1 instructs the computer to place the 80 numbers in column 1, and HIST C1 12 5 instructs it to print a histogram with 12 as the first class mark and 5 as the class interval for the data in column 1. Note that the frequency scale is horizontal, so that the histogram is, so to speak, on its side. Strictly speaking, the result shown in Figure 2.4 is not a histogram, at least not as we have defined this term. However, combining some of the features of Figures 2.5 and 2.9, it conveys the same idea.

[†] In the printout of Figure 2.4 and in others appearing in this book, words and numbers appear that relate to the technical aspects of operating the particular computer program employed. If a computer is available, the reader should refer to the appropriate manuals for operating instructions and for a list of problems that can be solved with existing programs.

```
MTB > SET C1
DATA> 23 24 18 14 20 24 24 26 23 21
DATA> 16 15 19 20 22 14 13 20 19 27
DATA> 29 22 38 28 34 32 23 19 21 31
DATA> 16 28 19 18 12 27 15 21 25 16
DATA> 30 17 22 29 29 18 25 20 16 11
DATA> 17 12 15 24 25 21 22 17 18 15
DATA> 21 20 23 18 17 15 16 26 23 22
DATA> 11 16 18 20 23 19 17 15 20 10
DATA> END
MTB > HIST C1 12 5

Histogram of C1 N=80
Midpoint  Count
    12.00      8 ********
    17.00     28 ****************************
    22.00     27 ***************************
    27.00     12 ************
    32.00      4 ****
    27.00      1 *
```

FIGURE 2.4 MINITAB printout for the histogram of the amounts of time that 80 students devoted to leisure activities during a typical school week.

Similar to histograms are **bar charts**, such as the one shown in Figure 2.5. The heights of the rectangles, or bars, again represent the class frequencies, but there is no pretense of having a continuous horizontal scale.

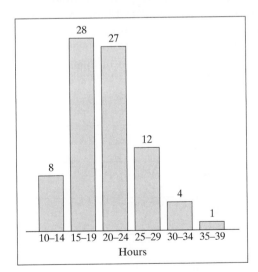

FIGURE 2.5 Bar chart of the distribution of the amounts of time that 80 students devoted to leisure activities during a typical school week.

Another, less widely used, form of graphical presentation is the **frequency polygon** (see Figure 2.6). In this form the class frequencies are plotted at the class marks, and the successive points are connected by means of straight lines. Note that we added classes with zero frequencies at both

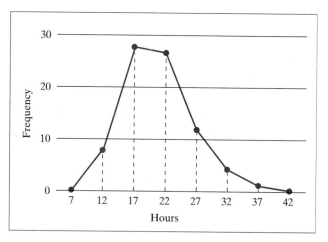

FIGURE 2.6 Frequency polygon of the distribution of the amounts of time that 80 students devoted to leisure activities during a typical school week.

ends of the distribution to "tie down" the graph to the horizontal scale. If we apply the same technique to a cumulative distribution, we obtain what is called an **ogive**, which rhymes with live or jive. However, the cumulative frequencies are plotted at the corresponding class limits instead of the class marks. It stands to reason that "less than or equal to 19," for example, should be plotted at 19. This value actually includes everything up to 19. Figure 2.7 shows an ogive of the cumulative "less than" distribution, which we constructed previously.

Although the visual appeal of histograms, bar charts, frequency polygons, and ogives exceeds that of tables, there are various ways in which distributions can be presented more dramatically and more effectively. Two kinds of such pictorial presentations (often seen in newspapers, magazines, and reports of various sorts) are illustrated by the **pictograms** of Figures 2.8 and 2.9.

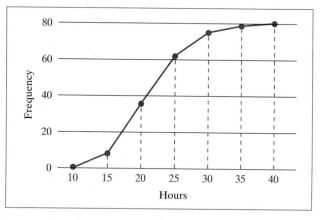

FIGURE 2.7 Ogive of the distribution of the amounts of time that 80 students devoted to leisure activities during a typical school week.

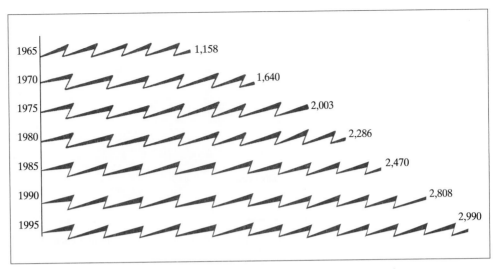

FIGURE 2.8 Electric energy production in the United States (billions of kilowatt hours).

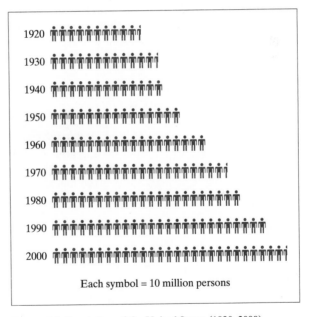

Each symbol = 10 million persons

FIGURE 2.9 Population of the United States (1920–2000).

Categorical (or qualitative) distributions are often presented graphically as **pie charts**, such as the one shown in Figure 2.10, where a circle is divided into sectors (pie-shaped pieces) that are proportional in size to the corresponding frequencies or percentages. To construct a pie chart, we first convert the distribution into a percentage distribution. Then, since a complete circle corresponds to 360°, we obtain the central angles of the various sectors by multiplying the percentages by 360. The central angles for Figure 2.10 were obtained by multiplying 360° by 19%, 17%, 43%, and 21%, obtaining 68°, 61°, 155°, and 76°, respectively.

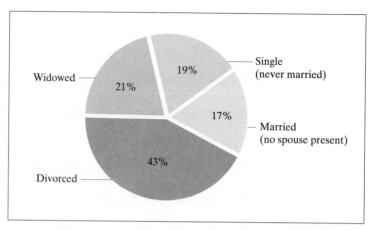

FIGURE 2.10 Marital status of white female heads of households with no spouse present, 1995.

■ **EXAMPLE**   To study their attitudes toward social issues, 1,200 persons were asked (among other things) whether we are spending "too little," "about right amount," or "too much" on social welfare programs. Draw a pie chart to display the results shown in the following table:

	NUMBER OF PERSONS
Too little	296
About right amount	360
Too much	544
Total	1,200

Solution   The percentages corresponding to the three categories are

$$\frac{296}{1,200} \cdot 100\% \approx 24.7\%, \quad \frac{360}{1,200} \cdot 100\% = 30.0\%,$$

$$\text{and} \quad \frac{544}{1,200} \cdot 100\% \approx 45.3\%$$

Multiplying by 360, we find that the central angles of the three sectors are

$$0.247(360) \approx 88.92°, \quad 0.300(360) = 108.00°,$$

$$\text{and} \quad 0.453(360) \approx 163.08°$$

The symbol $\approx$ means *approximately equal* (its use will be illustrated further in Section 3.1).

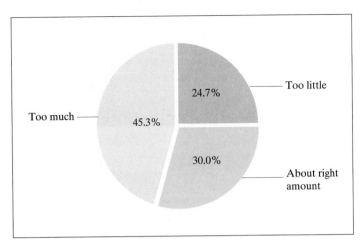

FIGURE 2.11 Responses to questions about welfare expenditures.

Rounding the angles to the nearest degree and measuring the angles with a protractor, we get the pie chart shown in Figure 2.11.   ■

Many computer programs can produce pie charts directly on their display screens.

Research on the ability to read graphical displays has suggested that untrained users have trouble interpreting pie charts. Comparing information from two adjacent pie charts is particularly difficult. The comparison of bar heights in histograms is rather easy. It is recommended that pie charts be used with great care.

Exercises

Exercises 2.38 and 2.42 are practice exercises; their complete solutions are given on pages 41 and 42.

2.38 The following is the distribution of the total finance charges that 200 customers paid on their budget accounts at a department store:

AMOUNT (DOLLARS)	FREQUENCY
1–20	18
21–40	62
41–60	63
61–80	43
81–100	14
Total	200

(a) Draw a histogram of this distribution.
(b) Draw a bar chart of this distribution.

2.39 Convert the distribution of Exercise 2.38 into a cumulative "less than" distribution and draw an ogive.

2.40 The following is the distribution of the weights of the 140 freshmen women entering a certain college:

WEIGHT (POUNDS)	FREQUENCY
90–99	4
100–109	23
110–119	49
120–129	38
130–139	17
140–149	6
150–159	3
Total	140

(a) Draw a histogram of this distribution.
(b) Draw a frequency polygon of this distribution.

2.41 Convert the distribution of Exercise 2.40 into a cumulative "less than" distribution and draw an ogive.

2.42 The following table shows how workers in Phoenix, Arizona, get to work:

MEANS OF TRANSPORTATION	PERCENTAGE
Ride alone	81
Car pool	14
Ride bus	3
Varies or work at home	2
Total	100

Construct a pie chart of this percentage distribution.

2.43 Data from the Institute of International Education shows that foreign students in U.S. colleges originate from the following regions:

Region	Enrollment in thousands
Africa	21
Asia	290
Europe	67
Latin America	47
North America	24
Oceania	4

Construct a pie chart of this categorical distribution.

2.44 Data from the U.S. Bureau of the Census shows that the resident populations of the East North Central states are as follows:

State	Population in thousands
Indiana	5,841
Illinois	11,847
Michigan	9,594
Wisconsin	5,160

Construct a pie chart of this categorical distribution.

2.45 According to the American Automobile Manufacturers Association, the numbers of domestic and imported cars sold at retail in the United States during 1996 were as follows:

SOURCE	CAR SALES IN MILLIONS
United States	6.6
Imports from:	
Japan	1.9
Germany	0.3
All other countries	0.2
Total	9.0

Construct a pie chart of this categorical distribution.

2.46 Here, again, are the grades from Exercise 2.32 that 50 students obtained on an accounting test:

```
73  65  82  70  45  50  70  54  32  75
75  67  65  60  75  87  83  40  72  64
58  75  89  70  73  55  61  78  89  93
43  51  59  38  65  71  75  85  65  85
49  97  55  60  76  75  69  35  45  63
```

Use a computer package to construct a histogram with the classes 30–39, 40–49, 50–59,…, and 90–99. Construct also a histogram with the classes 20–39, 40–59, 60–79, and 80–99. Which histogram do you prefer?

2.47 Here, again, are the weights of Exercise 2.30 on page 29, which are the body weights of 50 immature rats used in a study of vitamin deficiencies. Use a computer package to construct a histogram of these weights with the classes 80–89, 90–99, 100–109,…, and 140–149.

```
136   92  115  118  121  137  132  120  104  125
119  115  101  129   87  108  110  133  135  126
127  103  110  126  118   82  104  137  120   95
146  126  119  119  105  132  126  118  100  113
106  125  117  102  146  129  124  113   95  148
```

2.48 Here, again, are the mileages of Exercise 2.34 on page 30, which are the miles per gallon obtained with 40 tankfuls of gas. Use a computer package to construct a histogram with classes 23.0–23.4, 23.5–23.9, 24.0–24.4,…, and 25.5–25.9.

```
24.1  25.0  24.8  24.3  24.2  25.3  24.2  23.6  24.5  24.4
24.5  23.2  24.0  23.8  23.8  25.3  24.5  24.6  24.0  25.2
25.2  24.4  24.7  24.1  24.6  24.9  24.1  25.8  24.2  24.2
24.8  24.1  25.6  24.5  25.1  24.6  24.3  25.2  24.7  23.3
```

✓ Solutions to Practice Exercises

2.1

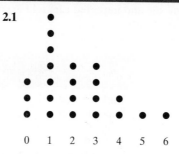

2.4

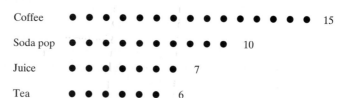

2.10

16·	9			
17∗	0			
17·	5			
18∗	1	3		
18·	6	7	7	
19∗	0	2	4	4
19·	6	8	9	
20∗	3	4		
20·	5	7		
21∗	2			
21·	6	8		
22∗	3			
22·	6			

2.15 The heights cover a range of 84 stories. It is reasonable that we use the five classes 20–39, 40–59, 60–79, 80–99 and 100–119.

2.18 (a) Yes; (b) no; (c) yes; (d) no; (e) no.

2.21 (a) The lower class limits are 0, 15, 30, and 45.
(b) The upper class limits are 14, 29, 44, and 59.
(c) The class marks are

$$\frac{0 + 14}{2} = 7, \quad \frac{15 + 29}{2} = 22, \quad \frac{30 + 44}{2} = 37, \quad \text{and} \quad \frac{45 + 59}{2} = 52$$

(d) The class interval of the distribution is $22 - 7 = 15$.
or $29.5 - 14.5 = 15$, $14.5 (-0.5) = 15$.

2.25   (a) The boundary between the first two classes is

$$\frac{4 + 11}{2} = 7.5$$

the boundary between the second and third classes is

$$\frac{11 + 18}{2} = 14.5$$

and the boundary between the third and fourth classes is

$$\frac{18 + 25}{2} = 21.5$$

Since the differences between successive class boundaries are $14.5 - 7.5 = 7$, the lower boundary of the first class is $7.5 - 7 = 0.5$, and the upper boundary of the fourth class is $21.5 + 7 = 28.5$.

(b)   The limits of the first class are the smallest and largest integers falling between 0.5 and 7.5, that is, 1 and 7; the limits of the second class are the smallest and largest integers falling between 7.5 and 14.5, that is, 8 and 14; similarly, the limits of the third class are 15 and 21, and the limits of the fourth class are 22 and 28.

2.28   There is an ambiguity because values from $15.00 to $15.99 can be put into the second or third class; also, there is no place to put values from $24.91 to $24.99 and no place to put $30.00.

2.38   (a)

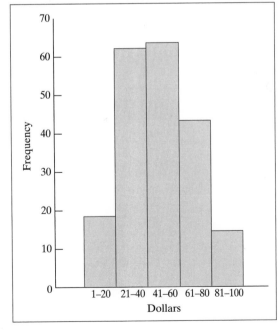

FIGURE 2.12 Histogram of the distribution.

(b)

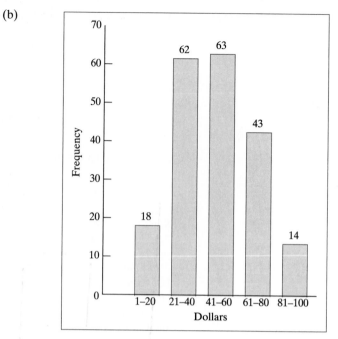

FIGURE 2.13 Bar chart of the distribution.

2.42   Multiplying 81%, 14%, 3%, and 2% by 360, we get the central angles of 291.6°, 50.4°, 10.8°, and 7.2°.

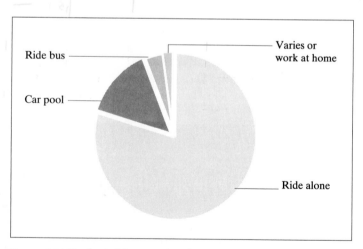

FIGURE 2.14 Pie chart of the central angles.

3

Summarizing Data: Statistical Descriptions

In the introduction to Chapter 1 we said that statistics is the subject that deals with all aspects of the handling of data. Actually, "statistics" has several other meanings. It is also used to refer to the data themselves—for instance, when we refer to birth and death records as **vital statistics**. Furthermore, yet another connotation of "statistics" is the plural of "statistic," that is, a quantity (such as an average) that is computed from sample data. In this sense, a **statistic** is also referred to as a **statistical measure** or simply as a **statistical description**.

It is customary to classify statistical measures according to the particular features of a set of data that they are intended to describe. In this chapter we shall be concerned mainly with **measures of location**, especially **measures of central location**, and **measures of variation**. The former may be described crudely as *averages*, in the sense that they are generally indicative of the "center," "middle," or the "most typical" of a set of data; the latter are indicative of the variability, spread, or dispersion of a set of data. Other statistical measures will be taken up later, as needed.

For most statistical descriptions, the calculations are fairly simple. However, for large sets of data we may well want to use a computer, a graphing calculator, a statistical calculator, or other type of calculating device.

Measures of location are discussed in Sections 3.1 through 3.4, and measures of variation in Sections 3.5 through 3.7. The description of grouped data is treated in Section 3.8, and some further kinds of descriptions are presented in Section 3.9.

Since the term *statistic* was introduced in connection with sample data, let us add that there is also a name for statistical descriptions of populations—we call them **parameters**. As we shall see, the distinction between statistics and parameters will serve to simplify and clarify our language. Indeed, we shall even use different symbols for statistical measures, depending on whether they are used to describe samples or populations.

3.1
Measures of Location: The Mean

The most popular measure of central location is what the layperson calls an "average" and what the statistician calls a **mean**.[†] We put the word "average" in quotes because in everyday language it has all sorts of connotations—we speak of a baseball player's batting average, we talk about the average suburban family, we describe a holdup man's appearance as average, and so on. Formally, then, the mean is defined as follows:

The mean of n numbers is their sum divided by n.

■ **EXAMPLE**   On a certain day, nine students received, respectively, 1, 3, 2, 0, 1, 5, 2, 1, and 3 pieces of mail. Find the mean.

Solution   The total number of pieces of mail that the nine students received is

$$1 + 3 + 2 + 0 + 1 + 5 + 2 + 1 + 3 = 18$$

Since $\frac{18}{9} = 2$, the mean number of pieces of mail per student is 2.   ■

■ **EXAMPLE**   A supermarket manager, who wants to study the traffic in her store, finds that 295, 1,002, 941, 768, and 1,283 persons entered the store during the past five days. Find the mean number of persons who entered the supermarket during these five days.

Solution   Altogether the number of persons who entered the supermarket during the past five days is

$$295 + 1,002 + 941 + 768 + 1,283 = 4,289$$

Since $\dfrac{4,289}{5} = 857.8$, this is the mean (or average) number of persons who entered the store per day.   ■

Since we shall have occasion to calculate the means of many different sets of data, it will be convenient to have a formula that is always applicable. This requires that we represent the numbers to be averaged by using symbols such as x, y, or z; the number of values in a sample, the **sample size**, is

---

[†]The mean is also referred to as the *arithmetic mean*, to distinguish it from the *geometric mean* and the *harmonic mean*, two other kinds of averages that are used only in very special situations and will not be discussed in this book.

usually denoted by the letter n. Choosing the letter x, we can refer to the n values in a sample as x_1 (which is read "x sub-one"), x_2 (which is read "x sub-two"), $x_3, \ldots$, and x_n, and write the **sample mean** as

$$\frac{x_1 + x_2 + x_3 + \cdots + x_n}{n}$$

This formula will take care of any set of sample data, but it can be made more compact by assigning the sample mean the symbol $\bar{x}$ (which is read "x bar") and by using the Σ **notation**. The symbol Σ is capital *sigma*, the Greek letter for S. In this notation, we let Σx stand for "the sum of the x's" (that is, $\Sigma x = x_1 + x_2 + x_3 + \cdots + x_n$), so that we can write

Sample mean

$$\bar{x} = \frac{\Sigma x}{n}$$

If we refer to the measurements as y's or z's, we write their mean as $\bar{y}$ or $\bar{z}$ and substitute Σy or Σz for Σx.

In the formula for $\bar{x}$, the expression Σx does not state explicitly which x's are to be added; let it be understood, therefore, that Σx always stands for the sum of all the x's under consideration in a given situation. In Section 3.10, the use of the sigma notation is discussed in some detail.

The mean of a population of N items is defined in the same way. It is the sum of the N items, $x_1 + x_2 + x_3 + \cdots + x_N$, or Σx, divided by the **population size N**. Assigning the **population mean** the symbol μ (lowercase mu, the Greek letter for m), we write

Population mean

$$\mu = \frac{\Sigma x}{N}$$

with the reminder that Σx is now the sum of all N values that constitute the population. To distinguish between parameters and statistics, namely, descriptions of populations and samples, it is common practice to denote the former with Greek letters. We note also that uppercase N is the common symbol for the population size, while lowercase n is used for the sample size.

The division step in finding the sample mean or population mean generally yields an answer which needs to be rounded. In our work in this book, we try to retain a reasonable number of figures after the rounding, but it must be admitted that one cannot form a dogmatic rule for dealing with rounding. We will use the *approximately equal* sign $\approx$ to indicate that a result has been rounded. As an illustration, we will write $\bar{x} = \frac{38}{7} \approx 5.43$. We will generally use the $\approx$ sign only at the step in which the arithmetic occurs; for example, a later reference to the same information would note $\bar{x} = 5.43$.

The popularity of the mean as a measure of the *middle* or *center* of a set of data is not accidental. Any time we use a single number to describe some aspect of a set of data, there are certain requirements, or desirable features, that should be kept in mind. Aside from the fact that the mean is a simple and familiar measure, the following are some of its noteworthy properties:

1. **The mean can be calculated for any set of numerical data.**
2. **For any set of numerical data, the mean is a unique, unambiguous value.**
3. **The mean lends itself to further statistical treatment; for example, the means of several sets of data can always be combined into the overall mean of all the data. Most other statistics do not have this property.**
4. **If each value in a sample were replaced by the mean, then $\sum x$ would remain unchanged.**
5. **The mean takes into account the value of each item in a set of data.**
6. **The mean is relatively reliable in the sense that means of many samples drawn from the same population generally do not fluctuate, or vary, as widely as other statistics used to estimate the mean of a population. (See Exercises 3.37 and 3.38.)**

This last property is of fundamental importance in statistical inference, and we shall study it in some detail in Chapter 8.

Sometimes a set of data may contain very small or very large values that are so far removed from the main body of the data that the appropriateness of including them when describing the data is questionable. Such values may be due to chance, or they may be due to gross errors in recording the data, gross errors in calculations, malfunctioning of equipment, or other identifiable sources of contamination. When such values are averaged in with the other values, they can affect the mean to such an extent that it is debatable whether it really provides a useful description of the middle of the data. In such cases, we may trim the data by deleting the upper and lower 5% and refer to the mean of the remaining data as the **trimmed mean**.

■ **EXAMPLE**   Five lightbulbs burned out after lasting, respectively, for 867, 849, 840, 852, and 822 hours of continuous use. Find the mean and also determine what the mean would have been if the second value had been recorded incorrectly as 489 instead of 849.

Solution   For the original data we get

$$\bar{x} = \frac{867 + 849 + 840 + 852 + 822}{5} = \frac{4{,}230}{5} = 846$$

and with 489 instead of 849 we get

$$\bar{x} = \frac{867 + 489 + 840 + 852 + 822}{5} = \frac{3{,}870}{5} = 774$$

This shows that a very small or very large value, here due to a careless error in recording the data, can have a pronounced effect on the mean.   ■

To avoid the possibility of being misled by very small or very large values, we sometimes describe the middle or center of a set of data with other kinds of statistical measures, for instance, those given in Sections 3.3 and 3.4.

3.2
Measures of Location: The Weighted Mean

When averaging quantities, it is often necessary to account for the fact that not all of them are equally important in the phenomenon being described. For instance, in 1996 the production of milk, in thousands of pounds per year per cow, was 17.0 in Michigan, 15.4 in Wisconsin, and 15.8 in Minnesota. The mean of these three milk production numbers is

$$\frac{17.0 + 15.4 + 15.8}{3} = \frac{48.2}{3} \approx 16.1$$

but we cannot very well say that this is the average milk production for the three states. The three figures do not carry equal weight because there are not equally many cows in the three states (see Exercise 3.19).

To give quantities being averaged their proper degree of importance, it is necessary to assign them **weights** (measures of their relative importance) and then calculate a **weighted mean**. In general, the weighted mean $\bar{x}_w$ of a set of numbers $x_1, x_2, x_3, \ldots$, and x_n, whose relative importance is expressed numerically by a corresponding set of numbers $w_1, w_2, w_3, \ldots$, and w_n, is given by

Weighted mean

$$\bar{x}_w = \frac{w_1 x_1 + w_2 x_2 + \cdots + w_n x_n}{w_1 + w_2 + \cdots + w_n} = \frac{\sum w \cdot x}{\sum w}$$

Here $\sum w \cdot x$ stands for the sum of the products obtained by multiplying each x by the corresponding weight, and $\sum w$ is simply the sum of the weights. If the weights are all equal, the formula for the weighted mean becomes

$$\bar{x}_w = \frac{wx_1 + wx_2 + \cdots + wx_n}{w + w + \cdots + w}$$

$$= \frac{w(x_1 + x_2 + \cdots + x_n)}{n \cdot w}$$

$$= \frac{\sum x}{n}$$

and thus reduces to that of the (ordinary) mean.

■ **EXAMPLE**   In a recent year, cod, flounder, haddock, and ocean perch brought commercial fishermen 33.0, 57.9, 39.4, and 28.3 cents per pound. Given that they caught 100 million pounds of cod, 201 million pounds of flounder, 55 million pounds of haddock, and 19 million pounds of ocean perch, what is the overall average price that they received per pound?

Solution   Substituting $x_1 = 33.0$, $x_2 = 57.9$, $x_3 = 39.4$, $x_4 = 28.3$, $w_1 = 100$, $w_2 = 201$, $w_3 = 55$, and $w_4 = 19$ into the formula for $\bar{x}_w$, we get

$$\bar{x}_w = \frac{(100)(33.0) + (201)(57.9) + (55)(39.4) + (19)(28.3)}{100 + 201 + 55 + 19}$$

$$= \frac{17{,}642.6}{375} \approx 47.0$$

Note that the figure in the denominator, 375, is the total catch in millions of pounds, and that the figure in the numerator, 17,642.6, is the total value of the catch in millions of cents, that is, in units of $10,000. Also, if we had averaged 33.0, 57.9, 39.4, and 28.3 without using weights, we would have obtained

$$\bar{x} = \frac{33.0 + 57.9 + 39.4 + 28.3}{4} = \frac{158.6}{4} \approx 39.6$$

Can you explain why this is much less than the actual average price of 47.0 cents?   ∎

A special application of the formula for the weighted mean arises when we must find the overall mean, or **grand mean**, of k sets of data having the means $\bar{x}_1, \bar{x}_2, \bar{x}_3, \ldots,$ and $\bar{x}_k$, consisting, respectively, of $n_1, n_2, n_3, \ldots,$ and n_k measurements or observations. The result is given by

Grand mean of combined data

$$\bar{\bar{x}} = \frac{n_1\bar{x}_1 + n_2\bar{x}_2 + \cdots + n_k\bar{x}_k}{n_1 + n_2 + \cdots + n_k} = \frac{\Sigma n \cdot \bar{x}}{\Sigma n}$$

where the weights are the sizes of the respective samples, the numerator is the total of all the measurements or observations, and the denominator is the sample size of the combined data.

∎ **EXAMPLE**   In a biology class there are 20 freshmen, 18 sophomores, and 12 juniors. If the freshmen averaged 68 on an examination, the sophomores averaged 75, and the juniors averaged 86, find the mean grade for the entire class.

Solution   Substituting $n_1 = 20$, $n_2 = 18$, $n_3 = 12$, $\bar{x}_1 = 68$, $\bar{x}_2 = 75$, and $\bar{x}_3 = 86$ into the formula for the grand mean of combined data, we get

$$\bar{\bar{x}} = \frac{20 \cdot 68 + 18 \cdot 75 + 12 \cdot 86}{20 + 18 + 12}$$

$$= \frac{3,742}{50} = 74.84 \approx 75$$ ∎

Exercises

Exercises 3.1, 3.4, 3.10, 3.17, and 3.18 are practice exercises; their complete solutions are given on pages 94 and 95.

   3.1 Which of the following are statistics, and which are parameters?
   (a) A statistical measure that is calculated from or used to describe a population.

(b) A statistical measure that is calculated from or used to describe a sample.

(c) A statistical measure that is calculated from or used to describe a large sample.

(d) A statistical measure that is calculated from or used to describe a small population.

3.2 Which of the following are statistics, and which are parameters?

(a) The mean height of a sample of basketball players in the National Basketball Association.

(b) The mean weight of the population of football players on the Washington Redskins team.

(c) The mean blood pressure of the population of patients in the cardiac ward of a hospital.

(d) The mean grade of students in a certain elementary algebra section at a college, which is used to estimate the mean grade of all students in all sections of elementary algebra at the college.

3.3 Suppose that we are given complete information about the number of lunches that the executives of an insurance company charged to their expense accounts during the first six months of 1998. Give one illustration each of a situation where the mean of these data would be looked upon as

(a) a parameter;   (b) a statistic.

3.4 The following are the ages of 20 persons empaneled for jury duty by a court:

$$
\begin{array}{ccccc}
48 & 58 & 33 & 42 & 57 \\
31 & 52 & 25 & 46 & 60 \\
61 & 49 & 38 & 53 & 30 \\
47 & 52 & 63 & 41 & 34
\end{array}
$$

Find their mean age.

3.5 According to the U.S. Bureau of the Census, the expectation of life at birth (in years) in the year 1997 for selected nations was as follows:

Country	Years	Country	Years
Argentina	74.4	India	60.2
Australia	79.6	Italy	78.2
Austria	76.7	Japan	79.7
Belgium	77.2	Mexico	74.0
Brazil	61.4	Netherlands	77.9
Canada	79.3	Russia	63.8
Chile	74.7	Spain	78.5
China	70.0	Sweden	78.2
Egypt	61.8	Switzerland	77.8
France	78.6	Ukraine	67.1
Germany	76.1	United Kingdom	76.6
Greece	78.3	United States	76.0

Find their mean expectation of life at birth.

3.6 According to the Federal Highway Administration, the state gasoline tax in the Mountain States in cents per gallon on July 1, 1997, was as follows:

State	Cents per gallon
Montana	27.0
Idaho	25.0
Wyoming	9.0
Colorado	22.0
New Mexico	18.875
Arizona	18.0
Utah	24.5
Nevada	24.0

Find their mean state tax on gasoline.

3.7 In January 1998, the 15 mutual funds that delivered the highest returns (over the preceding five years) had annual expenses, in percent, of

1.55 0.93 1.10 1.54 0.73 1.10 0.88 1.17
1.31 1.70 1.10 1.75 1.73 1.05 1.10

Find their mean annual expenses, in percent.

3.8 According to the Bureau of Labor Statistics, the average annual pay (in thousands of dollars) paid in the top 15 metropolitan areas in the United States in 1995 was

42.4 42.3 38.0 37.9 37.5 37.2 36.6 35.7
34.9 34.7 34.6 34.0 33.9 33.6 33.4

Find their mean annual pay in thousands of dollars. Note that data include approximately 97% of wage and salary civilian employment.

3.9 The records of 15 persons convicted of various crimes showed that, respectively, 4, 3, 0, 0, 2, 4, 4, 3, 1, 0, 2, 0, 2, 1, and 4 of their grandparents were foreign born. Find the mean and discuss whether it can be used to support the contention that the "average criminal" has two foreign-born grandparents.

3.10 An elevator in a hotel is designed to carry a maximum load of 2,000 pounds. Is it overloaded if at one time it carries eight women whose mean weight is 124 pounds and six men whose mean weight is 165 pounds?

3.11 An airplane cargo bay is designed to carry a maximum load of 12,000 pounds. If at a given moment it is loaded with 168 crates having a mean weight of 64 pounds, is there any real danger that the cargo bay is overloaded?

3.12 By mistake, an instructor has erased the grade that one of ten students in her class received in a final examination. However, she knows that the students averaged (had a mean grade of) 71 on the examination, and that the other nine students received grades of 96, 44, 82, 70, 47, 74, 94, 78, and 56. What must have been the grade that the teacher erased?

3.13 The hours of sleep that a student has had during each of the last 10 nights are 6, 8, 2, 13, 9, 7, 6, 5, 10, and 7. Find the mean and discuss its usefulness in describing the middle of the data.

3.14 A bill was introduced in a state legislature to repeal the sales tax on prescription drugs. Comment on the argument of the state finance director that the average (mean) per capita prescription bill for the past four years was a trifling $7.20, which is not really a burden to anyone.

3.15 Records show that in Phoenix, Arizona, the normal daily maximum temperature for each month is 65, 69, 74, 84, 93, 102, 105, 102, 98, 88, 74, and 66 degrees Fahrenheit. Verify that the mean of these figures is 85 and comment on the claim that, in Phoenix, the average daily maximum temperature is a very comfortable 85 degrees.

3.16 Careful measurements show that the actual coffee content of six jars of instant coffee is 6.03, 5.98, 6.04, 6.00, 5.99, and 6.02 ounces of coffee. What would have been the error in the mean coffee content of the six jars if the third value had been recorded incorrectly as 6.40?

3.17 If somebody invests $1,000 at 7%, $3,000 at 7.5%, and $16,000 at 8%, what is the overall percentage yield of these investments?

3.18 In a recent year, the average salaries of elementary school teachers in Oregon, Washington, and Alaska were $39,600, $37,900, and $49,600. Given that there were 17,800, 27,300, and 4,900 elementary school teachers in these states, find the average salary of all the elementary school teachers in the three states.

3.19 Previously, we stated that in 1996 the production of milk, in thousands of pounds per year per cow, was 17.0 in Michigan, 15.4 in Wisconsin, and 15.8 in Minnesota. Given that there were 320,000 cows in Michigan, 1,449,000 cows in Wisconsin, and 598,000 cows in Minnesota, what is the average annual milk production for the three states combined?

3.20 An instructor counts the final examination in a course three times as much as each of the three one-hour examinations. What is the average grade of a student who received grades of 75, 77, and 58 on the three one-hour examinations and 82 on the final examination?

3.21 In a recent season, a baseball team's five best hitters had batting averages of 0.307, 0.299, 0.297, 0.291, and 0.283. What is their combined batting average if they had, respectively, 488, 137, 646, 533, and 502 at bats?

3.22 During a special promotion, a discount chain sold 575, 410, and 520 microwave ovens in three of its stores at average prices of $495, $525, and $500, respectively. What is the mean price of the ovens sold?

3.23 In a study of home replacement costs, it was found that 12 houses in one city were underinsured on the average by $5,000, 24 homes in another city were underinsured on the average by $6,200, and 14 homes in a third city were underinsured on the average by $6,900. On the average, by how much were these 50 homes underinsured?

3.24 The following ordered data relates to the average travel time to work, in minutes, of 100 employees in a large downtown office building.

11.4	12.2	12.6	12.6	12.6	14.0	15.8	16.2	16.2	18.0
18.0	19.4	20.2	20.4	22.2	22.6	23.0	23.6	24.2	24.8
27.0	27.4	28.0	28.2	28.4	28.8	29.2	29.8	30.0	30.2
30.4	31.0	31.8	32.4	33.6	35.2	35.4	35.4	36.2	37.0
37.0	37.4	37.6	37.6	37.8	38.0	38.4	38.4	39.0	39.0
39.2	39.4	39.4	39.4	39.4	39.8	40.0	40.4	40.6	40.8
41.0	41.0	41.4	42.0	42.0	42.2	42.6	43.0	43.2	44.0
44.0	44.0	45.0	45.6	45.8	46.0	47.0	47.8	49.2	49.4
49.6	50.6	52.4	52.6	53.6	56.2	56.4	56.4	60.0	64.6
74.0	86.4	89.6	90.2	90.4	100.0	103.4	108.2	123.8	125.4

Find the trimmed mean for the average travel time of these 100 employees.

3.3
Measures of Location: The Median and Other Fractiles

To avoid the possibility of being misled by a few very small or very large values, as demonstrated previously, it is sometimes preferable to describe the *middle* or *center* of a set of data with statistical measures other than the mean.

The definition of one of these, the **median of *n* values**, requires that we arrange the data according to size. Then,

When n is odd, the median is the value of the item that is in the middle.

When n is even, the median is the mean of the two items that are nearest to the middle.

The symbol that we use for the median of a set of *x*'s is $\tilde{x}$ (and, hence, $\tilde{y}$ or $\tilde{z}$ if we refer to the measurements as *y*'s or *z*'s).

■ **EXAMPLE** In a recent month, a state's Game and Fish Department reported 53, 31, 67, 53, and 36 hunting or fishing violations for five different regions. Find the median number of violations for these regions.

Solution The median is not 67, the third (or middle) item, because the figures must first be arranged according to size. Thus, we get

$$31 \quad 36 \quad 53 \quad 53 \quad 67$$

and it can be seen that the median is 53. ■

Note that in this example there are two 53s among the data and that we do not refer to either of them as the median; the median is a number and not necessarily a particular measurement or observation.

■ **EXAMPLE** Ten persons, sent out to interview 50 students at each of 10 different campuses, found that 18, 13, 15, 12, 8, 21, 7, 11, 16, and 3 of the students sampled jog regularly. Find the median.

Solution Arranging these figures according to size, we get

$$3 \quad 7 \quad 8 \quad 11 \quad 12 \quad 13 \quad 15 \quad 16 \quad 18 \quad 21$$

and it can be seen that the median is $\dfrac{12 + 13}{2} = 12.5$, that is, the mean of the two items nearest the middle. ■

We cannot give a general formula for the value of the median, but we can give one for the **median position**. With data ranked from low to high or high to low and counting from either end,

The median is the value of the $\dfrac{n + 1}{2}$ th item.

When n is odd, $\dfrac{n+1}{2}$ is an integer and it gives the position of the median; when n is even, $\dfrac{n+1}{2}$ is midway between two integers and the median is the mean of the values of the corresponding items.

■ **EXAMPLE**   Find the median position for (a) $n = 11$, (b) $n = 25$, and (c) $n = 75$.

Solution   With the data arranged according to size and counting from either end,

(a) $\dfrac{n+1}{2} = \dfrac{11+1}{2} = 6$, so that for $n = 11$ the median is the value of the 6th item;

(b) $\dfrac{n+1}{2} = \dfrac{25+1}{2} = 13$, so that for $n = 25$ the median is the value of the 13th item;

(c) $\dfrac{n+1}{2} = \dfrac{75+1}{2} = 38$, so that for $n = 75$ the median is the value of the 38th item.   ■

■ **EXAMPLE**   Find the median position for (a) $n = 8$, (b) $n = 20$, and (c) $n = 100$.

Solution   With the data arranged according to size and counting from either end,

(a) $\dfrac{n+1}{2} = \dfrac{8+1}{2} = 4.5$, so that for $n = 8$ the median is the mean of the values of the 4th and 5th items;

(b) $\dfrac{n+1}{2} = \dfrac{20+1}{2} = 10.5$, so that for $n = 20$ the median is the mean of the values of the 10th and 11th items;

(c) $\dfrac{n+1}{2} = \dfrac{100+1}{2} = 50.5$, so that for $n = 100$ the median is the mean of the values of the 50th and 51st items.   ■

It is important to remember that $\dfrac{n+1}{2}$ is a formula for the median position, and not a formula for the value of the median. To simplify the determination of a median, it sometimes helps to utilize the grouping provided by a stem-and-leaf display.

■ **EXAMPLE**   The following are the numbers of passengers on 25 runs of a ferryboat:

$$
\begin{array}{ccccc}
52 & 84 & 40 & 57 & 61 \\
65 & 77 & 64 & 62 & 35 \\
82 & 58 & 50 & 78 & 103 \\
71 & 75 & 41 & 51 & 66 \\
60 & 95 & 58 & 49 & 89 \\
\end{array}
$$

Construct a stem-and-leaf display with one-digit leaves and use it to find the median.

Solution   First constructing the stem-and-leaf display and sorting the leaves, we get

$$
\begin{array}{r|llllll}
3 & 5 \\
4 & 0 & 1 & 9 \\
5 & 0 & 1 & 2 & 7 & 8 & 8 \\
6 & 0 & 1 & 2 & 4 & 5 & 6 \\
7 & 1 & 5 & 7 & 8 \\
8 & 2 & 4 & 9 \\
9 & 5 \\
10 & 3 \\
\end{array}
$$

Since the median position is $\dfrac{25 + 1}{2} = 13$ and 10 values are found on the first three stems, we must find the third smallest value on the fourth stem. As can be seen by inspection, it is 62.   ■

If we had calculated the mean in the example on page 52 that dealt with the hunting and fishing violations, we would have obtained

$$
\frac{53 + 31 + 67 + 53 + 36}{5} = \frac{240}{5} = 48
$$

and it should not come as a surprise that it differs from the median, which was 53. Each of these averages describes the middle or center of the data in its own way. The median is average in the sense that it splits the data into two parts so that as many of the values are to the left of the median position as there are to its right. If the values in a set of data are all unequal, we can substitute "below" for "to the left of" and "above" for "to the right of." When some of the values are equal, this may or may not be done. For instance, it cannot be done for 1, 2, 3, 3, 3, 3, and 5, where the median is 3 and there are two values below the median and only one value above it.

The mean, on the other hand, is typical in the sense that if each value in a set of data is replaced by the same number while the total remains unchanged, this number will have to be the mean. This is the fourth of the six properties listed in Section 3.1, and it follows from the fact that

$$
n \cdot \bar{x} = n \cdot \frac{\sum x}{n} = \sum x
$$

The median shares the first two properties of the mean as listed in Section 3.1; that is, it can be determined for any set of numerical data and it is always unique.[†] Also like the mean, the median is simple enough to find once the data has been arranged according to size, but it should be kept in mind that ordering a large set of data manually can be a very tedious job. As far as the third and sixth properties are concerned, the medians of several sets of data can generally not be combined into the overall median of all the data, and in problems of inference the median is generally not as reliable as the mean: that is, it is subject to greater chance fluctuations, as is illustrated by Exercises 3.37 and 3.38. When it comes to the fifth property, the median is actually preferable to the mean, because it is not so easily affected by extreme (very large or very small) values.

■ **EXAMPLE**   In a previous example on page 46 we showed that the mean lifetime of five lightbulbs is $\bar{x} = 846$ hours, but if one of the values, 849, is recorded incorrectly as 489, the mean lifetime is 774, which is off by $846 - 774 = 72$ hours. What would have been the corresponding error if we had used the medians?

Solution   Since the original values, sorted, are 822, 840, 849, 852, and 867, the correct value of the median is $\tilde{x} = 849$. With 489 substituted for 849, the values, arranged according to size, are 489, 822, 840, 852, and 867, and $\tilde{x} = 840$, which is off by $849 - 840 = 9$ hours. This is much smaller than the error we made when we used the mean.   ■

Also, unlike the mean, the median can be used to define the middle of a number of objects, properties, or attributes. It is possible, for example, to rank a number of tasks according to their difficulty and then describe the middle (or median) one as being of *average* difficulty; also, we might rank samples of chocolate fudge according to their consistency and then describe the middle (or median) one as having average consistency.

The median is but one of many different **fractiles** that divide a set of data into two or more equal parts. Also of importance in statistics are **quartiles** and **percentiles**; but since percentiles are used mainly in connection with large sets of data, we shall discuss them for grouped data in Section 3.8. Thus, let us treat here the three quartiles Q_1, Q_2, and Q_3, which are defined as follows: With the data arranged from left to right in an increasing order of magnitude,

The first quartile, Q_1, is the median of all the values to the left of the median position for the whole set of data.

The second quartile, Q_2, is the median.

The third quartile, Q_3, is the median of all the values to the right of the median position for the whole set of data.

[†]There are other definitions for the median that allow it to be nonunique when n is even. The definition used in this book leads to a unique median.

Generally, we find Q_3 by counting as many places as for Q_1, starting at the other end, that is, from the right rather than the left for data ordered from low to high.

Note that with this definition there are always as many values to the left of the Q_1 position as there are between the Q_1 position and the median position, between the median position and the Q_3 position, and to the right of the Q_3 position.

■ **EXAMPLE**   Verify the preceding sentence for $n = 23$.

Solution   The median position is $\dfrac{23 + 1}{2} = 12$, so there are 11 values to its left and 11 values to its right. Thus, the Q_1 position is $\dfrac{11 + 1}{2} = 6$ and, correspondingly, Q_3 is the 6th value counting from the other end; that is, the Q_3 position is 18 (see Figure 3.1). It follows that there are five values to the left of the Q_1 position, five values between the Q_1 position and the median position (between 6 and 12), five values between the median position and the Q_3 position (between 12 and 18), and five values to the right of the Q_3 position.   ■

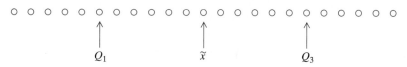

FIGURE 3.1 The median and quartile positions for $n = 23$.

■ **EXAMPLE**   Use the following stem-and-leaf display constructed on page 54 to find Q_1 and Q_3 for the data on the passengers of the 25 runs of a ferryboat. The median position is $\dfrac{25 + 1}{2} = 13$.

3	5					
4	0	1	9			
5	0	1	2	7	8	8
6	0	1	2	4	5	6
7	1	5	7	8		
8	2	4	9			
9	5					
10	3					

Solution   Since we have already shown that the median position is 13, we find that the Q_1 position is $\dfrac{12 + 1}{2} = 6.5$. Thus, Q_1 is the mean of the 6th and 7th values, and it can be seen from the stem-and-leaf plot that it equals $\dfrac{51 + 52}{2} = 51.5$.

To find Q_3, we count 6.5 values starting at the other end. Since the 6th value is 78 and the 7th value is 77, we get $\dfrac{78 + 77}{2} = 77.5$.   ■

■ **EXAMPLE** The following are the numbers of minutes that a woman, on her way to work, had to wait for the bus on 14 working days: 10, 2, 17, 6, 8, 3, 10, 2, 9, 5, 9, 13, 1, and 10. Find the median, Q_1, and Q_3.

Solution For $n = 14$, the median position is $\dfrac{14 + 1}{2} = 7.5$, so that the Q_1 position is $\dfrac{7 + 1}{2} = 4$, and Q_3 is the fourth value from the other end. Since the data, arranged according to size, are

1   2   2   3   5   6   8   9   9   10   10   10   13   17

it can be seen that the median is $\dfrac{8 + 9}{2} = 8.5$, Q_1 is 3, and $Q_3 = 10$.   ■

The information provided by the median, the two quartiles, and also the smallest and largest values, is sometimes presented in the form of a **box-and-whisker plot**, often just called a **box plot**. Such a plot, reproduced from the display screen of a TI-83 graphing calculator is shown in Figure 3.2 for the data of the preceding example. There is no numerical scale on the horizontal axis, but note that the rectangle extends from $Q_1 = 3$ to $Q_3 = 10$; the whisker on the left extends to the smallest value, which is 1 equals; the whisker on the right extends to the largest value, which is 17; the vertical line that divides the box is at the median, which is 8.5. Box plots provide useful information for the exploratory analysis of the overall shape of distribution (see also Section 3.8). They are sometimes embellished with other features of the data, but the simple form given here is adequate for most purposes.

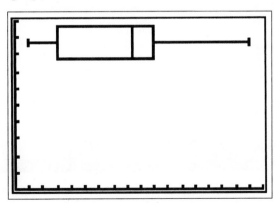

FIGURE 3.2 Box-and-whisker plot reproduced from the display screen of a TI-83 graphing calculator.

Optional directions for graphing the box plot with a TI-83 graphing calculator are provided on page 58. These are intended primarily for users of graphing calculators and can be omitted by other readers without loss of continuity.

Two other measures of central location, the midrange and the midquartile, are introduced in Exercises 3.41 and 3.48.

USING TECHNOLOGY : *The Graphing Calculator*

How to graph the box plot shown in Figure 3.2 using the TI-83.

(Tip: Before beginning, be sure to clear all lists and plots—see Appendix A: TI-83 Tips if you need help.)

Step 1.   Input the example data into list L1.
(Questions? See Appendix A: TI-83 Tips.)

Step 2.   Select the type of graph.
Press 2nd, Y= to choose the STAT PLOTS function.
Select PLOT 1, press ENTER.
Highlight ON.
Scroll to the box plot icon (the fifth graph icon), press ENTER.
Set   Xlist: L1
      Freq: 1

Step 3.   Set the window.
(See Appendix A: TI-83 Tips for more information on setting windows.)

> Press WINDOW
> Set   Xmin = 0
>   Xmax = 18
>   Xscl = 1
>   Ymin= 1
>   Ymax = 10
>   Yscl = 1
>   Xres = 1

Step 4.   View the graph.
Press GRAPH.

Exercises

Exercises 3.25, 3.28, 3.37, 3.42, 3.45, and 3.48 are practice exercises; their complete solutions are given on pages 95 and 96.

3.25 Find the median position for (a) $n = 15$ and (b) $n = 40$.

3.26 Find the median position for (a) $n = 17$ and (b) $n = 30$.

3.27 Find the median position for (a) $n = 39$ and (b) $n = 150$.

3.28 The value of investments, by country, held by the Janus Fund at the end of October 1997 were as follows:

Country	Market value (millions of dollars)
Argentina	123
France	87
Germany	174
Italy	97
Japan	59
Netherlands	1,215
Switzerland	68
United Kingdom	1,710
United States	15,613

(a) Find the median of this list of nine values.
(b) Find the mean.
(c) Indicate whether the mean or median is more appropriate for this list.

3.29 The following are the numbers of restaurant meals that 13 persons ate during a given week:

3 10 5 1 8 5 6 12 15 1 0 6 5

Find the median.

3.30 The melting points, in degrees Fahrenheit, of 16 selected metals are as follows:

Metals	Degrees Fahrenheit	Metals	Degrees Fahrenheit
Aluminum	1,220	Pewter	563
Chromium	3,430	Platinum	3,224
Copper	1,981	Plutonium	1,384
Gold	1,945	Silver	1,761
Iron	2,802	Tin	449
Lead	621	Titanium	3,300
Magnesium	1,202	Tungsten	6,170
Nickel	2,651	Zinc	787

Find
(a) the mean;   (b) the median.

3.31 The following list gives the duration in minutes of 24 power failures:

18	125	44	96	31	53
26	80	49	125	63	58
45	33	89	12	103	127
75	40	80	61	28	129

Find the median.

3.32 Civilian employment in the year 1995 as a percent of the civilian working age population in selected countries was as follows:

Nation	Percent
Australia	50.3
Canada	52.1
France	40.6
Germany	41.2
Italy	28.4
Japan	47.7
Sweden	54.9
United Kingdom	49.0
United States	55.6

Find the median percentage.

3.33 The numbers of full-length movies observed by a group of 400 persons were as follows:

NUMBER OF MOVIES	NUMBER OF PERSONS
0	72
1	106
2	153
3	40
4	18
5	7
6	3
7	0
8	1
Total	400

Find the median.

3.34 Twenty-five National Basketball Association (NBA) games lasted the following times (in minutes):

138	142	113	126	135
142	159	157	140	157
121	128	142	164	155
139	143	158	140	118
142	146	123	130	137

Find the median length of these games
(a) directly;   (b) by first constructing a stem-and-leaf display.

3.35 Using the following stem-and-leaf display concerning the scores obtained by 20 students on a physical coordination test (see Section 2.2),

5	2	7	5					
6	9	1	5	3	4	7	1	8
7	4	9	2	4	7			
8	4	8	2					
9	3							

determine the median score.

3.36 According to the U.S. Bureau of the Census, the 10 fastest-growing states in the United States from 1990 to 1996 were Washington, Oregon, Idaho, Nevada, Utah, Colorado, Arizona, New Mexico, Texas, and Georgia. The percentage population increase in these states was, respectively, 13.7, 12.7, 18.1, 33.4, 16.1, 16.0, 20.8, 13.1, 12.6, and 13.5. Find

    (a) the mean;   (b) the median.

    (c) Explain whether the mean or median is a better indicator of the states' average growth.

3.37 To verify the claim that the mean is generally more reliable than the median (meaning that it is subject to smaller chance fluctuations), a student conducted an experiment consisting of 15 times drawing 3 cards from a modified deck of cards. This modified deck had 40 cards and was obtained by removing the jacks, queens, and kings from an ordinary deck of 52. The aces were assigned the numeric value 1. Here are the results:

Trial number	Cards selected			Trial number	Cards selected		
1	8	3	2	9	4	5	9
2	4	4	10	10	1	8	8
3	7	10	4	11	3	5	9
4	8	10	3	12	10	1	4
5	3	6	7	13	5	1	9
6	6	6	7	14	5	3	8
7	8	3	8	15	10	5	6
8	1	3	1				

    (a) Calculate the 15 medians and 15 means.

    (b) Group the medians and the means obtained in part (a) into separate distributions having the classes 0.5–2.5, 2.5–4.5, 4.5–6.5, and 6.5–8.5. (There can be no ambiguities since the medians and the means of three whole numbers cannot equal 2.5, 4.5, or 6.5.)

    (c) Draw histograms of the two distributions obtained in part (b) and explain how they illustrate the claim that the mean is generally more reliable than the median.

3.38 To verify the claim that the mean is generally more reliable than the median (meaning that it is subject to smaller chance fluctuations), a student conducted an experiment consisting of 12 tosses of three dice. The results were:

Toss number	Dice values			Toss number	Dice values		
1	2	4	6	7	5	5	2
2	5	2	3	8	3	3	3
3	3	1	4	9	4	5	3
4	1	6	2	10	3	2	1
5	5	3	5	11	3	3	4
6	6	1	5	12	4	5	3

    (a) Calculate the 12 medians and the 12 means.

    (b) Group the medians and the means obtained in part (a) into separate distributions having the classes 1.5–2.5, 2.5–3.5, 3.5–4.5, and 4.5–5.5. (There will be no ambiguities since the medians and means of three whole numbers cannot equal 2.5, 3.5, or 4.5.)

(c) Draw histograms of the two distributions obtained in part (b), and explain how they illustrate the claim that the mean is generally more reliable than the median.

3.39 Repeat Exercise 3.38 with your own data by repeatedly rolling three dice (or one die three times) and construct corresponding distributions for the 12 medians and the 12 means. (If no dice are available, simulate the experiment mentally or by drawing numbered slips of paper out of a hat.)

3.40 A consumer testing service obtained the following miles per gallon in five test runs performed with each of three compact cars:

Car A:	27.9	30.4	30.6	31.4	31.7
Car B:	31.2	28.7	31.3	28.7	31.3
Car C:	28.6	29.1	28.5	32.1	29.7

(a) If the manufacturers of car A want to advertise that their car performed best in this test, which of the averages discussed in this text could they use to substantiate their claim?

(b) If the manufacturers of car B want to advertise that their car performed best in this test, which of the averages discussed in this text could they use to substantiate their claim?

3.41 Suppose that the manufacturers of car C of Exercise 3.40 hire an unscrupulous statistician and instruct him to find some kind of "average" that will show that their car performed best in the test. Show that the **midrange**, which is defined as the mean of the smallest and largest values, will serve their purpose. The midrange is occasionally used as a measure of central location.

3.42 Find the positions of the median, Q_1, and Q_3 when $n = 21$, and verify that there are as many values to the left of the Q_1 position as there are between the Q_1 position and the median position, between the median position and the Q_3 position, and to the right of the Q_3 position.

3.43 Rework Exercise 3.42 with $n = 18$.

3.44 Rework Exercise 3.42 with $n = 24$.

3.45 In Exercise 3.29, 13 persons ate 3, 10, 5, 1, 8, 5, 6, 12, 15, 1, 0, 6, and 5 restaurant meals during a week. Calculate the values of Q_1 and Q_3 for the numbers of restaurant meals.

3.46 The number of minutes spent shopping by 20 randomly selected persons in a certain department store was

31	75	45	26	63
125	33	80	18	103
89	49	44	80	61
40	96	125	12	28

(a) Calculate the values of Q_1 and Q_3 for the lengths of the shopping times.

(b) Construct a box-and-whisker plot for the lengths of the shopping times.

3.47 In Exercise 3.34, twenty-five National Basketball Association games lasted the following numbers of minutes:

$$
\begin{array}{ccccc}
138 & 142 & 113 & 126 & 135 \\
142 & 159 & 157 & 140 & 157 \\
121 & 128 & 142 & 164 & 155 \\
139 & 143 & 158 & 140 & 118 \\
142 & 146 & 123 & 130 & 137 \\
\end{array}
$$

Find Q_1 and Q_3 for the lengths of the games
(a) directly from the data;
(b) by first constructing a stem-and-leaf display.
(c) Then construct a box-and-whisker plot for the lengths of the NBA games.

3.48 The **midquartile** is occasionally used as a measure of central location, and its formula is $\dfrac{Q_1 + Q_3}{2}$. Previously, we gave the numbers of minutes that a woman had to wait for her bus, and we determined that Q_1 is 3 minutes and Q_3 is 10 minutes. The midquartile is therefore

$$
\frac{Q_1 + Q_3}{2} = \frac{3 + 10}{2} = 6.5 \text{ minutes}
$$

which is a measure of central location.

It may be noted that the mean, median, and midquartile are usually different values. Previously, we found that the median of this data is 8.5 minutes, and we can readily calculate that the mean is

$$
\frac{10 + 2 + 17 + 6 + 8 + 3 + 10 + 2 + 9 + 5 + 9 + 13 + 1 + 10}{14} = \frac{105}{14}
$$

$$
= 7.5 \text{ minutes}
$$

Finally, the midrange of the data is the mean of the smallest and largest values, which is $\dfrac{1 + 17}{2} = \dfrac{18}{2} = 9$ minutes.

(a) Exercise 3.46 relates to the length of time spent shopping. Given that $Q_1 = 32$ minutes, $Q_3 = 84.5$ minutes, the shortest shopping time is 12 minutes, and the longest is 125 minutes, find the value of the midquartile and also that of the midrange.
(b) Exercise 3.47 relates to the length of NBA games. Given that $Q_1 = 129$ minutes, $Q_3 = 150.5$ minutes, the shortest game is 113 minutes, and the longest is 164 minutes, find the value of the midquartile and also that of the midrange.

3.49 In a study of the breeding habits of the phoebe, a small grayish-brown bird, an ornithologist watched 15 nests plastered to agricultural barns and found that the number of eggs laid was, respectively,

$$
4 \quad 3 \quad 3 \quad 5 \quad 8 \quad 5 \quad 4 \quad 2 \quad 4 \quad 5 \quad 4 \quad 4 \quad 7 \quad 4 \quad 3
$$

Find
(a) the mean;
(b) the median;
(c) Q_1 and Q_3;
(d) the midquartile;
(e) the midrange.

<table>
<tr><td>

3.4
Measures of
Location: The Mode

</td></tr>
</table>

The **mode** is another measure of location that is sometimes used to describe the middle of a set of data. It is defined as follows:

The mode is the value that occurs with the highest frequency.

In this sense it is "most typical" of a set of data; its two main advantages are that it requires no calculations, only counting, and that it can be determined for qualitative as well as quantitative data.

■ **EXAMPLE**   U.S. Bureau of the Census data shows that the percentages of persons not covered by health insurance in 1995 in the 50 states and the District of Columbia were as follows:

$$
\begin{array}{ccccccccccc}
14 & 12 & 20 & 18 & 21 & 15 & 9 & 16 & 17 & 18 & 18 \\
9 & 14 & 11 & 13 & 11 & 12 & 15 & 20 & 14 & 15 & 11 \\
10 & 8 & 20 & 15 & 13 & 9 & 19 & 10 & 14 & 26 & 15 \\
14 & 8 & 12 & 19 & 12 & 10 & 13 & 15 & 9 & 15 & 24 \\
12 & 13 & 14 & 12 & 15 & 7 & 16 & & & &
\end{array}
$$

Find the mode.

Solution   Among these numbers, the value 15 appears eight times, which is more often than any other number occurs. Thus, 15 is the modal percentage.   ■

■ **EXAMPLE**   Asked to name the best collegiate basketball team in the country, 25 sports-writers named the following teams: Duke, Kansas, UCLA, Arizona, North Carolina, Indiana, Kansas, Kentucky, Kansas, UCLA, Kansas, Duke, Michigan, North Carolina, Kentucky, Washington, UCLA, North Carolina, Arizona, Kansas, Kansas, Duke, UCLA, Kentucky, and North Carolina. Find the mode.

Solution   Since Indiana, Michigan, and Washington are named once, Arizona is named twice, Duke and Kentucky are named three times, UCLA and North Carolina are named four times, and Kansas is named six times, Kansas is the modal choice.   ■

In a great many sets of data, all values are different. Certainly, the mode is useless in such a situation. A serious disadvantage of the mode is that its value is badly altered when the values in the set of data are rounded. Exercise 3.58 illustrates this point. We recommend that the mode be avoided for sets of values presented to several significant figures.

■ **EXAMPLE**   A sample from the records of a motor vehicle bureau shows that 16 drivers in a certain age group received 2, 3, 3, 1, 0, 0, 2, 1, 0, 3, 4, 0, 3, 2, 3, and 0 tickets in a recent year. Find the mode.

Solution   Since 0 occurs five times, 1 occurs two times, 2 occurs three times, 3 occurs five times, and 4 occurs once, 0 and 3 each occurs with the maximum frequency of five. Thus, there are two modes. We might infer from this that there are many very good drivers, many very poor drivers, and fewer drivers in the categories between these two extremes.   ■

There are many other measures of location besides the mean, the median, and the mode, and the question of what particular average should be used in a given situation is not always easily answered. The fact that the selection of statistical descriptions is to some extent arbitrary has led some persons to believe that the magic of statistics can be used to prove almost anything. A famous nineteenth-century British statesman is often quoted as having said that there are three kinds of lies: lies, damned lies, and statistics. Exercises 3.40 and 3.41 described a situation where this kind of criticism would well be justified.

Exercises

Exercises 3.50 and 3.55 are practice exercises; their complete solutions are given on pages 96 and 97.

3.50 The following data provided by the U.S. National Center for Health Statistics are the number of marriages per 1,000 population, year 1995, in 17 eastern states and also in the nation's capitol. These are Maine, New Hampshire, Vermont, Massachusetts, Rhode Island, Connecticut, New York, New Jersey, Pennsylvania, Delaware, Maryland, District of Columbia, Virginia, West Virginia, North Carolina, South Carolina, Georgia, and Florida. The values are, respectively,

9 8 10 7 8 7 8 7 6 8 9 6 10 6 9 12 8 10

Find the mode.

3.51 The following data provided by the U.S. National Center for Health Statistics are the numbers of divorces per 1,000 population, year 1995, for the corresponding eastern states and the District of Columbia as listed in Exercise 3.50:

4 4 5 2 4 3 3 3 3 5 3 3 4 5 5 4 5 6

Find the mode.

3.52 A buyer for a chain of food stores opens 14 identical cans of mixed nuts to count the numbers of different types of nuts in each can. The number of unbroken filberts (sometimes called hazelnuts or cobnuts) in each can was

6 20 10 2 16 10 12 24 30 2 0 12 3 10

Find the mode.

3.53 Exercise 3.29 gives the numbers of restaurant meals that 13 persons ate during a given week as

3 10 5 1 8 5 6 12 15 1 0 6 5

Find the mode.

3.54 The receiving clerk at a warehouse recorded the following number of deliveries received during 22 business days of a month:

4 7 9 3 5 5 6 5 8 3 7 8 6 10 5 4 5 5 6 9 4 8

Find the mode.

3.55 The following are the amounts of time (in minutes) that 16 persons spent standing in line waiting to buy tickets for a concert:

8 2 9 1 16 5 7 11 9 1 14 12 9 0 8 4

Find the mode.

3.56 Asked for their favorite color, 50 persons gave the following responses:

red	blue	blue	green	yellow
blue	brown	red	blue	red
red	green	white	blue	red
green	blue	red	green	green
purple	white	yellow	blue	blue
blue	red	red	brown	orange
white	green	blue	blue	black
red	blue	red	yellow	green
yellow	blue	blue	orange	red
green	white	purple	blue	red

What was their modal choice?

3.57 Forty registered voters were asked whether they considered themselves Democrats, Republicans, or Independents. Use the following results to determine their modal choice:

Democrat	Republican	Independent	Independent
Democrat	Independent	Republican	Republican
Independent	Democrat	Democrat	Independent
Democrat	Independent	Republican	Independent
Independent	Independent	Democrat	Democrat
Republican	Independent	Independent	Republican
Republican	Democrat	Republican	Democrat
Independent	Independent	Democrat	Democrat
Independent	Republican	Independent	Independent
Democrat	Independent	Republican	Democrat

3.58 The following figures are a local bakery's daily flour utilization, in pounds, for 20 consecutive weekdays:

452 677 481 690 707 514 671 488 483 534
611 638 572 514 623 664 631 570 484 612

(a) Find the mean, median, and mode for this set of values.
(b) Round the 20 values to the nearest 10 pounds. Then again give the mean, median, and mode.
(c) Round the 20 values to the nearest 100 pounds. Provide the mean, median, and mode.

(d) State a conclusion about the effect of rounding on the mean, median, and mode.

3.5
Measures of Variation: The Range

An important feature of almost any kind of data is that the values are not all alike, and the extent to which they are unalike or vary among themselves is of basic importance in statistics. The following are some examples that illustrate the importance of measuring the variability of statistical data.

The chocolate chip ice cream produced by one company averages 360 chocolate chips per quart, with all quarts containing anywhere from 340 to 380 chocolate chips. Another company's chocolate chip ice cream also averages 360 chocolate chips per quart, but some quarts contain as few as 20 chips and others as many as 740. If the ice cream produced by the two companies is the same in all other respects, it stands to reason that most persons would prefer the chocolate chip ice cream made by the first company—this should give them a better chance of getting ice cream with a "desirable" number of chocolate chips.

Suppose that in a hospital each patient's pulse rate is taken three times a day and that on a certain day the records of two patients show

Patient A:	72	76	74
Patient B:	59	92	71

The mean pulse rates are the same, as can easily be checked, but observe the difference in variability. Whereas patient A's pulse rate is quite stable, that of patient B fluctuates widely, and this should be taken into account when treatments are prescribed.

Finally, suppose that we have a coin that is slightly bent and we wonder whether it is still balanced or *fair*. Suppose that we toss the coin 200 times. If we get 100 heads and 100 tails, there would be no doubt. What if we get 130 heads, however, or only 92? Are the discrepancies due to chance or due to the coin's being bent? Clearly, we need some indication of how much variability can be attributed to chance.

These examples show the need for statistical descriptions that measure the extent to which data is dispersed, or spread out, namely, the need for measures of variation.

To introduce one of the simplest ways of measuring variability, let us refer to the second of the preceding examples, and let us observe that the pulse rate of patient A varied from 72 to 76, while that of patient B varied from 59 to 92. These extreme (smallest and largest) values tell us something about the variability of the respective sets of data, and so do their differences. Thus, we define the **range** as follows:

The range of a set of data is the largest value minus the smallest.

For patient A we get a range of 76 − 72 = 4, and for patient B we get a range of 92 − 59 = 33.

The range is easy to calculate and easy to understand, and there is a natural curiosity about the minimum and maximum values. Nonetheless, it is not generally a useful measure of variation. Its main shortcoming is that it does not tell us anything about the dispersion of the values that fall between the two extremes.

    Each of the following sets of data

Set 1:	5	20	20	20	20	20	20	20
Set 2:	5	5	5	5	20	20	20	20
Set 3:	5	7	9	12	15	17	19	20

has a range of $20 - 5 = 15$, but the dispersion is entirely different in each case. Thus, the range is used mainly as a quick-and-easy indication of variability. It is used, for instance, in industrial quality control to keep a close check on raw materials or products by observing and charting the range of small samples taken at regular intervals of time.

3.6
Measures of Variation: The Standard Deviation

To introduce the **standard deviation**, by far the most generally useful measure of variation, let us observe that the dispersion of a set of data is small if the values are closely bunched about their mean, and that it is large if the values are scattered widely about their mean. It would seem reasonable, therefore, to measure the variation of a set of data in terms of the amounts by which the individual values differ from their mean. If a set of numbers $x_1, x_2, x_3, \ldots,$ and x_N, constituting a population, has the mean μ, the differences

$$x_1 - \mu, \quad x_2 - \mu, \quad x_3 - \mu, \quad \ldots, \quad \text{and} \quad x_N - \mu$$

are called the **deviations from the mean**. It is plausible that we might use their average (that is, their mean) as a measure of the variation in the population. Unfortunately, this will not do. Unless the x's are all equal, some of the deviations will be positive and some will be negative, and, as the reader will be asked to show in Exercise 3.120, the sum of the deviations from the mean, $\Sigma(x - \mu)$, and consequently also the mean of the deviations is always zero.

    Since we are really interested in the magnitude of the deviations, and not in whether they are positive or negative, we might simply ignore the signs and define a measure of variation in terms of the absolute values of the deviations from the mean. Indeed, if we add the deviations from the mean as if they were all positive or zero and divide by N, we obtain the statistical measure called the **mean deviation**. This measure has intuitive appeal, but because of the absolute values it leads to theoretical difficulties in problems of inference, and it is rarely used.

    An alternative approach is to work with the squares of the deviations from the mean, as this will also eliminate the effect of the signs. Squares of real numbers cannot be negative; in fact, squares of the deviations from a mean are always positive unless a value happens to coincide with the mean, in which case both $x - \mu$ and $(x - \mu)^2$ are equal to zero. Then, if we average the squared deviations from the mean and take the square root of the result (to compensate for the fact that the deviations were squared), we get the **population standard deviation**.

Population standard deviation

$$\sigma = \sqrt{\frac{\Sigma(x - \mu)^2}{N}}$$

This measure of variation is denoted by σ (lowercase sigma, the Greek letter for s), and, expressing literally what we have done here mathematically, it is also called the **root-mean-square deviation**. The square of σ is called the **population variance**. The formula for the population variance is obtained by removing the square root symbol from the formula for the population standard deviation, getting

Population variance

$$\sigma^2 = \frac{\Sigma(x - \mu)^2}{N}$$

It may seem logical to use the same formula, with n and $\bar{x}$ substituted for N and μ, for the standard deviation of a sample, but this is not quite what we do. Instead of dividing the sum of the squared deviations from the mean by n, we divide it by $n - 1$ and define the **sample standard deviation**, denoted by s, as

Sample standard deviation

$$s = \sqrt{\frac{\Sigma(x - \bar{x})^2}{n - 1}}$$

Its square, s^2, is called the **sample variance**. The formula for the sample variance is obtained by removing the square root symbol from the formula for the sample standard deviation, getting

Sample variance

$$s^2 = \frac{\Sigma(x - \bar{x})^2}{n - 1}$$

In dividing by $n - 1$ instead of n, we are not just being arbitrary. There is a good theoretical reason for doing this; if we divided by n and used s^2 as an estimate of σ^2 (that is, used the variance of a sample to estimate the variance of the population from which it came), our result would tend to be too small, and we correct for this by dividing by $n - 1$ instead of n. If n is large, it generally will not matter much whether we divide by $n - 1$ or n, but it is common practice to define σ and s as we did.

In calculating the sample standard deviation using the formula by which it is defined, we must (1) find $\bar{x}$, (2) determine the n deviations from the mean $x - \bar{x}$, (3) square these deviations, (4) add all the squared deviations, (5) divide by $n - 1$, and (6) take the square root of the result obtained in step 5. In actual practice, this formula is rarely used, but we shall illustrate it here to emphasize what is really measured by σ and s.

■ **EXAMPLE**   On six consecutive Sundays, a tow-truck operator received 9, 7, 11, 10, 13, and 7 service calls. Calculate s.

Solution   First calculating the mean, we get

$$\bar{x} = \frac{9 + 7 + 11 + 10 + 13 + 7}{6} = \frac{57}{6} = 9.5$$

and the work required to find $\sum (x - \bar{x})^2$ may be arranged as in the following table:

x	$x - \bar{x}$	$(x - \bar{x})^2$
9	−0.5	0.25
7	−2.5	6.25
11	1.5	2.25
10	0.5	0.25
13	3.5	12.25
7	−2.5	6.25
Total 57	0.0	27.50

Then, dividing by $6 - 1 = 5$ and taking the square root, we get

$$s = \sqrt{\frac{27.50}{5}} = \sqrt{5.5} \approx 2.3$$

Note in this table that the total for the middle column is zero; since this must always be the case, it provides a check on the calculations.   ■

It was easy to calculate s in this example because the data were whole numbers and the mean was exact to one decimal. Otherwise, the calculations required by the formula defining s can be quite tedious, and, unless we can get s directly with a statistical calculator or a computer, it helps to use the computing formula:

Computing formula for the sample standard deviation

$$s = \sqrt{\frac{S_{xx}}{n - 1}} \quad \text{where} \quad S_{xx} = \sum x^2 - \frac{(\sum x)^2}{n}$$

The advantage of this computing formula is that we merely have to get $\sum x$ and $\sum x^2$ from the data, without having to determine the value of $\bar{x}$ and work with the deviations from the mean. In Exercises 3.64 and 3.65, the reader is asked to verify that the computing formula is, indeed, equivalent to the one defining s^2.

■ **EXAMPLE**   Use the computing formula for s to rework the preceding example.

Solution   First, we calculate $\sum x$ and $\sum x^2$, getting

x	x^2
9	81
7	49
11	121
10	100
13	169
7	49
Total 57	569

Then, substituting $\sum x = 57$ and $\sum x^2 = 569$, together with $n = 6$, we find that

$$S_{xx} = 569 - \frac{(57)^2}{6} = 27.50$$

This gives us

$$s = \sqrt{\frac{27.50}{6-1}} = \sqrt{5.5} \approx 2.3$$

which agrees with the result we obtained before. ■

Since beginners often seem to confuse $\sum x^2$ with $(\sum x)^2$, let us emphasize the point that for $\sum x^2$ we first square each x and then add all the squares; for $(\sum x)^2$, on the other hand, we first add all the x's and then square their sum.

The sum $\sum x$ can be obtained by merely pressing calculator keys; it is not necessary to write down intermediate steps. On most calculators it is possible to compute $\sum x^2$ without writing down the squares of the individual numbers, though the specific techniques vary from one calculator brand to another.

The usual method for verifying a calculation is to repeat it and see if the same result is obtained. When different results arise, the calculation must be redone. This checking can be time consuming for the sample standard deviation. Some calculators have keys that compute the standard deviation directly from the list of values; even with such calculators the checking is inconvenient. Accordingly, we recommend that hand-held calculators be used only for small problems, say $n \leq 20$; computers should be used for larger problems.

Exercises

Exercises 3.59 and 3.64 are practice exercises; their complete solutions are given on page 97.

3.59 The 1994 live birth rates per thousand population in the Mountain States of Montana, Idaho, Wyoming, Colorado, New Mexico, Arizona, Utah, and Nevada were 12.9, 15.5, 13.5, 14.8, 16.7, 17.4, 20.1, and 16.4, respectively. What is their range?

3.60 Total student enrollment (in millions) in two- and four-year colleges and universities for the 10-year period 1986–1995 was, respectively,

$$12.7 \quad 12.7 \quad 13.1 \quad 13.2 \quad 13.6 \quad 14.1 \quad 14.0 \quad 13.9 \quad 15.0 \quad 14.7$$

Find the range.

3.61 The following are the closing prices of two stocks on five consecutive Fridays:

Stock A: $18\frac{1}{16}$ $17\frac{5}{8}$ $18\frac{1}{4}$ $17\frac{3}{8}$ $18\frac{1}{8}$

Stock B: $20\frac{1}{4}$ $20\frac{3}{16}$ $19\frac{7}{8}$ $20\frac{1}{2}$ $20\frac{3}{8}$

Calculate the range for each stock and decide which is more stable, that is, less variable.

3.62 The 42 employees of a company are given an intensive course in CPR. The following are the numbers of correct answers that they gave on a test administered after the completion of the course:

13	9	18	15	14	21	7	10	11	20
5	18	23	16	17	8	11	18	20	9
18	17	14	21	21	18	9	15	10	21
12	14	20	18	19	7	9	13	11	20
22	16								

Find the range of these figures.

3.63 The densities of selected gases in kilograms per cubic meter are

air	1.17	neon	0.90
carbon dioxide	1.98	nitrogen	1.25
carbon monoxide	1.25	nitrous oxide	2.00
helium	0.18	oxygen	1.43
hydrogen	0.09	propane	2.02
methane	0.72	sulfur dioxide	2.93

Find the range of the densities of the selected gases.

3.64 In five attempts, it took a person 11, 15, 12, 8, and 14 minutes to change a tire on a car. Calculate s using
  (a) the formula that defines s.
  (b) Rework this exercise, using the computing formula.
  (c) Verify that the solutions of parts (a) and (b) are identical.

3.65 Annual unemployment rates in the United States, from 1986 to 1996, were, respectively,

$$7.0 \quad 6.2 \quad 5.5 \quad 5.3 \quad 5.6 \quad 6.8 \quad 7.5 \quad 6.9 \quad 6.1 \quad 5.6 \quad 5.4$$

Calculate s using
  (a) the formula that defines s;   (b) the computing formula.

3.66 A filling machine in a high-production bakery is set to fill open-face pies with 16 fluid ounces of fill. A sample of four pies from a large production lot shows fills of 16.2, 15.9, 15.8, and 16.1 fluid ounces. Calculate s using
  (a) the formula that defines s;   (b) the computing formula.

3.67 The following are the times it took eight cars to accelerate from 0 to 60 mph: 15, 12, 15, 18, 19, 14, 17, and 15 seconds. Use the computing formula to determine the value of s.

3.68 The following are the numbers of hours that 12 students studied for a final examination: 7, 14, 22, 19, 20, 13, 25, 28, 32, 11, 20, and 24. Use the computing formula to determine s^2 and s. For these values,

$$\Sigma x = 235 \quad \text{and} \quad \Sigma x^2 = 5{,}189$$

3.69 The numbers of congressional bills vetoed by Presidents Kennedy, Johnson, Nixon, Ford, Carter, Reagan, Bush, and Clinton (to the year 1996) are, respectively, 21, 30, 43, 66, 31, 78, 44, and 17. Use the computing formula to determine s^2 and s. For these values

$$\Sigma x = 330 \quad \text{and} \quad \Sigma x^2 = 16{,}816$$

3.70 According to the National Bureau of Economic Research, the lengths of business cycles from 1919 to 1995, measured from trough to trough (a trough is the low point) in months, were

51   28   36   40   64   63   88   48   55   47   34   117   52   64   28   100

Find s. For these values

$$\Sigma x = 915 \quad \text{and} \quad \Sigma x^2 = 62{,}057$$

3.71 In 1995 the electric energy generated, in units of billions of kilowatt-hours, in the six New England states was 2.7, 13.9, 4.8, 27.0, 0.7, and 26.9. Modify the computing formula for the standard deviation so that it applies to populations (that is, replace the $n-1$ by n, and then replace each n by N), and then use it to calculate σ for the given data.

3.72 With reference to Exercise 3.62, determine s for the number of correct answers that the company's employees got on the test. For these values,

$$\Sigma x = 628 \quad \text{and} \quad \Sigma x^2 = 10{,}366$$

3.73 It has been claimed that for samples of size $n=4$ the range should be roughly twice as large as the standard deviation. Check this claim with reference to the following data, representing the number of times that four students were absent from a course in anthropology: 4, 7, 3, and 7.

3.74 It has been claimed that for samples of size $n=10$ the range should be roughly three times as large as the standard deviation. Check this claim with reference to the following test scores, which 10 students obtained on an examination: 77, 90, 72, 71, 91, 84, 64, 80, 75, and 86.

3.75 The following list gives the debt service ratios, meaning foreign debt divided by export values (in percents), for 55 foreign countries in 1995.

Argentina	22	Honduras	25	Poland	7
Bangladesh	12	Hungary	27	Portugal	11
Bolivia	23	India	22	Senegal	7
Brazil	23	Indonesia	18	Slovenia	3
Bulgaria	14	Jamaica	14	Sri Lanka	6
Cameroon	13	Jordan	11	Syria	2
Chile	15	Kenya	23	Tanzania	15
China	7	Macedonia	2	Thailand	4
Colombia	27	Madagascar	7	Tunisia	16
Congo	11	Malaysia	4	Turkey	22
Costa Rica	13	Mali	11	Uganda	16
Cote d'Ivoire	17	Mexico	12	Uruguay	21
Croatia	2	Morocco	31	Venezuela	15
Dominican Rep.	10	Nicaragua	36	Yemen	3
Ecuador	24	Oman	7	Zambia	29
Egypt	13	Pakistan	19		
Ethiopia	13	Panama	2		
Gabon	13	Papua–N. Guinea	9		
Ghana	14	Peru	12		
Guatemala	9	Philippines	13		

Use a computer package to find $\bar{x}$ and s for these percentages.

3.76 The following are the ignition times of certain upholstery materials exposed to a flame (in seconds):

2.50	4.50	5.11	9.70	5.62	6.77	3.49	4.90	10.21	8.76
9.33	4.12	3.85	4.97	5.04	2.97	3.81	10.60	7.95	7.41
8.64	5.33	3.90	11.25	1.92	1.42	12.80	9.45	6.25	4.71
7.86	2.65	4.79	6.20	1.52	1.38	3.87	4.54	5.12	5.15
11.75	7.35	2.80	6.85	1.20	9.20	1.76	5.21	3.40	7.29
8.66	5.04	10.25	6.43	2.97	4.45	5.50	5.92	4.56	2.46
6.90	1.47	2.11	2.32	4.19	2.20	4.32	1.58	6.43	4.04
2.51	2.58	3.78	3.75	3.10	6.43	1.70	6.40	3.24	1.79
8.75	2.46	3.62	4.72	7.40	8.81	5.83	6.75	7.65	8.79
10.92	9.65	5.09	4.11	6.37	5.40	2.51	10.28	5.49	3.76

Use a computer package to find the mean and variance of the sample data.

3.7
Some Applications of the Standard Deviation

After this chapter, we shall use sample standard deviations primarily as estimates of population standard deviations in problems of inference. To get more of a feeling for what a standard deviation really measures, let us devote this section to some applications.

In the argument that led to the definition of the standard deviation, we observed that the dispersion of a set of data is small if the values are bunched closely about their mean, and that it is large if the values are scattered widely about their mean. Correspondingly, we can now say that, if the standard deviation of a set of data is small, the values are concentrated near the mean, and if the standard deviation is large, the values are scattered widely about the mean. This idea is expressed more formally by **Chebyshev's theorem**, named for the Russian mathematician P. L. Chebyshev (1821–1894):

Chebyshev's theorem

> **For any set of data (population or sample) and any constant k greater than 1, at least $1 - 1/k^2$ of the data must lie within k standard deviations on either side of the mean.**

Thus, we can be sure that at least

$$1 - \frac{1}{2^2} = 1 - \frac{1}{4} = \frac{3}{4} \quad \text{or} \quad 75\%$$

of the values in any set of data must lie within 2 standard deviations on either side of the mean; at least

$$1 - \frac{1}{3^2} = 1 - \frac{1}{9} = \frac{8}{9} \quad \text{or} \quad \text{about } 89\%$$

must lie within 3 standard deviations on either side of the mean; and at least

$$1 - \frac{1}{10^2} = 1 - \frac{1}{100} = \frac{99}{100} \quad \text{or} \quad 99\%$$

must lie within 10 standard deviations on either side of the mean.

■ **EXAMPLE**   If all the 1-pound cans of coffee filled by a food processor have a mean weight of 16.00 ounces with a standard deviation of 0.02 ounce, at least what percentage of the cans must contain anywhere from 15.95 to 16.05 ounces of coffee?

Solution   The range of values 15.95 to 16.05 represents the mean plus or minus 0.05. This 0.05 is to be interpreted as k standard deviations, so that

$$k = \frac{0.05}{0.02} = 2.5$$

Thus, at least

$$1 - \frac{1}{2.5^2} = 1 - \frac{1}{6.25} = 0.84 \quad \text{or} \quad 84\%$$

of the cans must contain anywhere from 15.95 to 16.05 ounces of coffee.   ■

Chebyshev's theorem applies to any kind of data, but it tells us only "at least what percentage" must lie between certain limits. For nearly all sets of data, the actual percentage of data lying between the limits is much greater than that specified by Chebyshev's theorem. For many distributions, we can make the following much stronger statement:

For distributions having the general shape of the cross section of a bell (see Figure 3.3),

(1) *about 68% of the values will lie within 1 standard deviation of the mean, that is, between $\bar{x} - s$ and $\bar{x} + s$;*

(2) *about 95% of the values will lie within 2 standard deviations of the mean, that is, between $\bar{x} - 2s$ and $\bar{x} + 2s$;*

(3) *about 99.7% of the values will lie within 3 standard deviations of the mean, that is, between $\bar{x} - 3s$ and $\bar{x} + 3s$.*

This result is sometimes referred to as the **empirical rule**, although actually it is a theoretical result based on the normal distribution, which we shall discuss in Chapter 7.

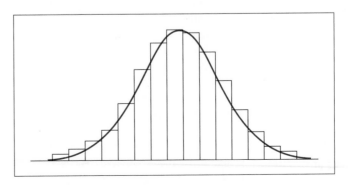

FIGURE 3.3 Bell-shaped distribution.

There are many ways in which knowledge of the variability of a set of data can be of importance, and some of them were previously shown in Section 3.6. Another application arises in the comparison of numbers belonging to different sets of data.

■ **EXAMPLE**   In an English class the final examination grades average 60 with a standard deviation of 16, and in a mathematics class the final examination grades average 58 with a standard deviation of 10. If a student gets a 72 on the English examination and a 68 on the mathematics examination, how many standard deviations is each of her grades above the average of the respective class? What does this tell us about her performance in the two subjects?

Solution   On the English examination her grade is $72 - 60 = 12$ points and, hence, $\frac{12}{16} = 0.75$ standard deviation above average; on the mathematics examination her grade is $68 - 58 = 10$ points and, hence, $\frac{10}{10} = 1.00$ standard deviation above average. Even though her English grade is higher than her mathematics grade, and her English grade is 12 points above average, while her mathematics grade is only 10 points above average, her performance relative to the two classes is better in mathematics than it is in English. Being 1.00 standard deviation above average in mathematics puts her in a relatively better position than being 0.75 standard deviation above average in English.   ■

What we have done here consisted of converting the grades into **standard units** or **z-scores**. In general, if x is a measurement belonging to a set of data having the mean $\bar{x}$ (or μ) and the standard deviation s (or σ), then its value in standard units, denoted by z, is

Formula for converting to standard units

$$z = \frac{x - \bar{x}}{s} \quad \text{or} \quad z = \frac{x - \mu}{\sigma}$$

depending on whether the data constitute a sample or a population. The z-score tells us how many standard deviations a value lies above or below the mean of the set of data to which it belongs. Standard units will be used frequently in later chapters. The standard deviation depends on the units of measurement. For instance, the weights of certain objects may have a standard deviation of 0.1 ounce or 2,835 milligrams, which is the same, but neither value really tells us whether it reflects a great deal of variation or very little variation. If the objects we are weighing are the eggs of small birds, either figure would reflect a great deal of variation, but this would not be the case if the objects that we are weighing are 100-pound bags of potatoes. What we need in a situation like this is a **measure of relative variation**, such as the **coefficient of variation**,

Coefficient of variation

$$V = \frac{s}{\bar{x}} \cdot 100\% \quad \text{or} \quad V = \frac{\sigma}{\mu} \cdot 100\%$$

which expresses the standard deviation as a percentage of what is being measured, at least on the average. Thus, if V is smaller for one set of data than another, there is relatively less variability in the first set of data, and we say that it is "more consistent."

■ **EXAMPLE**   During the past few months, one runner averaged 12 miles per week with a standard deviation of 2 miles, while another runner averaged 25 miles per week with a standard deviation of 3 miles. Which of the two runners is relatively more consistent in his weekly running habits?

Solution   The two coefficients of variation are, respectively,

$$\frac{2}{12} \cdot 100\% \approx 16.7\% \quad \text{and} \quad \frac{3}{25} \cdot 100\% = 12.0\%$$

since 12.0% is less than 16.7%, the second runner is relatively more consistent in his weekly running habits.   ■

Exercises

Exercises 3.77, 3.78, 3.84, and 3.88 are practice exercises; their complete solutions are given on page 97.

 3.77 According to Chebyshev's theorem, what can we assert about the percentage of any set of data that must lie within k standard deviations on either side of the mean when
(a) $k = 4$;   (b) $k = 12$?

 3.78 An airline's records show that its flights between two cities arrive on the average 5.4 minutes late with a standard deviation of 1.4 minutes. At least what percentage of its flights between the two cities arrive anywhere between
(a) 2.6 minutes late and 8.2 minutes late;
(b) 1.6 minutes early and 12.4 minutes late?

3.79 According to Chebyshev's theorem, what can we assert about the percentage of any set of data that must lie within k standard deviations on either side of the mean when
(a) $k = 5$;   (c) $k = 10$;
(b) $k = 8$;   (d) $k = 20$?

3.80 With reference to Exercise 3.78, at least what percentage of the flights between the two cities must arrive anywhere between
(a) 4.0 minutes late and 6.8 minutes late;
(b) 2.6 minutes late and 8.2 minutes late;
(c) 1.2 minutes late and 9.6 minutes late;
(d) 0.5 minute late and 10.3 minutes late?

3.81 Consider again Exercise 3.78. Using the empirical rule on page 75, about what percentage of the flights between the two cities will arrive between
(a) 4.0 minutes late and 6.8 minutes late;
(b) 2.6 minutes late and 8.2 minutes late;
(c) 1.2 minutes late and 9.6 minutes late?

3.82 A study of the nutritional value of a certain kind of bread shows that on the average one slice contains 0.260 milligram of thiamine (vitamin B_1) with a standard deviation of 0.005 milligram. According to Chebyshev's theorem, between what values must the thiamine content be of

(a) at least $\frac{35}{36}$ of all slices of this bread;

(b) at least $\frac{80}{81}$ of all slices of this bread?

3.83 With reference to Exercise 3.82, at least what percentage of the slices of the given kind of bread must have a thiamine content between 0.245 and 0.275 milligram? What can we say about this percentage if it can be assumed that the distribution of the thiamine content of the slices of bread is bell shaped?

3.84 In a large city the average retail price of a head of lettuce is $1.09 (with a standard deviation of $0.15), the average retail price of a pound of tomatoes is $0.88 (with a standard deviation of $0.06), and the average retail price of a cucumber is $0.32 (with a standard deviation of $0.04). If a certain food market charges $1.39 for a head of lettuce, $0.99 for a pound of tomatoes, and $0.35 for a cucumber, which of these food items is relatively the most overpriced?

3.85 For each stock that it lists, an investment service reports the price at which it is currently selling, its average price over a period of time, and a measure of its variability. Stock C, it reports, has a normal (mean) price of $58 with a standard deviation of $11, and it is currently selling at $76.50; stock D sells normally for $38, has a standard deviation of $4, and is currently selling at $50. Other things being equal, if a person owns both stocks and wants to dispose of one, which one should he sell and why?

3.86 Of two persons on a reducing diet, the first belongs to an age group for which the mean weight is 146 pounds with a standard deviation of 14 pounds, and the second belongs to an age group for which the mean weight is 160 pounds with a standard deviation of 17 pounds. If their respective weights are 178 pounds and 193 pounds, which of the two is more seriously overweight for his or her age group?

3.87 The applicants to one state university have an average ACT mathematics score of 21.4 with a standard deviation of 3.1, while the applicants to another state university have an average ACT mathematics score of 22.1 with a standard deviation of 2.8. With respect to which of these two universities is a student in a relatively better position if he or she scores

(a) 26 on this test;   (b) 31 on this test?

3.88 To compare the precision of two measuring instruments, a laboratory technician studies recent measurements made with both instruments. The first was recently used to measure the diameter of a ball bearing and the measurements had a mean of 4.92 mm with a standard deviation of 0.018 mm; the second was recently used to measure the unstretched length of a spring, and the measurements had a mean of 2.54 in. with a standard deviation of 0.012 in. Which of the two measuring instruments is relatively more precise?

3.89 On five tests, one student averaged 63.2 with a standard deviation of 3.3, while another student averaged 78.8 with a standard deviation of 5.3. Which student is relatively more consistent?

3.90 One patient's systolic blood pressure, measured daily over several weeks, averaged 202 with a standard deviation of 12.5, while that of another patient averaged 124 with a standard deviation of 8.1. Which patient's blood pressure is relatively more variable?

3.8
The Description of Grouped Data*

In the past, considerable attention was paid to the description of grouped data, because it was generally advantageous to group data before calculating the appropriate descriptions. This was true, particularly, in connection with large sets of data, where the manual determination of quantities such as Σx and Σx^2 entailed a considerable amount of work. Today, this is no longer the case, since such quantities can be determined quickly with a computer. Nevertheless, we shall devote this section to the description of grouped data, since some data (published government figures, for example) are available only in the form of frequency distributions. As we saw in Chapter 2, the grouping of data entails some loss of information. Each item loses its exact identity by being placed into a class. We only know the number of items in each class, so we must be satisfied with approximations. In the case of the mean and the standard deviation, we can get excellent approximations by proceeding as follows:

We assign to each item falling into a class the value of the class mark.

For intance, restating the distribution that pertains to the amount of time that 80 students devoted to leisure activities, we have the following:

HOURS	FREQUENCY
10–14	8
15–19	28
20–24	27
25–29	12
30–34	4
35–39	1
Total	80

To calculate the mean or the standard deviation of the distribution, we treat the 8 values falling into the class 10–14 as if they were all equal to 12, the 28 values falling into the class 15–19 as if they were all equal to 17, ..., and the value falling into the class 35–39 as if it were equal to 37. This procedure is usually quite satisfactory, since the errors that are thus introduced into the calculations will tend to average out.

To give general formulas for the mean and the standard deviation of a distribution with k classes, let us denote the successive class marks by x_1, x_2, x_3, ..., and x_k, and the corresponding class frequencies by f_1, f_2, f_3, ..., and f_k. The number of values is

$$f_1 + f_2 + f_3 + \cdots + f_k = \Sigma f = n$$

The sum of all the measurements or observations is represented by

$$x_1 f_1 + x_2 f_2 + x_3 f_3 + \cdots + x_k f_k = \Sigma x \cdot f$$

The sum of their squares is represented by

$$x_1^2 f_1 + x_2^2 f_2 + x_3^2 f_3 + \cdots + x_k^2 f_k = \Sigma x^2 \cdot f$$

The formula for $\bar{x}$ and the computing formula for s can be written as

$$\bar{x} = \frac{\sum x \cdot f}{n} \quad \text{and} \quad s = \sqrt{\frac{S_{xx}}{n-1}}$$

where

$$S_{xx} = \sum x^2 \cdot f - \frac{(\sum x \cdot f)^2}{n}$$

To get the corresponding formulas for a population, we replace n by N, and in the formula for the standard deviation we replace $n - 1$ by N. That is, we write

$$\mu = \frac{\sum xf}{N} \quad \text{and} \quad \sigma = \sqrt{\frac{S_{xx}}{N}}.$$

■ **EXAMPLE**   Find the mean and the standard deviation of the previous distribution pertaining to the amounts of time that 80 college students devoted to leisure activities.

Solution   To obtain $\sum x \cdot f$ and $\sum x^2 \cdot f$, we perform the calculations shown in the following table:

CLASS MARK x	x^2	FREQUENCY f	$x \cdot f$	$x^2 \cdot f$
12	144	8	96	1,152
17	289	28	476	8,092
22	484	27	594	13,068
27	729	12	324	8,748
32	1,024	4	128	4,096
37	1,369	1	37	1,369
	Total	80	1,655	36,525

Then, substitution into the formulas yields

$$\bar{x} = \frac{1,655}{80} = 20.6875 \approx 20.69$$

and

$$S_{xx} = 36,525 - \frac{(1,655)^2}{80} \approx 2,287.19$$

and then

$$s = \sqrt{\frac{2,287.19}{79}} \approx 5.38$$

To check the grouping error for this set of data, that is, the error introduced by first grouping the data, let us refer to the MINITAB printout of Figure 3.4. It shows that the mean of this ungrouped data is 20.613 rounded to three decimals and that the standard deviation is 5.56 rounded to two decimals. Thus, both errors are fairly small.   ■

```
MTB > SET C1
DATA> 23 24 18 14 20 24 24 26 23 21
DATA> 16 15 19 20 22 14 13 20 19 27
DATA> 29 22 38 28 34 32 23 19 21 31
DATA> 16 28 19 18 12 27 15 21 25 16
DATA> 30 17 22 29 29 18 25 20 16 11
DATA> 17 12 15 24 25 21 22 17 18 15
DATA> 21 20 23 18 17 15 16 26 23 22
DATA> 11 16 18 20 23 19 17 15 20 10
DATA> END
MTB > MEAN C1
   MEAN    =       20.613
MTB > STDEV C1
   ST.DEV. =       5.5632
```

FIGURE 3.4 MINITAB printout for the mean and the standard deviation of the amounts of time that 80 students devoted to leisure activities.

Once a set of data has been grouped, we can still determine (at least approximately) most other statistical measures besides the mean and the standard deviation, but we must modify their definitions. Although we can no longer determine the exact value of the median, we can find an approximate value as follows:

For grouped data, the median is such that half the total area of the rectangles of a histogram of their distribution lies to its left and the other half lies to its right (see Figure 3.5).

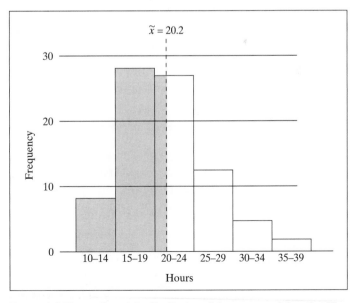

FIGURE 3.5 The median of the distribution of the amounts of time that 80 students devoted to leisure activities.

This definition is equivalent to the assumption that the values in the class containing the median are distributed evenly (that is, spread out evenly) throughout that class. To find the dividing line between the two halves of a histogram (each of which represents $\frac{n}{2}$ of the items grouped), we must somehow count $\frac{n}{2}$ of the items, starting at either end of the distribution. How this is done is illustrated by the following example.

■ **EXAMPLE**   Find the median of the distribution of the amounts of time that the students devoted to leisure activities.

Solution   Since $\frac{n}{2} = \frac{80}{2} = 40$, we must count 40 of the items, starting at either end. Starting at the bottom of the distribution (beginning with the smallest values), we find that $8 + 28 = 36$ of the values fall into the first two classes and that $8 + 28 + 27 = 63$ of the values fall into the first three classes. Therefore, we must count $40 - 36 = 4$ of the values beyond the 36 that fall into the first two classes. On the assumption that the 27 values in the third class are spread evenly throughout that class, we add $\frac{4}{27}$ of the class interval of 5 to the lower boundary of the third class. This gives us

$$\tilde{x} = 19.5 + \frac{4}{27} \cdot 5 \approx 20.2$$

for the median of the distribution.   ■

In general, if L is the lower boundary of the class into which the median must fall, f is its frequency, c is its interval, and j is the number of items we still lack when we reach L, then the median of the distribution is given by

Median of grouped data

$$\tilde{x} = L + \frac{j}{f} \cdot c$$

If we prefer, we can find the median of a distribution by starting to count at the other end (beginning with the largest values) and subtracting an appropriate fraction of the class interval from the upper boundary U of the class into which the median must fall. The corresponding formula is

Alternative formula for the median of grouped data

$$\tilde{x} = U - \frac{j'}{f} \cdot c$$

where j' is the number of items that we still lack when we reach U.

■ **EXAMPLE**   Use the alternative formula to find the median of the distribution of the amounts of time that the 80 students devoted to leisure activities.

Solution   Since $1 + 4 + 12 = 17$ of the values fall into the three classes at the top of the distribution, we need $40 - 17 = 23$ of the 27 values in the next class to reach the median, and we write

$$\tilde{x} = 24.5 - \frac{23}{27} \cdot 5 \approx 20.2$$

The result is, of course, the same.   ■

Note that the median of a distribution can be found by this method regardless of whether the class intervals are all equal; in fact, it can usually be found even when either or both of the classes at the top and at the bottom of a distribution are open (see Exercise 3.100).

The method by which we found the median of a distribution can also be used to determine other fractiles. For instance, the quartiles Q_1 and Q_3 of a distribution are defined so that 25% of the total area of the rectangles of the histogram lies to the left of Q_1 and 25% to the right of Q_3. To find them, we use either of the two formulas for the median of grouped data.

■ **EXAMPLE**   Find Q_1 and Q_3 for the distribution of the amounts of time that the 80 students devoted to leisure activities.

Solution   Since $\frac{n}{4} = \frac{80}{4} = 20$, we must count 20 of the items, starting at the bottom of the distribution, to find Q_1. Since there are eight values in the first class, we must count $20 - 8 = 12$ of the 28 values in the second class to reach Q_1, and we write

$$Q_1 = 14.5 + \frac{12}{28} \cdot 5 \approx 16.6$$

To find Q_3 we must count 20 of the items starting at the other end of the distribution. Since $1 + 4 + 12 = 17$ of the values fall into the three classes at the top of the distribution, we must count $20 - 17 = 3$ of the 27 values in the next class to reach Q_3, and we write

$$Q_3 = 24.5 - \frac{3}{27} \cdot 5 \approx 23.9$$ ■

The results that we have obtained here for the median, Q_1, and Q_3, together with the fact that the smallest value is 10 and the largest value is 38 (as shown in the ungrouped data on page 24), are summarized in the box-and-whisker plot of Figure 3.6.

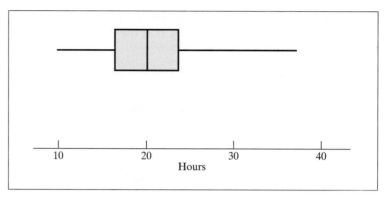

FIGURE 3.6 Box-and-whisker plot for the amounts of time that 80 students devoted to leisure activities.

Another set of useful fractiles consists of the **percentiles**. The first percentile is defined so that 1% of the total area of the rectangles of the histogram lies to its left. The second percentile is defined so that 2% of the total area of the rectangles of the histogram lies to its left, and the other percentiles are defined in an analogous manner. They are determined by using either of the two formulas on page 82.

■ **EXAMPLE**   Find P_{33} and P_{85} for the distribution of the amounts of time that the 80 students devoted to leisure activities.

Solution   To find P_{33}, we must count $0.33(80) = 26.4$ of the items starting at the bottom of the distribution. Since there are 8 values in the first class, we must count $26.4 - 8 = 18.4$ of the 28 values in the second class to reach P_{33}, and we write

$$P_{33} = 14.5 + \frac{18.4}{28} \cdot 5 \approx 17.8$$

To find P_{85}, we must count $0.15(80) = 12$ of the items starting at the top of the distribution. Since $1 + 4 = 5$ of the values fall into the two classes at the top of the distribution, we must count $12 - 5 = 7$ of the 12 values in the next class to reach P_{85}, and we get

$$P_{85} = 29.5 - \frac{7}{12} \cdot 5 \approx 26.6$$ ■

3.9
Some Further
Descriptions *

So far we have discussed statistical descriptions that come under the general heading of *measures of location* and *measures of variation*. Actually, there is no limit to the number of ways in which statistical data can be described, and statisticians continually develop new methods of describing characteristics of numerical data that are of interest in particular problems. In this section we shall consider briefly the problem of describing the overall shape of a distribution.

Although frequency distributions can take on almost any shape or form, most of the distributions that we meet in practice can be described fairly well by one or another of a few standard types. Among these, foremost in impor-

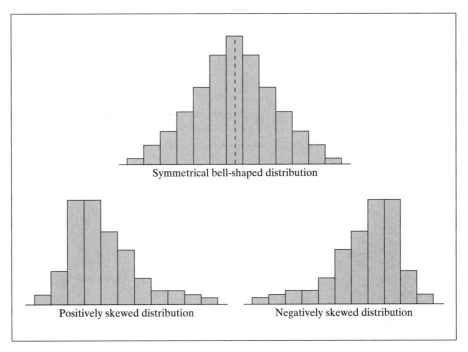

Symmetrical bell-shaped distribution

Positively skewed distribution   Negatively skewed distribution

FIGURE 3.7 Bell-shaped distributions.

tance is the aptly described **symmetrical bell-shaped distribution** shown at the top of Figure 3.7. Indeed, there are theoretical reasons why, in many instances, distributions of actual data can be expected to follow this general pattern. The other two distributions of Figure 3.7 can still, by a stretch of the imagination, be called **bell shaped**, but they are not symmetrical. Distributions like these, having a *tail* on one side or the other, are said to be **skewed**; if the tail is on the left, we say that they are **negatively skewed**, or skewed to the left, and if the tail is on the right, we say that they are **positively skewed** or skewed to the right. In general, we say that a distribution is skewed in the direction of the tail. Distributions of incomes or wages are often positively skewed because of the presence of some relatively high values that are not offset by correspondingly low values.

The symmetry or skewness of a distribution can also be judged visually by inspection of a box-and-whisker plot. For a symmetrical distribution, the median line divides the box into equal halves. It is moved to the left of center when a distribution is positively skewed and to the right of center when a distribution is negatively skewed. In Figure 3.6, the median line is slightly to the left of center, and this reflects the mild positive skewness, which is apparent from the histogram of Figure 3.5. Note also the long whisker extending from Q_3 to the largest value, which was 38.

There are several ways of actually measuring the extent to which a distribution is skewed. A relatively easy one is based on the fact that, for a perfectly symmetrical bell-shaped distribution such as the one at the top of Figure 3.7, the values of the median and the mean coincide. The presence of some relatively high values that are not offset by correspondingly low values will tend to make the mean greater than the median. Similarly, the presence of some relatively low values that are not offset by correspondingly high values will tend to make the mean less than the median. We can use this relationship between the mean and the median to define a relatively simple measure of the extent to which a distribution is skewed. It is called the **Pearsonian coefficient of skewness** and is given by

Pearsonian coefficient of skewness

$$SK = \frac{3(\text{mean} - \text{median})}{\text{standard deviation}}$$

For a perfectly symmetrical distribution, the value of SK is 0, and in general its values must fall between −3 and 3.

■ **EXAMPLE**   Find the Pearsonian coefficient of skewness for the distribution of the amounts of time the 80 students devoted to leisure activities.

Solution   Substituting into the formula the values obtained earlier for the mean, the median, and the standard deviation, $\bar{x} = 20.7$, $\tilde{x} = 20.2$, and $s = 5.4$, we get

$$SK = \frac{3(20.7 - 20.2)}{5.4} \approx 0.3$$

This reflects the fact that the distribution has a slight positive skewness.   ■

Two other kinds of distributions that sometimes arise in practice are the **reverse J-shaped** and **U-shaped distributions** shown in Figure 3.8. As can be seen from the diagram, the names of such distributions quite literally describe their shape. Examples of such distributions may be found in Exercises 3.110 and 3.111.

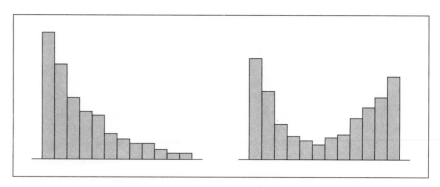

FIGURE 3.8 Reverse J-shaped and U-shaped distributions.

Exercises

Exercises 3.91, 3.95, 3.99, 3.104, and 3.107 are practice exercises; their complete solutions are given on page 98.

*3.91 Find $\bar{x}$ and s for the following distribution of the weekly earnings of 125 wage earners. Observe that the class marks are $124.995, $134.995, $144.995, and so on. The computations will be somewhat easier if you raise these by $0.005 to $125, $135, $145, and so on. The effect of this on the mean will be negligible, and it will not affect the standard deviation.

WEEKLY EARNINGS (DOLLARS)	FREQUENCY
120.00–129.99	9
130.00–139.99	20
140.00–149.99	36
150.00–159.99	30
160.00–169.99	15
170.00–179.99	11
180.00–189.99	4
Total	125

*3.92 The following is a summary of the gasoline tax rates that were in effect in each of the states in the United States at the end of the year 1995.

CENTS PER GALLON	NUMBER OF STATES
5–9	3
10–14	3
15–19	21
20–24	18
25–29	4
30–34	1
Total	50

Find the mean, $\bar{x}$, and the standard deviation, s, of this distribution.

*3.93 The per capita tax revenues of the Organization for Economic Development member countries for the year 1994, in thousands of dollars, were as follows:

THOUSANDS OF DOLLARS PER CAPITA[a]	NUMBER OF OECD COUNTRIES
0.0–2.9	6
3.0–5.9	5
6.0–8.9	5
9.0–11.9	8
12.0–14.9	3
15.0–17.9	1
Total	28

[a] U.S. per capita tax revenues were $7,234.

Find the mean, x, and standard deviation, s, of this distribution. You may find it helpful to use the same modification in the class marks as suggested in Exercise 3.91.

*3.94 The prices (in dollars) of premium gasoline in 28 countries, including taxes, for the time period closest to January 1, 1996, were as follows:

Dollars[a]	Number of Countries
0.00–0.99	2
1.00–1.99	6
2.00–2.99	6
3.00–3.99	7
4.00–4.99	5
5.00–5.99	2
Total	28

[a] The U.S. price was $1.28. Venezuela and Saudi Arabia had the lowest prices. Netherlands and Norway had the highest prices.

Find the mean, $\bar{x}$, and standard deviation, s, of this distribution.

*3.95 Find the median of the distribution of Exercise 3.91.

*3.96 Find the median of the following distribution of the numbers of raffle tickets sold by the 70 members of a social-service organization:

Number of Raffle Tickets	Frequency
Fewer than 30	28
30–44	19
45–59	10
60–74	8
75–89	5
Total	70

*3.97 According to the Central Intelligence Agency, *The World Factbook, 1995* estimates of the gross domestic product of the 40 countries with the highest per capita GDPs were as follows:

Dollars[a]	Number of Contries
Less than 13,000	5
13,000–15,000	5
16,000–18,000	8
19,000–21,000	11
22,000–24,000	8
More than 25,000	3
Total	40

[a] U.S. GDP is $27,900 per capita.

Find the median of this distribution.

*3.98 According to the U.S. Department of Defense, the dollar values of defense contracts received by the 50 largest suppliers for the year 1996 were as follows:

MILLIONS OF DOLLARS	NUMBER OF COMPANIES
0–400	22
500–900	14
1,000–1,400	4
1,500–1,900	3
Over 2,000	7
Total	50

Find the median.

*3.99 If possible, find Q_1 and Q_3 for the distribution of Exercise 3.96.

*3.100 Is it possible to find the mean and the median of each of the following distributions? Explain your answers.

(a)

GRADE	FREQUENCY
40–49	5
50–59	18
60–69	27
70–79	15
80–89	6
Total	71

(b)

PERCENT OF SALES QUOTA ACHIEVED	FREQUENCY
Less than 90	3
90–99	14
100–109	22
110–119	19
More than 119	7
Total	65

(c)

WEIGHT	FREQUENCY
100 or less	41
101–110	13
111–120	8
121–130	3
131–140	1
Total	66

*3.101 Find Q_1 and Q_3 for the gasoline tax distribution of Exercise 3.92.

*3.102 A geologist weighs his rock specimens and determines that the heaviest weighs 17.2 pounds, the lightest weighs 0.5 pound, and the value of the median, Q_1, and Q_3 are 9.27, 5.95, and 13.00 pounds, respectively. Use this information to draw a box-and-whisker plot.

3.103 Determine, by inspection, whether the box plots of Figures 3.9, 3.10, and 3.11 are positively skewed, negatively skewed, or symmetrical. Justify your answers.

(a)

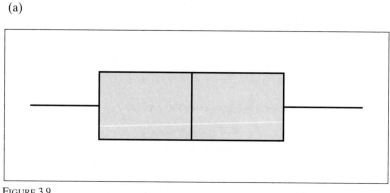

FIGURE 3.9

(b)

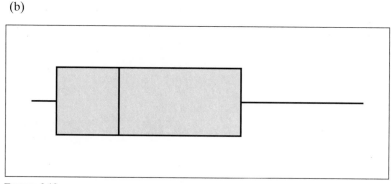

FIGURE 3.10

(c)

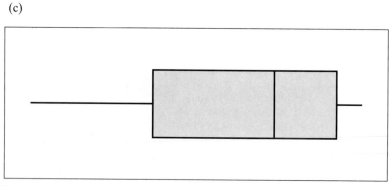

FIGURE 3.11

*3.104 If possible, find P_{10}, P_{40}, and P_{90} for the distribution of Exercise 3.96.

*3.105 Find P_{20} and P_{80} for the following distribution of grades for 180 students enrolled in a physics class.

GRADES IN PERCENT	NUMBER OF STUDENTS
40–49	8
50–59	16
60–69	26
70–79	64
80–89	45
90 and above	21
Total	180

*3.106 **Deciles** are calculated in the same manner as quartiles and percentiles. Nine deciles divide a set of data into 10 parts, and $D_1, D_2, D_3, \ldots$, and D_9 correspond to $P_{10}, P_{20}, P_{30}, \ldots, P_{90}$. Find D_2 and D_8 for the distribution of grades in Exercise 3.105 in this set of data.

*3.107 In a study of the weekly earnings of a large number of wage earners, we find that the mean, $\bar{x}$, is \$210.54, the median, $\tilde{x}$, is \$204.38, and the standard deviation, s, is \$9.43. Calculate the Pearsonian coefficient of skewness.

*3.108 On a final examination in an English literature course, the mean grade, $\bar{x}$, is 79.9, the median grade, $\tilde{x}$, is 81.4, and the standard deviation, s, is 3.1. Calculate the Pearsonian coefficient of skewness.

*3.109 In a tree nursery, a nursery attendant measures the heights of all the blue spruce trees to measure their growth in a month and finds that the mean, $\bar{x}$, growth is 2.25 inches, their median, $\tilde{x}$, growth is 1.96 inches, and their variance, s^2, is 0.23 inch. Find the Pearsonian coefficient of skewness.

*3.110 Roll a pair of dice 120 times and construct a distribution showing how many times there were 0 sixes, how many times there was 1 six, and how many times there were 2 sixes. Draw a histogram of this distribution and describe its shape.

*3.111 If a coin is flipped five times, the result may be represented by means of a sequence of H's and T's (for example, HHTTH), where H stands for *heads* and T for *tails*. Having obtained such a sequence of H's and T's, we can then check after each successive flip whether the number of heads exceeds the number of tails. For example, for the sequence HHTTH, heads is ahead after the first flip, after the second flip, after the third flip, not after the fourth flip, but again after the fifth flip; altogether it is ahead four times. Repeat this experiment 50 times, and construct a histogram showing in how many cases heads was ahead altogether 0 times, 1 time, 2 times, ..., and 5 times. Explain why the resulting distribution should be U shaped.

*3.112 The following are the numbers of minutes that a doctor kept 60 patients waiting beyond their appointment times:

12.1	9.8	10.5	5.6	8.2	0.5	6.8	10.1	17.2	4.2
8.3	1.3	7.9	11.3	6.3	7.2	9.3	9.9	7.2	12.7
1.2	4.6	10.3	8.5	10.0	12.8	9.6	13.5	10.8	5.1
12.7	11.5	3.8	12.9	13.0	3.9	7.5	16.1	11.1	8.3
9.6	6.4	15.7	5.8	9.7	11.9	2.4	5.2	8.4	16.7
2.5	13.0	4.8	10.7	11.4	9.3	4.7	6.0	9.5	14.6

If the data is grouped into a distribution with the classes 0.0–2.9, 3.0–5.9, 6.0–8.9, 9.0–11.9, 12.0–14.9, and 15.0–17.9, and if we solve for the mean, $\bar{x}$, and standard deviation, s, we get 8.90 minutes and 4.04 minutes, respectively. Use a computer package to find the mean and standard deviation of the raw data (ungrouped data). Determine the respective grouping errors by comparing these results with those obtained from the grouped data.

3.10
Technical Note:
Summations

In the abbreviated notation introduced in Section 3.1, the expression $\sum x$ does not make it clear which, or how many, values of x we have to add. This is taken care of by the more explicit notation

$$\sum_{i=1}^{n} x_i = x_1 + x_2 + \cdots + x_n$$

where it is made clear that we are adding the x's whose subscripts i are 1, 2, ..., and n. Generally, we shall not use the more explicit notation in this text to simplify the overall appearance of the formulas, assuming that it is clear in each case what x's we are referring to and how many there are.

Using the $\sum$ notation, we shall also have occasion to write such expressions as $\sum x^2$, $\sum xy$, $\sum x^2 f, \ldots$, which (more explicitly) represent the sums

$$\sum_{i=1}^{n} x_i^2 = x_1^2 + x_2^2 + \cdots + x_n^2$$

$$\sum_{j=1}^{m} x_j y_j = x_1 y_1 + x_2 y_2 + \cdots + x_m y_m$$

$$\sum_{i=1}^{n} x_i^2 f_i = x_1^2 f_1 + x_2^2 f_2 + \cdots + x_n^2 f_n$$

Working with two subscripts, we shall also have occasion to evaluate **double summations** such as

$$\sum_{j=1}^{3} \sum_{i=1}^{4} x_{ij} = \sum_{j=1}^{3} \left(x_{1j} + x_{2j} + x_{3j} + x_{4j} \right)$$

$$= x_{11} + x_{21} + x_{31} + x_{41}$$
$$+ x_{12} + x_{22} + x_{32} + x_{42}$$
$$+ x_{13} + x_{23} + x_{33} + x_{43}$$

To verify some of the formulas involving summations that are stated but not proved in the text, the reader will find it helpful to use the following rules:

Rules for summations

$$\text{Rule A:} \quad \sum_{i=1}^{n} \left(x_i \pm y_i \right) = \sum_{i=1}^{n} x_i \pm \sum_{i=1}^{n} y_i$$

$$\text{Rule B:} \quad \sum_{i=1}^{n} \left(k \cdot x_i \right) = k \cdot \sum_{i=1}^{n} x_i$$

$$\text{Rule C:} \quad \sum_{i=1}^{n} k = k \cdot n$$

The first of these rules states that the summation of the sum (or difference) of two terms equals the sum (or difference) of the individual summations, and it can be extended to the sum or difference of more than two terms. The second rule states that we can, so to speak, factor a constant out of a summation. The third rule states that the summation of a constant is simply n times that constant. All these rules can be proved by actually writing out in full what each of the summations represents. For instance, for Rule B we can write

$$\sum_{i=1}^{n} kx_i = kx_1 + kx_2 + \cdots + kx_n$$

$$= k(x_1 + x_2 + \cdots + x_n)$$

$$= k \cdot \sum_{i=1}^{n} x_i$$

Exercises

Exercise 3.115 is a practice exercise; its complete solution is given on page 98.

3.113 Write each of the following in full, that is, without summation signs:

(a) $\sum_{i=1}^{6} x_i$;

(c) $\sum_{i=1}^{3} x_i y_i$;

(e) $\sum_{i=3}^{7} x_i^2$;

(b) $\sum_{i=1}^{5} y_i$;

(d) $\sum_{j=1}^{8} x_j f_j$;

(f) $\sum_{j=1}^{4} (x_j + y_j)$.

3.114 Write each of the following as summations:

(a) $z_1 + z_2 + z_3 + z_4 + z_5$;

(b) $x_5 + x_6 + x_7 + x_8 + x_9 + x_{10} + x_{11} + x_{12}$;

(c) $x_1 f_1 + x_2 f_2 + x_3 f_3 + x_4 f_4 + x_5 f_5 + x_6 f_6$;

(d) $y_1^2 + y_2^2 + y_3^2$.

 3.115 Given $x_1 = 1$, $x_2 = 3$, $x_3 = 5$, $x_4 = 7$, $x_5 = 9$, $f_1 = 1$, $f_2 = 5$, $f_3 = 10$, $f_4 = 3$, $f_5 = 2$,

find

(a) $\sum_{i=1}^{5} x_i$;

(c) $\sum_{i=1}^{5} x_i \cdot f_i$;

(b) $\sum_{i=1}^{5} f_i$;

(d) $\sum_{i=1}^{5} x_i^2 \cdot f_i$.

3.116 Given $x_1 = -2$, $x_2 = 3$, $x_3 = 4$, and $x_4 = 4$, find

(a) $\sum_{i=1}^{4} x_i$;

(b) $\sum_{i=1}^{4} x_i^2$.

3.117 Given $x_1 = 2$, $x_2 = 3$, $x_3 = 4$, $x_4 = 5$, $x_5 = 6$, $x_6 = 7$, $f_1 = 3$, $f_2 = 12$, $f_3 = 10$, $f_4 = 6$, $f_5 = 3$, $f_6 = 1$, find

(a) $\displaystyle\sum_{i=1}^{6} x_i;$　　　　　　　(c) $\displaystyle\sum_{i=1}^{6} x_i \cdot f_i;$

(b) $\displaystyle\sum_{i=1}^{6} f_i;$　　　　　　　(d) $\displaystyle\sum_{i=1}^{6} x_i^2 \cdot f_i.$

3.118 Given $x_1 = 2,\ x_2 = 1,\ x_3 = 5,\ x_4 = 3\ \ y_1 = 1,\ y_2 = 3,\ y_3 = 2,\ y_4 = 1$

find

(a) $\displaystyle\sum_{i=1}^{4} x_i;$　　　　　　　(d) $\displaystyle\sum_{i=1}^{4} y_i^2;$

(b) $\displaystyle\sum_{i=1}^{4} y_i;$　　　　　　　(e) $\displaystyle\sum_{i=1}^{4} x_i \cdot y_i.$

(c) $\displaystyle\sum_{i=1}^{4} x_i^2;$

3.119 Given

$$
\begin{array}{lll}
x_{11} = 2, & x_{12} = \ \ 3, & x_{13} = -1, \\
x_{21} = 1, & x_{22} = \ \ 2, & x_{23} = \ \ 2, \\
x_{31} = 2, & x_{32} = -2, & x_{33} = \ \ 2, \\
x_{41} = 3, & x_{42} = -4, & x_{43} = -3,
\end{array}
$$

find

(a) $\displaystyle\sum_{i=1}^{4} x_{ij}$　for $j = 1, 2,$ and 3;

(b) $\displaystyle\sum_{j=1}^{3} x_{ij}$　for $i = 1, 2, 3,$ and 4;

(c) $\displaystyle\sum_{i=1}^{4}\sum_{j=1}^{3} x_{ij}.$

3.120 Show that $\displaystyle\sum_{i=1}^{n}(x_i - \bar{x}) = 0$ for any set of x's whose mean is $\bar{x}$.

3.121 Is it true in general that $\left(\displaystyle\sum_{i=1}^{n} x_i\right)^2 = \displaystyle\sum_{i=1}^{n} x_i^2$?

(*Hint*: Check whether the equation holds for $n = 2$.)

✓ Solutions to Practice Exercises

3.1 (a) parameter;　(b) statistic;　(c) statistic;　(d) parameter.

3.4 The total of the 20 values is $\Sigma x = 48 + 58 + \cdots + 34 = 920$, so that

$$\bar{x} = \frac{920}{20} = 46 \text{ years old}$$

3.10 The total weight of the women is $8(124) = 992$, and the total weight of the men is $6(165) = 990$, so that their combined weight is $992 + 990 = 1{,}982$ pounds. Since this does not exceed 2,000 pounds, the elevator is not overloaded (but the numbers are close enough to make one nervous).

3.17 Substituting $w_1 = 1,000$, $w_2 = 3,000$, $w_3 = 16,000$, $x_1 = 7$, $x_2 = 7.5$, and $x_3 = 8$ into the formula for the weighted mean, we get

$$\bar{x}_w = \frac{1,000(7) + 3,000(7.5) + 16,000(8)}{1,000 + 3,000 + 16,000}$$

$$= \frac{157,500}{20,000} = 7.875\%$$

3.18 Substituting $x_1 = \$39,600$, $x_2 = \$37,900$, $x_3 = \$49,600$, $n_1 = 17,800$, $n_2 = 27,300$, and $n_3 = 4,900$ into the formula for the grand mean of the combined data, we get

$$\bar{\bar{x}} = \frac{17,800(39,600) + 27,300(37,900) + 4,900(49,600)}{17,800 + 27,300 + 4,900} = \frac{1,982,590,000}{50,000}$$

$$\approx \$39,652$$

3.25 (a) $\dfrac{15 + 1}{2} = 8$, so that the median is the 8th value.

(b) $\dfrac{40 + 1}{2} = 20.5$, so that the median is the mean of the 20th and 21st values.

3.28 (a) Arranged according to size, the data is

$$59 \quad 68 \quad 87 \quad 97 \quad 123 \quad 174 \quad 1,215 \quad 1,710 \quad 15,613$$

Since $\dfrac{9 + 1}{2} = 5$, the median is the 5th value, or $123 million (Argentina).

(b) The mean is $\dfrac{123 + 87 + 174 + \cdots + 15,613}{9} = \dfrac{19,146}{9} \approx \$2,127$ million.

(c) Because of the large investment in the United States, the value of the mean is larger than the investment in eight out of the nine countries. The median is a better choice.

3.37 (a) The medians are

$$3 \quad 4 \quad 7 \quad 8 \quad 6 \quad 6 \quad 8 \quad 1 \quad 5 \quad 8 \quad 5 \quad 4 \quad 5 \quad 5 \quad 6$$

The means are

$$4\tfrac{1}{3} \quad 6 \quad 7 \quad 7 \quad 5\tfrac{1}{3} \quad 6\tfrac{1}{3} \quad 6\tfrac{1}{3} \quad 1\tfrac{2}{3} \quad 6 \quad 5\tfrac{2}{3} \quad 5\tfrac{2}{3} \quad 5 \quad 5 \quad 5\tfrac{1}{3} \quad 7$$

(b) Grouping the medians and the means into the indicated classes, we get the following:

Medians	Frequency	Means	Frequency
0.5–2.5	1	0.5–2.5	1
2.5–4.5	3	2.5–4.5	1
4.5–6.5	7	4.5–6.5	10
6.5–8.5	4	6.5–8.5	3

(c) The histograms of these distributions are shown in Figure 3.12, and it can be seen that there is less variability among the means than among the medians. In other words, the means are concentrated more closely about $\mu = 5.5$, the actual mean of the numbers on all the cards.

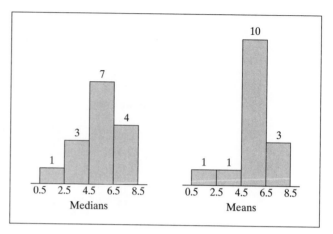

FIGURE 3.12 Histograms for part (c) of Exercise 3.37.

3.42 The median position is $\dfrac{21 + 1}{2} = 11$, the Q_1 position is $\dfrac{10 + 1}{2} = 5.5$, and the Q_3 position is the 5.5th from the other end, or 16.5. Thus, there are five values to the left of the Q_1 position, five values between the Q_1 and median positions, five values between the median and Q_3 positions, and five values to the right of the Q_3 position.

3.45 The median position is $\dfrac{13 + 1}{2} = 7$, the Q_1 position is $\dfrac{6 + 1}{2} = 3.5$, and the Q_3 position is 3.5 from the other end, or 10.5. Arranged according to size, the data is

$$0 \quad 1 \quad 1 \quad 3 \quad 5 \quad 5 \quad 5 \quad 6 \quad 6 \quad 8 \quad 10 \quad 12 \quad 15$$

so that

$$Q_1 = \frac{1 + 3}{2} = 2 \quad \text{and} \quad Q_3 = \frac{8 + 10}{2} = 9$$

3.48 (a) Midquartile is $\dfrac{Q_1 + Q_3}{2} = \dfrac{32 + 84.5}{2} = \dfrac{116.5}{2} = 58.25$ minutes.

Midrange is the mean of the smallest and largest values, or $\dfrac{12 + 125}{2} = \dfrac{137}{2} \approx 68.5$ minutes.

(b) Midquartile is $\dfrac{129 + 150.5}{2} = \dfrac{279.5}{2} = 139.75$ minutes.

Midrange is $\dfrac{113 + 164}{2} = \dfrac{277}{2} \approx 138.5$ minutes.

3.50 Since 12 occurs once, 6, 7, 9, and 10 occur three times each, and 8 occurs 5 times, the mode is 8 marriages per 1,000 population.

3.55 Most of the numbers occur only once, 1 and 8 occur two times, and 9 occurs three times. The mode is 9 minutes.

3.59 The range is $20.1 - 12.9 = 7.2$ live births per thousand population.

3.64 (a) The mean is $\bar{x} = \dfrac{11 + 15 + 12 + 8 + 14}{5} = \dfrac{60}{5} = 12$, so that

$$s = \sqrt{\frac{(11-12)^2 + (15-12)^2 + (12-12)^2 + (8-12)^2 + (14-12)^2}{5-1}}$$

$$= \sqrt{\frac{1 + 9 + 0 + 16 + 4}{4}} = \sqrt{7.5} \approx 2.7 \text{ minutes}$$

(b) Substituting $n = 5$, $\Sigma x = 60$, and $\Sigma x^2 = 11^2 + 15^2 + 12^2 + 8^2 + 14^2 = 750$ into the computing formula for S_{xx}, we get

$$S_{xx} = 750 - \frac{(60)^2}{5} = 30$$

Then

$$s = \sqrt{\frac{30}{4}} \approx 2.7 \text{ minutes}$$

(c) $2.7 = 2.7$.

3.77 (a) At least $1 - \dfrac{1}{4^2} = \dfrac{15}{16} = 0.9375$, or 93.75% of the data must lie within $k = 4$ standard deviations on either side of the mean.

(b) At least $1 - \dfrac{1}{12^2} = \dfrac{143}{144} \approx 0.9931$, or 99.31%, of the data must lie within $k = 12$ standard deviations on either side of the mean.

3.78 (a) $\dfrac{8.2 - 5.4}{1.4} = 2$; at least $1 - \dfrac{1}{2^2} = 0.75$, or 75%, of the planes arrive anywhere between 2.6 minutes late and 8.2 minutes late.

(b) $\dfrac{12.4 - 5.4}{1.4} = 5$; at least $1 - \dfrac{1}{5^2} = 0.96$, or 96%, of the planes arrive anywhere between 1.6 minutes early and 12.4 minutes late.

3.84 In standard units, the price of the head of lettuce is

$$\frac{1.39 - 1.09}{0.15} = 2$$

the price of the pound of tomatoes is

$$\frac{0.99 - 0.88}{0.06} \approx 1.83$$

and the price of the cucumber is

$$\frac{0.35 - 0.32}{0.04} = 0.75$$

Therefore, the head of lettuce is relatively most overpriced.

3.88 For the first measuring instrument, the coefficient of variation is

$$\frac{0.018}{4.92} \cdot 100\% \approx 0.37\%$$

and for the second measuring instrument, it is

$$\frac{0.012}{2.54} \cdot 100\% \approx 0.47\%$$

Since the first measuring instrument has the smaller coefficient of variation, it is relatively more precise.

3.91 It helps to elevate the class marks by $0.005. This gives

x	f	$x \cdot f$	$x^2 \cdot f$
125	9	1,125	140,625
135	20	2,700	364,500
145	36	5,220	756,900
155	30	4,650	720,750
165	15	2,475	408,375
175	11	1,925	336,875
185	4	740	136,900
Total	125	18,835	2,864,925

Then

$$\bar{x} = \frac{18,835}{125} = \$150.68$$

Next find

$$S_{xx} = 2,864,925 - \frac{(18,835)^2}{125} = 26,867.20$$

Then

$$s = \sqrt{\frac{26,867.20}{124}} \approx \$14.72$$

3.95 The median is

$$\tilde{x} = 139.995 + \frac{33.5}{36} \cdot 10 \approx \$149.30$$

3.99 For Q_1, we must count $\frac{70}{4} = 17.5$ values from the bottom of the distribution, and since this falls into the open class "less than 30," Q_1 cannot be found. For Q_3, we must count 17.5 values from the top of the distribution, and we get

$$Q_3 = 59.5 - \frac{4.5}{10} \cdot 15 = 52.75$$

3.104 P_{10} cannot be found as it falls into the "less than 30" open class, but $(0.40)(70) = 28$.

$$P_{40} = 29.5 + \frac{0}{19} \cdot 15 = 29.5 \quad \text{and} \quad P_{90} = 74.5 - \frac{2}{8} \cdot 15 = 70.75$$

3.107 $SK = \dfrac{3(210.54 - 204.38)}{9.43} \approx 1.96.$

3.115 (a) $1 + 3 + 5 + 7 + 9 = 25$;
(b) $1 + 5 + 10 + 3 + 2 = 21$;
(c) $1 \cdot 1 + 3 \cdot 5 + 5 \cdot 10 + 7 \cdot 3 + 9 \cdot 2 = 105$;
(d) $1^2 \cdot 1 + 3^2 \cdot 5 + 5^2 \cdot 10 + 7^2 \cdot 3 + 9^2 \cdot 2 = 605.$

Review: Chapters 1, 2, & 3

Having read and studied these chapters and worked a good portion of the exercises, you should be able to:

1. Explain the difference between descriptive statistics and statistical inference.
2. Construct frequency distributions of both numerical and categorical data.
3. Determine the class limits, class boundaries, class marks, and class interval of a distribution of numerical data.
4. Convert frequency distributions into percentage distributions.
5. Present ungrouped data in the form of dot diagrams and in other types of graphs.
6. Convert frequency distributions into cumulative distributions and percentage distributions into cumulative percentage distributions.
7. Present frequency distributions in the form of histograms, bar charts, frequency polygons, ogives, or pie charts.
8. Construct stem-and-leaf displays.
9. Explain the difference between samples and populations and the difference between statistics and parameters.
10. Determine the mean of a set of data.
11. Determine the median of a set of data.
12. Determine the mode of a set of data.
13. List some of the desirable and undesirable features of the various measures of location.
14. Calculate weighted means.
15. Determine the grand mean of combined data.
16. Determine the quartiles of a set of data.
17. Draw a box-and-whisker plot.
18. Determine the range of a set of data.
19. Determine the standard deviation (or variance) of a set of data.
20. List some of the desirable and undesirable features of the various measures of variation.
21. Explain Chebyshev's theorem.
22. Convert measurements into standard units and explain the advantage of using such units.

23. Calculate the coefficient of variation and explain its significance.

*24. Determine the mean and the standard deviation of a distribution.

*25. Determine the median of a distribution.

*26. Determine the quartiles of a distribution.

*27. Determine the percentiles of a distribution.

*28. Describe the shape of a distribution as symmetrical or skewed, as bell shaped, as U shaped, and so forth.

*29. Determine the Pearsonian coefficient of skewness.

30. Work with the Σ (summation) notation.

Checklist of Key Terms (with page references to their definitions)

Bar chart, 33
Bell-shaped distribution, 85
Bias, 4
Biased data, 4
Box-and-whisker plot, 57
Box plot, 57
Categorical data, 2
Categorical distribution, 22
Chebyshev's theorem, 74
Class boundary, 25
Class frequency, 25
Class interval, 25
Class interval of the distribution, 25
Class limit, 25
Class mark, 25
Coefficient of variation, 76
Cumulative distribution, 26
Cumulative percentage distribution, 27
Data, 1
Decile, 91
Decision theory, 9
Descriptive statistics, 7
Deviation from mean, 68
Dot diagram, 14
Double-stem display, 20
Double summation, 92
Empirical rule, 75
Exploratory data analysis, 10
Fractile, 55
Frequency distribution, 22
Frequency polygon, 33
Grand mean of k sets of data, 48
Grouping, 14
Histogram, 31

Inequality signs, 2
Interval data, 2
J-shaped distribution, 86
Leaf, 16
Listing, 13
Lower class limit, 25
Mean, 44
Mean deviation, 68
Measures of central location, 43
Measures of location, 43
Measures of relative variation, 76
Measures of variation, 43
Median of n values, 52
Median position, 52
Midquartile, 63
Midrange, 62
Mode, 64
Negatively skewed distribution, 85
Nominal data, 2
Numerical data, 2
Numerical distribution, 22
Ogive, 34
Open class, 23
Ordinal data, 2
Parameter, 44
Pareto diagram, 15
Pearsonian coefficient of skewness, 86
Percentage distribution, 26
Percentile, 55, 83
Pictogram, 35
Pie chart, 35
Population, 3
Population mean, 45
Population size, 45

Population standard deviation, 68
Population variance, 69
Positively skewed distribution, 85
Probability theory, 8
Quartile, 55
Range, 67
Ratio data, 3
"Real" class limits, 25
Reverse J-shaped distribution, 86
Root-mean-square deviation, 69
Sample, 3
Sample mean, 45
Sample size, 44
Sample standard deviation, 69
Sample variance, 69
Sigma notation (Σ notation), 45
Skewed distribution, 85
Standard deviation, 68
Standard unit, 76
Statistic, 43
Statistical description, 43
Statistical inference, 7
Statistical measure, 43
Statistics, 1
Stem, 16
Stem-and-leaf display, 16
Symmetrical bell-shaped distribution, 85
Trimmed mean, 46
Upper class limit, 25
U-shaped distribution, 86
Vital statistics, 43
Weighted mean, 47
Weights, 47
z-score, 76

Review Exercises

R.1 The following are the air miles between Chicago and 12 selected cities:

City	Air miles
Beijing	6,604
Berlin	4,414
Cairo	6,141
Cape Town	8,491
Caracas	2,495
Hong Kong	7,797
London	3,958
Madrid	4,189
Melbourne	9,673
Moscow	4,987
Paris	4,143
Tokyo	6,317

Find
  (a) the mean;
  (b) the median.

R.2 According to Chebyshev's theorem, what can we assert about the percentage of any set of data that must lie within k standard deviations on either side of the mean when (a) $k = 1.2$;   (b) $k = 6.25$;   (c) $k = 11$?

R.3 On three consecutive days, a meter maid issued 28, 20, and 32 parking tickets. On the same three days, she issued 7, 4, and 5 tickets for expired inspection stickers. Which of the following conclusions can be obtained from these data by purely descriptive methods and which require generalizations? Explain your answers.
  (a) The meter maid issued 80 parking tickets on the three days.
  (b) On any given day, this meter maid will seldom issue more than 10 tickets for expired inspection stickers.
  (c) The meter maid issued the fewest number of tickets on the second day because she took an extended lunch hour.
  (d) On these three days, the meter maid issued five times as many parking tickets as tickets for expired inspection stickers.

R.4 Repeating the weekly earnings data of Exercise 3.91, we have the following:

WEEKLY EARNINGS (DOLLARS)	FREQUENCY
120.00–129.99	9
130.00–139.99	20
140.00–149.99	36
150.00–159.99	30
160.00–169.99	15
170.00–179.99	11
180.00–189.99	4
Total	125

Convert the data into
  (a) a cumulative "less than" distribution;
  (b) a cumulative "or more" distribution.

R.5 The dean of a college has complete records on how many failing grades each faculty member gave to his or her students during the academic year 1999–2000. Give one example each of a situation in which the dean would look upon these data as   (a) a population;   (b) a sample.

R.6 During the five business days of a recent week, the closing prices per share of General Motors common stock on the New York Stock Exchange were as follows:

Day	Dollars
Monday	61.0625
Tuesday	62.2500
Wednesday	62.7500
Thursday	63.2500
Friday	61.9375

Find the mean closing price per share of General Motors common stock for this business week. (*Note:* Stocks are generally valued in sixteenths of a dollar per share, where $\frac{1}{16}$ = \$0.0625. Prices are therefore calculated to four decimal places.)

R.7 In the year 1996 the number of representatives to the House of Representatives of the Congress from the Pacific states of Washington, Oregon, California. Alaska, and Hawaii were 9, 5, 51, 1, and 2, respectively. Comment on the (misleading?) argument that the "average" Pacific state has 13.6 representatives.

R.8 It can be determined from data published by the Center for the American Woman and Politics that the mean number of women holding office in the state legislatures of the six New England states was 60.33 in the year 1996. If the numbers of women in the legislatures of five of the states were 36, 50, 49, 125, and 47, what must have been the number of women in the legislature of the sixth state (which was Vermont)?

R.9 The following is the distribution of the number of mistakes 150 students made in translating a certain passage from German to English:

Number of mistakes	Number of students
17–19	5
20–22	63
23–25	39
26–28	24
29–31	17
32–34	2

Find
  (a) the mean;   (b) the standard deviation.

R.10 Draw a histogram and also draw a bar chart of the distribution of Exercise R.9.

R.11 The following are the systolic blood pressures of 20 hospital patients:

165 135 151 153 155 182 142 158 146 149
124 162 173 204 159 130 177 162 141 156

Construct a stem-and-leaf display with the stem labels 12, 13, ... , and 20.

R.12 The costs for sending parcels overnight by a certain firm are to be placed in a frequency distribution having the following categories:

> Up to $2.00
> $2.01 to $5.00
> $5.00 to $10.00
> $10.01 to $20.00
> Over $25

Explain where difficulties might arise.

R.13 A fishery expert found the following concentrations of mercury, in parts per million, in eight fish caught in a certain stream:

> 0.058 0.062 0.070 0.063
> 0.068 0.061 0.068 0.060

(a) Find $\bar{x}$ and s.
(b) Find the median and the range.

R.14 According to the Martindale–Hubbell Law Directory, 192,353 of all lawyers reporting in 1996 were engaged in private practice. Of these lawyers, 116,911 practiced as individuals; 60,709 practiced as partners; and 14,733 were designated as employees by their employers. Construct a pie chart of this categorical distribution.

R.15 Given $x_1 = 6$, $x_2 = 8$, $x_3 = 5$, $x_4 = -1$, $x_5 = 3$, and $x_6 = 4$, find

(a) $\sum_{i=1}^{6} x_i$;   (b) $\sum_{i=1}^{6} x_i^2$.

R.16 Explain why each of the following samples may well yield misleading information:
(a) To ascertain facts about tooth-brushing habits, a sample of the residents of a community are asked how many times they brush their teeth each day.
(b) To study the spending patterns of families in a certain income group, a sample survey is conducted during the first three weeks of December.

∗R.17 If a distribution has the mean 112.8, the median 96.5, and the standard deviation 24.7, find the Pearsonian coefficient of skewness.

R.18 Based on many years' information, it is known that the bus that leaves Valley Center at 8:02 A.M. takes 32 minutes, with a standard deviation of 2.25 minutes, to arrive at the downtown bus terminal. At least what percentage of the time will this bus arrive at the downtown bus terminal between 8:25 A.M. and 8:43 A.M.?

R.19 On five runs, bus A carried 15, 24, 19, 12, and 20 passengers, and bus B carried 18, 21, 16, 14, and 16 passengers.
(a) Calculate the respective means to decide which bus averaged more passengers on the five runs.
(b) Use the formula that defines the standard deviation to calculate s for each of the two sets of data.
(c) Calculate the two coefficients of variation to determine whether the number of passengers is relatively more variable for bus A or bus B.

R.20 The following are the numbers of accidents that occurred at 18 intersections without left-turn arrows:

18   39   41   24   45   38   22   28   34   23   16   42   9   20   36   32   42   35

Find
   (a) the median;   (b) Q_1 and Q_3.

R.21 Use the results of Exercise R.20 to draw a box-and-whisker plot for the accident data.

R.22 For fresh salmon sold at four stores, the prices per pound and numbers of pounds sold on a particular date were as follows:

Store	Price	Quantity
Eastside Fish Mart	$5.95	80
Murray's	$5.25	120
Thompson Specialty	$7.00	40
Bayside Fish	$6.50	60

Find the mean price of the salmon sold in the four stores.

R.23 The daily ticket sales of a movie theater are grouped into a distribution having the classes 50–99, 100–149, 150–199, 200–249, 250–299, 300–349, and 350–399. Determine
   (a) the lower class limits;
   (b) the upper class limits;
   (c) the class boundaries;
   (d) the class marks;
   (e) the class interval of the distribution.

R.24 Specify the values that occur in the following stem-and-leaf display:

0.00	2
0.01	35
0.02	668
0.03	7

R.25 Ten athletes ran a 440-yard race in 46.2, 46.9, 48.3, 46.7, 46.3, 46.5, 47.2, 46.1, 46.9, and 46.3 seconds. Calculate the standard deviation.

R.26 On four Saturdays, a person jogged for 46, 50, 52, and 60 minutes.
   (a) Find the mean, the range, and the standard deviation of these four sample values.
   (b) Subtract 50 minutes from each of the times, recalculate the mean, the range, and the standard deviation, and compare the results with those obtained in part (a).
   (c) Divide each of the original values by 2, recalculate the mean, the range, and the standard deviation, and compare the results with those obtained in part (a).
   (d) Based on these results, what effect does (1) adding or subtracting a constant and (2) multiplying or dividing by a constant have on the mean, the range, and the standard deviation of a set of data?

R.27 Mr. Ames lives in a neighborhood where the average family income is $30,000 with a standard deviation of $4,000, and Mr. Brown lives in a neighborhood where average family income is $36,000 with a standard deviation of $6,000. If

Mr. Ames and family make $40,000 and Mr. Brown and family make $48,000, which of the two families is relatively better off with respect to the families in their neighborhoods?

R.28 Find the median position for
    (a) $n = 31$;   (b) $n = 62$.

R.29 The following is the distribution of the number of vehicles that received emission inspections during a recent month at 80 service stations in a certain county:

NUMBER OF VEHICLES	FREQUENCY
0–19	4
20–39	11
40–59	28
60–79	24
80–99	13
Total	80

    (a) Draw a histogram of this distribution.
    (b) Convert this distribution into a cumulative "less than" distribution and draw an ogive.

*R.30 With reference to Exercise R.29, find
    (a) the median;   (b) the quartiles Q_1 and Q_3.

*R.31 Regarding the distribution of Exercise R.29, suppose that the smallest value is 5 and the largest value is 95. Use the results of Exercise R.30 to draw a box-and-whisker plot.

*R.32 Regarding the distribution of Exercise R.29, find the percentiles $P_{0.20}$ and $P_{0.80}$.

*R.33 A church bulletin shows that on the five preceding Sundays the attendance was 352, 314, 3,360, 375, and 328.
    (a) Calculate the mean and the median of these attendance figures.
    (b) Assuming that the third value was printed incorrectly and should have been 360 instead of 3,360, recalculate the mean and the median.
    (c) Compare the effects of this printing error on the mean and on the median.

R.34 The 25 teachers of an elementary school are given an intensive course in first aid. Find the range of the following numbers of correct answers that they gave in a test administered after the completion of the course:

18	12	15	9	11	16	20
15	14	18	18	15	10	17
13	17	19	8	19	20	16
12	18	11	14			

R.35 Asked whether they ever accept social invitations from their students, 40 tennis pros replied as follows: occasionally, rarely, rarely, never, rarely, frequently, occasionally, occasionally, never, rarely, rarely, never, occasionally, occasionally, frequently, occasionally, rarely, rarely, occasionally, never, occasionally, occasionally, never, never, rarely, rarely, occasionally, occasionally, frequently, occasionally, never, occasionally, rarely, rarely, never, frequently, occasionally, rarely, rarely, occasionally.

(a) Construct a categorical distribution and display it in the form of a bar chart.

(b) What is the tennis pros' modal reply?

R.36 The following are the numbers of deer observed in 60 sections of land in a wildlife count:

11	14	21	15	16	4	18	11	17	11	7	12
21	0	16	12	20	17	13	10	16	5	10	14
5	19	10	6	15	1	26	8	18	19	2	14
17	6	15	14	22	7	7	13	19	0	15	17
2	16	11	18	10	28	15	4	32	6	20	7

(a) Group these figures into a distribution having the classes 0–4, 5–9, 10–14, 15–19, 20–24, 25–29, and 30–34, and draw a bar chart.

(b) Convert the distribution obtained in part (a) into a cumulative "less than" distribution and draw an ogive.

R.37 The following are the May 15 average low temperatures in degrees Fahrenheit at 40 selected cities in the United States:

55	62	69	34	70	43	48	57	53	64
38	45	46	57	61	61	72	41	48	54
67	31	28	58	63	52	36	75	60	69
52	40	53	57	26	43	64	39	71	51

Construct a stem-and-leaf display for these data.

R.38 Use the stem-and-leaf display obtained in Exercise R.37 to find the median of the 40 average low temperatures.

R.39 Construct a box-and-whisker plot for the data given in Exercise R.37.

R.40 U.S. Department of Agriculture data shows that the per capita consumption of citrus fruits (oranges, grapefruit, and other) in pounds for the years 1982 to 1995 was as follows:

24.8	29.5	24.0	22.6	26.1	25.8	26.4
23.5	21.4	19.1	24.4	26.0	25.0	24.4

Construct a box-and-whisker plot for this information.

4 Possibilities and Probabilities

In everyday language we use words such as "probable," "likely," and other terms expressing uncertainties as a matter of course. "The flight from Denver to Chicago will probably arrive on time." "She is an odds-on favorite to be elected to the school board." "The chances are that a person with malaria will fully recover." "More likely than not, the tennis player we favor will win this year's Paris Open Tournament."

Statements like these are all very vague, but they can be quantified (assigned numerical measures) through the concept of **probability**, which was first introduced in connection with games of chance.

Here we shall see how probabilities can, or may, be interpreted and how they can be used to make choices (among different courses of action) that promise to be the most profitable, or otherwise most desirable. Sections 4.1, 4.2, and 4.3 are devoted to the problem of determining *what is possible* in given situations, and in Section 4.4 we learn how to judge also *what is probable*. In Sections 4.5 and 4.6 we present the concept of a mathematical expectation and its application to problems of decision making.

**4.1
Counting**

There are essentially two ways to answer a "what is possible" question. One way consists of listing everything that can happen in a given situation. The second way determines how many different things can happen, without actually constructing a complete list. In connection with our work in this book, the second kind of answer is especially important; in most cases we will not need a complete list and hence can save ourselves a great deal of work. Although the first kind of answer, that of listing everything that can happen in a given situation, may seem easy, this is not always the case.

■ **EXAMPLE**   A woman is considering the replacement of flooring in two rooms of her apartment, but has not yet decided whether to replace the flooring in two rooms, one room, or neither room. Her choices of flooring materials are carpeting, ceramic tile, or wood. The entire floor of each room must be done using one material. List the different ways in which she could replace (or not replace) the flooring.

Solution   There are clearly many possibilities. Both rooms could be carpeted; one could be carpeted and the other one tiled; one could be tiled and the other one wooden; one could be carpeted and the other not replaced; and so forth. Continuing in this manner, we may be able to complete the list, but it is likely that we will omit at least one of the possibilities.

This problem can be handled systematically by drawing a **tree diagram** such as that of Figure 4.1(a). This figure shows that at first there are 3 possibilities (3 branches) corresponding to 0, 1, or 2 of the rooms being carpeted.

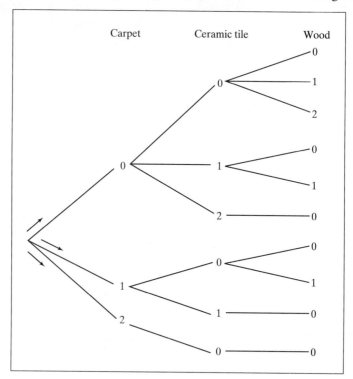

FIGURE 4.1(a) Tree diagram showing in each case how the flooring may be replaced.

Then for ceramic tile flooring, there are three branches emanating from the top branch, two from the middle branch, and one from the bottom branch. Observe that there are 3 possibilities (0, 1, or 2) for the number of tiled floors when none of the floors are to be carpeted; there are 2 possibilities (0 or 1) for the number of tiled floors when one room will be carpeted; and there is only 1 possibility (0) for the number of tiled floors when two rooms will be carpeted. For the number of wooden floors, the reasoning is similar. In going from left to right in Figure 4.1(a), we find that there are altogether 10 different paths along the branches of the tree. In other words, there are 10 possibilities. From top to bottom, on the right-hand side of the diagram, we can see that the choices are the following: do not replace either floor; replace only one floor using wood; replace two floors with wood; replace only one floor with tile; replace one floor with tile and the other with wood; replace both floors with tile; replace only one floor with carpeting; replace one floor with carpeting and the other with wood; replace one floor with carpeting and the other with tile; and, finally, replace both floors with carpeting. ■

■ **EXAMPLE**   After completing the previous example, suppose that we learn that the woman considering changes in the flooring of two rooms in an apartment decides also to consider using other types of flooring, such as concrete, stone, terrazzo, cork, linoleum, or plastic tile. Let us combine these into a catchall category called *other flooring*. Suppose that we also learn that she has decided to go ahead with the job and will definitely replace the flooring in both rooms. Change Figure 4.1(a) to reflect this additional information.

Solution   In Figure 4.1(b), we extend each branch of Figure 4.1(a) to include the new category, other flooring. Since the flooring in both rooms will be replaced, the

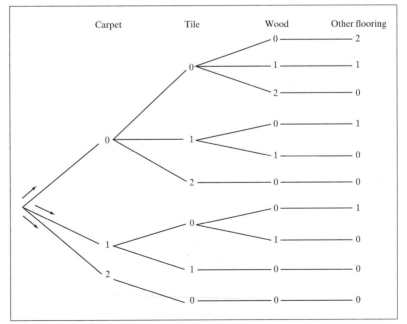

FIGURE 4.1(b) Continuation of Figure 4.1 (a).

totals of the values on each of the 10 branches must now equal 2. On the uppermost branch of Figure 4.1(a), for instance, the sum of the choices is $0+0+0=0$. We must therefore insert $2-0=2$ for the new category of other flooring. The completed changes are shown in Figure 4.1(b). As in Figure 4.1(a), where we had 10 branches, Figure 4.1(b) also has 10 branches.   ■

■ **EXAMPLE**   In a medical study, patients are classified according to whether they have blood type A, B, AB, or O and also according to whether their blood pressure is low, normal, or high. In how many different ways can a patient thus be classified according to blood type and blood pressure?

Solution   As is apparent from the tree diagram of Figure 4.2, the answer is 12. Starting at the top, the first path along the branches corresponds to a patient having blood type A and low blood pressure, the second path corresponds to a patient having blood type A and normal blood pressure, …, and the twelfth path corresponds to a patient having blood type O and high blood pressure.   ■

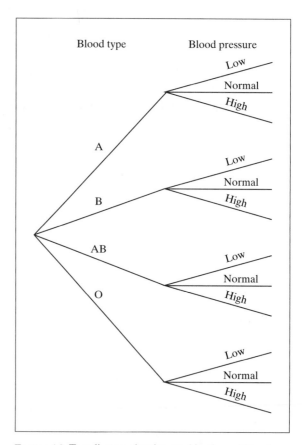

FIGURE 4.2 Tree diagram showing combinations of blood types and blood pressures.

The answer we obtained in the preceding example is $4 \cdot 3 = 12$, that is, the product of the number of blood types and the number of blood pressure levels. Generalizing from this example, let us state the following rule:

Multiplication of choices

> If a choice consists of two steps, of which the first can be made in m ways and for each of these the second can be made in n ways, then the whole choice can be made in $m \cdot n$ ways.

To prove this rule, we have only to draw a tree diagram similar to that of Figure 4.2. First, there are m branches corresponding to the possibilities in the first step, and then there are n branches emanating from each of these branches to represent the possibilities in the second step. This leads to $m \cdot n$ paths along the branches of the tree diagram, and hence to $m \cdot n$ possibilities.

■ **EXAMPLE**   A restaurant offers nine different desserts, which it serves with coffee, decaffeinated coffee, tea, milk, or hot chocolate. In how many different ways can one order a dessert and a drink?

Solution   Since $m = 9$ and $n = 5$, the answer is $9 \cdot 5 = 45$.   ■

■ **EXAMPLE**   If an English department schedules four lecture sections and 12 discussion groups for a course in modern literature, in how many different ways can a student choose a lecture section and a discussion group?

Solution   Since $m = 4$ and $n = 12$, the answer is $4 \cdot 12 = 48$.   ■

By using appropriate tree diagrams, we can readily generalize the foregoing rule for the **multiplication of choices** so that it will apply to choices involving more than two steps. For k steps, where k is a positive integer, we get the following rule:

Multiplication of choices (generalized)

> If a choice consists of k steps, of which the first step can be made in n_1 ways, for each of these the second step can be made in n_2 ways, for each combination of choices made in the first two steps the third can be made in n_3 ways, ..., and for each combination of choices made in all the preceding steps the kth can be made in n_k ways, then the whole choice can be made in $n_1 \cdot n_2 \cdot n_3 \cdot \ldots \cdot n_k$ ways.

We simply keep multiplying the numbers of ways in which the different steps can be made.

■ **EXAMPLE**   A new-car buyer has the choice of 4 body styles, 3 different engines, and 10 colors.

(a) In how many different ways can a person order one of these cars?

(b) If a person also has the options of ordering the car with or without air conditioning, with or without an automatic transmission, and with or without bucket seats, in how many different ways can he or she order one of these cars?

Solution   (a) Since $n_1 = 4$, $n_2 = 3$, and $n_3 = 10$, there are $4 \cdot 3 \cdot 10 = 120$ different ways in which a person can order one of the cars.

(b) Since $n_1 = 4$, $n_2 = 3$, $n_3 = 10$, $n_4 = 2$, $n_5 = 2$, and $n_6 = 2$, there are

$$4 \cdot 3 \cdot 10 \cdot 2 \cdot 2 \cdot 2 = 960$$

different ways in which a person can order one of the cars.   ■

■ **EXAMPLE**   A test consists of 12 multiple-choice questions, with each question having four possible answers. In how many different ways can a student check off one answer to each question?

Solution   Since $n_1 = n_2 = \cdots = n_{12} = 4$, there are

$$4 \cdot 4 \cdot 4 \cdot 4 \cdot 4 \cdot 4 \cdot 4 \cdot 4 \cdot 4 \cdot 4 \cdot 4 \cdot 4 = 16{,}777{,}216$$

ways in which a student can check off one answer to each question. Only in one of these cases will all the answers be correct, and in

$$3 \cdot 3 \cdot 3 \cdot 3 \cdot 3 \cdot 3 \cdot 3 \cdot 3 \cdot 3 \cdot 3 \cdot 3 \cdot 3 = 531{,}441$$

cases all the answers will be wrong.   ■

■ **EXAMPLE**   A bettor at a race track wants to place a single $2 win bet on each of the nine races. (A win bet pays only if the selected horse finishes first.) If there are eight horses in each of the first six races and seven horses in each of the final three races, in how many ways can this bettor place his bets?

Solution   Since $n_1 = n_2 = \cdots = n_6 = 8$ and $n_7 = n_8 = n_9 = 7$, there are

$$8 \cdot 8 \cdot 8 \cdot 8 \cdot 8 \cdot 8 \cdot 7 \cdot 7 \cdot 7 = 89{,}915{,}392$$

ways in which he can place his bets.   ■

Exercises

Exercises 4.1, 4.7, and 4.8 are practice exercises; their complete solutions are given on page 139.

4.1 In a baseball World Series (in which the winner is the first team to win four games), suppose that the National League champion leads the American League champion by three games to one. Construct a tree diagram to show the different ways in which these teams can win or lose the remaining game or games.

4.2 There are four routes, A, B, C, and D, between a businessman's home and his office, but route B is one-way, so that he cannot take it on the way to work, and route D is one-way, so that he cannot take it on the way home.
  (a) Draw a tree diagram showing the various ways by which he can go to and from work.
  (b) Draw a tree diagram showing the various ways by which he can go to and from work, given that he never goes by the same route both ways.

4.3 A person with $1 in his pocket bets $1, even money, on the flip of a coin, and he continues to bet $1 as long as he has any money. Draw a tree diagram to show the various things that can happen during the first three flips of the coin. In how many of the cases will he be exactly $1 ahead?

4.4 A student can study 0, 1, or 2 hours for a history test on any given night. Draw a tree diagram to show that there are seven different ways in which she can study altogether 3 hours for the test on three consecutive nights.

4.5 In a union election, Mr. Brown, Ms. Green, and Ms. Jones are running for president, while Mr. Adams, Ms. Roberts, and Mr. Smith are running for vice-president. Construct a tree diagram showing the nine possible outcomes, and use it to determine the number of ways in which these union officials will not both be of the same sex.

4.6 A purchasing agent places his orders by phone, fax, or e-mail, requesting in each case that his order be confirmed by fax or e-mail. Draw a tree diagram to show the various ways in which one of his orders can be placed and confirmed.

4.7 There are three trails to the top of a mountain. In how many different ways can a person hike up and down the mountain if
  (a) she wants to take the same trail both ways;
  (b) she can, but need not, take the same trail both ways;
  (c) she does not want to take the same trail both ways?

4.8 A basketball team consists of a center, two forwards, and two guards. If a basketball team has five centers, nine forwards, and eight guards, in how many ways can a most valuable player award be given to one of the centers, one of the forwards, and one of the guards?

4.9 A college has eight academic departments, but employs only four secretaries. In how many different ways can one of the secretaries be assigned to one of the department heads?

4.10 A display of greeting cards in a store has 18 different birthday cards and 11 different get-well cards. In how many ways can a customer buy one birthday card and one get-well card?

4.11 In a certain restaurant, a customer can order a hamburger rare, medium rare, medium, or well done, and also with or without cheese. In how many ways can a customer order a hamburger in this restaurant?

4.12 In the waiting room of a dental office there are five issues of *Time* magazine, seven issues of *National Geographic* magazine, and three issues of *Money* magazine. In how many different ways can a patient waiting to see the dentist glance at one of each kind if the order in which she looks at these magazines does not matter?

4.13 The menu of a restaurant lists 4 soups, 5 salads, 10 entrees, and 5 desserts. In how many different ways can one choose a soup, a salad, an entree, and a dessert?

4.14 A test consists of 10 multiple-choice questions, with each question having three possible answers. In how many different ways can a student check off one answer to each question?

4.15 If a test consists of 10 true–false questions, in how many different ways can a student answer all the questions?

4.16 In a census survey, families are classified into six categories according to income, five categories according to family size, three categories according to education of the head of the household, three categories with regard to home ownership, and eight categories with regard to ownership of major appliances. In how many different ways can a family thus be classified?

4.17 A computer retail store offers a basic desktop computer in a choice of four processor speeds, three sizes of read-only memory (ROM), five sizes of fixed disk, and three monitor designs. In how many different ways can a person select a computer from this store?

4.18 Trailer license plates in one state consist of three digits, the first of which cannot be 0, followed by two letters of the alphabet, the first of which cannot be I, O, Q, or X. How many different plates are possible with this scheme?

4.19 A housepainter has taken a job that requires the spackling and painting of a kitchen and dining room. This involves four tasks: spackle the kitchen; paint the kitchen; spackle the dining room; and paint the dining room. Spackling a room must precede painting. Construct a tree diagram to show the number of ways in which the painter can schedule the four tasks.

4.20 With reference to Exercise 4.19, suppose that the painter must also install fixtures in the kitchen after it is painted. Construct a tree diagram to show the number of ways in which the painter can schedule the five tasks.

4.2 Permutations

The rule for the multiplication of choices and its generalization are often used when several selections are made from one set of objects, items, or persons, and the order in which they are selected is important.

■ **EXAMPLE**   If 16 entries are submitted to an essay contest, in how many different ways can the judges award a first prize and a second prize?

Solution   Since the first prize can be awarded in 16 ways and the second prize must go to one of the other 15 entries, there are altogether $16 \cdot 15 = 240$ possibilities.   ■

■ **EXAMPLE**   In how many different ways can the 42 members of a union elect a president, a vice-president, a secretary, and a treasurer?

Solution   Assuming that the officers are selected in the order treasurer, secretary, vice-president, and president, we find that there are $n_1 = 42$ ways in which they can elect the treasurer, $n_2 = 41$ ways in which they can elect the secretary, $n_3 = 40$ ways in which they can elect the vice-president, and $n_4 = 39$ ways in which they can elect the president. Thus, there are altogether

$$42 \cdot 41 \cdot 40 \cdot 39 = 2{,}686{,}320$$

different possibilities. The solution would be the same for any other selection order.   ∎

In general, if r objects are selected from a set of n distinct objects, any particular arrangement (order) of these objects is called a **permutation**. For instance, Maine, Vermont, and Connecticut, in that order, constitute a permutation (a particular ordered arrangement) of three of the six New England states. Also, Purdue, Illinois, Michigan State, and Wisconsin, in that order, constitute a permutation of four of the universities in the Big 10 Conference.

■ **EXAMPLE**   Determine the number of different permutations of two of the five vowels a, e, i, o, and u, and list them all.

Solution   Since $n_1 = 5$ and $n_2 = 4$, there are $5 \cdot 4 = 20$ different permutations, and they are

$$ae \quad ai \quad ao \quad au \quad ei \quad eo \quad eu \quad io \quad iu \quad ou$$
$$ea \quad ia \quad oa \quad ua \quad ie \quad oe \quad ue \quad oi \quad ui \quad uo$$ ∎

To find a formula for the total number of permutations of r objects selected from a set of n distinct objects, observe that the first selection is made from the whole set of n objects, the second selection is made from the $n - 1$ objects that remain after the first selection has been made, the third selection is made from the $n - 2$ objects that remain after the first two selections have been made, ..., and the rth and final selection is made from the

$$n - (r - 1) = n - r + 1$$

objects that remain after the first $r - 1$ selections have been made. Therefore, direct application of the generalized rule for the multiplication of choices yields the result that the total number of permutations of r objects selected from a set of n distinct objects, which we shall denote by $_nP_r$, is

$$n(n - 1)(n - 2) \cdot \ldots \cdot (n - r + 1)$$

Since products of consecutive integers arise in many problems relating to permutations and other kinds of special arrangements or selections, it is convenient to introduce here what is called the **factorial notation**. In this notation, the product of all positive integers less than or equal to the positive integer n is called n *factorial* and denoted by $n!$. Thus,

$$1! = 1$$

$$2! = 2 \cdot 1 = 2$$

$$3! = 3 \cdot 2 \cdot 1 = 6$$

$$4! = 4 \cdot 3 \cdot 2 \cdot 1 = 24$$

$$5! = 5 \cdot 4 \cdot 3 \cdot 2 \cdot 1 = 120$$

$$6! = 6 \cdot 5 \cdot 4 \cdot 3 \cdot 2 \cdot 1 = 720$$

. . . .

and, in general,

$$n! = n(n - 1)(n - 2) \cdot \ldots \cdot 3 \cdot 2 \cdot 1$$

The factorials grow rapidly; for instance, 20! exceeds $2.0 \cdot 10^{18}$ (2 quintillion). Also, to make various formulas more generally applicable, we let $0! = 1$ by definition.

To express the formula for $_nP_r$ in terms of factorials, observe that

$$12 \cdot 11 \cdot 10! = 12!$$

$$7 \cdot 6 \cdot 5 \cdot 4 \cdot 3! = 7!$$

$$35 \cdot 34 \cdot 33 \cdot 32! = 35!$$

and, similarly,

$$_nP_r \cdot (n - r)! = n(n - 1)(n - 2) \cdot \ldots \cdot (n - r + 1) \cdot (n - r)! = n!$$

so that

$$_nP_r = \frac{n!}{(n - r)!}$$

To summarize,

Number of permutations of n distinct objects taken r at a time

> **The number of permutations of r objects selected from a set of n distinct objects is**
>
> $$_nP_r = n(n - 1)(n - 2) \cdot \ldots \cdot (n - r + 1)$$
>
> **or, in factorial notation,**
>
> $$_nP_r = \frac{n!}{(n - r)!}$$

In the first formula there are r factors; the first factor is n, and each successive factor is 1 less than the preceding factor.

■ **EXAMPLE**   Find the number of permutations of 4 objects selected from a set of 12 distinct objects (say, the number of ways in which 4 of 12 basketball teams can be ranked first, second, third, and fourth by a panel of coaches).

Solution   For $n = 12$ and $r = 4$, the first formula yields

$$_{12}P_4 = 12 \cdot 11 \cdot 10 \cdot 9 = 11{,}880$$

and the second formula yields

$$_{12}P_4 = \frac{12!}{(12 - 4)!} = \frac{12!}{8!} = \frac{12 \cdot 11 \cdot 10 \cdot 9 \cdot 8!}{8!} = 11{,}880$$

Essentially, the work is the same, but the second formula requires a few extra steps.   ■

abc	acb	bac	bca	cab	cba
abd	adb	bad	bda	dab	dba
acd	adc	cad	cda	dac	dca
bcd	bdc	cbd	cdb	dbc	dcb

If we are not concerned with the order in which the three letters are chosen from the four letters a, b, c, and d, there are only four ways in which the selection can be made: abc, abd, acd, and bcd, the values shown in the first column. Note that each row contains the $3! = 6$ different permutations of the corresponding letters shown in the first column. In general, there are $r!$ permutations of any r objects selected from a set of n distinct objects, so the $_nP_r$ permutations of r objects selected from a set of n distinct objects contain each set of r objects $r!$ times. Therefore, to write a formula for the number of ways in which r objects can be selected from a set of n distinct objects, also called the number of **combinations** of n objects taken r at a time and denoted by $_nC_r$, we must divide $_nP_r$ by $r!$, and we get

Number of combinations of n objects taken r at a time

The number of ways in which r objects can be selected from a set of n distinct objects is
$$_nC_r = \frac{n(n-1)(n-2)\cdot\ ...\ \cdot(n-r+1)}{r!}$$

or, in factorial notation,
$$_nC_r = \frac{n!}{r!(n-r)!}$$

Although the first of these formulas is generally easier to use, the one in factorial notation is easier to remember.

Instead of $_nC_r$, we often use the symbol $\binom{n}{r}$, and in this notation we refer to it as a **binomial coefficient**. The reason for this is explained in Exercise 4.46. For $n = 0$ to $n = 20$, the values of $\binom{n}{r}$ may be read from Table X at the end of the book.

■ **EXAMPLE**   In how many different ways can a person invite three of her eight closest friends to a party?

Solution   For $n = 8$ and $r = 3$, the first formula yields
$$\binom{8}{3} = \frac{8\cdot 7\cdot 6}{3!} = 8\cdot 7 = 56$$

and the second formula yields
$$\binom{8}{3} = \frac{8!}{3!\ 5!} = \frac{8\cdot 7\cdot 6\cdot 5!}{3!\ 5!} = \frac{8\cdot 7\cdot 6}{3\cdot 2\cdot 1} = 56$$

Basically, the work is the same, but the first formula requires fewer steps. ■

■ **EXAMPLE**   In how many different ways can a committee of four be selected from the 72 staff members of a hospital?

Solution   For $n = 72$ and $r = 4$, we get

$$\binom{72}{4} = \frac{72 \cdot 71 \cdot 70 \cdot 69}{4!} = \frac{72 \cdot 71 \cdot 70 \cdot 69}{24} = 1,028,790 \qquad ■$$

Observe that the result of the first example, but not that of the second, can be checked in Table X.

■ **EXAMPLE**   In how many different ways can a student select two of six mathematics courses together with three of seven English courses?

Solution   The student can select the two mathematics courses in $\binom{6}{2}$ ways, the three English courses in $\binom{7}{3}$ ways, and, hence, by the multiplication of choices, all five in

$$\binom{6}{2} \cdot \binom{7}{3} = 15 \cdot 35 = 525 \text{ ways}$$

The values of the binomial coefficients were obtained from Table X.   ■

When we take 7 objects from a set of 10 distinct objects, then $10 - 7 = 3$ of the objects are left. Thus, there are as many ways of leaving (or selecting) 3 objects from a set of 10 distinct objects as there are ways of selecting 7 objects, and we can write

$$\binom{10}{7} = \binom{10}{3}$$

In general, when r objects are selected from a set of n distinct objects, $n - r$ of the objects are left; consequently, there are as many ways of leaving (or selecting) $n - r$ objects from a set of n distinct objects as there are ways of selecting (or leaving) r objects. Symbolically, we write

Rule for binomial coefficients

$$\binom{n}{r} = \binom{n}{n - r} \qquad \text{for } r = 0, 1, 2, \dots, n$$

Sometimes this rule serves to simplify calculations and sometimes it is needed in connection with the use of Table X.

■ **EXAMPLE**   Determine the value of $\binom{75}{72}$.

Solution   To avoid having to write down the product

$$75 \cdot 74 \cdot 73 \cdot \dots \cdot 4$$

and cancel

$$72 \cdot 71 \cdot 70 \cdot \ldots \cdot 4$$

we write directly

$$\binom{75}{72} = \binom{75}{3} = \frac{75 \cdot 74 \cdot 73}{3!} = 67{,}525 \qquad ■$$

■ **EXAMPLE**   Find the value of $\binom{18}{13}$.

Solution $\binom{18}{13}$ cannot be looked up directly in Table X, but making use of the fact that

$$\binom{18}{13} = \binom{18}{18-13} = \binom{18}{5}$$

we look up $\binom{18}{5}$ and get 8,568.   ■

Exercises

Exercises 4.35 and 4.36 are practice exercises; their complete solutions are given on page 140.

4.35 A university catalogue lists 15 graduate courses in physics. Calculate the number of ways in which a graduate student can select
  (a) 4 courses;   (c) 2 courses;
  (b) 3 courses;   (d) 1 course.

4.36 Among the 16 candidates for four positions on a city council, 9 are Democrats, 5 are Republicans, and 2 are Independents. In how many ways can the 4 councilmen be chosen so that
  (a) 3 are Democrats and 1 is a Republican;
  (b) 2 are Democrats, 1 is a Republican, and 1 is an Independent?

4.37 Calculate the number of ways in which a motel chain can select 4 of 11 sites for the construction of new motels, and check your answer in Table X.

4.38 Draw poker is a card game played with an ordinary deck of 52 cards in which each player is dealt 5 cards. How many different 5-card hands are there?

4.39 A travel brochure lists 10 museums in the city of London. In how many ways can a tourist visit four museums
  (a) if the order in which the museums are visited does not matter;
  (b) if the order in which the museums are visited does matter?

4.40 Determine the number of ways in which the IRS can select 6 of 39 tax returns for a special audit.

4.41 A carton of 12 transistor batteries contains 4 that are defective. In how many different ways can one choose 3 of these batteries so that
  (a) none of the defective batteries is included;
  (b) exactly 1 of the defective batteries is included;
  (c) exactly 2 of the defective batteries are included;
  (d) exactly 3 of the defective batteries are included?

4.42 A health maintenance organization lists a number of medical doctors, among whom are eight internists and four urologists. In how many ways can a member of the HMO first get the opinions of two of the internists and then get the opinion of one of the urologists
    (a) if the order matters;
    (b) if the order does not matter?

4.43 A student committee must consist of three juniors and four seniors. If seven juniors and eight seniors are willing to serve on the committee, in how many different ways can it be selected?

4.44 To fill a number of vacancies, the personnel manager of a company has to choose three secretaries from among 10 applicants and two bookkeepers from among 4 applicants. In how many different ways can she fill the 5 vacancies?

4.45 Grouping the 100 U.S. senators in the 104th Congress (1995) by seniority, we have exactly 50 senators with 9 years or less of seniority and exactly 50 with 10 years or more of seniority. Altogether, in how many different ways could 2 senators be selected from each of the two seniority groups to form a four-member special committee?

4.46 The quantity $\binom{n}{r}$ is called a *binomial coefficient* because it is, in fact, the coefficient of $a^{n-r}b^r$ in the binomial expansion of $(a + b)^n$. Verify this for $n = 2, 3,$ and 4 by expanding $(a + b)^2, (a + b)^3,$ and $(a + b)^4$ and comparing the coefficients with the corresponding entries for $n = 2, n = 3,$ and $n = 4$ in Table X.

4.47 A table of binomial coefficients is easy to construct by following the pattern known as **Pascal's triangle**:

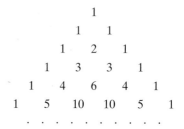

In this arrangement, each row begins with a 1 and ends with a 1, and each other entry is the sum of the nearest two values in the row immediately above. Construct the next three rows of Pascal's triangle and verify from Table X that they are, respectively, the binomial coefficients corresponding to $n = 6, n = 7,$ and $n = 8$.

4.48 In Pascal's triangle, denote the row 1 2 1 as row 2, the row 1 3 3 1 as row 3, and in general the row beginning 1 n ... as row n. Within row n, identify the $n + 1$ positions as 0, 1, 2, ..., n. Observe that $\binom{n}{r}$ appears in position r of row n.
    (a) What is row 7?
    (b) Give the value of $\binom{7}{3}$.
    (c) The triangle is constructed by using the relationship

$$\binom{n}{r} = \binom{n-1}{r-1} + \binom{n-1}{r}$$

By expressing the binomial coefficients in factorial form, verify that this relationship is correct.

4.4
Probability

Historically, the oldest way of measuring uncertainty is the **classical concept of probability**. It was developed originally in connection with games of chance, and it lends itself most readily to bridging the gap between possibilities and probabilities. This concept applies only when all possible outcomes are equally likely, in which case we say that

The classical probability concept

If there are n **equally likely possibilities, of which one must occur and s are** regarded as favorable, or as a *success*, then the probability of a success is $\dfrac{s}{n}$.

In the application of this rule, the terms *favorable* and *success* are used rather loosely—what is favorable to one player is unfavorable to his opponent, and what is a success from one point of view is a failure from another. Thus, these terms can be applied to any kind of outcome, even if favorable means that a house gets struck by lightning, or success means that a person catches pneumonia. This usage dates back to the days when probabilities were quoted only in connection with games of chance.

■ **EXAMPLE**   What is the probability of drawing an ace from a well-shuffled deck of 52 playing cards?

Solution   There are $s = 4$ aces among the $n = 52$ cards, so that we get

$$\frac{s}{n} = \frac{4}{52} = \frac{1}{13}$$ ■

■ **EXAMPLE**   What is the probability of rolling a 6 with a well-balanced die?

Solution   In this case, $s = 1$ and $n = 6$, so that the probability is

$$\frac{s}{n} = \frac{1}{6}$$ ■

■ **EXAMPLE**   Find the probability that 2 cards drawn from an ordinary deck of 52 playing cards will both be black.

Solution   According to what we learned in Section 4.3, the total number of possibilities is

$$n = \binom{52}{2} = \frac{52 \cdot 51}{2} = 1,326$$

and the number of favorable possibilities is

$$s = \binom{26}{2} = \frac{26 \cdot 25}{2} = 325$$

since half of the 52 playing cards are black and the others are red. It follows that the probability of drawing two black cards is

$$\frac{s}{n} = \frac{325}{1,326} \approx 0.245 \qquad\blacksquare$$

Although equally likely possibilities are found mostly in games of chance, the classical probability concept applies also to a great variety of situations where gambling devices are used to make random selections, say, when offices are assigned to research assistants by lot, when laboratory animals are chosen for an experiment (perhaps by the method described in Section 8.1) so that each has the same chance of being selected, when each family in a large apartment complex has the same chance of being included in a sample survey, or when machine parts are chosen for inspection so that each part has the same chance of being selected.

■ **EXAMPLE**   A tire manufacturer requires that 4 of the 20 tires in each production lot be inspected before they are shipped. If the tires are all satisfactory, the whole lot is shipped, but if they are not all satisfactory, the remaining 16 tires in the lot are also inspected. What is the probability that such a production lot will pass the inspection when actually 2 of the tires are defective?

Solution   There are $n = \binom{20}{4} = 4,845$ ways of choosing 4 of the 20 tires, and it will be assumed that they are all equally likely. The number of ways in which 4 of the 18 good tires can be selected is

$$s = \binom{18}{4} = 3,060$$

and it follows that the desired probability is

$$\frac{s}{n} = \frac{3,060}{4,845} = \frac{12}{19} \approx 0.632$$

The values of the two binomial coefficients, $\binom{20}{4}$ and $\binom{18}{4}$, were obtained directly from Table X.   ■

A major shortcoming of the classical probability concept is its limited applicability, for there are many situations in which the various possibilities cannot all be regarded as equally likely (that is, as having the same probability). This would be the case, for instance, if we are concerned with the question of whether it will rain on the next day. Surely, it would be nonsensical to say that either it will rain or it will not rain, and hence the probability for rain is $\frac{1}{2}$; or that there will be no precipitation, rain, hail, or snow, and hence the probability for rain is $\frac{1}{4}$. Also, the various possibilities cannot be regarded as equally likely when we wonder whether a person will get a raise, when we want to predict the outcome of an election or the score of a football game, or when we want to judge whether food prices will go up, go down, or remain the same. Among the various probability concepts, the most widely held is the **frequency interpretation**, according to which

The frequency interpretation of probability

> **The probability of an event (happening or outcome) is the proportion of the time that events of the same kind will occur in the long run.**

If we say that the probability is 0.88 that a jet from Denver to Seattle will arrive on time, we mean that such flights arrive on time 88% of the time. Also, if the Weather Service predicts that there is a 40% chance for rain (that is, the probability is 0.40 that it will rain), this is meant to imply that under the same weather conditions it will rain 40% of the time. More generally, we say that an event has a probability of, say, 0.90, in the same sense in which we might say that in cold weather our car will start 90% of the time. We cannot guarantee what will happen on any particular occasion—the car may start and then it may not—but if we kept records over a long period of time, we should find that the proportion of successes is very close to 0.90. In accordance with the frequency interpretation of probability, we estimate the probability of an event by observing what fraction of the time similar events have occurred in the past.

■ **EXAMPLE**   If data kept by a government agency shows that (over a period of time) 528 of 600 jets from Denver to Seattle arrived on time, what is the probability that any one jet from Denver to Seattle will arrive on time?

Solution   Since in the past $\frac{528}{600} = 0.88$ of the flights arrived on time, we use this figure as an estimate of the desired probability; or we say that there is an 88% chance that such a flight will arrive on time.   ■

■ **EXAMPLE**   If records show that 506 of 814 automatic dishwashers sold by a large retailer required repairs within the warranty year, what is the probability that an automatic dishwasher sold by this retailer will not require repairs within the warranty year?

Solution   Since $814 - 506 = 308$ of the dishwashers did not require repairs, we estimate the desired probability as $\frac{308}{814} \approx 0.38$.   ■

When probabilities are thus estimated, it is only reasonable to ask whether the estimates are any good. Later we shall answer this question in some detail, but for now let us refer to an important theorem called the **Law of Large Numbers**. Informally, this theorem may be stated as follows:

The Law of Large Numbers

If a situation, trial, or experiment is repeated again and again, the proportion of successes will tend to approach the probability that any one outcome will be a success.

We can illustrate this law by repeatedly flipping a balanced coin and recording the accumulated proportion of heads after each fifth flip. We can also use the MINITAB simulation of Figure 4.3, where the 1's and 0's denote heads and tails.

Reading across successive rows, we find that among the first 5 simulated flips there are 3 heads, among the first 10 there are 6 heads, among the first 15 there are 8 heads, among the first 20 there are 12 heads, among the first 25 there are 14 heads, ..., and among all 100 there are 51 heads. The corresponding proportions, plotted in Figure 4.4, are $\frac{3}{5} = 0.60$, $\frac{6}{10} = 0.60$, $\frac{8}{15} \approx 0.53$, $\frac{12}{20} = 0.60$, $\frac{14}{25} = 0.56$, ..., and $\frac{51}{100} = 0.51$. Observe that the proportion of heads fluctuates but comes closer and closer to 0.50, the probability of heads for each flip of a coin. Theoretical support for the Law of Large Numbers will be given in Chapter 6.

In the frequency interpretation, the probability of an event is defined in terms of what happens to similar events in the long run, so let us examine briefly whether it is at all meaningful to talk about the probability of an event that can occur only once. For instance, can we assign a probability to the event that Ms. Bertha Jones will be able to leave the hospital within four days after having an appendectomy or to the event that a certain major-party candidate will win an upcoming gubernatorial election? If we put ourselves in the position of Ms. Jones's doctor, we might check medical records, discover that patients left the hospital within four days of an appendectomy in, say, 61% of hundreds of cases, and apply this figure to Ms. Jones. This may not be of much

```
MTB > RANDOM 100 C1;
SUBC> INTEGERS 0 1.
MTB > PRINT C1

C1
    0  0  1  1  1  1  1  0  0  1
    1  0  0  1  0  1  1  1  0  1
    0  0  1  0  1  1  0  1  0  0
    1  1  0  1  0  0  1  1  1  0
    1  0  1  0  0  0  0  1  0  0
    1  1  0  0  0  0  0  1  0  0
    1  1  0  0  1  1  1  0  1  1
    1  0  1  1  0  1  1  0  0  0
    0  0  0  1  0  0  1  0  1  1
    1  0  1  1  1  1  0  1  0  1
MTB > MEAN C1
    MEAN = 0.51000
```

FIGURE 4.3 MINITAB simulation of 100 flips of a balanced coin.

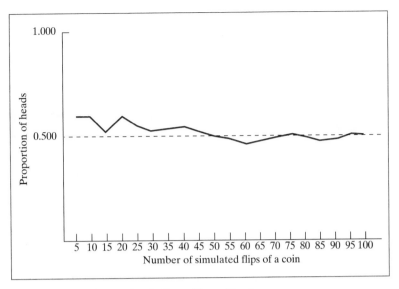

FIGURE 4.4 Graph illustrating the Law of Large Numbers.

comfort to Ms. Jones, but it does provide a meaning for a probability statement about her leaving the hospital within four days—the probability is 0.61. This illustrates that when we make a probability statement about a specific (nonrepeatable) event, the frequency interpretation of probability leaves us no choice but to refer to a set of similar events. As can well be imagined, however, this can easily lead to complications, since the choice of "similar" events is generally neither obvious nor straightforward. With reference to Ms. Jones's appendectomy, we might consider as "similar" only female patients, only patients of the same age as Ms. Jones, or only cases in which the patients were of the same height and weight as Ms. Jones. Ultimately, the choice of similar events is a matter of personal judgment, often motivated by the nature of available data, and it is by no means contradictory that we can arrive at different probabilities, all valid, concerning the same event.

Regarding the question of whether a certain major-party candidate will win an upcoming gubernatorial election, suppose that we ask the persons who have conducted a poll "how sure" they are that the candidate will win. If they say they are "95 percent sure" (that is, if they assign a probability of 0.95 to the candidate's winning the election), this is not meant to imply that he would win 95% of the time if he ran for office a great number of times. Rather, it means that the prediction is based on a method that "works" 95% of the time. It is in this way that we must interpret many of the probabilities attached to statistical results.

Finally, let us mention an alternative concept in which probabilities are interpreted as **personal** or **subjective evaluations**; they measure one's belief with regard to the uncertainties that are involved. Such probabilities apply especially when there is little or no direct evidence, so that there is really no choice but to consider collateral (indirect) information, "educated guesses," and perhaps intuition and other subjective factors. Subjective probabilities can be very useful to a person who is trying to sort out the consequences of his or her point of view, but they generally do not help to resolve differences among persons with very different points of view. Subjective probabilities can sometimes be determined by putting the issue on a "put up or shut up" basis, as will be explained in Sections 4.5 and 5.4.

Exercises

Exercises 4.49, 4.50, and 4.61 are practice exercises; their complete solutions are given on page 140.

4.49 What is the probability of rolling a total of 7 with a pair of well-balanced dice?

4.50 A car rental agency has 20 intermediate-sized cars and 10 compact cars. If 4 of these cars are randomly selected for a safety check, what is the probability that 2 of them will be intermediate-sized cars and 2 will be compact cars?

4.51 If 1 card is drawn from a well-shuffled deck of 52 playing cards, what are the probabilities of getting

    (a) a red king;   (c) a 3, 4, 5, or 6;   (e) not a diamond;

    (b) a black card;   (d) a diamond;   (f) not an ace?

4.52 If H stands for heads and T for tails, the four possible outcomes for two flips of a coin are HH, HT, TH, and TT. If it can be assumed that these four possibilities are equally likely, what are the probabilities of getting 0, 1, or 2 heads?

4.53 If H stands for heads and T for tails, the eight possible outcomes for three flips of a coin are HHH, HHT, HTH, THH, HTT, THT, TTH, and TTT. If it can be assumed that these eight possibilities are equally likely, what are the probabilities of getting 0, 1, 2, or 3 heads?

4.54 A bowl contains 21 red beads, 9 white beads, 11 blue beads, and 9 black beads. If one bead is drawn at random, what are the probabilities of getting

    (a) a red bead;

    (b) a bead that is white or black;

    (c) a bead that is neither red nor black?

4.55 If we roll a well-balanced die, what are the probabilities of getting

    (a) a 1 or a 6;   (b) an even number?

4.56 If 2 cards are drawn from a well-shuffled deck of 52 playing cards, what are the probabilities of getting

    (a) two spades;   (b) two aces;   (c) a king and a queen?

4.57 Among the 15 applicants for three identical positions at a bank, 12 have college degrees. What are the probabilities that the three positions will be randomly filled with

    (a) three of the applicants with college degrees;

    (b) two of the applicants with college degrees and one applicant without a college degree?

4.58 A carton of 12 lightbulbs includes 3 that are defective. If 2 bulbs are chosen at random, what are the probabilities that

    (a) neither bulb will be defective;

    (b) exactly 1 bulb will be defective;

    (c) both bulbs will be defective?

4.59 An apartment building has 24 two-bedroom apartments and 12 three-bedroom apartments. If 3 apartments are chosen at random to be redecorated, what are the probabilities that
   (a) all 3 will be two-bedroom apartments;
   (b) 2 will be two-bedroom apartments and 1 will be a three-bedroom apartment?

4.60 On a tray there are six pieces of chocolate cake and four pieces of walnut cake. If a waiter randomly picks two pieces of cake from the tray and gives them to customers who ordered chocolate cake, what is the probability that he will be making a mistake?

4.61 Data compiled by the manager of a department store shows that 1,018 of 1,956 persons who entered the store on a weekday morning made at least one purchase. Estimate the probability that a person who enters the store on a weekday morning will make at least one purchase.

4.62 According to the National Center for Health Statistics, there were 733,864 deaths from heart disease in 1996 from a total of 2,322,421 people who died of all causes. Find the probability that the cause of death of a person who died in 1996 was heart disease.

4.63 Bureau of Labor Statistics data shows that in the year 1997 there were 70,536,000 workers who were paid hourly rates; and 29,687,000 of these workers were paid $10 per hour or more. What is the probability that a randomly selected worker in this group received an hourly rate of $10 or more?

4.64 According to the American Medical Association, there were 737,764 physicians in the United States in the year 1996, of whom 580,377 were male and 157,387 were female. Of these physicians, 27,831 were male psychiatrists and 10,586 were female psychiatrists.
   (a) What is the probability that a randomly selected physician is male?
   (b) What is the probability that a randomly selected physician is female?
   (c) What is the probability that a randomly selected physician is a psychiatrist?
   (d) What is the probability that a randomly selected physician is not a psychiatrist?
   (e) What is the probability that a randomly selected psychiatrist is male?

4.65 In the state of Arizona, 1,159,000 out of a population of 4,801,000 persons were not covered by health insurance in 1996. What is the probability that a person residing in Arizona during 1996 did not have health insurance?

4.66 Use a computer to simulate 100 flips of a balanced coin and plot the accumulated proportion of heads after each fifth flip (as in Figure 4.4) to illustrate the Law of Large Numbers. If a suitable computer package is not available, actually flip a coin 100 times.

4.67 Use a computer to simulate 200 flips of a balanced coin and plot the accumulated proportion of heads after each fifth flip (as in Figure 4.4) to illustrate the Law of Large Numbers. If a suitable computer package is not available, actually flip a coin 200 times.

4.68 Use a computer to simulate 150 flips of a pair of coins, recording a 1 or a 0, depending on whether or not both coins come up heads. The probability for this is 0.25 (see Exercise 4.52). Then, to illustrate the Law of Large Numbers, plot the accumulated proportion of successes (both coins come up heads) after each tenth flip. If a suitable computer package is not available, actually flip a pair of coins 150 times.

4.5
Mathematical
Expectation

If an insurance agent tells us that in the United States a 45-year-old woman can expect to live 33 more years, this does not mean that anyone really expects a 45-year-old woman to live until her 78th birthday and then pass away the next day. Similarly, if we read that a person living in the United States can expect to eat 276 eggs per year and 128.9 pounds of beef, or that a child in the age group from 6 to 16 can expect to visit a dentist 1.92 times a year, it must be obvious that the word *expect* is not being used in its colloquial sense. A child cannot go to the dentist 1.92 times, and it would be surprising, indeed, if we found somebody who has actually eaten 276 eggs and 128.9 pounds of beef in a given year. Insofar as 45-year-old women are concerned, some will live another 15 years, some will live another 25 years, some will live another 40 years, ..., and the life expectancy of "33 more years" will have to be interpreted as an average or, as we shall call it here, a **mathematical expectation**.

Originally, the concept of a mathematical expectation arose in connection with games of chance, and in its simplest form it is the product of the amount a player stands to win and the probability that he or she will win.

■ **EXAMPLE**   What is our mathematical expectation if we will receive $20 if and only if a coin comes up heads?

Solution   If we assume that the coin is balanced and randomly tossed, that is, the probability of heads is $\frac{1}{2}$, our mathematical expectation is $20 \cdot \frac{1}{2} = \$10$.   ■

■ **EXAMPLE**   What is our mathematical expectation if we buy 1 of 2,000 raffle tickets issued for a television set worth $540?

Solution   The probability that we will win is $\dfrac{1}{2,000}$, so our mathematical expectation is

$$540 \cdot \frac{1}{2,000} = \$0.27 \quad \text{or} \quad 27 \text{ cents}$$

Thus, it would be unwise to spend more than 27 cents for the ticket, unless, of course, the proceeds of the raffle go to a worthy cause (or the difference can be credited to whatever pleasure a person might derive from placing a bet).   ■

In both of these examples there is a single prize, but in each case there are two possible payoffs—$20 and $0 in the first example and $540 and $0 in the other. Indeed, in the second example we can argue that 1,999 of the tickets will pay $0 and one of the tickets will pay $540 (or the equivalent in merchandise). Altogether, the 2,000 tickets will thus pay $540 or, on the average,

$$\frac{540}{2,000} = \$0.27 \text{ per ticket}$$

and this is the mathematical expectation. To generalize the concept of a mathematical expectation, let us consider the following change in the raffle of the preceding example.

■ EXAMPLE   What is our mathematical expectation if we buy 1 of 2,000 raffle tickets issued for a first prize of a television set worth \$540, a second prize of a tape recorder worth \$180, and a third prize of a pocket radio worth \$40?

Solution   Now we can argue that 1,997 of the raffle tickets will not pay anything at all, one ticket will pay \$540 (in merchandise), another will pay \$180 (in merchandise), and a third will pay \$40 (in merchandise). Altogether the 2,000 tickets will thus pay $540 + 180 + 40 = \$760$ (in merchandise) or, on the average,

$$\frac{760}{2,000} = \$0.38 \text{ per ticket}$$

As before, this is the mathematical expectation for each ticket. Looking at the problem in a different way, we could argue that if the raffle were repeated many times, we would lose

$$\frac{1,997}{2,000} \cdot 100\% = 99.85\% \text{ of the time}$$

(or with probability 0.9985); and win each of the prizes

$$\frac{1}{2,000} \cdot 100\% = 0.05\% \text{ of the time}$$

(or with probability 0.0005). On average, we would thus win

$$0(0.9985) + 540(0.0005) + 180(0.0005) + 40(0.0005) = \$0.38$$

which is the sum of the products obtained by multiplying each amount by the corresponding proportion or probability.   ■

Generalizing from this example, let us now give the following general definition:

Mathematical expectation

> If the probabilities of obtaining the amounts $a_1, a_2, \ldots,$ or a_k are, respectively, $p_1, p_2, \ldots,$ and p_k, then the mathematical expectation is
> $$E = a_1 p_1 + a_2 p_2 + \cdots + a_k p_k$$

Each amount is multiplied by the corresponding probability, and the mathematical expectation, E, is the sum of all these products. In the Σ notation,

$$E = \Sigma a \cdot p$$

Insofar as the a's are concerned, it is important to keep in mind that they are positive when they represent profits, winnings, or gains (amounts that we receive) and that they are negative when they represent losses, penalties, or deficits (amounts that we have to pay).

■ **EXAMPLE**   What is our mathematical expectation if we win $6 when a die comes up 1 or 2 and lose $3 when the die comes up 3, 4, 5, or 6?

Solution   The amounts are $a_1 = 6$ and $a_2 = -3$, and if we assume that the die is balanced and randomly tossed, $p_1 = \frac{2}{6}$ and $p_2 = \frac{4}{6}$. Thus, the mathematical expectation is

$$E = 6 \cdot \frac{2}{6} + (-3)\frac{4}{6} = 0$$ ■

This example illustrates an **equitable** or **fair game**. It is a game that does not favor either player; that is, each player's mathematical expectation is zero.

■ **EXAMPLE**   The probabilities are 0.22, 0.36, 0.28, and 0.14 that an investor will be able to sell a piece of property at a profit of $5,000, that he will be able to sell it at a profit of $2,000, that he will break even, or that he will sell it at a loss of $3,000. What is his expected profit?

Solution   If we substitute $a_1 = 5,000$, $a_2 = 2,000$, $a_3 = 0$, $a_4 = -3,000$, $p_1 = 0.22$, $p_2 = 0.36$, $p_3 = 0.28$, and $p_4 = 0.14$ into the formula for E, we get

$$E = 5,000(0.22) + 2,000(0.36) + 0(0.28) - 3,000(0.14) = \$1,400$$ ■

Although we referred to the quantities $a_1, a_2, \ldots,$ and a_k as *amounts*, they need not be cash winnings, losses, penalties, or rewards. When we said previously that a child in the age group from 6 to 16 can expect to visit a dentist 1.92 times a year, we referred to the result that was obtained by multiplying $0, 1, 2, 3, \ldots,$ by the corresponding probabilities that a child in this age group will visit a dentist that many times a year and then adding all these products (see Exercise 4.77).

■ **EXAMPLE**   The police chief of a city knows that the probabilities for 0, 1, 2, 3, 4, or 5 car thefts on any given day are, respectively, 0.21, 0.37, 0.25, 0.13, 0.03, and 0.01. How many car thefts can he expect per day?

Solution   Substituting into the formula for a mathematical expectation, we get

$$E = 0(0.21) + 1(0.37) + 2(0.25) + 3(0.13) + 4(0.03) + 5(0.01) = 1.43$$ ■

In all the examples in this section, we were given the values of a and p (or the value of the a's and p's) and calculated E. Now let us consider an example in which we are given the values of a and E to arrive at some result about p.

■ **EXAMPLE**   To handle a liability suit, a lawyer has to decide whether to charge a straight fee of $2,400 or a contingent fee of $9,600, which she will get only if her client wins. How does she feel about her client's chances if she prefers the straight fee of $2,400?

Solution   If she feels that the probability is p that her client will win and she decided on the contingent fee, her mathematical expectation would be $9,600p$. Since she feels that the certainty of $2,400 is preferable to the mathematical expectation of 9,600p, we can write

$$2{,}400 > 9{,}600p$$

which yields $p < \dfrac{2{,}400}{9{,}600}$ and, hence, $p < 0.25$. ∎

Note that this is one of the two methods of determining subjective probabilities referred to earlier. To pin it down further, we might ask the lawyer if she would still prefer the straight fee of $2,400 if the contingent fee were, say, $19,200 (see Exercise 4.80).

4.6 A Decision Problem

When we are faced with uncertainties, mathematical expectations can often be used to great advantage in making decisions. In general, if we have to choose between two or more alternatives, it is considered rational to select the one with the *most promising* mathematical expectation: the one that maximizes expected profits, minimizes expected costs, maximizes expected tax advantages, minimizes expected losses, and so on.

■ **EXAMPLE** A clothing manufacturer must decide whether to spend a considerable sum of money to build a new factory. He knows that if the new factory is built and the clothing business has a good sales year, there will be a $451,000 profit; if the new factory is built and the clothing business has a poor sales year, there will be a deficit of $110,000; if the new factory is not built and the clothing business has a good sales year, there will be a $220,000 profit; and if the new factory is not built and the clothing business has a poor year, there will be a $22,000 profit (mostly because of lower overhead cost). If the clothing manufacturer feels that the probabilities for a good sales year or a poor sales year are, respectively, 0.40 and 0.60, would building the new factory maximize his expected profit?

Solution In problems like this, it usually helps to present the information about profits and deficits in the following kind of table, called a **payoff table**.

	New factory built	New factory not built
Good sales year	$451,000	$220,000
Poor sales year	−$110,000	$22,000

As can be seen from this table, it will be advantageous to build the new factory only if the clothing business is going to have a good sales year, and the decision whether to build the new factory will, therefore, have to depend on the chances that this will be the case. Using the manufacturer's probabilities of 0.40 and 0.60 for a good sales year and a poor sales year, we find that if the new factory is built, his expected profit is

$$451,000(0.40) - 110,000(0.60) = \$114,400$$

If the new factory is not built, his expected profit is

$$220,000(0.40) + 22,000(0.60) = \$101,200$$

Since the first of these two figures is larger, it follows that building the new factory maximizes the manufacturer's expected profit.   ■

The way in which we have studied this problem is called a **Bayesian analysis**. In this kind of analysis, probabilities are assigned to the alternatives about which uncertainties exist (the **states of nature**, which in our example were a good sales year and a poor sales year); then we choose whichever alternative promises the greatest expected profit or the smallest expected loss. This approach to decision making has great intuitive appeal, but it is not without complications. It may be quite tricky to sort out complex situations in a logical way. In the preceding example, is it adequate to consider sales as simply "good" or "poor"? Also, it is essential that the appraisals of all relevant probabilities be close to correct and also that the values of the payoffs associated with the various possibilities be close to correct. (See Exercises 4.87 through 4.90).

■ **EXAMPLE**   A dealer for luxury boats is concerned about the number of a very expensive line of boats to order for the April–June sales season. He would like to have available exactly the number that will be ordered. If demand for this boat exceeds supply, then the extra customers will either buy the boat elsewhere or make no purchase at all. If supply exceeds demand, then the extra boats are not likely to sell at all until the next sales season and must be returned to the manufacturer for a substantial service charge. The profit is $10,000 on each boat sold, and the loss for an unsold boat is $3,500.

The number of boats to be demanded is virtually certain to be at most 5. Therefore, the dealer must decide whether to order 0, 1, 2, 3, 4, or 5 boats for the coming season.

It is believed that the number of boats demanded will be 0 with probability 0.05, 1 with probability 0.20, 2 with probability 0.30, 3 with probability 0.25, 4 with probability 0.15, and 5 with probability 0.05.

How many boats should the dealer order to maximize his expected profit?

Solution   Construct a chart showing the profit to be made under each combination of order size and number demanded. In this chart, the amounts are given in thousands of dollars.

		Order				
	0	1	2	3	4	5
0	0	−3.5	−7	−10.5	−14	−17.5
1	0	10	6.5	3	−0.5	−4
2	0	10	20	16.5	13	9.5
3	0	10	20	30	26.5	23
4	0	10	20	30	40	36.5
5	0	10	20	30	40	50

Number of boats demanded (row labels 0–5)

For example, if zero boats are ordered, the profit will certainly be zero. If one boat is ordered, the profit will be $10,000 so long as the demand is one or greater; if the demand is zero, the profit will be −$3,500. If two boats are ordered and the demand is one boat, the dealer will make $10,000 on the boat that is sold and lose $3,500 on the boat that is not sold; the amount

$$\$10,000 - \$3,500 = \$6,500$$

is presented as 6.5 in the box (Order 2, Demand 1).

If one boat is ordered, the profit will be −3.5 thousand dollars with probability 0.05 and 10 thousand dollars with probability 0.95. The expected profit is then

$$-3.5 \cdot 0.05 + 10 \cdot 0.95 = 9.325 \text{ thousand dollars}$$

If two boats are ordered, the profit will be −7 thousand dollars with probability 0.05. It will be 6.5 thousand dollars with probability 0.20, and it will be 20 thousand dollars with probability

$$0.30 + 0.25 + 0.15 + 0.05 = 0.75$$

The expected profit is then

$$-7 \cdot 0.05 + 6.5 \cdot 0.20 + 20 \cdot 0.75 = 15.95 \text{ thousand dollars}$$

Similarly, the expected profits for ordering three, four, and five boats are 18.525, 17.725, and 14.900 thousand dollars, respectively.

Ordering three boats will maximize the expected profit. ■

Exercises

Exercises 4.69, 4.70, 4.79, and 4.85 are practice exercises; their complete solutions are given on pages 140 and 141.

4.69 At a bazaar held to raise money for a charity, it costs 50 cents to try one's luck in drawing an ace from an ordinary deck of 52 playing cards. What is the bazaar's expected profit per card drawn by a customer if the prize is $4.00 if and only if the customer draws an ace?

4.70 A union wage negotiator feels that the probabilities are 0.25, 0.50, 0.20, and 0.05 that the members of the union will get a $1.20 raise in their hourly wage, an 80-cent raise, a 40-cent raise, or no raise at all. What is the corresponding expected raise?

4.71 If a service club sells 500 raffle tickets for a cash prize of $100, what is the mathematical expectation of a person who buys one of the tickets?

4.72 In a friendly game, if we receive 10 cents each time that we roll a 6 with a balanced die, how much should we pay when we roll a 1, 2, 3, 4, or 5 so as to make the game equitable?

4.73 In the game of roulette a metal ball is dropped into a spinning wheel so that it lands in one of 38 compartments. The compartments are numbered 00, 0, 1, 2, 3,..., 36. The compartments for 00 and 0 are colored green; 18 of the remaining compartments are red and 18 are black.

(a) You can bet $1 on red. If a red number comes up, then your bet is returned to you with another $1 so that your net winning is +$1. If the number that comes up is not red, your bet is lost, and your net winning is −$1. What is your expected net winning?

(b) You can bet $1 on the set of numbers 1 through 12. If one of these numbers comes up, then your bet is returned to you with another $2 so that your net winning is +$2. If the number that comes up is not one of these, your bet is lost, and your net winning is −$1. What is your expected net winning?

(c) You can bet $1 on any individual number. If this number comes up, then your bet is returned to you with another $35 so that your net winning is +$35. If your chosen number does not come up, your bet is lost, and your net winning is −$1. What is your expected net winning?

4.74 In the finals of a tennis tournament, the winner gets $48,000 and the loser gets $24,000. What are the two finalists' mathematical expectations if
(a) they are evenly matched;
(b) their respective probabilities of winning are $\frac{2}{3}$ and $\frac{1}{3}$?

4.75 A stockbroker feels that the probabilities are 0.40, 0.30, 0.20, and 0.10 that the value of a stock will increase by $1.50, $1.00, 50 cents, or not at all. What is the expected increase in the value of the stock?

4.76 An importer is offered a shipment of pineapples for $12,000, and the probabilities that he will be able to sell them for $16,000, $15,000, $14,000, or $13,000 are, respectively, 0.15, 0.41, 0.33, and 0.11. What is the importer's expected profit?

4.77 If the probabilities are 0.15, 0.28, 0.27, 0.17, 0.08, 0.03, and 0.02 that a child in the age group from 6 to 16 will visit a dentist 0, 1, 2, 3, 4, 5, or 6 times, respectively, a year, how many times can a child in this age group expect to visit a dentist in any given year?

4.78 The probabilities that a person who enters "The Department Store" will make 0, 1, 2, 3, 4, or 5 purchases are 0.11, 0.33, 0.31, 0.12, 0.09, and 0.04, respectively. How many purchases can a person entering this store be expected to make?

4.79 A salesperson has to choose between a straight salary of $24,000 and a salary of $20,000 plus a bonus of $8,000 if her sales exceed a certain quota. How does she assess her chances of exceeding the quota if she chooses the lower salary with the possibility of the bonus?

4.80 Suppose that a lawyer in a liability suit is asked whether she would prefer a fee of $2,400 or a contingent fee of $19,200, which she will get only if her client wins. How does she feel about her client's chances if she prefers the fee of $2,400?

4.81 An insurance company agrees to pay the promoter of a drag race $15,000 if the race has to be canceled because of rain. If the company's actuary feels that a fair net premium for this risk is $2,400, what does this tell us about his assessment of the probability that the race will have to be canceled because of rain?

4.82 A shopping-center manager finds a contractor to do a road repair job for $45,000. Another contractor offers to do the job for $50,000 with a penalty of $12,500 if the job is not finished on time. The shopping center will not experience a monetary loss if the repair is not completed on time. If the manager prefers the second offer, what does this tell us about her assessment of the probability that the second contractor will not finish the job on time?

*4.83 A dealer in luxury boats needs to place his order for his stock of Narwhal cruisers for the coming spring buying season. Each boat that he stocks and sells will generate a profit of $15,000. Each boat that he stocks but does not sell costs him $4,000 in storage and financing costs. Ideally, he would like to order exactly the number that he will be able to sell, but the number that he can sell cannot be perfectly predicted. He feels that the probabilities that he can sell 0, 1, 2, 3, or 4 of the Narwhal cruisers are 0.10, 0.25, 0.40, 0.20, and 0.05, respectively. How many Narwhal cruisers should be ordered?

4.84 Referring to Exercise 4.80 in which a lawyer was asked whether she would prefer a fee of $2,400 or a contingent fee of $19,200, which she would get only if her client wins, the lawyer chose the fee of $2,400. Now let us assume that she chose the contingent fee of $19,200. How did she feel about the probability that her client would win?

*4.85 A truck driver has to deliver a load of building materials to one of two construction sites and return to the lumberyard. These sites are 16 and 20 miles from the lumberyard. He has misplaced the order telling him where the load should go, and the telephone at the lumberyard is out of order. The two construction sites are 6 miles apart. If the driver feels that the probability is $\frac{1}{5}$ that the load should go to the site that is 16 miles from the lumberyard and the probability is $\frac{4}{5}$ that the load should go to the other site, where should he go first to minimize the expected distance that he will have to drive?

*4.86 With reference to Exercise 4.85, where should the driver go first so as to minimize the expected distance that he will have to drive if he reassessed the probabilities as $\frac{1}{8}$ and $\frac{7}{8}$ instead of $\frac{1}{5}$ and $\frac{4}{5}$?

*4.87 The management of a mining company must decide whether to continue an operation at a certain location. If they continue and are successful, they will make a profit of $4,500,000; if they continue and are not successful, they will lose $2,700,000; if they do not continue but would have been successful if they had continued, they will lose $1,800,000 (for competitive reasons); and if they do not continue and would not have been successful if they had continued, they will make a profit of $450,000 (because funds allocated to the operation remain unspent). What decision would maximize the company's expected profit if it is felt that there is a 50–50 chance for success?

*4.88 Show that it does not matter what they decide to do in Exercise 4.87 if it is felt that the probabilities for and against success are, respectively, $\frac{1}{3}$ and $\frac{2}{3}$.

4.89 With reference to the following payoff table, which was previously used in the example on page 133,

	New factory built	New factory not built
Good sales year	$451,000	$220,000
Poor sales year	-$110,000	$22,000

show that building the new factory would not maximize the expected profit if the manufacturer felt that the probabilities for a good sales year and a poor sales year are, respectively, $\frac{1}{6}$ and $\frac{5}{6}$.

4.90 With reference to Exercise 4.89 and using the probabilities of 0.40 and 0.60, would the clothing manufacturer's decision not to build the new factory still maximize his expected profit if it is found that the -$110,000 figure is in error and should be a loss of $165,000?

*4.91 A realty company owns a newly constructed but unoccupied strip shopping center that has locations for four stores. The realty company needs to know the number of these stores for which they should install indoor furnishings (plumbing, counter space, display shelves, and so on). If the realty company finds more tenants than there are finished stores, the tenants will be lost. If the realty company has more finished stores than tenants, the dollar loss will be represented by the money tied up in the indoor furnishings. The value of a tenant is computed at $30,000 over the next year, while the cost of installing indoor furnishings is set at $12,000. An expert in the local real estate market believes that the following probabilities govern the number of tenants that can be found for this shopping center:

Number of tenants	Probability
0	0.1
1	0.4
2	0.3
3	0.1
4	0.1

In order to maximize the expected return, how many stores should be given indoor furnishings?

*4.92 The realty company in Exercise 4.91 consulted a second expert. The second expert gave these probabilities:

Number of tenants	Probability
0	0.1
1	0.1
2	0.1
3	0.3
4	0.4

According to the second expert, how many stores should be given indoor furnishings?

Solutions to Practice Exercises

4.1 The tree diagram is shown in Figure 4.5, where N denotes a win by the National League champion and A denotes a win by the American League champion.

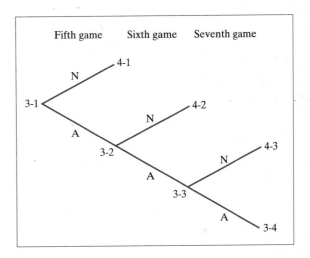

FIGURE 4.5 Tree diagram for Exercise 4.1.

4.7 (a) There are three choices corresponding to the three trials.
　　 (b) $m = 3, n = 3$, so there are $3 \cdot 3 = 9$ possibilities.
　　 (c) $m = 3, n = 2$, so there are $3 \cdot 2 = 6$ possibilities.

4.8 $n_1 = 5, n_2 = 9$, and $n_3 = 8$ and there are

$$n_1 \cdot n_2 \cdot n_3 = 5 \cdot 9 \cdot 8 = 360 \text{ possibilities}$$

4.21 (a) $8! = 8 \cdot 7 \cdot (6 \cdot 5 \cdot 4 \cdot 3 \cdot 2 \cdot 1) = 8 \cdot 7 \cdot 6!$ is true; in general,

$$n! = n(n-1)! = n(n-1)(n-2)! = n(n-1)(n-2)(n-3)!$$

and so forth.
　　 (b) $3! = 6, 2! = 2, 6! = 720$, and $6 \cdot 2$ does not equal 720.
　　 (c) $\dfrac{12!}{12 \cdot 11 \cdot 10} = \dfrac{12 \cdot 11 \cdot 10 \cdot 9!}{12 \cdot 11 \cdot 10} = 9!$ is true.

4.24 $7 \cdot 6 \cdot 5 = 210$ ways.

4.25 $6! = 6 \cdot 5 \cdot 4 \cdot 3 \cdot 2 \cdot 1 = 720$ ways.

4.35   (a) $\binom{15}{4} = \dfrac{15 \cdot 14 \cdot 13 \cdot 12}{4!} = \dfrac{32{,}760}{24} = 1{,}365$ ways.

     (b) $\binom{15}{3} = \dfrac{15 \cdot 14 \cdot 13}{3!} = \dfrac{2{,}730}{6} = 455$ ways.

     (c) $\binom{15}{2} = \dfrac{15 \cdot 14}{2!} = \dfrac{210}{2} = 105$ ways.

     (d) $\binom{15}{1} = \dfrac{15}{1!} = 15$ ways.

4.36   (a) The three Democrats can be chosen in $\binom{9}{3} = 84$ ways, the Republican can be chosen in 5 ways, so that, by the multiplication of choices, there are $84 \cdot 5 = 420$ possibilities.

     (b) The two Democrats can be chosen in $\binom{9}{2} = 36$ ways, the Republican can be chosen in 5 ways, and the Independent can be chosen in 2 ways, so that by the generalized rule for the multiplication of choices, there are $36 \cdot 5 \cdot 2 = 360$ possibilities.

4.49 Altogether there are $n = 6 \cdot 6 = 36$ possibilities, and in $s = 6$ of these (1 and 6, 6 and 1, 2 and 5, 5 and 2, 3 and 4, 4 and 3) the total is 7; therefore, the probability is $\frac{6}{36} = \frac{1}{6}$.

4.50 Altogether there are

$$n = \binom{30}{4} = \frac{30 \cdot 29 \cdot 28 \cdot 27}{4!} = 27{,}405 \text{ possibilities}$$

The two intermediate-sized cars can be chosen in

$$\binom{20}{2} = \frac{20 \cdot 19}{2} = 190 \text{ ways}$$

and the two compact cars can be chosen in

$$\binom{10}{2} = \frac{10 \cdot 9}{2} = 45 \text{ ways}$$

so that by the multiplication of choices, $s = 190 \cdot 45 = 8{,}550$; it follows that the probability is

$$\frac{s}{n} = \frac{8{,}550}{27{,}405} \approx 0.312$$

4.61 The estimate is $\dfrac{1{,}018}{1{,}956} \approx 0.520$.

4.69 The mathematical expectation of a person who draws a card is $4 \cdot \frac{1}{13} \approx 0.31$, so that the bazaar's approximate expected profit is $0.50 - 0.31 = \$0.19$.

4.70 Substituting $a_1 = 120$, $a_2 = 80$, $a_3 = 40$, $a_4 = 0$, $p_1 = 0.25$, $p_2 = 0.50$, $p_3 = 0.20$, and $p_4 = 0.05$ into the formula for a mathematical expectation, we get

$$120(0.25) + 80(0.50) + 40(0.20) + 0(0.05) = 78 \text{ cents}$$

4.79 If she feels that the probability of her exceeding the quota is p, the mathematical expectation corresponding to the salary plus bonus alternative is $20,000 + 8,000p$. Since this exceeds 24,000, we get $20,000 + 8,000p > 24,000$, and hence

$$8,000p > 4,000 \quad \text{and} \quad p > \frac{4,000}{8,000} = \frac{1}{2}$$

Thus, she feels that there is a better than 50–50 chance that she will exceed the quota.

4.85 Letting I and II denote the construction sites that are, respectively, 16 and 20 miles from the lumberyard, we can show the distances corresponding to the various alternatives (depending on where the driver goes first and where he should go) as in the following table:

		Driver goes first to	
		I	II
Driver should go to	I	16 + 16 = 32	20 + 6 + 16 = 42
	II	16 + 6 + 20 = 42	20 + 20 = 40

Thus, if he first goes to site I, the expected distance is

$$32 \cdot \frac{1}{5} + 42 \cdot \frac{4}{5} = \frac{32 + 168}{5} = \frac{200}{5} = 40.0 \text{ miles}$$

and if he goes first to site II, the expected distance is

$$42 \cdot \frac{1}{5} + 40 \cdot \frac{4}{5} = \frac{42 + 160}{5} = \frac{202}{5} = 40.4 \text{ miles}$$

Thus, if the driver wants to minimize the expected distance that he has to drive, he should go first to site I, that is, the construction site that is 16 miles from the lumberyard.

5

Some Rules of Probability

There are three fundamental questions in the study of probability: (1) What do we mean when we say that the probability of an event is 0.60, 0.88, or 0.02? (2) How are the numbers we call probabilities determined, or measured, in actual practice? (3) What are the mathematical rules that probabilities must obey?

Since the first two kinds of questions have already been studied in Chapter 4, we shall concentrate here on some of the rules which probabilities must obey, namely, on the **theory of probability** or, as it is sometimes called, the **mathematics of chance**.

Probability theory was originally developed in connection with games of chance. Nowadays, it has become an indispensable tool for business managers who must plan inventories without knowing with certainty what products will sell, for military strategists who must commit personnel and equipment to the hazards of battle, for doctors who risk their lives in combating disease, for anyone who sends a message by mail without an assurance that it will be delivered on time, and so on.

In this chapter, after some preliminaries in Sections 5.1 and 5.2, we shall study some of the basic rules in Section 5.3, the relationship between probabilities and odds in Section 5.4, addition rules in Section 5.5, conditional probabilities and related problems in Sections 5.6, 5.7, and 5.8, and Bayes' theorem in Section 5.9.

5.1
The Sample Space

In statistics, the word *experiment* is used in a very wide and unconventional sense. For lack of a better term, it refers to any process of observation or measurement. Thus, an **experiment** may consist of counting how many times a student has been absent; it may consist of the simple process of noting whether a light is on or off, or whether a person is single or married; or it may consist of the very complicated process of obtaining and evaluating data to predict trends in the economy, to find the source of social unrest, or to study the cause of a disease. The results that one obtains from an experiment, whether they are instrument readings, counts, yes or no answers, or values obtained through extensive calculations, are called the **outcomes** of the experiment.

For each experiment, the set of all possible outcomes is called the **sample space**, and it is usually denoted by the letter S. For instance, if a zoologist must choose 3 of his 24 guinea pigs for an experiment, the sample space consists of the

$$\binom{24}{3} = 2{,}024 \text{ ways}$$

in which the selection can be made; if the dean of a college has to assign 2 of her 84 faculty members as advisors to a political science club, the sample space consists of the

$$\binom{84}{2} = 3{,}486 \text{ ways}$$

in which this can be done. Also, if we are concerned with the number of days that it rains in Chicago during the month of January, the sample space is the set

$$S = \{0, 1, 2, 3, 4, \ldots, 30, 31\}$$

When we study the outcomes of an experiment, we usually identify the various possibilities with numbers, points, or some other kinds of symbols so that we can treat all questions about them mathematically, without having to go through long verbal descriptions of what has taken place, is taking place, or will take place. For instance, if there are eight candidates for a scholarship and we let a, b, c, d, e, f, g, and h denote that it is awarded to Ms. Adam, or Mr. Bean, or Ms. Clark, and so on, then the sample space for this experiment is the set

$$S = \{a, b, c, d, e, f, g, h\}$$

The use of points rather than letters or numbers has the advantage that it makes it easier to visualize the various possibilities and perhaps discover some special features that several of the outcomes may have in common.

■ **EXAMPLE**   A used car dealer has two 1997 Chevrolet Camaros on his lot, and we are interested in how many of them each of two salespersons will sell in a given week.

(a) Use two coordinates so that (0, 1), for example, represents the outcome

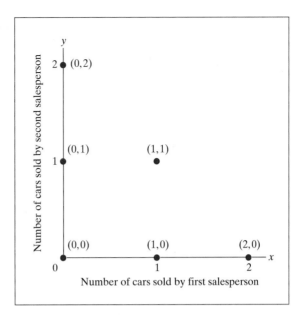

FIGURE 5.1 Sample space for two-salespersons example.

that the first salesperson will sell neither of the Camaros and the second salesperson will sell one, $(1, 1)$ represents the outcome that each of the two salespersons will sell one of the Camaros, and $(2, 0)$ represents the outcome that the first salesperson will sell them both. Now list all possible outcomes of this experiment.

(b) Draw a figure showing the corresponding points of the sample space.

Solution   (a) The six possible outcomes are $(0, 0), (1, 0), (0, 1), (2, 0), (1, 1)$, and $(0, 2)$.

(b) The corresponding points are shown in Figure 5.1, from which it is apparent, for instance, that they sell equally many 1997 Camaros in two of the six possibilities, and that they sell both cars in three of the six possibilities.   ■

Generally, we classify sample spaces according to the number of elements, or points, that they contain. The ones we have studied so far in this section contained, respectively, 2,024, 3,486, 32, 8, and 6 elements, and we refer to them all as **finite**. In this chapter we shall consider only sample spaces that are finite, but in later chapters we shall consider also sample spaces that are **infinite**. An infinite sample space arises, for example, when we throw a dart at a target and there is a continuum of points that we may hit.

In general, if the elements of a sample space are obtained by a measuring process, then the sample space is infinite. If the elements of a sample space are obtained by counting, then the sample space may be finite (as in the preceding used-car example), or it may be infinite (as in recording the number of tails preceding the first head in repeated flips of a coin).

5.2
Events

In statistics, any subset of a sample space is called an **event**. By subset, we mean any part of a set, including the set as a whole and, trivially, a set called the **empty set** and denoted by ∅, which has no elements at all. For instance, with reference to the example on page 144 dealing with the number of days that it rains in Chicago during the month of January, the subset

$$M = \{10, 11, 12, \ldots, 19, 20\}$$

is the event that there will be anywhere from 10 to 20 rainy days, and the subset

$$N = \{18, 19, 20, \ldots, 30, 31\}$$

is the event that there will be at least 18 rainy days.

■ **EXAMPLE**   With reference to Figure 5.1, express in words the events represented by
(a) $C = \{(0, 0), (1, 1)\}$;
(b) $D = \{(1, 0), (1, 1)\}$;
(c) $E = \{(0, 2)\}$.

Solution   (a) C is the event that the two salespersons will sell equally many of the 1997 Chevrolet Camaros.
(b) D is the event that the first salesperson will sell one and only one of the two cars.
(c) E is the event that the second salesperson will sell both cars.   ■

As is the custom, in both of these examples we denoted events by capital letters. In many probability problems we must deal with events that are compounded by forming **unions**, **intersections**, and **complements**.

The union of two events A and B, denoted by A ∪ B, is the event that consists of all the elements (outcomes) contained in event A, in event B, or in both.

The intersection of two events A and B, denoted by A ∩ B, is the event that consists of all the elements (outcomes) contained in both A and B.

The complement of event A, denoted by A', is the event that consists of all the elements (outcomes) of the sample space that are not contained in A.

The symbol ∪ is commonly read as *or*. Similarly, ∩ is read as *and* while A' is read as *not A*. One sometimes encounters the symbols A^c or $\overline{A}$ for *not A*. We will use A'.

■ **EXAMPLE**   With reference to the example dealing with the number of days that it rains in Chicago in January, where we had

$$S = \{0, 1, 2, 3, 4, \ldots, 30, 31\}$$

$$M = \{10, 11, 12, \ldots, 19, 20\}$$

and
$$N = \{18, 19, 20, \ldots, 30, 31\}$$

express each of the following events symbolically and also in words:

(a) $M \cup N$; (c) M'; (e) $M' \cup N'$;

(b) $M \cap N$; (d) N'; (f) $M' \cap N'$.

Solution (a) Since $M \cup N$ contains all the elements that are in M, in N, or in both, we find that

$$M \cup N = \{10, 11, 12, \ldots, 30, 31\}$$

and this is the event that there will be at least 10 rainy days.

(b) Since $M \cap N$ contains all the elements that are in both M and N, we find that

$$M \cap N = \{18, 19, 20\}$$

and this is the event that there will be from 18 to 20 rainy days.

(c) Since M' contains all the elements of the sample space that are not in M, we find that

$$M' = \{0, 1, 2, \ldots, 8, 9, 21, 22, \ldots, 30, 31\}$$

and this is the event that there will be fewer than 10 or more than 20 rainy days.

(d) Since N' contains all the elements of the sample space that are not in N, we find that

$$N' = \{0, 1, 2, \ldots, 16, 17\}$$

and this is the event that there will be at most 17 rainy days.

(e) Since $M' \cup N'$ contains all the elements that are not in M, not in N, or in neither, we find that

$$M' \cup N' = \{0, 1, 2, \ldots, 16, 17, 21, 22, \ldots, 30, 31\}$$

and this is the event that there will be fewer than 18 or more than 20 rainy days (or the event that there will not be 18, 19, or 20 rainy days).

(f) Since $M' \cap N'$ contains all the elements that are in both M' and N', we find that

$$M' \cap N' = \{0, 1, 2, \ldots, 8, 9\}$$

and this is the event that there will be at most 9 rainy days. ■

Observe also that in the two-salespersons example on page 146 events D and E have no elements in common. Such events are called **mutually exclusive**.

Two events A and B are mutually exclusive if they cannot both occur at the same time, namely, when $A \cap B = \varnothing$.

■ **EXAMPLE**   With reference to the two-salespersons example, which of the following pairs of events are mutually exclusive:

(a) C and D;   (b) C and E?

Solution   (a) Since events C and D both contain $(1, 1)$, they are not mutually exclusive.

(b) Since $C \cap E = \varnothing$, events C and E are mutually exclusive.   ■

Sample spaces and events, particularly relationships among events, are often pictured by means of **Venn diagrams** such as those of Figures 5.2 and 5.3. In each case, the sample space is represented by a rectangle, while events are represented by regions within the rectangle, usually by circles or parts of circles. The tinted regions of the four Venn diagrams of Figure 5.2 represent

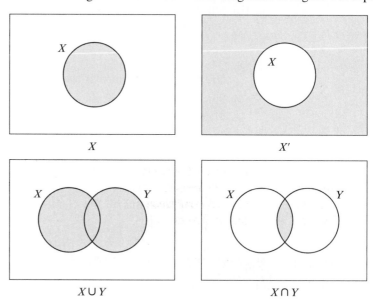

FIGURE 5.2 Venn diagrams.

the event X, the complement of event X, the union of two events X and Y, and the intersection of two events X and Y.

■ **EXAMPLE**   If X is the event that a given high school student will be accepted at Duke University and Y is the event that he or she will be accepted at the University of Virginia, what events are represented by the tinted regions of the four Venn diagrams of Figure 5.2?

Solution   (a) The tinted region of the upper left diagram represents the event that the student will be accepted at Duke University.

(b) The tinted region of the upper right diagram represents the event that he or she will not be accepted at Duke University.

(c) The tinted region of the lower left diagram represents the event that he or she will be accepted at Duke University or at the University of Virginia, or both.

(d) The tinted region of the lower right diagram represents the event that he or she will be accepted at both Duke University and at the University of Virginia. ∎

When we deal with three events, we draw the circles as in Figure 5.3. In this diagram, the circles divide the sample space into eight regions, numbered 1 through 8, and it is easy to determine whether each of the corresponding events is contained in X or in X', in Y or in Y', and in Z or in Z'.

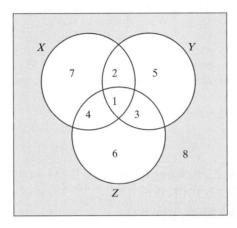

FIGURE 5.3 Venn diagram.

∎ **EXAMPLE** Suppose that the employees of a company are planning a picnic and that X is the event that the weather will be good, Y is the event that there will be enough food, and Z is the event that they will have a good time. With reference to the Venn diagram of Figure 5.3, express in words the events represented by the following regions:

(a) region 4;

(b) regions 1 and 3 together;

(c) regions 3, 5, 6, and 8 together.

Solution (a) Since this region is contained in X and in Z, but not in Y, it represents the event that the weather will be good and they will have a good time, but there will not be enough food.

(b) Since this is the region common to Y and Z, it represents the event that there will be enough food and that they will have a good time.

(c) Since this is the entire region outside X, it represents the event that the weather will not be good. ∎

Exercises

Exercises 5.1, 5.2, and 5.12 are practice exercises; their complete solutions are given on pages 182 and 183.

5.1 With reference to the illustration on page 144 concerning eight candidates for a scholarship, suppose that a, b, c, d, e, f, g, and h denote that Ms. Adams, Mr.

Bean, Ms. Clark, Ms. Daly, Mr. Earl, Ms. Fuentes, Ms. Gardner, or Mr. Hall will be awarded the scholarship. Suppose that

$$X = \{a, c, f, g\}, \quad Y = \{d, f, h\}, \quad \text{and} \quad Z = \{a, b, e, g\}$$

List the outcomes that comprise each of the following events, and also express the events in words:

(a) X';   (c) $X \cap Y$;   (e) $Y \cap Z$;
(b) $X \cup Y$;   (d) $X \cap Z$;   (f) $X' \cup Y$.

✓ **5.2** A small restaurant has two chefs and five waiters, and at least one chef and two waiters have to be present at all times.

(a) Using two coordinates so that (1, 3), for example, represents the event that one chef and three waiters are present and (2, 4) represents the event that two chefs and four waiters are present, draw a diagram similar to that of Figure 5.1, showing the eight points of the sample space.

(b) Describe in words the events represented by

$$Q = \{(1, 4), (2, 4)\}$$
$$R = \{(1, 2), (2, 3)\}$$
$$T = \{(1, 2), (1, 3), (1, 4), (1, 5)\}$$

(c) List the outcomes that comprise each of the following events and also express the events in words: $Q \cup T$, $R \cap T$, and T'.

(d) With reference to part (b), which pairs of events Q and R, Q and T, and R and T are mutually exclusive?

5.3 In an experiment we roll a balanced die and observe the total number of points, which are $1, 2, 3, 4, 5$, and 6. These are the elements of the sample space. If $D = \{3, 4, 5\}$ and $E = \{2, 4, 6\}$, list the outcomes that comprise each of the following events, and also express the events in words: D', $D \cup E$, and $D \cap E$.

5.4 A laboratory technician marks six mice with the numbers $1, 2, 3, 4, 5$, and 6, so that the sample space for selecting one of the mice is the set $S = \{1, 2, 3, 4, 5, 6\}$. Let

$$L = \{1, 2, 3, 4\}$$
$$A = \{3, 4, 5, 6\}$$
$$B = \{2, 3, 4, 5\}$$

If the technician selects one of the numbered mice for an experiment, list the outcomes that comprise each of the following events, and also express the events in words.

(a) $L \cap A$;
(b) $L \cup A$;
(c) $A' \cap B'$;
(d) $A \cap B'$.

5.5 If one card is drawn from an ordinary deck of 52 playing cards, the sample space may be written as

$$S = \{A\spadesuit, \quad 2\spadesuit, \quad 3\spadesuit, \quad \ldots, \quad K\spadesuit,$$
$$A\heartsuit, \quad 2\heartsuit, \quad 3\heartsuit, \quad \ldots, \quad K\heartsuit,$$
$$A\diamondsuit, \quad 2\diamondsuit, \quad 3\diamondsuit, \quad \ldots, \quad K\diamondsuit,$$
$$A\clubsuit, \quad 2\clubsuit, \quad 3\clubsuit, \quad \ldots, \quad K\clubsuit\}$$

where $\spadesuit$, $\heartsuit$, $\diamondsuit$, and $\clubsuit$ denote the suits spades, hearts, diamonds, and clubs. If

$$M = \{Q\spadesuit,\ K\spadesuit,\ Q\heartsuit,\ K\heartsuit,\ Q\diamondsuit,\ K\diamondsuit,\ Q\clubsuit,\ K\clubsuit\}$$

and

$$N = \{10\spadesuit,\ J\spadesuit,\ Q\spadesuit,\ K\spadesuit\}$$

list the outcomes which comprise each of the following events, and also express the events in words:

(a) $M \cup N$;   (c) M';   (e) $M' \cup N'$;
(b) $M \cap N$;   (d) N';   (f) $M' \cap N'$.

5.6 To construct sample spaces for experiments in which we deal with categorical data, we often code the various alternatives by assigning them numbers. For instance, if persons are asked whether their favorite color is red, yellow, blue, green, brown, white, purple, or some other color, we might assign these alternatives the codes 1, 2, 3, 4, 5, 6, 7, and 8, respectively. If $A = \{3, 4\}$, $B = \{1, 2, 3, 4, 5, 6, 7\}$, and $C = \{5\ 6, 7, 8\}$, list the outcomes that comprise each of the following events:

(a) B';   (c) $A \cap B$;   (e) $A' \cup C$;
(b) $A \cup B$;   (d) C';   (f) $B \cap C'$.

5.7 With reference to Exercise 5.6, which pairs of events A and B, A and C, and B and C are mutually exclusive?

5.8 A movie critic has two days in which to view some of the movies that have recently been released. She wants to see at least three of the movies, but not more than three on either day.

(a) Using two coordinates so that $(3, 1)$, for example, represents the event that she will see three of the movies on the first day and one on the second day, draw a diagram similar to that of Figure 5.1 showing the 10 points of the corresponding sample space.

(b) If T is the event that altogether she will see three of the movies, U is the event that she will see more of the movies on the second day than on the first, V is the event that she will see three of the movies on the first day, and W is the event that she will see equally many movies on both days, list the outcomes which comprise each of these four events.

5.9 With reference to Exercise 5.8, indicate which of the following six pairs of events are mutually exclusive:

(a) T and U;   (d) U and V;
(b) T and V;   (e) U and W;
(c) T and W;   (f) V and W.

5.10 A small marina has three fishing boats that are sometimes in dry dock for repairs.

(a) Using two coordinates so that $(2, 1)$, for example, represents the event that two of the fishing boats are in dry dock and one is rented out for the day and $(0, 2)$ represents the event that none of the boats is in dry dock and two are rented out for the day, draw a diagram similar to that of Figure 5.1 showing the 10 points of the corresponding sample space.

(b) If K is the event that at least two of the boats are rented out for the day, L is the event that more boats are in dry dock than are rented out for the day, and M is the event that all the boats that are not in dry dock are rented out for the day, list the outcomes that comprise each of these three events.

(c) With reference to part (b), list the outcomes that comprise K' and $L \cap M$, and also express these events in words.

(d) With reference to part (b), which of the three pairs of events K and L, K and M, and L and M are mutually exclusive?

5.11 Which of the following pairs of events are mutually exclusive? Explain your answers.
(a) Getting a passing grade or a failing grade in an English literature course.
(b) Being a medical doctor and being a psychiatrist.
(c) Flying to Chicago in an airplane and calling home on a telephone.
(d) Electing a Republican or a Democratic president of the United States.
(e) Being born in Los Angeles or in Boston.

5.12 In Figure 5.4, D is the event that a person vacationing in Southern California visits Disneyland, and U is the event that he visits Universal Studios. Explain in words the events represented by regions 1, 2, 3, and 4.

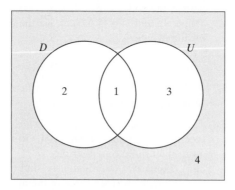

FIGURE 5.4 Venn diagram for Exercise 5.12.

5.13 With reference to Exercise 5.12, what events are represented by
(a) regions 1 and 2 together;
(b) regions 2 and 3 together;
(c) regions 2 and 4 together?

5.14 In Figure 5.5, G is the event that a murder suspect is guilty, and B is the event that he is allowed out on bail. Explain in words the events represented by regions 1, 2, 3, and 4.

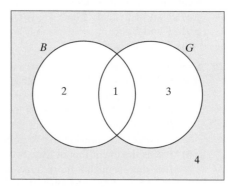

FIGURE 5.5 Venn diagram for Exercise 5.14.

5.15 With reference to Exercise 5.14, what events are represented by
(a) regions 1 and 3 together;
(b) regions 1 and 4 together;
(c) regions 3 and 4 together?

5.16 Suppose that Figure 5.6 pertains to a library where B is the event that the number of books is adequate, P is the event that the number of periodicals is adequate, and C is the event that the chairs are comfortable. What regions or combinations of regions represent the following events?
   (a) The number of books is adequate, and the chairs are comfortable.
   (b) The number of books is adequate, and the number of periodicals is adequate.
   (c) The number of books is inadequate, the number of periodicals is inadequate, and the chairs are uncomfortable.
   (d) The number of books is adequate, the number of periodicals is adequate, and the chairs are comfortable.

5.17 Continuing with Exercise 5.16, explain in words what the following regions represent:
   (a) region 1;
   (b) region 3;
   (c) region 6;
   (d) region 8;
   (e) regions 1 and 4 together;
   (f) regions 3 and 5 together;
   (g) regions 1, 3, 4, and 6 together;
   (h) regions 2, 5, 7, and 8 together.

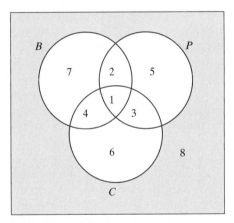

FIGURE 5.6 Venn diagram for Exercise 5.17.

5.18 With reference to Exercise 5.16 and Figure 5.6, list the regions or combinations of regions that represent the following events:
   (a) The number of periodicals is adequate, but the number of books is not adequate and the chairs are uncomfortable.
   (b) The number of books is adequate, but the number of periodicals is not adequate.
   (c) The number of books is adequate and the number of periodicals is adequate, or the chairs are comfortable.

**5.3
Some Basic Rules
of Probability**

Probabilities always pertain to the occurrence or nonoccurrence of events (whether a coin will come up heads, whether a letter will arrive on time, whether it will not rain, whether a project will succeed, …); now that we have learned how to deal with events, let us turn to the rules according to which probabilities must behave, that is, the mathematical rules that they must obey.

To express these rules symbolically, we shall denote events with capital letters as in Section 5.2, and write the probability of event A as $P(A)$, the probability of event B as $P(B)$, and so forth.

The following are some of the most basic rules of probabilities, and we shall justify them here with reference to the frequency interpretation; in Exercise 5.29, the reader will be asked to justify them with reference to the classical probability concept, and in Section 5.4 we shall see to what extent they are compatible with subjective probabilities.

Basic rules of probabilities

1. **Probabilities are real numbers between 0 and 1, inclusive.**
2. **If an event is certain to occur, its probability is 1, and if an event is certain not to occur, its probability is 0.**
3. **If two events are mutually exclusive, the probability that one or the other will occur equals the sum of their probabilities.**
4. **The sum of the probabilities that an event will occur and that it will not occur is equal to 1.**

With reference to the frequency interpretation, the first of these rules simply expresses the fact that an event cannot occur less than 0% or more than 100% of the time; that is, the proportion of successes, or the proportion of successes in the long run, cannot be negative or exceed 1. Symbolically, we write $0 \leq P(A) \leq 1$ for any event A. By the same token, the second rule expresses the fact that an event that is certain to occur will occur 100% of the time and an event that is certain not to occur will occur 0% of the time. Symbolically, we write $P(S) = 1$ for any sample space S, which expresses the certainty that one of the outcomes in any sample space S must occur, and $P(\emptyset) = 0$, which expresses the fact that an event which cannot occur (which does not include any of the outcomes) has zero probability.

To show that the third rule is satisfied by the frequency interpretation, we have only to observe that if one event occurs, say, 15% of the time, if another event occurs, say, 28% of the time, and if the two events cannot both occur at the same time (that is, they are mutually exclusive), then one or the other occurs 15% + 28% = 43% of the time. Of course, the same argument applies also to proportions. If the probabilities are, respectively, 0.45 and 0.17 that the weather will improve or that it will remain the same, then the probability is 0.45 + 0.17 = 0.62 that it will improve or remain the same. Symbolically, we write

$$P(A \cup B) = P(A) + P(B)$$

for any two mutually exclusive events A and B.

The fourth rule expresses the certainty that an event either will or will not occur. If someone is late for work 22% of the time, he is not late 78% of the time, the corresponding probabilities are 0.22 and 0.78, and their sum is 0.22 + 0.78 = 1. Symbolically, we write

$$P(A) + P(A') = 1 \quad \text{for any event } A$$

We can also write $P(A') = 1 - P(A)$ if we want to calculate the probability that an event will not occur in terms of the probability that it will occur.

The examples that follow illustrate how these rules are put to use in actual practice.

■ **EXAMPLE** If A is the event that a student will stay home to study and B is the event that she will instead go to a movie, $P(A) = 0.64$, and $P(B) = 0.21$, find

(a) $P(A')$;   (b) $P(A \cup B)$;   (c) $P(A \cap B)$.

Solution (a) Using the fourth rule, we find that the probability of A', the event that the student will not stay home to study, is

$$1 - P(A) = 1 - 0.64 = 0.36$$

(b) Since A and B are mutually exclusive, we can use the third rule and write

$$P(A \cup B) = P(A) + P(B) = 0.64 + 0.21 = 0.85$$

for the probability that the student will either stay home to study or go to a movie.

(c) Since A and B are mutually exclusive, they cannot possibly both occur, and hence

$$P(A \cap B) = P(\varnothing) = 0$$ ■

In problems like this, it often helps to draw a Venn diagram, fill in the probabilities associated with the various regions, and then read the answers directly off the diagram.

■ **EXAMPLE** If C is the event that at 9:30 A.M. a certain doctor is in her office and D is the event that she is in the hospital, $P(C) = 0.48$ and $P(D) = 0.27$, find $P(C' \cap D')$, the probability that she is neither in her office nor in the hospital.

Solution Drawing the Venn diagram as in Figure 5.7, we first put a 0 probability into region 1, because the events C and D are mutually exclusive. It follows that the 0.48 probability of event C must go into region 2, the 0.27 probability of event D must go into region 3, and since the probability for the entire sample space must total 1, we put $1 - (0.48 + 0.27) = 0.25$ into region 4. Since the

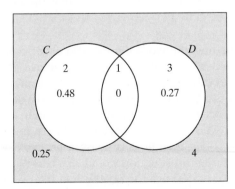

FIGURE 5.7 Venn diagram.

event $C' \cap D'$ is represented by the region outside both circles, namely, region 4, we find that the answer is

$$P(C' \cap D') = 0.25 \qquad ■$$

5.4
Probabilities and Odds

If an event is twice as likely to occur as not to occur, we say that the **odds** are 2 to 1 that it will occur; if an event is three times as likely to occur as not to occur, we say that the odds are 3 to 1; if an event is ten times as likely to occur as not to occur, we say that the odds are 10 to 1; and so forth. In general,

The odds that an event will occur are given by the ratio of the probability that it will occur to the probability that it will not occur.

Symbolically, if the probability of an event is p, the odds for its occurrence are a to b, where a and b are positive values such that

$$\frac{a}{b} = \frac{p}{1 - p}$$

It is customary to express odds in terms of positive integers having no common factors, if possible.

■ **EXAMPLE**   What are the odds for the occurrence of an event if its probability is

(a) $\frac{5}{9}$;   (b) 0.85?

Solution   (a) By definition, the odds are $\frac{5}{9}$ to $1 - \frac{5}{9} = \frac{4}{9}$, or 5 to 4.

(b) By definition, the odds are 0.85 to $1 - 0.85 = 0.15$, 85 to 15, or better, 17 to 3.   ■

If an event is more likely not to occur than to occur, it is customary to quote the odds that it will not occur, rather than the odds that it will occur.

■ **EXAMPLE**   What are the odds if the probability of an event is 0.20?

Solution   The odds for the occurrence of the event are 0.20 to $1 - 0.20 = 0.80$, or 1 to 4, but it is customary to say instead that the odds against the occurrence of the event are 4 to 1.   ■

In betting, the word *odds* is also used to denote the ratio of the wager of one party to that of another. For instance, if a gambler says that he will give 3 to 1 odds on the occurrence of an event, he means that he is willing to bet $3 against $1 (or perhaps $30 against $10 or $1,500 against $500) that the event *will* occur. If such **betting odds** actually equal the odds that the event will occur, we say that the betting odds are fair. If a gambler really believes that a bet is fair, then he is, at least in principle, willing to bet on either side. The gambler in this situation would also be willing to bet $1 against $3 (or $10 against $30 or $500 against $1,500) that the event *will not* occur.

■ **EXAMPLE**   Records show that one-twelfth of the trucks weighed at a certain checkpoint in Nevada carry too heavy a load. If someone offers to bet $40 against $4 that the next truck weighed at this checkpoint will not carry too heavy a load, are these betting odds fair?

Solution   Since the probability is $1 - \frac{1}{12} = \frac{11}{12}$ that the truck will not carry too heavy a load, the odds are 11 to 1. The bet would be fair if the person offered to bet $44 against $4 that the next truck weighed at the checkpoint will not carry too heavy a load. Thus, the $40 against $4 bet is not fair; it favors the person offering the bet.   ■

This discussion of odds and betting odds provides the groundwork for a way of measuring subjective probabilities. If a businessman feels that the odds for the success of a new clothing store are 3 to 2, this means that he is willing to bet (or considers it fair to bet) $300 against $200, or perhaps $3,000 against $2,000, that the new store will be a success. In this way he is expressing his belief regarding the uncertainties connected with the success of the store, and to convert it into a probability, we take the equation

$$\frac{a}{b} = \frac{p}{1 - p}$$

and solve it for p. Leaving the details to the reader in Exercise 5.37, let us merely state the result that

Formula relating probabilities to odds

> **If the odds are a to b that an event will occur, the probability of its occurrence is**
> $$p = \frac{a}{a + b}$$

■ **EXAMPLE**   Convert the businessman's 3 to 2 odds for the success of the new clothing store into a probability.

Solution   Substituting $a = 3$ and $b = 2$ into the formula for p, we get

$$p = \frac{3}{3 + 2} = \frac{3}{5} = 0.6$$ ■

■ **EXAMPLE**   If an applicant for a managerial position feels that the odds are 7 to 4 that she will get the job, what probability is she thus assigning to her getting the job?

Solution   Substituting $a = 7$ and $b = 4$ into the formula for p, we get

$$p = \frac{7}{7 + 4} = \frac{7}{11} \approx 0.64$$ ■

Let us now see whether subjective probabilities, determined in this way, behave in accordance with the four basic rules of probabilities on page 154. As far as the first and fourth rules are concerned, this is easy to see. Since a and b are positive quantities, $\dfrac{a}{a+b}$ is a fraction between 0 and 1; since the probabilities that an event will occur and that it will not occur are, respectively,

$$\frac{a}{a+b} \quad \text{and} \quad \frac{b}{a+b}$$

their sum is

$$\frac{a}{a+b} + \frac{b}{a+b} = \frac{a+b}{a+b} = 1$$

As for the second rule, observe that the surer we are that an event will occur, the better odds we should be willing to give, say, 100 to 1, 1,000 to 1, or perhaps even 1 million to 1.

The corresponding probabilities are

$$\frac{100}{100+1} \approx 0.9901, \qquad \frac{1,000}{1,000+1} \approx 0.999001,$$

and

$$\frac{1,000,000}{1,000,000+1} \approx 0.999999$$

and it can be seen that the surer we are that an event will occur, the closer its probability will be to 1. By the same token, it can be shown that the surer we are that an event will not occur, the closer its probability will be to 0.

This leaves only the third rule,

$$P(A \cup B) = P(A) + P(B)$$

for any two mutually exclusive events A and B, and this rule is not necessarily satisfied when it comes to subjective probabilities. Indeed, proponents of the subjectivist point of view impose it as a **consistency criterion**, and this provides a means of "policing" a person's subjective probabilities.

■ **EXAMPLE**   An economist feels that the odds are 2 to 1 that the price of beef will go up during the next month, 1 to 5 that it will remain unchanged, and 8 to 3 that it will go up or remain unchanged. Are the corresponding probabilities consistent?

Solution   According to this economist, the probability that the price of beef will go up during the next month is $\dfrac{2}{2+1} = \dfrac{2}{3}$. The probability that this price will remain unchanged is $\dfrac{1}{1+5} = \dfrac{1}{6}$, and the probability that this price will go up or remain unchanged is $\dfrac{8}{8+3} = \dfrac{8}{11}$. Since $\dfrac{2}{3} + \dfrac{1}{6} = \dfrac{5}{6}$, which is not equal to $\dfrac{8}{11}$, the probabilities are not consistent. Hence, the economist's judgment must be questioned.   ■

Exercises

Exercises 5.19, 5.22, 5.30, *and* 5.31 *are practice exercises; their complete solutions are given on pages* 183 *and* 184.

5.19 In a study of the future needs of a community, D stands for the event that there will be enough doctors, and H stands for the event that there will be enough hospital beds. State in words the probabilities expressed by

(a) $P(D')$;
(b) $P(H')$;
(c) $P(D \cup H)$;
(d) $P(D \cap H)$;
(e) $P(D' \cap H')$;
(f) $P(D \cap H')$.

5.20 In a study of the adequacy of fuel supplies, C stands for the event that a power plant will use coal, and E is the event that it will be able to provide enough electricity. State in words the probabilities expressed by

(a) $P(C')$;
(b) $P(E')$;
(c) $P(C \cup E)$;
(d) $P(C \cap E)$;
(e) $P(C' \cap E)$;
(f) $P(C' \cap E')$.

5.21 If J is the event that Harry will find a job and M is the event that he will make enough money to live comfortably, express symbolically the probabilities that he will

(a) not find a job;
(b) find a job, and make enough money to live comfortably;
(c) not find a job, and not make enough money to live comfortably;
(d) find a job, make enough money to live comfortably, or both.

5.22 Explain why there must be a mistake in each of the following statements:

(a) The probability that a certain basketball player will score on a free throw is 0.67, and the probability that he will miss is 0.23.
(b) A gold-panner knows that the probabily is 0.25 that a pan of gold-bearing soil will contain gold and the probability is 0.90 that it will not contain gold.
(c) The probability that a man will wear a tuxedo to a certain social event is 0.24, the probability that he will wear a business suit is 0.38, and the probability that he will wear a tuxedo or a business suit is 0.50.

5.23 Explain why there must be a mistake in each of the following statements:

(a) The probability that a new vaccine is effective is 1.09.
(b) The probability that Tom will take Mary to the senior prom is 0.65, and the probability that Bill will take the same woman to the senior prom is 0.45.
(c) The probability that an employer will hire Anne for a job is 0.30, the probability that he will hire Mildred for the same job is 0.40, and the probability that he will hire either Anne or Mildred is 0.90.
(d) Sam and George are scheduled for a boxing match in a college-sponsored tournament. The probability that Sam will win is 0.35, while the probability that George will win is 0.55. Assume that winning or losing are the only possible outcomes.

5.24 Given the *mutually exclusive* events Y and Z, for which $P(Y) = 0.28$ and $P(Z) = 0.47$, find

(a) $P(Y')$;
(b) $P(Z')$;
(c) $P(Y \cap Z)$;
(d) $P(Y \cup Z)$;
(e) $P(Y' \cap Z')$.

(*Hint:* Draw a Venn diagram and fill in the probabilities associated with the various regions.)

5.25 Given the mutually exclusive events Q and R for which $P(Q) = 0.45$ and $P(R) = 0.30$, find
(a) $P(Q')$;
(b) $P(R')$;
(c) $P(Q \cup R)$;
(d) $P(Q \cap R)$;
(e) $P(Q' \cap R)$;
(f) $P(Q \cap R')$.

5.26 Events D and E are mutually exclusive. It also happens that D' and E are mutually exclusive. Find $P(E)$.

5.27 The probabilities that the reactions of a newspaper's food critic to a new restaurant will be unfavorable or indifferent are, respectively, 0.45 and 0.30. Find the probabilities that his reaction
(a) will be favorable;
(b) will be favorable or indifferent;
(c) will be neither favorable nor indifferent.

5.28 The probabilities that a secretary will make fewer than two mistakes in a letter or anywhere from two to four mistakes are, respectively, 0.72 and 0.18. Find the probabilities that the secretary will make
(a) two or more mistakes;
(b) at most four mistakes;
(c) more than four mistakes.

5.29 Show that the four basic rules of probabilities on page 154 are satisfied if we interpret probabilities in accordance with the classical probability concept, which states: If there are n equally likely possibilities, of which one must occur and s are regarded as favorable, or as a success, then the probability of a success is $\dfrac{s}{n}$.

 5.30 Convert each of the following probabilities to odds or odds to probabilities:
(a) The probability of getting three heads and three tails in six flips of a balanced coin is $\frac{5}{16}$.
(b) The odds against rolling 7 or 11 with a pair of balanced dice are 7 to 2.
(c) If a pollster randomly selects 5 of 24 households to be included in a survey, the probability is $\frac{5}{24}$ that any particular household will be included.
(d) If 3 eggs are randomly chosen from a carton of 12 eggs of which 3 are cracked, the odds are 34 to 21 that at least one of them will be cracked.

 5.31 A taxpayer claims that the odds are 2 to 1 that he will complete his tax return on time and 2 to 1 that it will be late. Can these odds be right? Explain.

5.32 Convert each of the following odds to probabilities:
(a) The odds that a particular horse will lose a race are 7 to 1.
(b) The odds are 3 to 5 that a sequence of four coin tosses will result in two heads and two tails.
(c) If a secretary randomly places six letters into six addressed envelopes, the odds are 1 to 719 that all letters will end up in the correct envelopes.
(d) The odds are 2 to 17 for winning a roulette bet made by placing a token at the intersection of four number boxes.

5.33 A football coach claims that the odds are 2 to 1 that his team will win an upcoming game, and that the odds against his team's losing or tying are, respectively, 4 to 1 and 9 to 1. Can these odds be right? Explain.

5.34 If a student is anxious to bet $25 against $5 that she will pass a certain course, what does this tell us about the probability she assigns to her passing the course? (*Hint:* The answer should read "greater than ....")

5.35 A politician states that the odds are 5 to 1 that he will not run for the House of Representatives and 4 to 1 that he will not run for the Senate. Furthermore, he feels that the odds are 1 to 2 that he will run for either of these offices. Are the corresponding probabilities consistent?

5.36 An automobile salesperson feels that the odds are 9 to 1 against her selling a certain car and 4 to 1 against her being able to lease it. To be consistent, what should she consider to be fair odds that she will either sell the house or lease it?

5.37 Verify algebraically that the formula $\dfrac{a}{b} = \dfrac{p}{1 - p}$, solved for p, yields $p = \dfrac{a}{a + b}$.

5.38 A gambler believes that the true odds favoring team Y in a coming game are 5 to 2. Which of the following bets would be attractive to him?
   (a) Bet $5 on team Y to win $5.
   (b) Bet $7 on team Y to win $3.
   (c) Bet $3 against team Y to win $7.
   (d) Bet $30 against team Y to win $70.

5.39 A gambler feels that the probability is 0.60 that Punchy Joe will win a certain boxing match. Which of the following bets would be attractive to her?
   (a) Bet $6.00 on Punchy Joe to win $6.00.
   (b) Bet $6.00 on Punchy Joe to win $7.00.
   (c) Bet $6.00 on Punchy Joe to win $3.00.

5.5 Addition Rules

The third of our basic probabilities rules on page 154 is sometimes referred to as the **special addition rule**. It is special in that it applies only to mutually exclusive events, and it is an addition rule because we add the probabilities of the two events. As we shall see in this section, it can be generalized so that it will apply to more than two mutually exclusive events and also to events that need not be mutually exclusive.

Repeatedly applying the third rule on page 154 we can show that

Special addition rule for two or more events

> If k events are mutually exclusive, the probability that one of them will occur equals the sum of their respective probabilities. In symbols,
> $$P(A_1 \cup A_2 \cup \cdots \cup A_k) = P(A_1) + P(A_2) + \cdots + P(A_k)$$
> for any mutually exclusive events $A_1, A_2, \ldots,$ and A_k, where, again, $\cup$ is usually read *or*.

■ **EXAMPLE**   The probabilities that a woman will buy a new dress for a party at Bullock's, the Broadway Southwest department store, or the May Co. are 0.22, 0.18, and 0.35. What is the probability that she will buy the new dress at one of these stores?

Solution   Since the three possibilities are mutually exclusive, the answer is
$$0.22 + 0.18 + 0.35 = 0.75$$ ■

■ **EXAMPLE**   The probabilities that a consumer testing service will rate a new antipollution device for cars very poor, poor, fair, good, very good, or excellent are 0.07, 0.12, 0.17, 0.32, 0.21, and 0.11. What is the probability that it will rate the device poor, fair, good, or very good?

Solution   Since the four possibilities are mutually exclusive, we get
$$0.12 + 0.17 + 0.32 + 0.21 = 0.82$$ ■

The job of assigning probabilities to all possible events connected with a given situation can be very tedious, to say the least. If a sample space has 10 elements (outcomes), we can form more than a thousand different events, and if a sample space has 20 elements (outcomes), we can form more than a million different events.[†] Fortunately, it is seldom, if ever, necessary to find the probabilities of all possible events relating to a given situation; the following rule, which is a direct application of the special addition rule for two or more events, makes it easy to determine the probability of any event on the basis of the probabilities assigned to the individual elements (outcomes) that form the sample space.

Rule for calculating the probability of an event

> **The probability of any event A is given by the sum of the probabilities of the individual outcomes comprising A.**

In the special case where the outcomes are all equiprobable, this rule leads to the formula

$$P(A) = \frac{s}{n}$$

which we studied earlier in connection with the classical probability concept. Here, n is the total number of outcomes and s is the number of successes, that is, the number of outcomes in event A.

■ **EXAMPLE**   In our previous example dealing with two used-car salespersons trying to sell two used 1997 Chevrolet Camaros, suppose that the six points of the sample space have the probabilities shown in Figure 5.8. Find the probabilities that

(a) the first salesperson will not sell either of the two cars;

(b) both cars will be sold;

(c) the second salesperson will sell at least one of the two cars.

Solution   (a) Adding the probabilities associated with the points $(0, 0)$, $(0, 1)$, and $(0, 2)$, we get

$$0.06 + 0.17 + 0.22 = 0.45$$

(b) Adding the probabilities associated with the points $(2, 0)$, $(1, 1)$, and $(0, 2)$, we get

$$0.22 + 0.16 + 0.22 = 0.60$$

(c) Adding the probabilities associated with the points $(0, 1)$, $(1, 1)$, and $(0, 2)$, we get

$$0.17 + 0.16 + 0.22 = 0.55 \qquad ■$$

[†] In general, if a sample space has n elements, we can form 2^n different events. Each element is either included or excluded for a given event, so by the multiplication of choices there are $2 \cdot 2 \cdot 2 \cdot \ldots \cdot 2 = 2^n$ possibilities. Note that

$$2^{10} = 1,024 \quad \text{and} \quad 2^{20} = 1,048,576$$

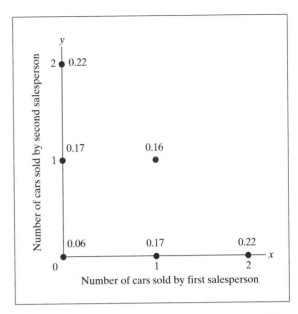

FIGURE 5.8 Sample space for two-salesperson example with probabilities assigned to the outcomes.

■ **EXAMPLE**   Assuming that the 44 points (outcomes) of the sample space of Figure 5.9 are all equiprobable, find $P(A)$.

Solution   Since there are $s = 10$ outcomes in A and the $n = 44$ outcomes are all equiprobable, it follows that

$$P(A) = \frac{10}{44} = \frac{5}{22}$$ ■

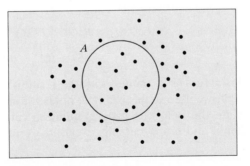

FIGURE 5.9 Sample space with 44 outcomes.

Since the addition rules that we have studied so far apply only to mutually exclusive events, they cannot be used, for example, to find the probability that at least one of two friends will pass a final examination; the probability that in an automobile accident the driver will break an arm, a rib, or a leg; or the probability that a customer will buy a shirt or a sweater at a given department store. Both friends can pass the final examination; the driver of the car can break an arm and a rib, and also a leg; and the customer of the department store can buy a sweater as well as a shirt.

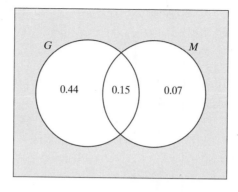

FIGURE 5.10 Venn diagram.

To find a formula for $P(A \cup B)$ that holds regardless of whether the events A and B are mutually exclusive, let us consider the Venn diagram of Figure 5.10, which concerns the election of a mayor. The letter G stands for the election of a person who is a college graduate, and the letter M stands for the election of a member of a certain minority group. It follows from the figures in the Venn diagram that

$$P(G) = 0.44 + 0.15 = 0.59, \qquad P(M) = 0.15 + 0.07 = 0.22,$$

and

$$P(G \cup M) = 0.44 + 0.15 + 0.07 = 0.66$$

where we were able to add the respective probabilities because they pertain to mutually exclusive events (to regions of the Venn diagram that do not overlap).

Had we erroneously used the special addition rule, the third of the four basic rules, to calculate $P(G \cup M)$, we would have obtained $P(G) + P(M) = 0.59 + 0.22 = 0.81$, which exceeds the correct value by 0.15. This error results from including $P(G \cap M) = 0.15$ twice, once in $P(G) = 0.59$ and once in $P(M) = 0.22$, and we could correct for it by subtracting 0.15 from 0.81. Symbolically, we could thus write

$$P(G \cup M) = P(G) + P(M) - P(G \cap M)$$

$$= 0.59 + 0.22 - 0.15 = 0.66$$

and this agrees, as it should, with the result obtained before. Since the argument that we used in this example holds for any two events A and B, we can now state the following **general addition rule**, which applies regardless of whether A and B are mutually exclusive events:

General addition rule

$$P(A \cup B) = P(A) + P(B) - P(A \cap B)$$

When A and B are mutually exclusive, note that $P(A \cap B) = 0$ and the general addition rule reduces to the special addition rule, the third of our basic rules.

■ **EXAMPLE**   If one card is drawn from an ordinary deck of 52 playing cards, what is the probability that it will be either a club or a face card (king, queen, or jack)?

Solution   If C denotes drawing a club and F denotes drawing a face card, then

$$P(C) = \frac{13}{52}, \quad P(F) = \frac{12}{52}, \quad \text{and} \quad P(C \cap F) = \frac{3}{52}$$

so that

$$P(C \cup F) = \frac{13}{52} + \frac{12}{52} - \frac{3}{52} = \frac{22}{52} = \frac{11}{26}$$ ■

■ **EXAMPLE**   If the probabilities are, respectively, 0.92, 0.33, and 0.29 that a person vacationing in Washington, D.C., will visit the Capitol building, the Smithsonian Institution, or both, what is the probability that a person vacationing there will visit at least one of these buildings?

Solution   Substituting into the formula, we get

$$0.92 + 0.33 - 0.29 = 0.96$$

Note that if we had incorrectly used the special addition rule (for mutually exclusive events), we would have obtained the impossible answer $0.92 + 0.33 = 1.25$.   ■

The general addition rule can be generalized further so that it will apply to more than two events, but we shall not go into that in this book.

Exercises

Exercises 5.40, 5.45, 5.49, and 5.52 are practice exercises; their complete solutions are given on page 184.

5.40 A police department needs new tires for its patrol cars, and the probabilities are, respectively, 0.22, 0.17, 0.21, and 0.09 that it will buy Uniroyal, Goodyear, Goodrich, or Michelin tires. What is the probability that it will buy one of these four kinds of tires?

5.41 If the probabilities that a personnel manager will evaluate an employee as superior, above average, or average are, respectively, 0.08, 0.23, and 0.45, what is the probability that the employee will get one of these evaluations?

5.42 In a practical nurse training program, the probabilities that a practical nurse trainee will be evaluated as superior, above average, or average are 0.17, 0.43, and 0.31, respectively. What is the probability that a practical nurse trainee will be evaluated as below average (assuming that this is the only remaining category)?

5.43 The probabilities that a TV station will receive 0, 1, 2, 3, ..., 7, or at least 8 complaints after showing a controversial program are, respectively, 0.02, 0.04, 0.07, 0.12, 0.15, 0.19, 0.18, 0.14, and 0.09. What are the probabilities that after showing such a program the station will receive
  (a) at least 5 complaints;
  (b) at most 3 complaints;
  (c) anywhere from 2 to 4 complaints?

5.44 The probabilities that the serviceability of a new laser printer will be rated very difficult, difficult, average, easy, or very easy are, respectively, 0.11, 0.16, 0.35, 0.28, and 0.10. Find the probabilities that the serviceability of the new laser printer will be rated
  (a) difficult or very difficult;
  (b) difficult, average, or easy;
  (c) average or better.

5.45 Figure 5.11 pertains to the number of persons who are invited to a conference and the number of persons who attend. If each of the 35 points of the sample space has the probability $\frac{1}{35}$, what are the probabilities that
  (a) all the persons who are invited will attend;
  (b) at most two persons will attend;
  (c) at least seven persons will be invited;
  (d) one or two of the persons who are invited will not attend?

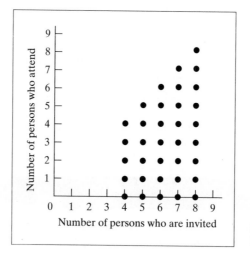

FIGURE 5.11 Sample space for Exercise 5.45.

5.46 If each point of the sample space of Figure 5.12 represents an outcome having the probability $\frac{1}{32}$, find
  (a) $P(A)$;
  (b) $P(B)$;
  (c) $P(A \cap B)$;
  (d) $P(A \cup B)$;
  (e) $P(A' \cap B)$;
  (f) $P(A' \cap B')$;
  (g) $P(A' \cup B)$;
  (h) $P(A \cup B')$.

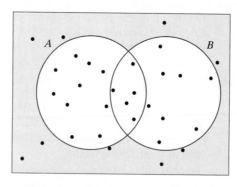

FIGURE 5.12 Sample space for Exercise 5.46.

5.47 With reference to the data of Figure 5.8, find the probabilities that
   (a) only one of the two cars will be sold;
   (b) either salesperson will sell both cars;
   (c) the two salespersons will sell equally many cars.

5.48 If H stands for heads and T for tails, the 16 possible outcomes for four flips of a coin are

HHHH	HHHT	HHTH	HTHH	THHH	HHTT
HTHT	HTTH	THHT	THTH	TTHH	HTTT
THTT	TTHT	TTTH	TTTT		

Assuming that these 16 possibilities are all equally likely, what are the probabilities of getting 0, 1, 2, 3, or 4 heads in four flips of a balanced coin?

5.49 An artist who has entered a large oil painting and a small oil painting in a show feels that the probabilities are, respectively, 0.15, 0.18, and 0.11 that she will sell the large oil painting, the small one, or both. What is the probability that she will sell
   (a) either or both of the two paintings;
   (b) neither of the two paintings?

5.50 The probabilities that a reckless driver will be fined, have his license revoked, or both are, respectively, 0.88, 0.60, and 0.55. What is the probability that he will be fined or have his license revoked?

5.51 The probabilities that a dentist's receptionist, his assistant, or both will be sick on a given day are, respectively, 0.04, 0.07, and 0.02. What is the probability that at least one of the two will be sick on that day?

5.52 Given $P(K) = 0.45$, $P(L) = 0.27$, and $P(K \cap L) = 0.13$, draw a Venn diagram, fill in the probabilities associated with the various regions, and thus determine
   (a) $P(K \cap L')$;   (d) $P(K' \cup L)$;
   (b) $P(K' \cap L)$;   (e) $P(K' \cap L')$;
   (c) $P(K \cup L)$;   (f) $P(K' \cup L')$;

5.53 Among the 64 doctors on the staff of a hospital, 58 carry malpractice insurance, 33 are surgeons, and 31 of the surgeons carry malpractice insurance. If one of these doctors is chosen by lot to represent the hospital staff at an AMA convention (that is, each doctor has a probability of $\frac{1}{64}$ of being selected), what is the probability that the one chosen is not a surgeon and does not carry malpractice insurance?

5.54 Given $P(A) = 0.59$, $P(B) = 0.46$, and $P(A \cap B) = 0.28$, draw a Venn diagram, fill in the probabilities associated with the various regions, and thus determine
   (a) $P(A' \cap B)$;   (c) $P(A \cup B)$;
   (b) $P(A \cap B')$;   (d) $P(A' \cap B')$.

5.6 Conditional Probability

Difficulties can easily arise when we ask for the probability of an event without specifying the sample space. For instance, the probability that a person will find a bargain depends on where she shops, the chances that a person will get a traffic ticket depend on how well and how carefully he drives, and the odds that a woman will get promoted depend on how well she is trained or how hard she is willing to work.

Since the choice of the sample space (the set of all possibilities under consideration) is rarely self-evident, it helps to use the symbol $P(A|S)$ to denote the **conditional probability** of event A relative to the sample space S or, as we often say, the probability of A given S. In many cases, S will be smaller than the original sample space that we started with. The symbol $P(A|S)$ makes it explicit that we are referring to a particular sample space S. It is preferable to the abbreviated notation $P(A)$ unless the tacit choice of S is clearly understood.

To elaborate on the idea of a conditional probability, suppose that a consumer research organization has studied the service provided by the 150 appliance repair persons in a certain city. Their findings are summarized in the following table:

	Good service	Poor service	Total
Factory trained	48	16	64
Not factory trained	24	62	86
Total	72	78	150

If one of these repair persons is randomly selected (that is, each has the probability $\frac{1}{150}$ of being selected), we find that the probability of choosing one who provides good service is

$$P(G) = \frac{72}{150} = 0.48$$

and the probability of choosing one who is factory trained is

$$P(F) = \frac{64}{150} \approx 0.43$$

The probability of choosing one who provides good service and is factory trained is

$$P(G \cap F) = \frac{48}{150} = 0.32$$

All these probabilities were calculated by means of the formula $\frac{s}{n}$ for equally likely possibilities.

Since the first of these probabilities is particularly disconcerting—there is less than a 50–50 chance of choosing a repair person who provides good service—let us see what will happen if we limit the choice to those who are factory trained. Looking at the reduced sample space represented by the first row of the table, we get

$$P(G|F) = \frac{48}{64} = 0.75$$

and this is quite an improvement over $P(G) = 0.48$, as might have been expected. Note that this conditional probability, 0.75, can also be written as

$$P(G|F) = \frac{\frac{48}{150}}{\frac{64}{150}} = \frac{P(G \cap F)}{P(F)}$$

which is the ratio of the probability of choosing a repair person who provides good service and is factory trained to the probability of choosing a repair person who is factory trained.

Generalizing from this example, let us now make the following definition of conditional probability, which applies to any two events A and B belonging to a given sample space S:

Definition of conditional probability

> If $P(B)$ is not equal to zero, then the conditional probability of A relative to B, namely, the probability of A given B, is
> $$P(A|B) = \frac{P(A \cap B)}{P(B)}$$

■ EXAMPLE   With reference to the appliance repair persons of the preceding illustration, what is the probability that one of them who is not factory trained will provide good service?

Solution   As can be seen from the table,

$$P(G \cap F') = \frac{24}{150} \quad \text{and} \quad P(F') = \frac{86}{150}$$

so that substitution into the formula yields

$$P(G|F') = \frac{P(G \cap F')}{P(F')} = \frac{\frac{24}{150}}{\frac{86}{150}} = \frac{24}{86} \approx 0.28$$

Of course, the fraction $\frac{24}{86}$ could have been obtained directly by considering only the second row of the table.   ■

Although we introduced the formula for $P(A|B)$ with an example in which the possibilities were all equally likely, this is not a requirement for its use. The only restriction is that $P(B)$ must not equal zero.

■ EXAMPLE   For burglaries in a certain city, police records show that the probability is 0.35 that an arrest will be made. The probability that an arrest and conviction will occur is 0.14. What is the probability that a person arrested for burglary will be convicted?

Solution   If A is the event that an arrest will follow a burglary and C is the event that a conviction will follow a burglary, then $P(A) = 0.35$ and $P(A \cap C) = 0.14$. The formula for conditional probability yields

$$P(C|A) = \frac{P(A \cap C)}{P(A)} = \frac{0.14}{0.35} = 0.40 \qquad ■$$

■ EXAMPLE   The probability that there will be a shortage of cement is 0.28, and the probability that there will not be a shortage of cement and a construction job will be finished on time is 0.64. What is the probability that the construction job will be finished on time given that there will not be a shortage of cement?

Solution   If we let N denote the event that there will not be a shortage of cement and F the event that the construction job will be finished on time, then $P(N) = 1 - 0.28 = 0.72$, $P(F \cap N) = 0.64$, and it follows that

$$P(F|N) = \frac{P(F \cap N)}{P(N)} = \frac{0.64}{0.72} = \frac{8}{9} \approx 0.89 \qquad \blacksquare$$

5.7 Independent Events

To introduce another concept that is important in the study of probability, let us consider the following problem:

■ **EXAMPLE**   The probabilities that a student will get passing grades in algebra, in litera- ture, or in both subjects are, respectively, $P(A) = 0.75$, $P(L) = 0.84$, and $P(A \cap L) = 0.63$. What is the probability that the student will get a passing grade in algebra given that he or she gets a passing grade in literature?

Solution   Substituting into the formula that defines conditional probability, we get

$$P(A|L) = \frac{P(A \cap L)}{P(L)} = \frac{0.63}{0.84} = 0.75 \qquad \blacksquare$$

What is special and interesting about this result is that $P(A|L) = P(A) = 0.75$; that is, the probability of event A is the same regardless of whether or not event L has occurred (occurs, or will occur).

In general, if $P(A|B) = P(A)$, we say that even A is **independent** of event B; that is,

> *Event A is independent of event B if the probability of event A is not affected by the occurrence or nonoccurrence of event B.*

Since it can be shown that event B is independent of event A whenever event A is independent of event B, it is customary to say simply that *events A and B are independent* whenever one is independent of the other (see Exercise 5.73). If two events A and B are not independent, we say that they are **dependent events**.

5.8 Multiplication Rules

So far we have used the formula

$$P(A|B) = \frac{P(A \cap B)}{P(B)}$$

only to calculate conditional probabilities; but if we multiply both sides of the equation by $P(B)$, we get the following formula, called the **general multipli- cation rule**, which enables us to calculate the probability that two events will both occur.

**General
multiplication rule**

$$P(A \cap B) = P(B) \cdot P(A|B)$$

In words, the probability that two events will both occur is the product of the probability that one event will occur and the conditional probability that the other event will occur given that the first event has occurred (occurs, or will occur). As it does not matter which event is denoted by A and which by B, the preceding formula can also be written as

**General
multiplication rule
(alternative form)**

$$P(A \cap B) = P(A) \cdot P(B|A)$$

■ **EXAMPLE**   A jury consists of nine persons who are native born and three persons who are foreign born. If two of the jurors are randomly picked for an interview, what is the probability that they will both be foreign born?

Solution   Let A be the event that the first juror is foreign born and B the event that the second juror is foreign born. If we assume equal probabilities for each choice (which is, in fact, what we mean by the selection being random), the probability that the first juror picked will be foreign born is $P(A) = \frac{3}{12}$. Then, if the first juror picked is foreign born, the probability that the second juror will also be foreign born is $P(B|A) = \frac{2}{11}$. Hence, the probability of getting two jurors who are both foreign born is

$$P(A \cap B) = P(A) \cdot P(B|A) = \frac{3}{12} \cdot \frac{2}{11} = \frac{1}{22} \qquad ■$$

When A and B are independent, we can substitute $P(A)$ for $P(A|B)$ in the first of the two formulas for $P(A \cap B)$, or $P(B)$ for $P(B|A)$ in the second, and we obtain the **special multiplication rule**:

**Special
multiplication rule
(independent events)**

If A and B are independent events, then
$$P(A \cap B) = P(A) \cdot P(B)$$

In words, the probability that two independent events will both occur is simply the product of their respective probabilities. This rule is sometimes used as the definition of independence; in any case, it may be used to check whether two given events are independent.

■ **EXAMPLE**   What is the probability of getting two heads in two honest flips of a balanced coin?

Solution   Since *honest* and *balanced* mean that the outcomes of the two flips are independent and that the probability of getting heads is $\frac{1}{2}$ for each flip of the coin, the answer is $\frac{1}{2} \cdot \frac{1}{2} = \frac{1}{4}$.   ■

■ **EXAMPLE**   If $P(C) = 0.65$, $P(D) = 0.40$, and $P(C \cap D) = 0.26$, are the events C and D independent?

Solution   Since

$$P(C) \cdot P(D) = (0.65)(0.40) = 0.26 = P(C \cap D)$$

the two events are independent.   ■

■ **EXAMPLE**   Select one card randomly from a standard deck of 52. Let R be the event that the card is red, and let Q be the event that the card is a queen. Are the events R and Q independent?

Solution   The event $R \cap Q$ is the selection of a red queen. Since there are two red queens in the deck,

$$P(R \cap Q) = \frac{2}{52}$$

Noting also that

$$P(R) = \frac{26}{52} \quad \text{and} \quad P(Q) = \frac{4}{52}$$

we find that $P(R) \cdot P(Q) = \dfrac{26}{52} \cdot \dfrac{4}{52} = \dfrac{2}{52} = P(R \cap Q)$, and therefore the two events are independent.   ■

The three previous examples illustrate common ways in which independence is asserted. In the first example, the two flips are physically unrelated, so that we simply claim independence and apply the formula. In the second example, we find $P(C \cap D) = P(C) \cdot P(D)$; in this case, the numbers allow us to say that C and D are independent. In the third example, the events R and Q are physically related (since they describe the same random selection from the deck), but the numbers still allow us to say that R and Q are independent.

■ **EXAMPLE**   What is the probability of getting two aces in a row when 2 cards are drawn from an ordinary deck of 52 playing cards, if

(a) the first card is replaced before the second card is drawn;

(b) the first card is not replaced before the second card is drawn?

Solution   (a) Since there are four aces among the 52 cards, we get

$$\frac{4}{52} \cdot \frac{4}{52} = \frac{1}{169} \approx 0.0059$$

(b) Since there are only three aces among the 51 cards that remain if one ace has been removed from the deck, the answer is

$$\frac{4}{52} \cdot \frac{3}{51} = \frac{1}{221} \approx 0.0045$$ ■

Observe the difference in the calculations and the results of the two parts of the preceding example. They serve to illustrate the distinction between **sampling with replacement** and **sampling without replacement**. In most statistical investigations, sampling is done without replacement.

The special multiplication rule can readily be generalized so that it applies to three or more independent events; again, we simply multiply all their probabilities.

■ **EXAMPLE**   If the probability is 0.25 that a person will name red as his or her favorite color, what is the probability that three totally unrelated persons will all name red as their favorite color?

Solution   Assuming independence, we get

$$(0.25)(0.25)(0.25) \approx 0.0156$$ ■

When three or more events are not independent, the multiplication rule becomes more complicated: we form the product of the probability that one of the events will occur; the conditional probability that a second event will occur given that the first event has occurred; the conditional probability that a third event will occur, given that the first two events have occurred; and so on.

■ **EXAMPLE**   Suppose that four cards are to be selected without replacement from a standard deck. What is the probability that the first two cards will be hearts and the next two cards will be clubs?

Solution   The probability that the first card will be a heart is $\frac{13}{52}$. The probability, thereafter, that the second card will be a heart is $\frac{12}{51}$. The probability that these draws will be followed by a club is $\frac{13}{50}$. Finally, the probability that the fourth draw will be a club is $\frac{12}{49}$. The probability then of getting two hearts followed by two clubs is

$$\frac{13}{52} \cdot \frac{12}{51} \cdot \frac{13}{50} \cdot \frac{12}{49} = \frac{24{,}336}{6{,}497{,}400} \approx 0.0037$$ ■

Exercises

Exercises 5.55, 5.56, 5.61, 5.62, and 5.80 are practice exercises; their complete solutions are given on page 185.

5.55   If F is the event that a student will get financial aid, J is the event that the student will find a part-time job, and G is the event that the student will graduate, express symbolically the probabilities that
(a)   a student who gets financial aid will graduate;
(b)   a student who gets no financial aid will find a part-time job;
(c)   a student who gets no financial aid will neither find a part-time job nor graduate;
(d)   a student who gets financial aid and finds a part-time job will not graduate.

5.56   If E is the event that a female applicant for a sales position has had prior experience, C is the event that she owns a car, and G is the event that she is a college graduate, state in words what probabilities are expressed by
(a) $P(C|G)$;
(b) $P(E|C')$;
(c) $P(C'|E)$;
(d) $P(G'|C')$;
(e) $P(C|E \cup G)$;
(f) $P(E \cap C'|G)$.

5.57 If W is the event that a worker is well trained and Q is the event that he or she meets the production quota, express symbolically the probabilities that
(a) a worker who is well trained will meet the production quota;
(b) a worker who meets the production quota is not well trained;
(c) a worker who is not well trained will not meet the production quota.

5.58 With reference to Exercise 5.57, state in words what probabilities are expressed by
(a) $P(W|Q)$;   (b) $P(Q'|W)$;   (c) $P(W'|Q')$.

5.59 If H is the event that a probation officer is honest, E is the event that he or she is easy going, and W is the event that he or she is well liked, express symbolically the probabilities that
(a) a probation officer who is easy going will be well liked;
(b) a probation officer who is dishonest will be easy going;
(c) a probation officer who is honest and easy going will be well liked.

5.60 With reference to Exercise 5.59, state in words what probabilities are expressed by
(a) $P(H|W')$;   (b) $P(W'|E')$;   (c) $P(W \cap E|H)$.

✓ **5.61** Among the 400 inmates of a prison, some are first offenders, some are repeat offenders, some are serving sentences of less than five years, and some are serving longer sentences. The exact breakdown is

	Sentences less than five years	Longer sentences
First offenders	120	40
Repeat offenders	80	160

One of the inmates is to be selected at random to be interviewed about prison conditions. Let R be the event that he is a repeat offender, and let L be the event that he is serving a longer sentence. Determine each of the following probabilities directly from the entries and the row and column totals of the table:
(a) $P(R)$;   (d) $P(R' \cap L)$;
(b) $P(L)$;   (e) $P(L|R)$;
(c) $P(L \cap R)$;   (f) $P(R'|L)$.

✓ **5.62** Use the results of Exercise 5.61 to verify that
(a) $P(L|R) = \dfrac{P(L \cap R)}{P(R)}$;   (b) $P(R'|L) = \dfrac{P(R' \cap L)}{P(L)}$.

5.63 In some of the delinquent charge accounts of a store, the amount owed is Tess than $100, in some it is $100 or more, some have been delinquent for less than a month, and some have been delinquent for a month or more. The exact breakdown is

	Less than a month	A month or more
Less than $100	168	72
$100 or more	57	43

One of these delinquent accounts is to be selected at random for a new credit check. Let L be the event that the amount owed is less than \$100, and let M be the event that the account has been delinquent for a month or more. Determine each of the following probabilities directly from the entries and the row and column totals of the table:

(a) $P(L)$;

(b) $P(M)$;

(c) $P(L \cap M')$;

(d) $P(M' \cap L')$;

(e) $P(L|M')$;

(f) $P(M'|L')$.

5.64 Use the results of Exercise 5.63 to verify that

(a) $P(L|M') = \dfrac{P(L \cap M')}{P(M')}$; (b) $P(M'|L') = \dfrac{P(M' \cap L')}{P(L')}$.

5.65 In the following table, 60 college students are classified according to their class standing and also according to their favorite pizza topping:

	Anchovies	Onions	Mushrooms	Hamburger
Freshman	7	6	7	3
Sophomore	1	9	0	9
Junior	3	2	5	8

If one of these students is selected at random, if F, S, and J denote the three classes, and if A, O, M, and H denote the four pizza toppings, find

(a) $P(M \cup J)$;

(b) $P(H|F)$;

(c) $P(O \cap S)$;

(d) $P(F'|A)$;

(e) $P(M \cup H|J')$;

(f) $P(J|A \cup M)$.

5.66 With reference to Exercise 5.65, find the probabilities that the student chosen will be

(a) a freshman whose favorite pizza topping is mushrooms;

(b) an anchovy pizza eater given that he or she is a junior;

(c) a sophomore given that he or she is not a junior.

5.67 If M is the event that a young child has received a low grade in a manual dexterity test and A is the event that the child cannot dress herself without assistance, express in words what probability is represented by

(a) $P(A|M)$;

(b) $P(M|A)$;

(c) $P(A|M')$;

(d) $P(M|A')$;

(e) $P(M'|A')$.

5.68 The probability that a certain concert will be well advertised is 0.80, and the probability that it will be well advertised and a great success is 0.76. What is the probability that the concert, if well advertised, will be a great success?

5.69 An employment recruiter figures that the probability is 0.82 that a résumé submitted to his office will contain all relevant facts needed for processing. The probability is 0.58 that a résumé submitted to his office will contain all relevant facts and also be reviewed by the entire placement staff. What is the probability that a résumé with all relevant facts will be reviewed by the entire placement staff?

5.70 Some special cards are prepared for an unusual game. There are four varieties: cards with red letter X, cards with green letter X, cards with red letter Y, and cards with green letter Y. Deck A has 40 of these cards, and the following table gives the contents of this deck.

Deck A

	Red	Green
X	5	14
Y	5	16

If a card is selected at random from this deck, it follows that

$$P(X|\text{Red, Deck } A) = \frac{5}{10} = 0.50$$

since 5 of the 10 red cards contain the letter X. Also,

$$P(X|\text{Green, Deck } A) = \frac{14}{30} \approx 0.47$$

Deck B also has 40 of these cards, and the following table gives the makeup of Deck B.

Deck B

	Red	Green
X	5	2
Y	20	13

(a) Show that

$$P(X|\text{Red, Deck } A) > P(X|\text{Green, Deck} A)$$

(b) Show that

$$P(X|\text{Red, Deck } B) > P(X|\text{Green, Deck } B)$$

(c) Suppose that the two decks are combined to make a single deck of 80 cards. How many cards of each of the four varieties are in this deck?

(d) Suppose that a single card is to be selected randomly from this deck of 80 cards. Is it true that

$$P(X|\text{Red, Combined Deck}) > P(X|\text{Green, Combined Deck})?$$

(*Note:* Conditional probabilities behave in peculiar ways. This particular problem is an example of **Simpson's paradox**.)

5.71 Abel and Baker are quarterbacks for a football team. The coach uses both players during each game. In the first game of the season, Abel got to throw 20 passes, and 11 of them were complete. Also in the first game, Baker threw 5 passes, completing 3 of them. In the second game, Abel threw 5 passes, completing only 1. Baker threw 16 passes, completing 6.

(a) Show that Baker's proportion of pass completions was better than Abel's proportion of pass completions in the first game.

(b) Show that Baker's proportion of pass completions was better than Abel's proportion of pass completions in the second game.

(c) A newspaper combined the data from the two games to get the season-to-date data. According to the combined information, which quarterback has the higher proportion of pass completions?

5.72 If the odds are 5 to 3 that event K will not occur, 2 to 1 that event L will occur, and 3 to 1 that they will not both occur, are these two events independent?

5.73 In the text example on page 170 concerning a student's grades in algebra and literature, we had

$$P(A) = 0.75$$

$$P(L) = 0.84$$

$$P(A \cap L) = 0.63$$

$$P(A|L) = 0.75$$

Show that event L is independent of event A; that is, $P(L|A) = P(L) = 0.84$.

5.74 If $P(A) = 0.80$, $P(B) = 0.35$, and $P(A \cap B) = 0.28$, check whether events A and B are independent.

5.75 For three rolls of a balanced die, find the probabilities of getting
   (a) three sixes;
   (b) first a six, then a five, then a four;
   (c) no sixes.

5.76 If two cards are drawn from an ordinary deck of 52 playing cards, what are the probabilities of getting two aces if the drawing is
   (a) with replacement;   (b) without replacement?

5.77 A fifth-grade class consists of 16 boys and 14 girls. If one pupil is chosen each week by lot to assist the teacher, what are the probabilities that a boy will be chosen two weeks in a row if
   (a) the same pupil can serve two weeks in a row;
   (b) the same pupil cannot serve two weeks in a row?

5.78 If a zoologist has four male guinea pigs and eight female guinea pigs and randomly chooses two of them for an experiment, what are the probabilities that
   (a) both will be males;
   (b) both will be females;
   (c) there will be one of each sex?

5.79 Find the probabilities of getting
   (a) six heads in a row with a balanced coin;
   (b) no fives in three rolls of a balanced die.

5.80 If 5 of a company's 15 delivery trucks do not meet emission standards, and 4 of them are randomly chosen for inspection, what is the probability that all of them will meet emission standards?

5.81 A carton contains 12 shirts of which 3 have blemishes and the rest are good. If 2 of the shirts are randomly selected from the carton, what is the probability that they will both have blemishes?

5.82 A jar contains 15 numbered balls that are identical except for the numbers. These numbers are 1 through 15. Suppose that you remove 4 balls randomly and without replacement.
   (a) What is the probability that all 4 balls will have even numbers?
   (b) What is the probability that the first 2 balls will have even numbers and the last 2 balls will have odd numbers?
   (c) What is the probability that the first ball will have an even number and the last 3 balls will have odd numbers?

5.83 In a certain city the probability that it will snow on a December day is 0.40, the probability that a December day on which it snows will be followed by a day on which it snows is 0.70, and the probability that a December day on which it does not snow will be followed by a day on which it snows is 0.20. What are the probabilities that on three consecutive December days it will

(a) snow on each day;

(b) not snow on the first day and then snow on the next two days?

5.84 With reference to Exercise 5.83, what is the probability that it will snow, snow, not snow, and snow on four consecutive December days?

5.9 Bayes' Theorem*

Although the symbols $P(A|B)$ and $P(B|A)$ may look very much alike, there is a great difference between the probabilities that they represent. For instance, we previously calculated the probability $P(G|F)$ that a factory-trained appliance repair person will provide good service, but what do we mean when we write $P(F|G)$? This is the probability that a repair person who provides good service will have been factory trained. Thus, we turned things around—cause becomes effect and effect becomes cause. To give another example, suppose that a person applying for a loan has his credit checked, and R denotes the event that the person is a bad risk, while J denotes the event that the person is judged a bad risk. Then $P(J|R)$ is the probability that a person who is a bad risk will be judged a bad risk, and $P(R|J)$ is the probability that a person who is judged a bad risk actually is a bad risk.

Since many problems involve such pairs of conditional probabilities, let us try to find a formula that expresses $P(B|A)$ in terms of $P(A|B)$ for any two events A and B. Fortunately, we do not have to look very far; all we have to do is equate the expressions for $P(A \cap B)$ in the two forms of the general multiplication rule on page 171, and we get

$$P(A) \cdot P(B|A) = P(B) \cdot P(A|B)$$

and hence

$$P(B|A) = \frac{P(B) \cdot P(A|B)}{P(A)}$$

after dividing by $P(A)$.

■ **EXAMPLE**   In a state where cars have to be tested for the emission of pollutants, 25% of all cars emit excessive amounts of pollutants. When tested, 99% of all cars that emit excessive amounts of pollutants will fail, but 17% of the cars that do not emit excessive amounts of pollutants will also fail. What is the probability that a car that fails the test actually emits excessive amounts of pollutants?

Solution   Letting A denote the event that a car fails the test and B the event that it emits excessive amounts of pollutants, we can translate the given percentages into probabilities and write

$$P(B) = 0.25, \quad P(A|B) = 0.99, \quad \text{and} \quad P(A|B') = 0.17$$

Before we can calculate $P(B|A)$ by means of the preceding formula, we will first have to determine $P(A)$, and to this end let us look at the tree diagram of Figure 5.13. Here A is reached either along the branch that passes through B or along the branch that passes through B', and the probabilities of this happening are, respectively,

$$(0.25)(0.99) = 0.2475 \quad \text{and} \quad (0.75)(0.17) = 0.1275$$

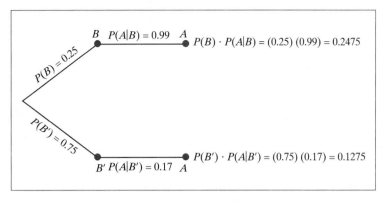

FIGURE 5.13 Tree diagram for emission testing example.

Since the alternatives represented by the two branches are mutually exclusive, we find that

$$P(A) = 0.2475 + 0.1275 = 0.3750$$

and substitution into the formula for $P(B|A)$ given previously yields

$$P(B|A) = \frac{P(B) \cdot P(A|B)}{P(A)} = \frac{(0.25)(0.99)}{0.3750} = 0.66$$

This is the probability that a car that fails the test actually emits excessive amounts of pollutants. ▪

With reference to the tree diagram of Figure 5.13, we can say that 0.2475 is the probability that event A was reached via the upper branch of the tree, and we showed that $P(B|A)$ is given by the ratio of the probability associated with that branch to the sum of the probabilities associated with both branches of the tree. This argument can be generalized to the case where there are more than two possible causes, namely, more than two branches leading to an event A. With reference to Figure 5.14, we can say that $P(B_i|A)$ is the probability that event A was reached via the ith branch of the tree (for $i = 1, 2, ..., $ or k), and it can be shown that its value is given by the ratio of

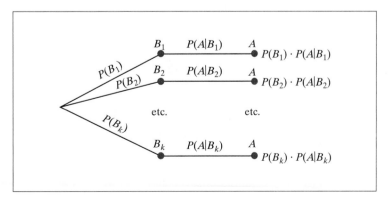

FIGURE 5.14 Tree diagram for Bayes' theorem.

the probability associated with the ith branch to the sum of the probabilities associated with all the branches. Symbolically, this result, called **Bayes' theorem**, is given by

Bayes' theorem

$$P(B_i|A) = \frac{P(B_i) \cdot P(A|B_i)}{P(B_1) \cdot P(A|B_1) + P(B_2) \cdot P(A|B_2) + \cdots + P(B_k) \cdot P(A|B_k)}$$

for $i = 1, 2, \dots,$ or k.

■ **EXAMPLE**   In a cannery, assembly lines I, II, and III account, respectively, for 37%, 42%, and 21% of the total output. If 0.6% of the cans from assembly line I are improperly sealed, while the corresponding percentages for assembly lines II and III are 0.4% and 1.2%, what is the probability that an improperly sealed can (discovered at the final inspection of outgoing products) came from assembly line III?

Solution   If we let A denote the event that a can is improperly sealed and B_1, B_2, and B_3 denote the events that a can comes from assembly lines I, II, or III, we can translate the given percentages into probabilities and write

$$P(B_1) = 0.37, \quad P(B_2) = 0.42, \quad P(B_3) = 0.21$$

$$P(A|B_1) = 0.006$$

$$P(A|B_2) = 0.004$$

$$P(A|B_3) = 0.012$$

Then, with reference to the tree diagram of Figure 5.15 we find that the probabilities associated with the three branches are, respectively,

$$(0.37)(0.006) = 0.00222, \qquad (0.42)(0.004) = 0.00168,$$

and

$$(0.21)(0.012) = 0.00252$$

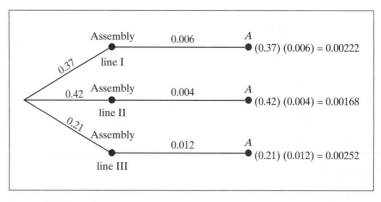

FIGURE 5.15 Tree diagram for cannery example.

Thus, the required probability is

$$P(B_3|A) = \frac{0.00252}{0.00222 + 0.00168 + 0.00252} \approx 0.393$$

If we had wanted to work this problem without referring to a tree diagram, direct substitution into the formula for Bayes' theorem would have yielded

$$P(B_3|A) = \frac{(0.21)(0.012)}{(0.37)(0.006) + (0.42)(0.004) + (0.21)(0.012)}$$

$$\approx 0.393$$

and all the details of the calculations are, of course, the same.   ■

As can be seen from our two examples, Bayes' theorem is a relatively simple mathematical rule. There can be no question about its validity, but criticism has frequently been raised about its applicability. This is because it involves a *backward* or *inverse* sort of reasoning, that is, reasoning from effect to cause. In our examples we used Bayes' theorem to determine the probability that a car that fails the emission test actually emits excessive amounts of pollutants, and the probability that an improperly sealed can was "caused" by assembly line III. It is precisely this aspect of Bayes' theorem that makes it play an important role in statistical inference, where our reasoning goes from sample data that is observed to the populations from which it came. Discussion of such inferences, appropriately called **Bayesian inferences**, may be found in more advanced texts.

Exercises

Exercise 5.85 is a practice exercise; its complete solution is given on page 185.

5.85 Suppose that 20% of the residents of an equatorial city have a certain tropical disease. Among afflicted individuals, failure in a medical diagnostic test accurately discloses the presence of the disease 95% of the time; but among persons who do not have this disease, the test falsely indicates its presence 10% of the time. Let A denote the event that a person fails the test, and let B denote the event that a person has the tropical disease; thus $P(A|B)$ denotes the probability that a person who has the disease will fail the test. Calculate

$$P(B|A) = \frac{P(B) \cdot P(A|B)}{P(A)}$$

to determine the probability that a person who fails the test actually has the disease.

*5.86 At a shoe factory, it is known from past experience that the probability is 0.82 that a worker who has attended the factory's training program will meet the production quota, and that the corresponding probability is 0.53 for a worker who has not attended the factory's training program. If 60% of the workers attend the factory's training program, what is the probability that a worker who meets the production quota will have attended the training program?

*5.87 The probability that a one-car accident is due to faulty brakes is 0.05, the probability that a one-car accident is correctly attributed to faulty brakes is 0.82, and the probability that a one-car accident is incorrectly attributed to faulty brakes is 0.03. What is the probability that a one-car accident attributed to faulty brakes was actually due to faulty brakes?

*5.88 In a T-maze, a rat is given food if it turns left and an electric shock if it turns right. On the first trial there is a 50–50 chance that a rat will turn either way; then if it receives food on the first trial, the probability is 0.72 that it will turn left on the next trial, and if it receives a shock on the first trial, the probability is 0.88 that it will turn left on the next trial. If a rat turns left on the second trial, what is the probability that it turned left also on the first trial?

5.89 A driver's license examiner knows that 75% of all applicants have attended a driving school. If an applicant has attended a driving school, the probability is 0.85 that he or she will pass the license examination, and if an applicant has not attended a driving school, the probability is 0.60 that he or she will pass the examination. If an applicant passes the license examination, what is the probability that he or she has attended a driving school?

5.90 In the example on page 180, we used Bayes' theorem to show that the probability that an improperly sealed can came from assembly line III was

$$P(B_3|A) = \frac{0.00252}{0.00222 + 0.00168 + 0.00252} = \frac{0.00252}{0.00642} \approx 0.393$$

Here A denotes the event that a can is improperly sealed, and B_3 denotes the event that it came from assembly line III. Making use of the results that $P(B_1) \cdot P(A|B_1) = 0.00222$, $P(B_2) \cdot P(A|B_2) = 0.00168$, and $P(B_3) \cdot P(A|B_3) = 0.00252$, determine the probabilities that

(a) an improperly sealed can came from assembly line I;
(b) an improperly sealed can came from assembly line II.
(c) Verify that the probability that an improperly sealed can came from assembly line I, II, or III is, in fact, equal to 1.

*5.91 A hotel gets cars for its guests from three rental agencies: 25% from agency X, 25% from agency Y, and 50% from agency Z. If 8% of the cars from X, 6% of the cars from Y, and 15% of the cars from Z need tune-ups, what is the probability that a car needing a tune-up that is delivered to a guest of the hotel came from rental agency Y?

*5.92 A retailer of automobile parts has four employees, K, L, M, and N, who make mistakes in filling an order one time in 100, four times in 100, two times in 100, and six times in 100. Of all the orders filled, K, L, M, and N fill, respectively, 20%, 40%, 30%, and 10%. If a mistake is found in a particular order, what are the probabilities that it was filled by K, L, M, or N?

Solutions to Practice Exercises

5.1 (a) $X' = \{b, d, e, h\}$ is the event that Mr. Bean, Ms. Daly, Mr. Earl, or Mr. Hall will be awarded the scholarship.
(b) $X \cup Y = \{a, c, d, f, g, h\}$ is the event that Ms. Adam, Ms. Clark, Ms. Daly, Ms. Fuentes, Ms. Garner, or Mr. Hall will be awarded the scholarship.
(c) $X \cap Y = \{f\}$ is the event that Ms. Fuentes will be awarded the scholarship.
(d) $X \cap Z = \{a, g\}$ is the event that Ms. Adams or Ms. Gardner will be awarded the scholarship.
(e) $Y \cap Z$ is the empty set.
(f) $X' \cup Y = \{b, d, e, f, h\}$ is the event that Mr. Bean, Ms. Daly, Mr. Earl, Ms. Fuentes, or Mr. Hall will be awarded the scholarship.

5.2 (a) The sample space is shown in Figure 5.16.
(b) Q is the event that four waiters are present, R is the event that the num-

ber of waiters exceeds the number of chefs by 1, and T is the event that only one chef is present.

(c) $Q \cup T = \{(1, 4), (2, 4), (1, 2), (1, 3), (1, 5)\}$ is the event that four waiters are present or one chef is present, $R \cap T = \{(1, 2)\}$ is the event that one chef and two waiters are present, and $T' = \{(2, 2), (2, 3), (2, 4), (2, 5)\}$ is the event that both chefs are present.

(d) Q and R are mutually exclusive, Q and T are not mutually exclusive, and R and T are not mutually exclusive.

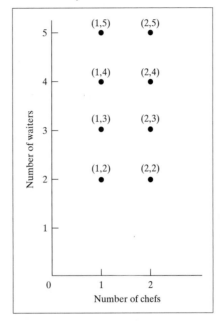

FIGURE 5.16 Diagram for Exercise 5.2.

5.12 Region 1 represents the event that a person vacationing in Southern California visits Disneyland and Universal Studios.

Region 2 represents the event that a person vacationing in Southern California visits Disneyland, but not Universal Studios.

Region 3 represents the event that a person vacationing in Southern California visits Universal Studios, but not Disneyland.

Region 4 represents the event that a person vacationing in Southern California visits neither Disneyland nor Universal Studios.

5.19 (a) The probability that there will not be enough doctors.
(b) The probability that there will not be enough hospital beds.
(c) The probability that there will be enough doctors, enough hospital beds, or both.
(d) The probability that there will be enough doctors and enough hospital beds.
(e) The probability that there will be neither enough doctors nor enough hospital beds.
(f) The probability that there will be enough doctors but not enough hospital beds.

5.22 (a) The sum of the two probabilities should equal 1, but they only equal 0.90.
(b) The sum of the two probabilities cannot exceed 1, and it is 1.15.

(c) Since the two events are mutually exclusive, the sum of the first two probabilities should equal the third.

5.30 (a) The odds against getting three heads and three tails in six flips of a balanced coin are 11 to 5.

(b) The probability of rolling 7 or 11 with a pair of balanced dice is

$$\frac{2}{2+7} = \frac{2}{9}$$

(c) If a pollster randomly selects 5 of 24 households to be included in a survey, the odds are 19 to 5 that any particular household will not be included.

(d) If 3 eggs are randomly chosen from a carton of 12 eggs of which 3 are cracked, the probability is $\dfrac{34}{34+21} = \dfrac{34}{55}$ that at least 1 of them will be cracked.

5.31 The corresponding probabilities are $\dfrac{2}{2+1} = 0.67$ and $\dfrac{2}{2+1} = 0.67$, and since the events are mutually exclusive, we can write $0.67 + 0.67 = 1.34$. Since a probability cannot exceed 1, this is an impossible value, and the original odds must have been wrong.

5.40 $0.22 + 0.17 + 0.21 + 0.09 = 0.69$.

5.45 (a) The event consists of $(4, 4)$, $(5, 5)$, $(6, 6)$, $(7, 7)$, and $(8, 8)$, so that the probability is $\frac{5}{35} = \frac{1}{7}$.

(b) The event consists of $(4, 0)$, $(4, 1)$, $(4, 2)$, $(5, 0)$, $(5, 1)$, $(5, 2)$, $(6, 0)$, $(6, 1)$, $(6, 2)$, $(7, 0)$, $(7, 1)$, $(7, 2)$, $(8, 0)$, $(8, 1)$, and $(8, 2)$, and the probability is $\frac{15}{35} = \frac{3}{7}$.

(c) The event consists of $(7, 0)$, $(7, 1)$, $(7, 2)$, $(7, 3)$, $(7, 4)$, $(7, 5)$, $(7, 6)$, $(7, 7)$, $(8, 0)$, $(8, 1)$, $(8, 2)$, $(8, 3)$, $(8, 4)$, $(8, 5)$, $(8, 6)$, $(8, 7)$, and $(8, 8)$, and the probability is $\frac{17}{35}$.

(d) The event consists of $(4, 2)$, $(4, 3)$, $(5, 3)$, $(5, 4)$, $(6, 4)$, $(6, 5)$, $(7, 5)$, $(7, 6)$, $(8, 6)$, and $(8, 7)$, and the probability is $\frac{10}{35} = \frac{2}{7}$.

5.49 (a) $0.15 + 0.18 - 0.11 = 0.22$;

(b) $1 - 0.22 = 0.78$.

5.52 The probabilities associated with the various regions are shown in the Venn diagram of Figure 5.17.

(a) 0.32;

(b) 0.14;

(c) $0.32 + 0.13 + 0.14 = 0.59$;

(d) $0.13 + 0.14 + 0.41 = 0.68$;

(e) 0.41;

(f) $1 - 0.13 = 0.87$.

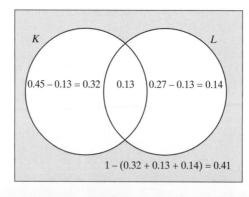

FIGURE 5.17 Venn diagram for Exercise 5.52.

5.55 (a) $P(G|F)$; (c) $P(J' \cap G'|F')$;

(b) $P(J|F')$; (d) $P(G'|F \cap J)$.

5.56 (a) The probability that a female applicant who is a college graduate will own a car.

(b) The probability that a female applicant who does not own a car will have had prior experience.

(c) The probability that a female applicant with prior experience will not own a car.

(d) The probability that a female applicant who does not own a car will not be a college graduate.

(e) The probability that a female applicant who has had prior experience or is a college graduate will own a car.

(f) The probability that a female applicant who is a college graduate will have had prior experience, but not own a car.

5.61 (a) $\dfrac{80 + 160}{400} = \dfrac{240}{400} = 0.60$; (d) $\dfrac{40}{400} = 0.10$;

(b) $\dfrac{40 + 160}{400} = \dfrac{200}{400} = 0.50$; (e) $\dfrac{160}{80 + 160} = \dfrac{2}{3} \approx = 0.67$;

(c) $\dfrac{160}{400} = 0.40$; (f) $\dfrac{40}{40 + 160} = 0.20$.

5.62 (a) $\dfrac{2}{3} = \dfrac{0.40}{0.60} \approx 0.67$, which checks; (b) $0.20 = \dfrac{0.10}{0.50} = 0.20$, which checks.

5.80 The probability that the first truck chosen will meet emission standards is $\frac{10}{15}$; the probability that the second will meet emission standards, given that the first met emission standards, is $\frac{9}{14}$; the probability that the third will meet emission standards, given that the first two met emission standards, is $\frac{8}{13}$; and the probability that the fourth will meet emission standards, given that the first three met emission standards, is $\frac{7}{12}$. Thus, the probability that all of them will meet emission standards is

$$\frac{10}{15} \cdot \frac{9}{14} \cdot \frac{8}{13} \cdot \frac{7}{12} = \frac{2}{13}$$

5.85 Translate the percentages into probabilities and write $P(B) = 0.20$, $P(A|B) = 0.95$, and $P(A|B') = 0.10$. To determine $P(A)$, we must add the results that $P(B) \cdot P(A|B) = (0.20)(0.95) = 0.19$ and $P(B') \cdot P(A|B') = (0.80)(0.10) = 0.08$, getting $P(A) = 0.19 + 0.08 = 0.27$. Finally, substituting these values into the formula for $P(B|A)$, we get

$$P(B|A) = \frac{P(B) \cdot P(A|B)}{P(A)} = \frac{(0.20)(0.95)}{0.27} \approx 0.704$$

In other words, the probability that a person has the tropical disease given that he or she fails the test is approximately 0.704. The test is not particularly convincing.

Review: Chapters 4 & 5

Having read and studied these chapters and having worked a good portion of the exercises, you should be able to:

1. Draw tree diagrams to determine all the alternatives that are possible in given situations.

2. Apply the formula for the multiplication of choices and its generalization.

3. Work with the factorial notation.

4. Determine the number of permutations of n distinct objects taken r at a time.

5. Determine the number of combinations of n distinct objects taken r at a time.

6. Use the table of binomial coefficients.

7. Explain the classical probability concept.

8. Explain the frequency interpretation of probability.

9. Calculate mathematical expectations.

*10. Use mathematical expectations as the basis for making rational decisions.

11. Explain what is meant by *experiment*, *outcome*, *sample space*, *event*, and *mutually exclusive events*.

12. Construct compound events by forming unions, intersections, and complements.

13. Picture events with the use of Venn diagrams.

14. List, justify, and apply the basic rules of probability.

15. Convert probabilities to odds and odds to probabilities.

16. Use betting odds to measure subjective probabilities.

17. State and apply the special addition rule and its generalization.

18. State and apply the general addition rule for events that need not be mutually exclusive.

19. Define *conditional probability* and apply the formula.

20. Explain what is meant by *independent events*, and check whether two events are independent or dependent.

21. State and apply the general multiplication rule.

22. State and apply the special multiplication rule.

23. State and apply generalizations of the multiplication rules.

*24. Use Bayes' theorem.

Checklist of Key Terms (with page references to their definitions)

Addition rules, 161
Bayesian analysis, 134
Bayesian inferences, 181
Bayes' theorem, 180
Betting odds, 156
Binomial coefficients, 119
Classical concept of probability, 123
Combinations, 119
Complement, 146
Conditional probability, 167
Consistency criterion, 158
Dependent events, 170
Empty set, 146
Equitable game, 132
Event, 146
Experiment, 144
Factorial notation, 115
Fair game, 132

Finite sample space, 145
Frequency interpretation of probability, 125
General addition rule, 164
General multiplication rule, 170
Independent events, 170
Infinite sample space, 145
Intersection, 146
Law of Large Numbers, 125
Mathematical expectation, 130
Mathematics of chance, 143
Multiplication of choices, 111
Multiplication rules, 170
Mutually exclusive events, 147
Odds, 156
Outcome, 144
Pascal's triangle, 122
Payoff table, 133

Permutations, 115
Personal evaluations, 127
Probability, 107
Sample space, 144
Sampling with replacement, 173
Sampling without replacement, 173
Simpson's paradox, 177
Special addition rule, 161
Special multiplication rule, 171
States of nature, 134
Subjective probabilty, 127
Theory of probabilty, 143
Tree diagram, 108
Union, 146
Venn diagram, 148

Review Exercises

R.41 A local newspaper reporter must cover local high school sports and is given his choice of three of six basketball games, one of four wrestling matches, and one of five swimming meets. In how many different ways can he choose the events to cover?

R.42 A luncheon special at a certain diner includes one of four different types of soup, one of eight different types of meat, and one of three different types of bread, along with two side dishes chosen from a group of five. How many different luncheon special choices are possible?

R.43 A small real estate office has four part-time salespersons. Using two coordinates so that $(3, 1)$, for example, represents the event that three of the salespersons are at work and one of them is busy with a customer, and $(2, 0)$ represents the event that two of the salespersons are at work but none of them is busy with a customer, draw a diagram similar to that of Figure 5.1, showing the 15 points of the corresponding sample space.

R.44 With reference to Exercise R.43, if each of the 15 points of the sample space has the probability $\frac{1}{15}$, find the probabilities that
    (a) all the salespersons that are at work are busy with customers;
    (b) at least three of the salespersons are at work;
    (c) at least three salespersons are busy with a customer;
    (d) none of the salespersons who are at work are busy with customers.

R.45 The following table specifies the probabilities that Al's Towing Service will receive 0, 1, 2, 3, or 4 calls for help during the evening rush hour:

Number of calls	0	1	2	3	4
Probability	0.08	0.15	0.30	0.35	0.12

What is the expected number of calls that Al's Towing will receive during the evening rush hour?

R.46 If $P(G) = 0.50$, $P(H) = 0.24$, and $P(G \cup H) = 0.62$, are events G and H dependent or independent?

R.47 Among six applicants for an executive job, A is a college graduate, is foreign born, and is single; B is not a college graduate, is foreign born, and is married; C is a college graduate, is native born, and is married; D is not a college graduate, is native born, and is single; E is a college graduate, is native born, and is married; and F is not a college graduate, is native born, and is married. One of these applicants is to get the job, and the event that the job is given to a college graduate, for example, is denoted $\{A, C, E\}$. State in a similar manner the event that the job is given to

    (a) a single person;

    (b) a native-born college graduate;

    (c) a married person who is foreign born.

R.48 If $P(A) = 0.48$, $P(B) = 0.50$, and $P(A \cap B) = 0.24$, are events A and B mutually exclusive or not mutually exclusive?

R.49 A grab bag contains 16 packages worth $1.00 apiece, 20 packages worth 40 cents apiece, and 14 packages worth 30 cents apiece. Is it worthwhile to pay 50 cents for the privilege of picking one of the packages at random?

R.50 If a student feels that the odds are 10 to 1 that she will pass a certain course, what is her personal probability that she will pass the course?

R.51 A real-estate agent has listings for ten single-family homes, five multifamily homes, and three condominium apartments. In how many ways can he choose three single-family homes, two multifamily homes, and one condominium apartment to show to a customer?

*R.52 During a time of national emergency, a country uses lie detectors to uncover security risks. Since lie detectors are not infallible, let us suppose that the probabilities are 0.10 and 0.04 that a lie detector will fail to detect a security risk or will incorrectly label a person a security risk. If 2% of the persons who are given lie detector tests are security risks, what is the probability that a person labeled a security risk by a lie detector actually is a security risk?

R.53 A large firm has four employees named Davis: Arthur Davis, Jennifer Davis, Paul Davis, and Wendy Davis.

    (a) In how many ways can the payroll department distribute their checks so that they each receive a check made out to one of the Davises?

    (b) In how many of the possibilities will none of them get the correct paycheck?

    (c) In how many of the possibilities will exactly one of them get the correct paycheck?

    (d) In how many of the possibilities will exactly two of them get the correct paycheck?

    (e) In how many of the possibilities will exactly three of them get the correct paycheck?

    (f) In how many of the possibilities will all four of them get the correct paycheck?

R.54 A bank teller has 20 ten-dollar bills, of which 4 are counterfeit and the rest are genuine. If 3 of the ten-dollar bills are randomly selected, what is the probability that they will all be counterfeit?

R.55 A psychologist preparing three-letter nonsense words for use in a memory test chooses the first letter from among the consonants q, w, x, and z; the second letter from among the vowels e, i, and u; and the third letter from among the consonants c, f, p, and v.

(a) How many different three-letter nonsense words can he construct?

(b) How many of these nonsense words will begin with the letter w?

(c) How many of these nonsense words will end either with the letter f or the letter p?

R.56 Five of the 25 men in an all-male chorus are tenors, and the others are baritones and basses. (The highest male voice is tenor, the next below is baritone, and the lowest is bass.) If the conductor randomly selects 3 men to sing a musical passage, what are the odds that the selected singers will not include tenors?

R.57 If B and M are the mutually exclusive events that a lepidopterist will catch a butterfly or a moth in a field, $P(B) = 0.43$, and $P(M) = 0.21$, find the probabilities that a lepidopterist will

(a) not catch a butterfly;

(b) catch a butterfly or a moth;

(c) not catch a butterfly or a moth.

R.58 The probabilities are, respectively, 0.32, 0.49, and 0.25 that a bird watcher will see a bobolink, a meadowlark, or both during a short walk in a certain grassy field. What is the probability that a birdwatcher will see either a bobolink or a meadowlark or both during a short walk in this location?

R.59 If there are six horses in a race, in how many different ways can they place first, second, and third?

R.60 The probabilities are 0.12, 0.27, and 0.07 that a family driving through a Western city will spend the night at one of its hotels, at one of its motels, or at its campground. What is the probability that a family driving through this city will spend the night at one of these kinds of facilities?

R.61 If 934 of 1,250 letters mailed by a government agency were delivered within 48 hours, estimate the probability that any one letter mailed by the agency will be delivered within 48 hours.

R.62 An appliance wholesaler distributes washing machines made in factories in Akron and Wheeling. Some of these washing machines receive, at customer request, an in-home warranted repair, with the exact breakdown being:

	Warranted repair	No warranted repair
Akron	20	160
Wheeling	50	150

If one of the washing machine customers is randomly selected to check on consumer satisfaction, and A is the event that the machine was made in Akron, and R is the event that it had an in-home warranted repair, determine each of the following probabilities:

(a) $P(A)$;

(b) $P(R')$;

(c) $P(A \cap R)$;

(d) $P(A|R)$;

(e) $P(R'|A)$;

(f) $P(A' \cup R)$.

R.63 Regarding Exercise R.62, verify that

$$P(A|R') = \frac{P(A \cap R')}{P(R')}$$

R.64 If the probability is 0.28 that any one woman will name yellow or orange as her favorite color, what is the probability that four women, selected at random, will all name yellow or orange as their favorite color?

R.65 If the probability is 0.40 that next year's inflation rate will exceed this year's, what are the corresponding odds?

*R.66 Ms. Kwan is planning to attend a convention in Denver, and she must send in her room reservations immediately. The convention is so large that the activities are held partly in hotel A and partly in hotel B, and Ms. Kwan does not know whether the particular session she wants to attend will be held in hotel A or hotel B. She is planning to stay only one day, which would cost her $80.00 at hotel A and $72.80 at hotel B, but it will cost her an extra $12.00 for cab fare if she stays at the wrong hotel. Where should she make her reservation if she feels that the probability is 0.75 that the session she wants to attend will be held at hotel A and she wants to minimize her expected cost?

R.67 If A is the event that a school's football team is rated among the top 20 by AP and U is the event that it is rated among the top 20 by UPI, what events are represented by the four regions of the Venn diagram of Figure R.1?

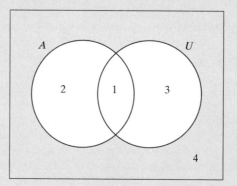

FIGURE R.1 Venn diagram for Exercise R.67.

R.68 A guidance department gives students various kinds of tests. If I is the event that a student scores high in intelligence, A is the event that a student rates high on a social adjustment scale, and N is the event that a student displays neurotic tendencies, express each of the following probabilities in symbolic form:

(a) The probability that a student who scores high in intelligence will display neurotic tendencies.

(b) The probability that a student who does not rate high on the social adjustment scale will not score high in intelligence.

(c) The probability that a student who displays neurotic tendencies will neither score high in intelligence nor rate high on the social adjustment scale.

(d) The probability that a student who scores high in intelligence and rates high on the social adjustment scale will not display any neurotic tendencies.

R.69 If an applicant for a job feels that the odds against getting the job are 12 to 7, what subjective probability does he assign to the probability that he will get the job?

R.70 The following are the probabilities that a policeman will issue 0 through 6 traffic tickets on any given day:

Number of traffic tickets	0	1	2	3	4	5	6
Probability	0.05	0.16	0.26	0.25	0.15	0.09	0.04

Find the probability that on a given day the policeman will issue
(a) at least three traffic tickets;
(b) more than five traffic tickets;
(c) no more than two traffic tickets;
(d) fewer than two traffic tickets.

R.71 Among the 10 applicants for four assistantships in an English department are 5 men and 5 women. In how many ways can these vacancies be filled
(a) with any 4 of the applicants;
(b) with any 4 of the female applicants;
(c) with 2 male applicants and 2 female applicants?

R.72 In Figure R.2, E, T, and N are the events that a car brought to a garage needs an engine overhaul, transmission repairs, or new tires. Express in words the events represented by
(a) region 1;   (d) regions 1 and 4 together;
(b) region 3;   (e) regions 2 and 5 together;
(c) region 7;   (f) regions 3, 5, 6, and 8 together.

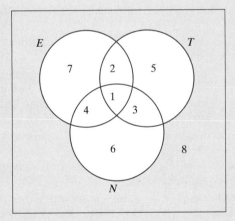

FIGURE R.2 Venn diagram for Exercise R.72.

R.73 With reference to Exercise R.72 and the Venn diagram of Figure R.2, list the regions or combinations of regions that represent the events that a car brought to the garage needs
(a) transmission repairs, but neither an engine overhaul nor new tires;
(b) an engine overhaul and transmission repairs;
(c) transmission repairs or new tires, but not an engine overhaul;
(d) new tires.

R.74 Explain why there must be a mistake in each of the following statements.
(a) The probability that an experiment will succeed is 0.95, and the probability that it will not succeed is −0.5.
(b) The probability that he will buy a Chevrolet is 0.42, the probability that he will buy a Ford is 0.33, and the probability that he will buy one or the other is 0.85.
(c) The probabilities that a student will get a grade of A, B, C, D, or F are, respectively, 0.10, 0.25, 0.40, 0.25, and 0.10.

(d) The probability that she will win a tennis match is 0.10, but the probability that she will lose is 10 times as great.

R.75 What is the probability of an event if the odds that it will occur are
  (a) 4 to 1;
  (b) 3 to 2;
  (c) 1 to 1.

R.76 An automobile salesman claims that the odds are 7 to 2 that a new-car customer will want air-conditioning, 1 to 2 that he or she will want power windows, and 8 to 1 that he or she will want at least one of these features. Find the odds that a new-car customer will want both air-conditioning and power windows.

R.77 A hospital wants to buy new magazines for several of its waiting rooms, and the mutually exclusive probabilities are 0.10, 0.20, 0.30, 0.13, 0.15, and 0.12 that it will buy *Better Homes and Gardens*, *National Geographic*, *Reader's Digest*, *People*, *Time*, or *Newsweek*, respectively. If these are their only choices, find the probabilities that the hospital will buy
  (a) *National Geographic* or *Time*;
  (b) *Better Homes and Gardens*, *People*, or *National Geographic*;
  (c) *Reader's Digest* or *Newsweek*;
  (d) *National Geographic*, *People*, or *Newsweek*.

R.78 If E is the event that an applicant for a home mortgage is employed, G is the event that he has a good credit rating, and A is the event that the application is approved, state in words the probabilities expressed by
  (a) $P(A|E)$;   (c) $P(A'|E')$;
  (b) $P(A|G)$;   (d) $P(A|E \cap G)$.

R.79 The probability that a woman shopping in a certain department store will visit the women's apparel department is 0.35, the probability that she will visit the children's apparel department is 0.23, and the probability that she will visit both is 0.09. What is the probability that a woman shopping will visit at least one of these departments?

R.80 David is willing to bet $20 against $30 that the Phoenix Cardinals will win their next football game, but he is not willing to bet $24 against $32. What does this tell us about his subjective probability that the Cardinals will win their next game?

R.81 Given $P(A) = 0.65$, $P(B) = 0.20$, and $P(A \cap B) = 0.13$, verify that
  (a) $P(A|B) = P(A)$;   (c) $P(B|A) = P(B)$;
  (b) $P(A|B') = P(A)$;   (d) $P(B|A') = P(B)$.

6 Probability Distributions

If a teacher records how many students are absent, or a social scientist determines what percentage of a panel of jurors is over 60 years old, or a geologist measures the hardness of a rock, or an engineer calculates the projected cost of a new engine, they are in each case concerned with a number that is associated with an *element of chance*, namely, with the value of a **random variable**. A random variable is a quantity that can take on the different values of a given set with specified probabilities. This definition may not be as rigorous as it could be since, strictly speaking, a random variable is a function. Most beginning students, however, find it easiest to think of random variables simply as quantities that can take on different values depending on chance. In addition to the examples cited, random variables can include such things as the number of speeding tickets issued each day on a freeway between two cities, the annual production of coffee in Brazil, the wind velocity at Kennedy airport, the size of the attendance at a baseball game, and the number of mistakes a person makes computing his or her income tax.

Random values are usually classified according to the number of values that they can assume. In this chapter we shall limit our discussion to **discrete random variables**, which can take on only a finite number of values, or a countable infinity of values (as many as there are whole numbers). For instance, the number of points that we roll with a pair of dice is a discrete random variable that can take on only the finite, or fixed, number of values 2, 3, 4, 5, 6, 7, 8, 9, 10, 11, or 12. In contrast, the number of the flip on which a coin comes up heads for the first time is a discrete random variable, which can take on the countable infinity of values 1, 2, 3, 4, 5, ... It is possible, though highly unlikely, that a coin will come up tails a million times, a billion times, or even more, before it finally comes up heads.

In the study of discrete random variables, we are usually interested primarily in the probabilities associated with all their values, that is, in their **probability distributions**. A general introduction to probability distributions in Section 6.1 is followed by discussions of various special probability distributions in Sections 6.2 through 6.5, and a description of their most important properties is given in Sections 6.6 through 6.8.

There are also **continuous random variables**, whose values are determined by means of continuous measuring scales (yardsticks, thermometers, stopwatches, and so on). These will be taken up later in Chapter 7.

**6.1
Probability
Distributions**

The tables in the two examples that follow illustrate what we mean by the probability distribution of a random variable—it is a correspondence that assigns probabilities to its values.

■ **EXAMPLE** Construct a table showing the probabilities of rolling a 1, 2, 3, 4, 5, or 6 with a balanced die.

Solution Since *balanced* means that the outcomes are all equiprobable, each value has the probability $\frac{1}{6}$, and we get

Number of points rolled with a die	Probability
1	$\frac{1}{6}$
2	$\frac{1}{6}$
3	$\frac{1}{6}$
4	$\frac{1}{6}$
5	$\frac{1}{6}$
6	$\frac{1}{6}$

■

■ **EXAMPLE** Construct a table showing the probabilities of getting 0, 1, 2, or 3 heads in three flips of a balanced coin.

Solution The eight equally likely possibilities are HHH, HHT, HTH, THH, HTT, THT, TTH, and TTT, where H stands for heads and T for tails. Counting the number of heads in each case and using the formula $\frac{s}{n}$ for equiprobable outcomes, we obtain the following probability distribution for the total number of heads:

Number of heads	Probability
0	$\frac{1}{8}$
1	$\frac{3}{8}$
2	$\frac{3}{8}$
3	$\frac{1}{8}$

■

Whenever possible, we try to express probability distributions by means of mathematical formulas that enable us to calculate the probabilities associated with the various values of a random variable.[†] For instance, for the number of points we roll with a balanced die, we can write

$$f(x) = \frac{1}{6} \quad \text{for } x = 1, 2, 3, 4, 5, \text{ and } 6$$

where $f(1)$ represents the probability of rolling a 1, $f(2)$ represents the probability of rolling a 2, and so on, in the usual functional notation. Most of the time we write $f(x)$ for the probability that a random variable takes on the value x, but we could as well write

$$g(x), \quad h(x), \quad p(x), \ldots$$

■ **EXAMPLE**   Verify that the probability distribution of the number of heads obtained in three flips of a balanced coin is given by

$$f(x) = \frac{\binom{3}{x}}{8} \quad \text{for } x = 0, 1, 2, \text{ and } 3$$

Solution   Looking up the binomial coefficients in Table X, we find that

$$\binom{3}{0} = 1, \quad \binom{3}{1} = 3, \quad \binom{3}{2} = 3, \quad \text{and} \quad \binom{3}{3} = 1$$

Thus, the probabilities for $x = 0, 1, 2,$ and 3 are $\frac{1}{8}, \frac{3}{8}, \frac{3}{8},$ and $\frac{1}{8}$, and this agrees with the results obtained previously.   ■

Note that each probability in these examples is between 0 and 1, inclusive, and that the sum of the probabilities is in each case equal to 1. These facts suggest the following two general rules, which apply to any probability distribution:

1. Since the values of a probability distribution are probabilities, they must be numbers between 0 and 1 inclusive.

2. Since a random variable has to take on one of its values, the sum of all the values of a probability distribution must be equal to 1.

[†] It would be necessary, for example, when a random variable can take on infinitely many different values, and we cannot possibly write down all the probabilities.

■ **EXAMPLE**   Check whether the following function can serve as the probability distribution of an appropriate random variable:

$$f(x) = \frac{x+2}{12} \quad \text{for } x = 1, 2, \text{ and } 3$$

Solution   Substituting $x = 1, 2,$ and 3, we get

$$f(1) = \frac{3}{12}, \quad f(2) = \frac{4}{12}, \quad \text{and} \quad f(3) = \frac{5}{12}$$

Since none of these values is negative or greater than 1, and since their sum is $\frac{3}{12} + \frac{4}{12} + \frac{5}{12} = 1$, the given function can serve as the probability distribution of a random variable.   ■

6.2
The Binomial Distribution

In many applied problems we are interested in the probability that an event will occur x times out of n. For instance, we may be interested in the probability of getting 75 responses to 200 mail questionnaires sent out as part of a sociological survey, or the probability that 12 of 50 tagged wild turkeys will be recaptured, or the probability that 45 of 300 drivers stopped at a road block will be wearing their seatbelts, or the probability that 66 of 200 television viewers (interviewed by a market research organization) will recall the products advertised on a given program. To borrow from the language of games of chance, we could say that in each of these examples we are interested in the probability of getting "x successes in n trials" or, in other words, "x successes and $n - x$ failures in n attempts." In the problems that we study in this section, we shall always make the following assumptions:

1. *There is a fixed number of trials, each resulting in success or failure.*
2. *The probability of a success is the same for each trial.*
3. *The trials are all independent.*

This means that the theory that we develop will not apply, for example, if we are interested in the number of dresses a woman may try on before she buys one (where the number of trials is not fixed), or if we check every hour whether traffic is congested at a certain intersection (where the probability of "success" is not constant), or if we are interested in the number of times that a person voted for the Republican candidate in the last five presidential elections (where the trials are not independent). As an example of a situation in which the three conditions apply, consider the plight of a student who is reduced to pure guessing on eight questions of a multiple-choice test. Suppose that each question has five choices, exactly one of which is correct. If C denotes a correct answer and I an incorrect answer, one way in which the student can get six correct answers and two incorrect answers is given by the sequence of letters

$$C\,C\,C\,C\,C\,C\,I\,I$$

Here, the first six questions are answered correctly and the last two are answered incorrectly. Since $P(C) = \frac{1}{5}$ and $P(I) = \frac{4}{5}$, and the trials are all independent, the probability of this particular sequence of correct and incorrect answers is

$$\frac{1}{5} \cdot \frac{1}{5} \cdot \frac{1}{5} \cdot \frac{1}{5} \cdot \frac{1}{5} \cdot \frac{1}{5} \cdot \frac{4}{5} \cdot \frac{4}{5} = \left(\frac{1}{5}\right)^6 \left(\frac{4}{5}\right)^2$$

However, there are many other sequences of six C's and two I's, for instance, *CCIICCCC, ICICCCCC,* and *CCICCCIC,* and, by the same argument as before, each has the probability

$$\left(\frac{1}{5}\right)^6 \left(\frac{4}{5}\right)^2$$

To get the probability of six C's and two I's in any order, we must count the number of ways of arranging six C's and two I's (the number of combinations of eight objects taken six at a time) and multiply by $\left(\frac{1}{5}\right)^6 \left(\frac{4}{5}\right)^2$. Thus, the result is given by

$$\binom{8}{6} \left(\frac{1}{5}\right)^6 \left(\frac{4}{5}\right)^2 = 0.00114688 \approx 0.0011$$

This suggests that, in general, if n is the number of trials, p is the probability of a success for each trial, and the trials are all independent, then the probability of x successes in n trials is

Binomial distribution

$$f(x) = \binom{n}{x} p^x (1 - p)^{n-x} \quad \text{for } x = 0, 1, 2, \ldots, \text{ or } n$$

It is customary to say here that the number of successes in n trials is a random variable having the **binomial probability distribution**, or simply the **binomial distribution**. The binomial distribution is called by this name because, for $x = 0, 1, 2, \ldots,$ and n, the values of the probabilities are the $n + 1$ successive terms of the binomial expansion of $\left[(1 - p) + p\right]^n$. Since

$$\left[(1 - p) + p\right]^n = 1^n = 1$$

the probabilities clearly sum to 1.

■ **EXAMPLE**   If the probability is 0.70 that a student with very high grades will get into law school, what is the probability that exactly three of five students with very high grades will get into law school?

Solution   Substituting $x = 3$, $n = 5$, and $p = 0.70$ into the formula for the binomial distribution, we get

$$f(3) = \binom{5}{3} (0.70)^3 (1 - 0.70)^{5-3}$$

$$= 10(0.70)^3 (0.30)^2 = 0.3087 \approx 0.31 \qquad ■$$

The following is an example in which we calculate all the probabilities of a binomial distribution:

■ **EXAMPLE**   The probability is 0.60 that a person shopping at a certain market will spend at least $50.

(a) Find the probability that, among five persons shopping at this market, none of them will spend at least $50. Similarly, find the separate probabilities that 1, 2, 3, 4, or 5 will spend at least $50.

(b) Use the results of part (a) to draw a histogram of the binomial distribution with $n = 5$ and $p = 0.60$.

Solution   (a) Substituting $n = 5$, $p = 0.60$, and $x = 0$ into the formula for the binomial distribution, we find the probability that none of them will spend at least $50. This probability is

$$f(0) = \binom{5}{0}(0.60)^0(1 - 0.60)^{5-0} = 0.01024 \approx 0.010$$

Similarly using $x = 1, 2, 3, 4$, and 5, we have

$$f(1) = \binom{5}{1}(0.60)^1(1 - 0.60)^{5-1} \approx 0.077$$

$$f(2) = \binom{5}{2}(0.60)^2(1 - 0.60)^{5-2} \approx 0.230$$

$$f(3) = \binom{5}{3}(0.60)^3(1 - 0.60)^{5-3} \approx 0.346$$

$$f(4) = \binom{5}{4}(0.60)^4(1 - 0.60)^{5-4} \approx 0.259$$

$$f(5) = \binom{5}{5}(0.60)^5(1 - 0.60)^{5-5} \approx 0.078$$

(b) The histogram is shown in Figure 6.1.   ■

In actual practice, binomial probabilities are seldom obtained by direct substitution into the formula. Sometimes we use the approximations discussed later in this chapter and in Chapter 7. In many cases we can refer to special tables such as Table I at the end of this book or the more detailed tables listed in the Bibliography. Table I gives the binomial probabilities for $n = 2$ to $n = 15$ and $p = 0.05, 0.1, 0.2, 0.3, 0.4, 0.5, 0.6, 0.7, 0.8, 0.9$, and 0.95, all rounded to three decimals. Values omitted in the table are 0.0005 or less and, therefore, 0.000 rounded to three decimals.

There are also specially programmed statistical calculators and an abundance of computer programs that give binomial probabilities; because these can deal with nearly any values of n and p, they turn out to be more useful than tables.

Many of the situations in which binomial random variables are used involve descriptions like *at most, at least, not more than, less than, more than,* and so on. Care is needed in dealing with these descriptions. For instance, *less than* 7 refers to the set of values

0, 1, 2, 3, 4, 5, 6

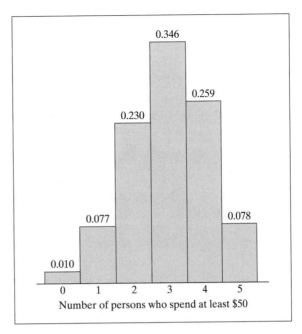

FIGURE 6.1 Histogram of binomial distribution with $n = 5$ and $p = 0.60$.

while *not more than* 7 refers to the set of values

$$0, \quad 1, \quad 2, \quad 3, \quad 4, \quad 5, \quad 6, \quad 7$$

■ **EXAMPLE**   If the probability is 0.40 that a divorcee will remarry within three years, find the probabilities that of ten divorcees

(a) at most three will remarry within three years;

(b) at least seven will remarry within three years;

(c) from two to five will remarry within three years;

(d) at least two will remarry within three years.

Solution   (a) From Table I for $n = 10$, the column headed 0.4, and the rows corresponding to $x = 0$, 1, 2, and 3, the probabilities are 0.006, 0.040, 0.121, and 0.215; thus, the probabilty that at most three of ten divorcees will remarry within three years is

$$0.006 + 0.040 + 0.121 + 0.215 = 0.382$$

(b) From Table I for $n = 10$, the column headed 0.4, and the rows corresponding to $x = 7$, 8, 9, and 10, the probabilities are 0.042, 0.011, 0.002, and 0.000; thus, the probability that at least seven of ten divorcees will remarry within three years is

$$0.042 + 0.011 + 0.002 + 0.000 = 0.055$$

(c) From Table I for $n = 10$, the column headed 0.4, and the rows corresponding to $x = 2$, 3, 4, and 5, the probabilities are 0.121, 0.215, 0.251, and 0.201; thus, the probability that from two to five of ten divorcees will remarry within three years is

$$0.121 + 0.215 + 0.251 + 0.201 = 0.788$$

(d) From Table I for $n = 10$, the column headed 0.4, and the rows corresponding to $x = 0$ and 1, the probabilities are 0.006 and 0.040; thus, the probability that at least two of ten divorcees will remarry within three years is

$$1 - (0.006 + 0.040) = 0.954$$ ∎

If the probability that a divorcee will remarry within three years had been 0.42 instead of 0.40, we could not have worked the preceding example by using Table I. In general, if n is greater than 15 and/or p takes on a value other than 0.05, 0.1, 0.2, …, 0.9, or 0.95, we will have to use a more detailed table (see the Bibliography), refer to the formula for the binomial distribution, or employ a computer as in the following example.

■ **EXAMPLE**   Use the computer printout of Figure 6.2 to rework the preceding example with $p = 0.42$ instead of $p = 0.40$.

Solution   (a) Adding the probabilities in the P(X = K) column corresponding to 0, 1, 2, and 3, we get

$$0.0043 + 0.0312 + 0.1017 + 0.1963 = 0.3335$$

Note, however, that since the printout also gives the cumulative *less than or equal* probabilities, the answer is the value in the P(X LESS OR = K) column corresponding to K = 3.

(b) Adding the probabilities in the P(X = K) column corresponding to 7, 8, 9, and 10, we get

$$0.0540 + 0.0147 + 0.0024 + 0.0002 = 0.0713$$

Since the probability of "at least 7" is 1 minus the probability of "6 or less," we get $1 - 0.9288 = 0.0712$, where 0.9288 is the value in the P(X LESS OR = K) column corresponding to K = 6. The difference between the two results is due to rounding.

(c) Adding the probabilities in the P(X = K) column corresponding to 2, 3, 4, and 5, we get

$$0.1017 + 0.1963 + 0.2488 + 0.2162 = 0.7630$$

Since the probability of "2 to 5" equals the difference between the probabilities of "5 or less" and "1 or less," the values in the P(X LESS OR = K) column yield $0.7984 - 0.0355 = 0.7629$. Again, the difference between the two results is due to rounding.

```
MTB > PDF;
SUBC> BINOMIAL N = 10 P = 0.42.


    BINOMIAL WITH N = 10 P = 0.420000
      K                 P( X = K)
      0                   0.0043
      1                   0.0312
      2                   0.1017
      3                   0.1963
      4                   0.2488
      5                   0.2162
      6                   0.1304
      7                   0.0540
      8                   0.0147
      9                   0.0024
     10                   0.0002


MTB > CDF;
SUBC> BINOMIAL N = 10 P = 0.42.


    BINOMIAL WITH N = 10 P = 0.420000
      K            P( X LESS OR = K)
      0                   0.0043
      1                   0.0355
      2                   0.1372
      3                   0.3335
      4                   0.5822
      5                   0.7984
      6                   0.9288
      7                   0.9828
      8                   0.9975
      9                   0.9998
     10                   1.0000
```

FIGURE 6.2 MINITAB printout for the binomial distribution with $n = 10$ and $p = 0.42$.

(d) Subtracting from 1 the probability of "1 or less," we get

$$1 - (0.0043 + 0.0312) = 0.9645$$

from the column headed P(X = K), and $1 - 0.0355 = 0.9645$ from the column P(X LESS OR = K). In this case, the results are the same. ∎

When we observe a value of a random variable having the binomial distribution—for instance, when we observe the number of heads in 50 flips of a coin, or the number of seeds (in a package of 24 seeds) that germinate, or the number of students (among 200 interviewed) who are opposed to an increase in tuition, or the number of automobile accidents (among 100 investigated) that are due to drunk driving—we say that we are **sampling a binomial population**. This terminology is widely used in statistics. Computers can also be used to simulate sampling binomial populations, that is, to simulate observations of random variables having given binomial distributions. In fact, that is what we did in the illustration on page 126, which illustrated the Law of Large Numbers. In Figure 4.3, instructions "RANDOM 100 C1;" and "INTEGERS 0 1." tell the computer to take 100 random observations of a random variable having the binomial distribution with $n = 1$ and $p = 0.50$.

Exercises

Exercises 6.1, 6.5, 6.6, 6.7, and 6.18 are practice exercises; their complete solutions are given on page 229.

6.1 In each case, determine whether the given values can be looked upon as the values of a probability distribution of a random variable that can take on the values 1, 2, and 3, and explain your answers.
  (a) $f(1) = 0.37, f(2) = 0.35, f(3) = 0.30$;
  (b) $f(1) = \frac{4}{9}, f(2) = \frac{4}{9}, f(3) = \frac{1}{9}$;
  (c) $f(1) = 0.57, f(2) = -0.59, f(3) = 0.16$.

6.2. Check in each case whether the given function can serve as the probability distribution of an appropriate random variable.

  (a) $f(x) = \dfrac{1}{5}$ for $x = 0, 1, 2, 3, 4, 5$;

  (b) $f(x) = \dfrac{x^2 - 1}{25}$ for $x = 0, 1, 2, 3, 4$;

  (c) $f(x) = \dfrac{x^2}{30}$ for $x = 0, 1, 2, 3, 4.$

6.3 In each case determine whether the given values can be looked upon as the values of a probability distribution of a random variable that can take on the values 1, 2, 3, and 4. Explain your answers.
  (a) $f(1) = 0.25, f(2) = 0.35, f(3) = 0.35, f(4) = 0.10$;
  (b) $f(1) = 0.01, f(2) = 0.02, f(3) = 0.03, f(4) = 0.90$;
  (c) $f(1) = 0.90, f(2) = 0.05, f(3) = 0.03, f(4) = 0.02$;
  (d) $f(1) = -0.35, f(2) = -0.25, f(3) = -0.20, f(4) = -0.20.$

6.4. Check in each case whether the given function can serve as the probability distribution of an appropriate random variable.

  (a) $f(x) = \dfrac{\binom{2}{x}}{4}$ for $x = 0, 1, 2$;

  (b) $f(x) = \dfrac{x - 2}{9}$ for $x = 1, 2, 3, 4, 5, 6$;

  (c) $f(x) = \dfrac{x^2 - 6x + 9}{10}$ for $x = 1, 2, 3, 4, 5$;

  (d) $f(x) = \dfrac{x^2 - 6x + 8}{5}$ for $x = 1, 2, 3, 4, 5.$

6.5 According to the Bureau of Labor Statistics, about 10% of the civilian labor force (both men and women) were 20 to 24 years of age in 1996. Using this figure and the formula for the binomial distribution, calculate the probabilities that for a randomly selected group of six members of the civilian labor force in 1996

(a) exactly two were 20 to 24 years of age;

(b) fewer than two were 20 to 24 years of age.

6.6 According to the College Board, 70% of high school graduates in the Commonwealth of Massachusetts took the SAT examination in a certain year. Find the probability that two out of four randomly selected high school graduates in Massachusetts took the SAT examination

(a) using the formula for the binomial distribution;

(b) referring to Table I.

6.7 A civil service examination is designed so that 80% of all high school graduates can pass. Use Table I to find the probabilities that among 14 high school graduates

(a) at least 12 will pass the test;   (b) at most 10 will pass the test.

6.8 In a given city, medical expenses are given as the reason for 70% of all personal bankruptcies. Use Table I to find the probabilities that in the next four personal bankruptcies filed in that city

(a) medical expenses will be given as the reason in three of the cases;

(b) medical expenses will be given as the reason in at least three of the cases.

6.9 If the probability is 0.20 that a set of tennis will go into a tie breaker, what is the probability that two of three sets will go into tie breakers?

6.10 A social scientist claims that only 70% of all high school seniors capable of doing college work actually go to college. If this is so, use the formula for the binomial distribution to calculate the probability (rounded to three decimals) that among eight high school seniors capable of doing college work, only four will go to college.

6.11 If the probability is 0.15 that a person at a license plate renewal line in a state motor vehicle office will pay extra for a special commemorative license plate, find the probability (rounded to three decimals) that only 2 of 20 persons, selected at random from this line, will pay extra for the special commemorative license plate.

6.12 If it is true that 80% of all industrial accidents can be prevented by paying strict attention to safety regulations, find the probability that five of seven industrial accidents can thus be prevented

(a) by using the formula for the binomial distribution;

(b) by referring to Table I.

6.13 According to the U.S. Bureau of Labor Statistics, approximately 20% of workers had five to nine years of tenure with their current employer in 1966. Using this value, find the probability that one of nine randomly selected workers had five to nine years of tenure with their employer

(a) by using the formula for the binomial distribution;

(b) by referring to Table I.

6.14 During the year 1995, 70% of Continental Airline's flights arrived on time (within 15 minutes of scheduled time). Use Table I to find the probabilities that among 10 randomly selected flights

(a) at least five arrived on time;

(b) at most five arrived on time;

(c) anywhere from four to six flights arrived on time.

6.15 A study conducted at a certain college shows that 60% of the school's graduates obtain a job in their chosen field within a year after graduation. Use Table I to find the probabilities that, within a year after graduation, among 14 randomly selected graduates of that college

(a) at least 6 will find a job in their chosen field;

(b) at most 3 will find a job in their chosen field;

(c) anywhere from 5 through 8 will find a job in their chosen field.

6.16 Research shows that 30% of all women taking a certain medication do not respond favorably. Use Table I to find the probabilities that among nine randomly selected women taking the medication

(a) at least four will respond favorably;

(b) at least four will not respond favorably.

6.17 A study has shown that 50% of the families in a certain large area have at least two cars. Find the probabilities that among 15 families randomly selected in this area

(a) 7 have at least two cars;

(b) more than 8 have at least two cars;

(c) 6 have fewer than two cars;

(d) more than 9 have fewer than two cars.

6.18 A quality control engineer wants to check whether (in accordance with specifications) 95% of the products shipped are in perfect condition. To this end, he randomly selects 10 items from each large lot ready to be shipped and passes it only if they are all in perfect condition; otherwise, each item in the lot is checked. Use Table I to find the probabilities that he will commit the error of

(a) holding a lot for further inspection even though 95% of the items are in perfect condition;

(b) letting a lot pass through even though only 90% of the items are in perfect condition;

(c) letting a lot pass through even though only 80% of the items are in perfect condition.

6.19 A food distributor claims that 80% of her 6-ounce cans of mixed nuts contain at least three pecans. To check on this, a consumer testing service decides to examine six of these 6-ounce cans of mixed nuts from a very large production lot and reject the claim if fewer than four of them contain at least three pecans. Use Table I to find the probabilities that the testing service will commit the error of

(a) rejecting the claim even though it is true;

(b) not rejecting the claim when in reality only 60% of the cans of mixed nuts contain at least three pecans;

(c) not rejecting the claim when in reality only 40% of the cans of mixed nuts contain at least three pecans.

6.20 With reference to Exercise 6.15, suppose that the study had shown that 61% of the school's graduates obtain a job in their chosen field within a year after graduation. Use a computer printout of the binomial distribution with $n = 14$ and $p = 0.61$ to rework the three parts of that exercise.

6.21 With reference to Exercise 6.16, suppose that the study had shown that 33% of all women taking the medication do not respond favorably. Use a computer printout of the binomial distribution with $n = 9$ and $p = 0.33$ to rework both parts of that exercise.

6.22 With reference to Exercise 6.17, suppose that the study had shown that 49% of the families in that area have at least two cars. Use a computer printout of the binomial distribution with $n = 15$ and $p = 0.49$ to rework all four parts of that exercise.

6.23 A florist needs 12 exotic potted ferns for a floral exhibition to take place in six months. He knows that the probability is 0.70 that a fern planted now will be suitable for the exhibition in six months. His concern is the minimum number of ferns that he should plant if he wants the probability of having 12 suitable ferns in six months to be at least 0.90.

    (a) Will 15 be enough? If he plants 15 ferns, find the probability that he will have at least 12 suitable ones.

    (b) Will 20 be enough? If he plants 20 ferns, find the probability that he will have at least 12 suitable ones.

    (c) Will 25 be enough? If he plants 25 ferns, find the probability that he will have at least 12 suitable ones.

    (d) Find the smallest number of ferns for which the probability of 12 or more suitable ones is 0.90 (or more).

6.3 The Hypergeometric Distribution

To illustrate another important kind of probability distribution, let us consider the following problem. A factory ships certain tape recorders in lots of 16, and when they arrive at their destination, three are randomly selected from each lot for inspection. Now suppose that four of the tape recorders in a lot are defective, and we are interested in the probability that among the three tape recorders selected, two will be all right and one will be defective. In other words, we are interested in the probability of "one success (defective tape recorder) in three trials," and we might be tempted to argue that since 4 of the 16 tape recorders are defective, the probability of x objects of type A and $n - x$ objects of type B is

$$f(1) = \binom{3}{1}\left(\frac{1}{4}\right)^1\left(1 - \frac{1}{4}\right)^{3-1} \approx 0.422$$

This result, obtained by means of the formula for the binomial distribution, would be correct if sampling is with replacement, that is, if each tape recorder is replaced in the lot before the next one is selected. In actual practice, we seldom, if ever, sample with replacement. To obtain the correct answer for our problem when sampling is without replacement, we might argue as follows. There are altogether $\binom{16}{3}$ ways of choosing 3 of the 16 tape recorders, and they are all equiprobable by virtue of the assumption that the selection is random. Since there are $\binom{4}{1}$ ways of choosing 1 of the defective tape recorders and $\binom{12}{2}$ ways of choosing 2 that are not defective, there are $\binom{4}{1} \cdot \binom{12}{2}$ ways of getting "1 success (defective tape recorder) in 3 trials." Thus, it follows by the special formula $\frac{s}{n}$ for equiprobable outcomes that the desired probability is

$$\frac{\binom{4}{1}\binom{12}{2}}{\binom{16}{3}} = \frac{4 \cdot 66}{560} \approx 0.471$$

This suggests that, in general, if n objects are chosen at random from a set consisting of a objects of type A and b objects of type B, and the selection is without replacement, then the probability of "x of type A and $n - x$ of type B" is

Hypergeometric distribution

$$f(x) = \frac{\binom{a}{x} \cdot \binom{b}{n-x}}{\binom{a+b}{n}} \quad \text{for } x = 0, 1, 2, \ldots, \text{or } n$$

where x cannot exceed a and $n - x$ cannot exceed b, since we cannot get more successes (or failures) than there are in the whole set. Note also that x cannot exceed n. This is the formula for the **hypergeometric distribution**.

■ **EXAMPLE**   A secretary is supposed to send 6 of 15 overseas letters by airmail, but she gets them all mixed up and randomly puts airmail stamps on 6 of the letters. What is the probability that only 3 of the letters that should go by airmail get an airmail stamp?

Solution   Since this problem involves sampling without replacement, the situation requires the formula for the hypergeometric distribution. Substituting $a = 6$, $b = 9$, $n = 6$, and $x = 3$ into this formula, we get

$$f(3) = \frac{\binom{6}{3}\binom{9}{3}}{\binom{15}{6}} = \frac{20 \cdot 84}{5,005} \approx 0.336 \qquad ■$$

The following is an example in which we calculate all the probabilities of a hypergeometric distribution.

■ **EXAMPLE**   Among a department store's 16 delivery trucks, 5 have worn brakes. If 3 of the trucks are randomly picked for a general overhaul, find the separate probabilities that these 3 will include 0, 1, 2, or 3 of the trucks with worn brakes.

Solution   Again, the problem involves sampling without replacement. Thus, substituting $a = 5$, $b = 11$, $n = 3$, and $x = 0, 1, 2,$ and 3 into the formula for the hypergeometric distribution, we get

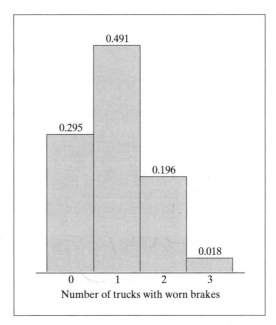

FIGURE 6.3 Histogram of hypergeometric distribution
with $a = 5$, $b = 11$, and $n = 3$.

$$f(0) = \frac{\binom{5}{0}\binom{11}{3}}{\binom{16}{3}} = \frac{1 \cdot 165}{560} \approx 0.295$$

$$f(1) = \frac{\binom{5}{1}\binom{11}{2}}{\binom{16}{3}} = \frac{5 \cdot 55}{560} \approx 0.491$$

$$f(2) = \frac{\binom{5}{2}\binom{11}{1}}{\binom{16}{3}} = \frac{10 \cdot 11}{560} \approx 0.196$$

$$f(3) = \frac{\binom{5}{3}\binom{11}{0}}{\binom{16}{3}} = \frac{10 \cdot 1}{560} \approx 0.018$$

A histogram of this probability distribution is shown in Figure 6.3.   ∎

This probability distribution answers many card game questions. Consider the following:

■ **EXAMPLE**   Suppose that 5 cards are dealt, without replacement, from a standard deck of 52. What is the probability that the 5 cards will include exactly two spades?

Solution   Here the spades correspond to type A and the nonspades correspond to type B. Therefore, $a = 13$, $b = 39$, $n = 5$, and $x = 2$. We find that

$$f(2) = \frac{\binom{13}{2}\binom{39}{3}}{\binom{52}{5}} = \frac{78 \cdot 9{,}139}{2{,}598{,}960} \approx 0.274 \qquad ■$$

■ **EXAMPLE**   A bridge hand consists of 13 cards dealt at random from the deck of 52. Find the probabilities that a bridge hand will have $0, 1, 2, \ldots, 13$ spades.

Solution   The probability that a bridge hand will have exactly x spades is

$$f(x) = \frac{\binom{13}{x}\binom{39}{13-x}}{\binom{52}{13}}$$

for $x = 0, 1, 2, \ldots,$ or 13.

This expression could be calculated directly for each value of x, but the task is much more easily done with a computer. As can be seen from the computer output in Figure 6.4, bridge hands commonly (about 73% of the time) have two, three, or four spades, while bridge hands with seven or more spades are very unlikely.   ■

x	Prob[X=x]	x	Prob[X=x]	x	Prob [X=x]	x	Prob[X=x]
0	.012791	5	.124692	10	.000004		
1	.080062	6	.041564	11	.000000		
2	.205873	7	.008817	12	.000000		
3	.286330	8	.001167	13	.000000		
4	.238608	9	.000093				

FIGURE 6.4 MINITAB printout of the hypergeometric distribution with $a = 13$, $b = 39$, and $n = 13$. The program used here is Econometrics Toolkit (ET).

As we have pointed out, the binomial distribution does not apply when sampling is without replacement, but sometimes it can be used as an approximation.

■ **EXAMPLE**   In a federal prison for women, 100 of the 240 inmates have radical political views. If 5 of them are randomly chosen to appear before a legislative committee, find the probability that only 1 of them will have radical political views by using

(a) the formula for the hypergeometric distribution;

(b) the formula for the binomial distribution with $n = 5$ and $p = \frac{100}{240} = \frac{5}{12}$ as an approximation.

Solution   (a) Substituting $a = 100$, $b = 140$, $n = 5$, and $x = 1$ into the formula for the hypergeometric distribution, we get

$$f(1) = \frac{\binom{100}{1}\binom{140}{4}}{\binom{240}{5}}$$

$$= \frac{100 \cdot \dfrac{140 \cdot 139 \cdot 138 \cdot 137}{4!}}{\dfrac{240 \cdot 239 \cdot 238 \cdot 237 \cdot 236}{5!}} \approx 0.2409$$

(b) Substituting $n = 5$, $p = \frac{5}{12}$, and $x = 1$ into the formula for the binomial distribution, we get

$$f(1) = \binom{5}{1}\left(\frac{5}{12}\right)^{1}\left(1 - \frac{5}{12}\right)^{5-1}$$

$$= 5 \cdot \frac{5}{12} \cdot \frac{2,401}{20,736} \approx 0.2412$$

Since the difference between the two values is only 0.0003, the approximation is very good.   ■

Observe that the hypergeometric situation involves a population of finite size $a + b$. The binomial approximation does not give an exact answer to problems with a finite population. It is generally agreed, though, that the approximation illustrated above may be used as long as the sample does not exceed 5% of the population, namely, when

Condition for binomial approximation to hypergeometric distribution

$$n \leq (0.05)(a + b)$$

In our example, $n = 5$ is less than $0.05(100 + 140) = 12$, so that the condition is satisfied. The main advantage of the binomial approximation to the hypergeometric distribution is that the binomial distribution appears much more extensively in published tables and computer programs. Also, the binomial calculations are generally less complicated. Observe that the binomial distribution is described by two parameters (n and p), while the hypergeometric distribution requires three (a, b, and n).

Exercises

Exercises 6.24, 6.25, 6.32, and 6.33 are practice exercises; their complete solutions are given on pages 229 and 230.

6.24 A produce market has 16 tomatoes of which 7 have minor blemishes. If 5 of these are randomly selected for a customer, what is the probability that exactly 3 of them will have minor blemishes?

6.25 Among the 15 used cars on a used car lot, 5 have faulty starters. If 2 of these cars are randomly selected for a newspaper advertisement, find the probabilities that
  (a) neither car will have a faulty starter;
  (b) only one of the cars will have a faulty starter;
  (c) both cars will have faulty starters.

6.26 Find the probability that an Internal Revenue Service auditor will catch two income tax returns with illegitimate deductions if she randomly selects five returns from among twenty returns of which nine contain illegitimate deductions.

6.27 A collection of 15 gold coins contains 4 counterfeits. If 2 of them are randomly selected to be sold at auction, what is the probability that neither of them is a counterfeit?

6.28 A supplier has 6 white mice and 12 black mice, of which 6 are randomly selected for sale to a laboratory. Use the formula for the hypergeometric distribution to determine the probabilities that
  (a) 5 of the 6 mice will be white;
  (b) all 6 of the mice will be white.

6.29 Among 15 applicants for nursing positions at a hospital are 11 registered nurses (who are graduates of three or four-year specialized nursing programs) and 4 practical nurses (who have less training). If 3 of these applicants are randomly selected for interviews, find the probability that 2 of them will be registered nurses and 1 will be a practical nurse.

6.30 To pass a quality control inspection, two cellular phones are randomly chosen from each lot of 10 cellular phones, and the lot is passed only if neither phone is defective. Use the formula for the hypergeometric distribution to determine the probabilities that a lot will
  (a) pass the inspection when only 1 of the 10 phones is defective;
  (b) pass the inspection when 2 of the 10 phones are defective;
  (c) pass the inspection when 3 of the 10 phones are defective;
  (d) fail the inspection when 4 of the 10 phones are defective.

6.31 Among a person's nine pairs of socks, four pairs need mending. If he randomly picks three pairs of these socks to take along on a trip, determine the probabilities that
  (a) none of the socks will need mending;
  (b) one pair will need mending;
  (c) two pairs will need mending;
  (d) all three pairs will need mending.

6.32 Check in each case whether the condition for the binomial approximation to the hypergeometric distribution is satisfied.
  (a) $a = 40$, $b = 160$, and $n = 15$;
  (b) $a = 380$, $b = 120$, and $n = 22$;
  (c) $a = 400$, $b = 240$, and $n = 30$.

6.33 A shipment of 200 burglar alarms contains 6 defectives. If 3 of these burglar alarms are randomly selected and shipped to a customer, find the probability that she will get 1 bad unit using
  (a) the formula for the hypergeometric distribution;
  (b) the binomial distribution as an approximation.

6.34 Check in each case whether the condition for the binomial approximation to the hypergeometric distribution is satisfied.
  (a) $a = 200$, $b = 200$, and $n = 32$;
  (b) $a = 150$, $b = 180$, and $n = 12$;
  (c) $a = 100$, $b = 120$, and $n = 14$.

6.35 Among 90 cities that a professional group is considering for its next three annual conventions, 36 are in the western part of the United States. If, to avoid arguments, the selection is left to chance (that is, random sampling without replacement), what is the probability that none of the three conventions will be held in the western part of the United States? How big an error would we make if we approximated this probability with the binomial probability of "zero successes in three trials" when $p = \frac{36}{90} = 0.40$?

6.36 Among the 120 employees of a company, 45 are members of a clerical workers union, and the others are members of the typesetters union. If 5 of the employees are chosen by lot to serve on a grievance committee, find the probability that 2 of them will be members of the clerical workers union and the other 3 members of the typesetters union, using

   (a) the formula for the hypergeometric distribution;
   (b) the binomial distribution as an approximation.

6.37 A commercial hatchery has 240 White Plymouth Rock chicks and 60 Rhode Island Red chicks. Use the binomial distribution (and Table I) to approximate the probability that if a poultryman randomly selects 12 chicks for examination, only one will be a Rhode Island Red.

6.4
The Poisson Distribution

When n is large and p is small, binomial probabilities are often approximated by means of the formula

Poisson approximation to binomial distribution

$$f(x) = e^{-np} \frac{(np)^x}{x!} \quad \text{for } x = 0, 1, 2, 3, \ldots$$

which is a special form of the **Poisson distribution**; the more general form is given on page 214. It is difficult to state precisely what we mean by "n is large and p is small," and there are other situations where the approximation may be used; but we shall use it here only when n is at least 100 and np is less than 10. Symbolically,

$$n \geq 100 \quad \text{and} \quad np < 10$$

To get some idea of the closeness of the Poisson approximation to the binomial distribution, consider the computer printout of Figure 6.5, which gives the binomial distribution with $n = 100$ and $p = 0.015$ and the corresponding Poisson distribution with $np = 100(0.015) = 1.5$.[†] Comparing the probabilities in the columns headed P(X = K), we find that the maximum difference, corresponding to K = 0, is

$$0.2231 - 0.2206 = 0.0025$$

In the formula for the Poisson distribution, e is the number 2.71828... used in connection with natural logarithms, and the values of $e^{-\lambda}$ may be obtained from Table XI. Most scientific calculators have keys that will calculate exponentials, and these are generally more convenient, and often more

---

[†] In the computer printout of Figure 6.5, the quantity MU = 1.5 is referred to as the **mean of the Poisson distribution**. This will be explained on pages 214 and 221.

```
MTB > PDF;
SUBC> BINOMIAL N = 100 P = 0.015

   BINOMIAL WITH N = 100 P = 0.015000
   K                P( X = K)
   0                 0.2206
   1                 0.3360
   2                 0.2532
   3                 0.1260
   4                 0.0465
   5                 0.0136
   6                 0.0033
   7                 0.0007
   8                 0.0001
   9                 0.0000

MTB > PDF;
SUBC> POISSON MU = 1.5

   POISSON WITH MEAN = 1.500
   K                P( X = K)
   0                 0.2231
   1                 0.3347
   2                 0.2510
   3                 0.1255
   4                 0.0471
   5                 0.0141
   6                 0.0035
   7                 0.0008
   8                 0.0001
```

FIGURE 6.5 MINITAB printout of the binomial distribution with $n = 100$ and $p = 0.015$ and the Poisson distribution with mean 1.5.

precise, than tables. For the Poisson distribution, x can take on the countably infinite set of values $0, 1, 2, 3, \ldots$; but this poses no problems, since the probabilities usually become negligible after relatively few values of x.

■ **EXAMPLE**   A very large shipment of books contains 2% with defective bindings. Use the Poisson approximation to the binomial distribution to find the probability that among 400 books taken at random from this shipment, only 5 will have defective bindings.

Solution   Substituting $x = 5$ and $np = 400(0.02) = 8$ into the formula for the Poisson distribution and getting $e^{-8} = 0.00034$ from Table XI, we obtain

$$f(5) = 0.00034 \frac{8^5}{5!} = 0.00034 \frac{32{,}768}{120} \approx 0.093$$

A calculator gives $e^{-8} \approx 0.000335462$, resulting in

$$f(5) \approx 0.091604 \approx 0.092$$

■

It would have been possible, though very cumbersome, to use the formula for the binomial distribution in this example.

■ **EXAMPLE**   Records show that the probability is 0.00005 that a car will have a flat tire while driving through a certain tunnel. Use the Poisson approximation to the binomial distribution to find the probabilities that among 10,000 cars passing through this tunnel

(a) at least 2 will have a flat tire;

(b) at most 2 will have a flat tire.

Solution   (a) For this probability, we shall subtract from 1 the probability that 0 or 1 of the cars will have a flat tire. Substituting

$$np = 10,000(0.00005) = 0.5$$

and, respectively, $x = 0$ and $x = 1$ into the formula for the Poisson distribution and getting $e^{-0.5} = 0.607$ from Table XI, we obtain

$$f(0) = 0.607 \frac{0.5^0}{0!} = 0.607$$

$$f(1) = 0.607 \frac{0.5^1}{1!} = 0.304$$

Thus, the answer is $1 - (0.607 + 0.304) = 0.089$.

(b) This probability is $f(0) + f(1) + f(2)$. If we substitute $np = 0.5$, $e^{-0.5} = 0.607$, and $x = 2$ into the formula for the Poisson distribution, we get

$$f(2) = 0.607 \frac{0.5^2}{2!} \approx 0.076$$

Combining this result with the two probabilities calculated in part (a), we find that the answer is

$$0.607 + 0.304 + 0.076 = 0.987$$

A calculator gives $e^{-0.5} \approx 0.606531$. This more precise figure would give 0.090 for (a) and 0.986 for (b).   ■

Since in some cases the hypergeometric distribution can be approximated by a binomial distribution, and the binomial distribution can in some cases be approximated by a Poisson distribution, there will exist situations in which the hypergeometric distribution can be approximated by a Poisson distribution. Consider the following example:

■ **EXAMPLE**   An auditor has been asked to investigate a collection of 4,000 sales invoices, of which 28 contain errors. A sample of 150 invoices is selected. What is the probability that this set of 150 invoices will contain exactly 2 with errors?

Solution   This is a hypergeometric problem with $a = 28$, $b = 3,972$, $n = 150$, and $x = 2$. Since $150 < 0.05(4,000) = 200$, it is reasonable to use the binomial approximation. This binomial distribution has

$$p = \frac{a}{a+b} = \frac{28}{4,000} = 0.007, \quad n = 150, \quad \text{and} \quad x = 2$$

Since $n > 100$ and $np = 150(0.007) = 1.05 < 10$, it would also be reasonable to use the Poisson approximation with $np = 1.05$. This yields

$$f(2) = e^{-1.05}\frac{1.05^2}{2!} \approx 0.3505\frac{1.1025}{2} \approx 0.193$$

where 0.3505 is halfway between 0.368 and 0.333, the values given for $e^{-1.0}$ and $e^{-1.1}$ in Table XI. ■

Since auditors deal frequently with situations with large values of n and even larger values of $a + b$, they find the Poisson distribution to be quite useful. Observe that, aside from x, the hypergeometric distribution involves three parameters (a, b, and n), while the binomial distribution involves two (n and p), and the Poisson distribution involves only one (np). In actual practice, Poisson probabilities are seldom obtained by direct substitution into the formula. Sometimes we refer to tables of Poisson probabilities, which may be found in handbooks of statistical tables, and more frequently we use a computer. For instance, had we used the computer printout of Figure 6.6 in the flat-tire example, we would have found that the answer to part (a) is $1 - 0.9098 = 0.0902$, where 0.9098 is the value in the right-hand column corresponding to K = 1. Similarly, the answer to part (b) is 0.9856, which is the value in the right-hand column corresponding to K = 2. In both instances, the differences between the results obtained here and before are due to rounding. The Poisson distribution also has many important applications that have no direct connection with the binomial distribution. In that case, np is replaced by the parameter λ (Greek lowercase lambda) and we calculate the probability of getting x "successes" by means of the formula

```
MTB > POISSON MU=.5

   POISSON PROBABILITIES FOR MEAN =     .500

   K           P(X = K)            P(X LESS OR = K)
   0             .6065                  .6065
   1             .3033                  .9098
   2             .0758                  .9856
   3             .0126                  .9982
   4             .0016                  .9998
   5             .0002                 1.0000
```

FIGURE 6.6 MINITAB printout of the Poisson distribution with $np = 0.5$.

Poisson distribution (parameter λ)

$$f(x) = e^{-\lambda}\frac{\lambda^x}{x!} \qquad \text{for } x = 0, 1, 2, 3, \ldots$$

where λ is interpreted as the expected, or average, number of successes (see discussion on page 221). This formula applies to many situations where we can expect a fixed number of successes per unit time (or for some other kind of unit), say, when 1.8 accidents can be expected per day at a busy intersection, or when eight small pieces of meat can be expected in a frozen meat pie, or when 4.5 imperfections can be expected per roll of cloth, or when 0.12 complaint per passenger can be expected by an airline, and so on.

■ **EXAMPLE**   If a bank receives on the average $\lambda = 6$ bad checks per day, what is the probability that it will receive 4 bad checks on any given day?

Solution   Substituting into the formula, we get

$$f(4) = e^{-6} \cdot \frac{6^4}{4!} \approx 0.0025 \cdot \frac{1{,}296}{24} \approx 0.135$$

where the value of e^{-6} was read from Table XI. Using $e^{-6} \approx 0.002478$ from a calculator would give the final answer as 0.134.   ■

In this chapter we have given a number of approximations relating the hypergeometric, binomial, and Poisson random variables. This simply means that the numerical values produced by the corresponding formulas will be nearly equal. The probability distributions still apply to very distinct situations: the hypergeometric distribution deals with a sample without replacement from a finite population, the binomial distribution deals with repeated independent trials, and the Poisson distribution deals with counted events with no fixed upper limit. The conditions under which the approximations are reasonable, noted on pages 209 and 211, are conservatively stated. The approximations may perform well even when these conditions are violated, but no promises can be made.

Exercises

Exercises 6.38, 6.40, and 6.46 are practice exercises; their complete solutions are given on page 230.

6.38 Check in each case whether the conditions for the Poisson approximation to the binomial distribution are satisfied.

(a) $n = 300$ and $p = \dfrac{1}{50}$;

(b) $n = 900$ and $p = \dfrac{1}{100}$;

(c) $n = 400$ and $p = \dfrac{1}{25}$.

6.39 Check in each case whether the conditions for the Poisson approximation to the binomial distribution are satisfied.

(a) $n = 90$ and $p = \dfrac{1}{10}$;

(b) $n = 200$ and $p = \dfrac{1}{5}$;

(c) $n = 500$ and $p = \dfrac{1}{60}$.

6.40 It is known from experience that 2% of the calls received by a switchboard are wrong numbers. Use the Poisson approximation to the binomial distribution to determine the probability that among 250 calls received by the switchboard, 4 will be wrong numbers.

6.41 According to 1995 data of the U.S. Immigration and Naturalization Service, 1.9% of foreigners visiting the United States for pleasure were from Australia. Use the Poisson approximation to the binomial distribution to determine the probability, in 1995, that in a random sample of 100 foreign visitors to the United States for pleasure, 3 were Australian.

6.42 According to the U.S. Bureau of Justice, 0.24% of the population of the United States in 1992 were full-time sworn police officers of state and local governments. Use the Poisson approximation to the binomial distribution to determine the probability that in a random sample of 2,000 inhabitants of the United States 4 or 5 were full-time sworn police officers of state and local governments in 1992.

6.43 Mental Health Services Administration data indicates that, in 1995, 1.3% of the population who were 18 to 25 years old were current users of cocaine. Use the Poisson approximation to the binomial distribution to assist drug enforcement authorities in determining the probability that in a random sample of 500 inhabitants who were 18 to 25 years old in 1995, at most 2 were users of the drug at that time.

6.44 An insurance adjuster has been asked to undertake a survey of automobile collision claims. He plans to take a random sample of 200 claims from a computerized record of 8,200 claims. He operates under the premise that exactly 1% of the computerized record consists of blatantly fraudulent claims. He wants to know the probability of getting at most three of these blatantly fraudulent claims in his sample.
  (a) Identify the values of a, b, and n for this hypergeometric situation.
  (b) The binomial approximation is appropriate here. Identify the values of n and p for the binomial approximation.
  (c) Use the Poisson approximation to the binomial distribution, and determine the probability that the sample will have at most three blatantly fraudulent claims.
  (d) Suppose that the adjuster decides to change the 1% premise to 1.5%. For this revised premise, use the same approximation as in part (c) to determine the probability that the sample will have at most three blatantly fraudulent claims.

6.45 The conditions for good approximations, as noted on pages 209 and 211, can be combined.
  (a) Under what conditions on a, b, and n would it be reasonable to approximate a hypergeometric distribution by a Poisson distribution?
  (b) In terms of a, b, and n, what is the value of np to be used in the Poisson calculation?

6.46 The number of minor injuries a football coach can expect during the course of a game is a random variable having the Poisson distribution with $\lambda = 4.4$. Find the probability that during the course of a game there will be at most three minor injuries.

6.47 If an automobile repair shop repairs, on the average, $\lambda = 4$ air conditioners per day, what is the probability that it will repair 5 air conditioners on any given day?

6.48 In the inspection of a fabric produced in continuous rolls, the number of imperfections spotted by an inspector during a 5-minute period is a random variable having the Poisson distribution with $\lambda = 3.4$. What are the probabilities that during a 5-minute period an inspector will spot more than one imperfection?

6.49 If the number of wild pigs seen on a 2-hour jeep trip in the Sonora desert is a random variable having the Poisson distribution with $\lambda = 0.8$, find the probabilities that on such a jeep trip one will see

    (a) no wild pigs;   (c) two wild pigs;

    (b) one wild pig;   (d) more than two wild pigs.

6.50 Use appropriate computer software or a graphing calculator to rework Exercise 6.41 by determining the binomial probability of "3 successes in 100 trials" with $p = 0.019$. Compare the result with 0.171, the value obtained in Exercise 6.41.

6.51 Use a computer printout of the binomial distribution with $n = 2,000$ and $p = 0.0024$ to determine the error made in Exercise 6.42, where the Poisson approximation to the binomial distribution yielded a probability of 0.34.

6.52 With reference to Exercise 6.46, use appropriate software or a graphing calculator to determine the probabilities that during the course of a game there will be

    (a) at most six minor injuries;

    (b) at least four minor injuries.

As we indicated in Exercise 6.46, the number of minor injuries during the course of a game is a random variable having the Poisson distribution with $\lambda = 4.4$.

6.53 With reference to Exercise 6.47, use a computer printout of the Poisson distribution with $\lambda = 4.0$ or a graphing calculator to determine the probabilities that on any given day the shop will repair

    (a) at most three air conditioners;

    (b) at least five air conditioners;

    (c) anywhere from two to six air conditioners.

Round your answers to six decimals.

6.54 With reference to Exercise 6.48, use a computer printout of the Poisson distribution with $\lambda = 3.4$ to determine the probabilities that during a 5-minute interval an inspector will spot

    (a) three or four imperfections;

    (b) at most five imperfections;

    (c) at least six imperfections;

    (d) anywhere from one to three imperfections.

Round your answers to four decimals.

6.5
The Multinomial Distribution

An important generalization of the binomial distribution arises when there are more than two possible outcomes for each trial, the probabilities of the various outcomes remain the same for each trial, and the trials are all independent. This is the case, for example, when we repeatedly roll a die, where each trial has six possible outcomes; or when students are asked whether they like a certain new recording, dislike it, or don't care; or when a U.S. Department of Agriculture inspector grades beef as prime, choice, good, commercial, or utility.

    If there are k possible outcomes for each trial and their probabilities are $p_1, p_2, \ldots,$ and p_k, it can be shown that the probability of x_1 outcomes of the first kind, x_2 outcomes of the second kind, ..., and x_k outcomes of the kth kind in n trials is given by

Multinomial distribution	$$\dfrac{n!}{x_1!x_2!\cdot\,\dots\,\cdot x_k!}\ p_1^{x_1}\cdot p_2^{x_2}\cdot\,\dots\,\cdot p_k^{x_k}$$

In using this expression, it is required that

$$x_1 + x_2 + \dots + x_k = n \quad \text{and} \quad p_1 + p_2 + \dots + p_k = 1$$

■ **EXAMPLE**   In a certain city, channel 3 has 50% of the viewing audience on Saturday nights, channel 12 has 30%, and channel 10 has 20%. What is the probability that among eight television viewers randomly selected in that city on a Saturday night, five will be watching channel 3, two will be watching channel 12, and one will be watching channel 10?

Solution   Substituting $n = 8$, $x_1 = 5$, $x_2 = 2$, $x_3 = 1$, $p_1 = 0.50$, $p_2 = 0.30$, and $p_3 = 0.20$ into the formula for the multinomial distribution, we get

$$\frac{8!}{5!\cdot 2!\cdot 1!}\ (0.50)^5(0.30)^2(0.20)^1 = 0.0945 \qquad ■$$

Exercises

Exercise 6.55 is a practice exercise; its complete solution is given on page 230.

6.55   The percentage distribution of money incomes in U.S. households in 1995 was

Dollars	Percentage of households
Under 14,999	21.0
15,000–49,999	30.1
50,000–74,999	34.0
75,000 and over	14.9

What is the probability that among twelve randomly chosen households, one has income under $14,999, three have incomes between $15,000 and $49,999, seven have incomes between $50,000 and $74,999, and one has income of $75,000 and over?

6.56   According to U.S. Bureau of Justice statistics for 1995, the victim–offender relationship in cases of criminal simple assault are

Victim–offender relationship	Percent
Relatives	10
Well-known	26
Casual acquaintance	18
Stranger	46

What is the probability that in eight randomly selected cases, six assaults are committed by offenders who are strangers, and two assaults are committed by offenders who are relatives? Round your answer to five decimals.

6.57 It can readily be shown that the probabilities of getting two heads, a head, and a tail, and two tails when flipping a pair of balanced coins are, respectively, $\frac{1}{4}$, $\frac{1}{2}$, and $\frac{1}{4}$. What is the probability of getting two heads once, a head and a tail twice, and two tails twice in five flips of a pair of balanced coins?

6.58 The probabilities are $0.60, 0.20, 0.10$, and 0.10 that a state income tax form will be filled out correctly, that it will contain only errors favoring the taxpayer, that it will contain only errors favoring the government, and that it will contain both kinds of errors. What is the probability that, among 10 such tax forms (randomly selected for audit), 7 will be filled out correctly, 1 will contain only errors favoring the taxpayer, 1 will contain only errors favoring the government, and 1 will contain both kinds of errors?

6.59 The probability that a randomly selected person contacted by telephone will respond to a set of questions about cosmetics purchases is 0.40.
   (a) Using the binomial distribution, find the probability that exactly 6 of 15 persons selected at random will respond to the set of questions.
   (b) Use the multinomial distribution to answer the same question. Assume that $k = 2$ and that the probabilities of response and nonresponse are 0.40 and 0.60.
   (c) Can you give a relationship between the binomial distribution and the multinomial distribution with $k = 2$?

6.6
The Mean of a Probability Distribution

Let us again direct our attention to the concept of a mathematical expectation, which we explained in Section 4.5. There we used several examples and illustrations of mathematical expectations and made the following statement: When we said that a child ... can *expect* to visit a dentist 1.92 times a year, we referred to the result that was obtained by multiplying $0, 1, 2, 3, \ldots$, by the corresponding probabilities that such a child will visit a dentist that many times a year and then adding all these products.

If we apply the same argument to the probability distribution of rolling a balanced die, we find that the number of points that we can expect in one roll of a die is

$$1 \cdot \frac{1}{6} + 2 \cdot \frac{1}{6} + 3 \cdot \frac{1}{6} + 4 \cdot \frac{1}{6} + 5 \cdot \frac{1}{6} + 6 \cdot \frac{1}{6} = 3\frac{1}{2}$$

Similarly, the number of heads that we can expect in three flips of a balanced coin is

$$0 \cdot \frac{1}{8} + 1 \cdot \frac{3}{8} + 2 \cdot \frac{3}{8} + 3 \cdot \frac{1}{8} = 1\frac{1}{2}$$

Of course, we cannot roll $3\frac{1}{2}$ with a die or get $1\frac{1}{2}$ heads; like all mathematical expectations, these figures must be looked upon as averages. In general, if a random variable takes on the values $x_1, x_2, x_3, \ldots$, or x_k, with the probabilities $f(x_1), f(x_2), f(x_3), \ldots$, or $f(x_k)$, its expected value (or its mathematical expectation) is given by

$$x_1 \cdot f(x_1) + x_2 \cdot f(x_2) + x_3 \cdot f(x_3) + \cdots + x_k \cdot f(x_k)$$

We refer to this quantity as the **mean of a probability distribution** of the random variable, and, in the Σ notation, we write

Mean of the probability distribution

$$\mu = \Sigma x \cdot f(x)$$

where the summation extends over all values taken on by the random variable. Like the mean of a population, it is denoted by the Greek letter μ (lowercase mu). The notation is the same, for as we pointed out in connection with the binomial distribution, when we observe any random variable, we use its probability distribution to identify the population that we are sampling. For instance, Figure 6.1 pictures the population that we are sampling when we observe a value of a random variable having the binomial distribution with $n = 5$ and $p = 0.60$, and Figure 6.3 pictures the population we are sampling when we observe a value of a random variable having the hypergeometric distribution with $a = 5$, $b = 11$, and $n = 3$.

■ **EXAMPLE**   Use the probabilities obtained in the shopping example on pages 198 and 199 to determine how many of five persons shopping at the given market can be expected to spend at least $50.

Solution   Substituting $x = 0, 1, 2, 3, 4$, and 5 and the corresponding probabilities into the formula for μ, we get

$$\mu = 0(0.010) + 1(0.077) + 2(0.230) + 3(0.346) + 4(0.259) + 5(0.078)$$

$$= 3.001$$

or approximately 3. Thus, three out of five shoppers can be expected to spend at least $50 at the given market.   ■

When a random variable can take on many different values, the calculation of μ becomes very laborious. For instance, if we want to know how many of 400 persons attending a movie can be expected to buy popcorn (when the probability is, say, 0.30 that any one of them will buy popcorn), we will first have to calculate the 401 probabilities corresponding to 0, 1, 2, ..., or 400 of them buying popcorn. We might argue, however, that if 30% of all moviegoers buy popcorn, then we can expect that 30% of our 400, or 400(0.30) = 120, will buy popcorn. Similarly, if a balanced coin is flipped 500 times, we can argue that in the long run heads will come up 50% of the time and, hence, that we can expect to get 500(0.50) = 250 heads in 500 flips of a balanced coin. These two values are, indeed, correct; both problems deal with random variables having binomial distributions, and it can be shown that in general

Mean of the binomial distribution

$$\mu = n \cdot p$$

for the **mean of the binomial distribution**. In words, the mean of a binomial distribution is the product of the number of trials and the probability of success on an individual trial.

■ **EXAMPLE**   With reference to the preceding example, use this formula to find the mean of the probability distribution of the number of persons, among five, who will spend at least $50 at the given market.

Solution   Since we are dealing with a binomial distribution with $n = 5$ and $p = 0.60$, we find that $\mu = 5(0.60) = 3$, and it should be apparent that the small difference of 0.001 between this exact value and the one obtained before is due to rounding the probabilities to three decimals.   ■

■ **EXAMPLE**   Find the mean of the probability distribution of the number of heads obtained in three flips of a balanced coin.

Solution   Since we are dealing with a binomial distribution with $n = 3$ and $p = \frac{1}{2}$, we get $\mu = 3 \cdot \frac{1}{2} = 1\frac{1}{2}$, and this agrees with the result obtained in the beginning of this section.   ■

It is important to remember that the formula $\mu = np$ applies only to binomial distributions. There are other formulas for other distributions; for instance, for the hypergeometric distribution, the formula for the mean is $\mu = n\dfrac{a}{a+b}$, and for the Poisson distribution it is $\mu = \lambda$. Proofs of these special formulas may be found in most textbooks on mathematical statistics.

**6.7
The Standard
Deviation of a
Probability
Distribution**

In Chapter 3 we saw that there are many situations in which we must not only calculate the mean or some other measure of location, but also describe the variability (spread or dispersion) of a set of data. As we indicated in that chapter, the most widely used measure of variation is the variance and its square root, the standard deviation. For probability distributions, we measure variability in almost the same way, but instead of averaging the squared deviations from the mean, we calculate their expected value. If x is a value of a random variable whose probability distribution has the mean μ, its deviation from the mean is $x - \mu$, and we define the **variance of a probability distribution** as the expected value (mathematical expectation) of the squared deviation from the mean, namely as,

**Variance of a
probability
distribution**

$$\sigma^2 = \Sigma(x - \mu)^2 \cdot f(x)$$

where the summation extends over all values taken on by the random variable. As in the preceding section and for the same reason, we denote descriptions of probability distributions with the same symbols as the corresponding descriptions of populations. The square root of the variance defines the **standard deviation of a probability distribution**, and we write

**Standard deviation of
a probability
distribution**

$$\sigma = \sqrt{\Sigma(x - \mu)^2 \cdot f(x)}$$

Note that this formula is like that of the standard deviation of a population as defined on page 68, with the probability $f(x)$ substituted for $\frac{1}{N}$.

■ **EXAMPLE**   Use the probabilities obtained in the shoppers' expenditures example on page 198, which are $f(0) = 0.010$, $f(1) = 0.077$, $f(2) = 0.230$, $f(3) = 0.346$, $f(4) = 0.259$, and $f(5) = 0.078$ to determine the standard deviation of the distribution of the number of persons, among five, who will spend at least $50 at the given market.

Solution   Since we have already shown that $\mu = 3$, we can arrange the necessary calculations as in the following table:

NUMBER OF PERSONS x	$x - \mu$	$(x - \mu)^2$	PROBABILITY $f(x)$	$(x - \mu)^2 \cdot f(x)$
0	−3	9	0.010	0.090
1	−2	4	0.077	0.308
2	−1	1	0.230	0.230
3	0	0	0.346	0.000
4	1	1	0.259	0.259
5	2	4	0.078	0.312
			Total	$\sigma^2 = 1.199$

The values in the column on the right were obtained by multiplying each squared deviation from the mean by its probability, and their sum is the variance of the distribution. To find the standard deviation, we must obtain the square root of 1.199. This is found readily on most calculators as

$$\sqrt{1.199} \approx 1.0949885 \approx 1.095$$

■

The calculations were easy in this example since the deviations from the mean were all small integers. When the deviations from the mean are large numbers, or when they are given to several decimal places, it is usually worthwhile to simplify the calculations by using a special computing formula for the variance, which will be given in Exercise 6.64.

As in the case of the mean, the calculation of the variance or the standard deviation of a probability distribution can generally be simplified when we deal with special kinds of probability distributions. For instance, for the binomial distribution we have the formula

Standard deviation of the binomial distribution

$$\sigma = \sqrt{np(1 - p)}$$

■ **EXAMPLE**   Use the preceding formula to verify the result obtained in the previous example.

Solution   Since we are dealing with a binomial distribution with $n = 5$ and $p = 0.60$, we get

$$\sigma = \sqrt{5(0.60)(0.40)} = \sqrt{1.20} \approx 1.095$$

This agrees with the results obtained previously.   ■

■ **EXAMPLE**   Find the variance of the probability distribution of the number of heads obtained in three flips of a balanced coin.

Solution   Since we are dealing with a binomial distribution with $n = 3$ and $p = \frac{1}{2}$, we note that

$$1 - p = \frac{1}{2} \quad \text{and} \quad \sigma^2 = 3 \cdot \frac{1}{2} \cdot \frac{1}{2} = \frac{3}{4}$$

In Exercise 6.74 the reader is asked to verify this result using the formula given previously that defines the variance of a probability distribution.   ■

For the hypergeometric distribution, the variance is given by

$$\sigma^2 = \frac{nab}{(a+b)^2} \cdot \frac{a+b-n}{a+b-1}$$

and the standard deviation is the square root of this. For the Poisson distribution, the standard deviation is simply $\sigma = \sqrt{\lambda}$. The derivations of these formulas may be found in more advanced texts.

Exercises

Exercises 6.60, 6.66, and 6.67 are practice exercises; their complete solutions are given on page 231.

6.60 The probabilities that a building inspector will observe 0, 1, 2, 3, 4, or 5 violations of the building code in a home built in a large development are given in the following table:

Number of violations	0	1	2	3	4	5
Probability	0.41	0.22	0.17	0.13	0.05	0.02

Find the mean of this probability distribution.

6.61 The probabilities that a police officer will issue 0, 1, 2, 3, 4, 5, or 6 parking tickets during his work shift are given in the following table:

Number of violations	0	1	2	3	4	5	6
Probability	0.20	0.30	0.25	0.10	0.08	0.06	0.01

Find the mean of this probability distribution.

6.62 Suppose that the probabilities are 0.4, 0.3, 0.2, and 0.1 that 0, 1, 2, or 3 hurricanes will hit a certain coast area in any given year.
(a) Find the mean of this probability distribution.
(b) Find the variance of this probability distribution.

6.63 The following table gives the probabilities that a probation officer will receive 0, 1, 2, 3, 4, 5, or 6 reports of probation violations on any given day:

Number of reports	0	1	2	3	4	5	6
Probability	0.01	0.13	0.20	0.29	0.20	0.11	0.06

Find the mean and the standard deviation of this probability distribution.

6.64 The calculation of the variance and the standard deviation was easy in the example on page 222 and in Exercise 6.62, mainly because μ, and hence the deviations from the mean, were all integers. When this is not the case, as in Exercise 6.63, the calculations can be simplified by using the computing formula

$$\sigma^2 = \Sigma x^2 \cdot f(x) - \mu^2$$

To calculate μ, we add the products obtained by multiplying each value of x by the corresponding probability. Then we add the products obtained by multiplying the square of each x by the corresponding probability and substitute these two sums into the formula for σ^2. This parallels what we did on page 222 in connection with the calculation of the sample variance (or the sample standard deviation).
(a) Use this computing formula for σ^2 to rework the standard deviation of Exercise 6.63.
(b) The probabilities of 0, 1, 2, 3, 4, or 5 armed robberies in a western city in any given month are, respectively, 0.10, 0.25, 0.35, 0.20, 0.08, and 0.02. First find the mean of this probability distribution, and then use the computing formula for σ^2 to determine its variance.

6.65 Previously, we showed that $\mu = 3.5$ for the number of points we roll with a balanced die, that is, for a random variable that takes on the values 1, 2, 3, 4, 5, or 6 with equal probabilities of $\frac{1}{6}$. Find the standard deviation of this probability distribution using
(a) the formula on page 221 that defines σ^2;
(b) the computing formula given in Exercise 6.64.

6.66 Find the mean of the binomial distribution with $n = 4$ and $p = 0.10$
(a) by looking up the probabilities in Table I and then using the formula that defines the mean of a probability distribution;
(b) by using the special formula for the mean of the binomial distribution.

6.67 Find the standard deviation of the binomial distribution with $n = 4$ and $p = 0.10$
(a) by using the value of μ obtained in part (b) of Exercise 6.66, the probabilities looked up in part (a) of that exercise, and the formula that defines the standard deviation of a probability distribution;
(b) by using the special formula for the standard deviation of the binomial distribution.

6.68 As can easily be verified by listing all possibilities or by using the formula for the binomial distribution, the probabilities of getting 0, 1, 2, 3, 4, or 5 heads in five flips of a balanced coin are given in this table:

Number of heads	0	1	2	3	4	5
Probability	$\frac{1}{32}$	$\frac{5}{32}$	$\frac{10}{32}$	$\frac{10}{32}$	$\frac{5}{32}$	$\frac{1}{32}$

Find the mean of this probability distribution using
(a) the formula that defines μ;
(b) the special formula for the mean of a binomial distribution.

6.69 With reference to Exercise 6.68, find the variance of the probability distribution using
(a) the computing formula given in Exercise 6.64;
(b) the special formula for the variance of a binomial distribution.

6.70 A traffic study shows that 90% of motor vehicles that come to a stop for a stop signal at a certain intersection have to wait at least 15 seconds before proceeding. Find the mean of the distribution of the number of motor vehicles among 10 randomly selected motor vehicles that come to this stop signal and have to wait at least 15 seconds before proceeding
(a) by looking up the probabilities in Table I and then using the formula that defines the mean of a probability distribution;
(b) by using the special formula for the mean of the binomial distribution.

6.71 With reference to Exercise 6.70, where $\mu = 9$ motor vehicles, find the standard deviation of the number of motor vehicles that have to wait at least 15 seconds before proceeding
(a) using the formula for the standard deviation of a probability distribution;
(b) using the special formula for the standard deviation of the binomial distribution.

6.72 If 95% of the bottles collected in a recycling bin are suitable for recycling, find the mean and the standard deviation of the distribution of the number of these bottles among 15 selected at random that are suitable for recycling, using the special formulas for the mean and the standard deviation of a binomial distribution.

6.73 Find the mean and the standard deviation of the distribution of each of the following binomial random variables:
(a) the number of heads obtained in 900 flips of a balanced coin;
(b) the number of 4's obtained in 405 rolls of a balanced die;
(c) the number of persons, among 756 invited, who will attend the opening of a new car dealership, if the probability is 0.30 that any one of them will attend.

6.74 Previously, we determined that the variance of the probability distribution of the number of heads obtained in three flips of a balanced coin is $\frac{3}{4}$. Use Table

I and the formula defining the variance of a probabilty distribution to verify the result of $\frac{3}{4}$ for the variance of the binomial distribution with $n = 3$ and $p = \frac{1}{2}$.

6.75 The probabilities for Figure 6.3 are

Number of trucks	0	1	2	3
Probability	0.295	0.491	0.196	0.018

Use these probabilities to calculate the mean of the given hypergeometric distribution. Verify the result by using the special formula

$$\mu = n \cdot \frac{a}{a + b}$$

6.76 The probabilities that there will be 0, 1, 2, 3, 4, or 5 fires caused by lightning during a summer storm are as follows:

Number of fires caused by lightning	0	1	2	3	4	5
Probability	0.449	0.360	0.144	0.038	0.008	0.001

Calculate the mean of this Poisson distribution with $\lambda = 0.8$, and use the result to verify the special formula $\mu = \lambda$ mentioned on page 221.

6.8 Chebyshev's Theorem

Intuitively speaking, the variance and the standard deviation of a probability distribution measure the expected size of the chance fluctuations of a corresponding random variable. When σ is small, the probability is high that we will get a value close to the mean, and when σ is large, we are more likely to get a value far away from the mean. This important idea is expressed formally by Chebyshev's theorem, which we introduced in Section 3.7, as it pertains to numerical data. For probability distributions, **Chebyshev's theorem** can be stated as follows:

Chebyshev's theorem

> **The probability that a random variable will take on a value within k standard deviations of the mean is at least $1 - \dfrac{1}{k^2}$.**

Thus, the probability of getting a value within two standard deviations of the mean (a value between $\mu - 2\sigma$ and $\mu + 2\sigma$) is at least $1 - \dfrac{1}{2^2} = \dfrac{3}{4}$, the probability of getting a value within five standard deviations of the mean (a value between $\mu - 5\sigma$ and $\mu + 5\sigma$) is at least $1 - \dfrac{1}{5^2} = \dfrac{24}{25}$, and so forth.

■ **EXAMPLE**   The number of telephone calls that an answering service receives between 9 A.M. and 10 A.M. is a random variable whose distribution has the mean

$\mu = 27.5$ and the standard deviation $\sigma = 3.2$. What does Chebyshev's theorem with $k = 3$ tell us about the number of telephone calls that the answering service may receive between 9 A.M. and 10 A.M.?

Solution   Since

$$\mu - 3\sigma = 27.5 - 3(3.2) = 17.9 \quad \text{and} \quad \mu + 3\sigma = 27.5 + 3(3.2) = 37.1$$

we can assert with a probability of at least

$$1 - \frac{1}{3^2} = \frac{8}{9}$$

or approximately 0.89, that the answering service will receive between 17.9 and 37.1 calls (that is, anywhere from 18 to 37 calls).   ■

■ **EXAMPLE**   What does Chebyshev's theorem with $k = 6$ tell us about the number of heads that we may get in 400 flips of a balanced coin?

Solution   Here $n = 400$ and $p = \frac{1}{2} = 0.5$, the mean and the standard deviation of the number of heads are

$$\mu = n \cdot p = 400 \cdot \frac{1}{2} = 200$$

and

$$\sigma = \sqrt{np(1 - p)} = \sqrt{400(0.5)(0.5)} = \sqrt{100} = 10$$

Then

$$\mu - 6\sigma = 200 - 6(10) = 140 \quad \text{and} \quad \mu + 6\sigma = 200 + 6(10) = 260$$

Thus, we can assert with a probability of at least

$$1 - \frac{1}{6^2} = \frac{35}{36}$$

or approximately 0.97, that we will get between 140 and 260 heads.   ■

If we convert the numbers of heads into proportions in the preceding example, we can assert with a probability of at least $\frac{35}{36}$ that the proportion of heads that we get in 400 flips of a balanced coin will lie between

$$\frac{140}{400} = 0.35 \quad \text{and} \quad \frac{260}{400} = 0.65$$

To continue this argument, the reader will be asked to show in Exercise 6.82 that the probability is at least $\frac{35}{36}$ that for 10,000 flips of a balanced coin the proportion of heads will lie between 0.47 and 0.53, and that for 1,000,000 flips of a balanced coin it will lie between 0.497 and 0.503. All this provides support for the Law of Large Numbers, which we mentioned in Section 4.5 in connection with the frequency interpretation of probability. In actual practice,

Chebyshev's theorem is very rarely used, since the probability $1 - \dfrac{1}{k^2}$ is usually unnecessarily small. For instance, in the preceding example we showed that the probability of getting a value within six standard deviations of the mean is at least 0.97, whereas the actual probability that this will happen for a random variable having the binomial distribution with $n = 400$ and $p = \frac{1}{2}$ is about 0.999999998. "At least 0.97" is not wrong, but for most practical purposes it does not tell us quite enough. In spite of the difference between these probabilities, Chebyshev's theorem cannot be improved, because sometimes it is exact, as will be noted in Exercise 6.82.

Exercises

Exercise 6.77 is a practice exercise; its complete solution is given on page 231.

6.77 The number of customers to whom a restaurant serves breakfast on a weekday morning is a random variable with $\mu = 142$ and $\sigma = 12$. According to Chebyshev's theorem, with what probability can we assert that between 94 and 190 customers will have breakfast there on a weekday morning?

6.78 A student answers the 144 questions on a true–false test by flipping a balanced coin.
 (a) Use the special formulas for the mean and the standard deviation of the binomial distribution to find the values of μ and σ for the distribution of the number of correct answers that he or she will get.
 (b) According to Chebyshev's theorem, what can we assert with a probability of at least 0.96 about the number of correct answers that he or she will get?

6.79 Use Chebyshev's theorem to determine the probability of getting a value
 (a) within 4 standard deviations of the mean;
 (b) within 2.5 standard deviations of the mean;
 (c) within 2.8 standard deviations.

6.80 The number of marriage licenses issued in a certain city during the month of June is a random variable with $\mu = 146$ and $\sigma = 7.5$.
 (a) What does Chebyshev's theorem with $k = 3$ tell us about the number of marriage licenses issued there during the month of June?
 (b) According to Chebyshev's theorem, with what probability can we assert that between 101 and 191 marriage licenses will be issued there during the month of June?

6.81 The annual number of rainy days in a certain city is a random variable with $\mu = 126$ and $\sigma = 9$.
 (a) What does Chebyshev's theorem with $k = 4$ tell us about the number of days that it will rain in the given city in any one year?
 (b) According to Chebyshev's theorem, with what probability can we assert that it will rain in the given city between 99 and 153 days in any one year?

6.82 Use Chebyshev's theorem to show that the probability is at least $\frac{35}{36}$ that
 (a) in 10,000 flips of a balanced coin there will be between 4,700 and 5,300 heads, and hence the proportion of heads will be between 0.47 and 0.53;
 (b) in 1,000,000 flips of a balanced coin there will be between 497,000 and 503,000 heads, and hence the proportion of heads will be between 0.497 and 0.503.

Solutions to Practice Exercises

6.1 (a) The values are all on the interval from 0 to 1, but since $0.37 + 0.35 + 0.30 = 1.02$, and not 1, the values cannot serve as the values of a probability distribution.

(b) The values are all on the interval from 0 to 1, and since their sum is 1, they can serve as the values of a probability distribution.

(c) Since one of the values is negative, they cannot serve as the values of a probability distribution.

6.5 (a) Substituting $n = 6, p = 0.10$, and $x = 2$ into the formula for the binomial distribution, we get

$$\binom{6}{2}(0.10)^2(0.90)^4 = 15(0.01)(0.6561) = 0.098415 \approx 0.098$$

(b) Following the procedure of part (a), we find that the probabilities for $x = 0$ and $x = 1$ are 0.531441 and 0.354294, respectively. Adding these two values, we get $0.885735 \approx 0.886$.

6.6 (a) Substituting $n = 4, p = 0.70$, and $x = 2$ into the formula for the binomial distribution, we get

$$f(2) = \binom{4}{2}(0.70)^2(1 - 0.70)^{4-2} = 6(0.49)(0.09) = 0.2646 \approx 0.265$$

(b) The value in Table I is 0.265.

6.7 (a) $0.250 + 0.154 + 0.044 = 0.448$;

(b) $0.002 + 0.009 + 0.032 + 0.086 + 0.172 = 0.301$.

6.18 (a) The probability of not getting 10 items in perfect condition is $1 - 0.599 = 0.401$.

(b) The probability of getting 10 items in perfect condition is 0.349.

(c) The probability of getting 10 items in perfect condition is 0.107.

6.24 $\dfrac{\binom{7}{3}\binom{9}{2}}{\binom{16}{5}} = \dfrac{35 \cdot 36}{4,368} = \dfrac{1,260}{4,368} \approx 0.288.$

6.25 (a) $\dfrac{\binom{5}{0}\binom{10}{2}}{\binom{15}{2}} = \dfrac{1 \cdot 45}{105} = \dfrac{45}{105} \approx 0.429;$

(b) $\dfrac{\binom{5}{1}\binom{10}{1}}{\binom{15}{2}} = \dfrac{5 \cdot 10}{105} = \dfrac{50}{105} \approx 0.476;$

(c) $\dfrac{\binom{5}{2}\binom{10}{0}}{\binom{15}{2}} = \dfrac{10 \cdot 1}{105} = \dfrac{10}{105} \approx 0.095.$

6.32 (a) Since $n = 15$ exceeds $0.05(40 + 160) = 10$, the condition is not satisfied.

(b) Since $n = 22$ does not exceed $0.05(380 + 120) = 25$, the condition is satisfied.

(c) Since $n = 30$ does not exceed $0.05(400 + 240) = 32$, the condition is satisfied.

6.33 (a) Substituting $a = 6$, $b = 194$, $n = 3$, and $x = 1$ into the formula for the hypergeometric distribution, we get

$$f(1) = \frac{\binom{6}{1}\binom{194}{2}}{\binom{200}{3}} = \frac{6 \cdot 18{,}721}{1{,}313{,}400} \approx 0.086$$

(b) Substituting $n = 3$, $p = \frac{6}{200} = 0.03$, and $x = 1$ into the formula for the binomial distribution, we get

$$f(1) = \binom{3}{1}(0.03)^1(1 - 0.03)^{3-1} = 3(0.03)(0.97)^2 \approx 0.085$$

6.38 (a) Since $n = 300$ is not less than 100 and $np = 300 \cdot \frac{1}{50} = 6$ is less than 10, the conditions are satisfied.

(b) Since $n = 900$ is not less than 100 and $np = 900 \cdot \frac{1}{100} = 9.0$ is less than 10, the conditions are satisfied.

(c) $n = 400$ is not less than 100, but since $np = 400 \cdot \frac{1}{25} = 16$ exceeds 10, the conditions are not satisfied.

6.40 The Poisson approximation with $np = 250(0.02) = 5.00$ yields

$$f(4) = 0.0067 \cdot \frac{5.0^4}{4!} = 0.0067\left(\frac{6.25}{24}\right) = 0.174$$

6.46 Substituting $\lambda = 4.4$ and, respectively, $x = 0, 1, 2,$ and 3 into the formula for the Poisson distribution, we get

$$f(0) = \frac{4.4^0 \cdot e^{-4.4}}{0!} = e^{-4.4} \approx 0.012$$

$$f(1) = \frac{4.4^1 \cdot 0.012}{1!} \approx 0.053$$

$$f(2) = \frac{4.4^2 \cdot 0.012}{2!} \approx 0.116$$

$$f(3) = \frac{4.4^3 \cdot 0.012}{3!} \approx 0.170$$

and the answer is $0.012 + 0.053 + 0.116 + 0.170 = 0.351$.

6.55 Substituting $n = 12$, $x_1 = 1$, $x_2 = 3$, $x_3 = 7$, and $x_4 = 1$; and $p_1 = 0.210$, $p_2 = 0.301$, $p_3 = 0.340$, and $p_4 = 0.149$ into the formula for the multinomial distribution, we get

$$\frac{12!}{1!3!7!1!} \cdot (0.210)^1(0.301)^3(0.340)^7(0.149)^1$$

$$= (15.840)(0.210)(0.02727)(0.000525)(0.149) \approx 0.0071$$

6.60 $\mu = 0(0.41) + 1(0.22) + 2(0.17) + 3(0.13) + 4(0.05) + 5(0.02) = 1.25.$

6.66 (a) $\mu = 0(0.656) + 1(0.292) + 2(0.049) + 3(0.004) = 0.402;$
(b) $\mu = 4(0.10) = 0.40.$

6.67 (a) $\sigma^2 = (0 - 0.4)^2(0.656) + (1 - 0.4)^2(0.292)$
$$+ (2 - 0.4)^2(0.049) + (3 - 0.4)^2(0.004)$$
$$= 0.36256$$
$$\sigma = \sqrt{0.36256} \approx 0.602;$$
(b) $\sigma = \sqrt{4(0.10)(0.90)} = 0.60.$

6.77 Since $142 - 12k = 94$ and $142 + 12k = 190$, it follows that $12k = 48$ and $k = 4$; thus, the probability is at least

$$1 - \frac{1}{4^2} = \frac{15}{16}, \quad \text{or approximately } 0.94$$

7 The Normal Distribution

Continuous sample spaces and continuous random variables arise whenever we deal with quantities that are measured on a continuous scale, for instance, when we measure the amount of alcohol in a person's blood, the net weight of a package of frozen food, the amount of tar in a cigarette, the speed of a car, and so forth.

In the continuous case, the place of histograms is taken by continuous curves. These may be pictured as in Figure 7.1, where we can think of histograms with narrower and narrower classes approaching the continuous curve.

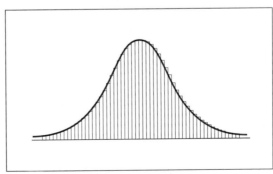

FIGURE 7.1 Continuous distribution curve.

233

Among the many continuous distribution curves used in statistics, by far the most important is the **normal distribution**, which is sometimes described simply as a bell-shaped distribution. This description leaves a good deal to be desired; it is true that normal distribution curves, often called **normal curves**, are bell shaped, but not all bell-shaped distribution curves are necessarily those of normal distributions. The normal distribution dates back to the work of the French-English mathematician Abraham de Moivre (1667–1745), who studied a curve which closely approximates the binomial distribution when the number of trials is very large.

In this chapter we shall study continuous random variables. The concept of a continuous distribution will be introduced in Section 7.1, followed by that of a normal distribution in Section 7.2. Various applications of the normal distribution will be discussed in Sections 7.3 and 7.4.

7.1 Continuous Distributions

In all the histograms shown in preceding chapters, the frequencies, percentages, or probabilities are represented by the heights of the rectangles; but as long as the class intervals are all equal, we can also say that the frequencies, percentages, or probabilities are represented by the areas of the rectangles. The latter is preferable for the work of this chapter, for in the continuous case we also represent probabilities by areas—not by areas of rectangles, but, as is illustrated in Figure 7.2, by areas under continuous curves. The diagram on the left shows a histogram of the probability distribution of a random variable that takes on only the values $0, 1, 2, \ldots$, and 10, and the probability of getting a 3, for example, is given by the area of the blue rectangle. The diagram on the right refers to a continuous random variable, which can take on any value on the interval from 0 to 10, and the probability of getting a value on the interval from 2.5 to 3.5 is given by the area of the blue region under the curve. Similarly, the area of the dark blue region under the curve gives the probability of getting a value greater than 8.

Continuous curves such as the one shown in the right-hand diagram of Figure 7.2 are the graphs of functions that we refer to as **probability densities** or, more informally, as **continuous distributions**. The first of these terms is borrowed from the language of physics, where the terms *weight* and *density* are used in very much the same way in which we use the terms *probability*

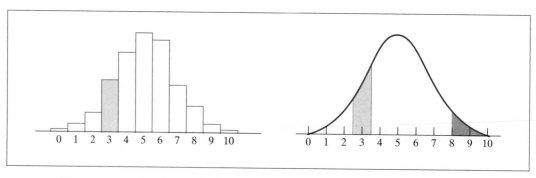

FIGURE 7.2 Histogram of a probability distribution and graph of a continuous distribution.

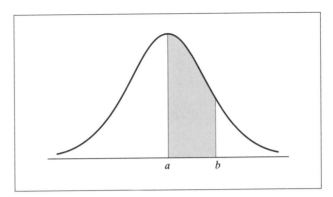

FIGURE 7.3 Probability density.

and *probability density* in statistics. What characterizes a probability density is the fact that

The area under the curve between any two values a and b (see Figure 7.3) gives the probability that a random variable having this continuous distribution will take on a value on the interval from a to b.

Observe that when a and b are very close together the heights of the density at a and at b, and at the midpoint $\dfrac{a+b}{2}$ as well, are nearly equal. Thus the area under the curve between a and b is very nearly equal to

$$\left(\text{height of the density at } \frac{a+b}{2}\right) \cdot (b-a)$$

It follows that the values of a probability density should not be negative and that the total area under the curve, representing the certainty that a random variable must take on one of its values, is always equal to 1. This corresponds to the two rules about probability distributions given near the end of Section 6.1. There is no requirement that the heights of a probability density be less than 1.

■ **EXAMPLE**   If a continuous random variable has the probability density shown in Figure 7.4, find the probabilities that it will take on a value

(a) between −2 and 3;

(b) greater than 1;

(c) greater than or equal to 1.

Solution   (a) Since this is the total area under the curve, it must be equal to 1; also, multiplying the base of the rectangle by its height, we get

$$[3 - (-2)] \cdot \frac{1}{5} = 5 \cdot \frac{1}{5} = 1$$

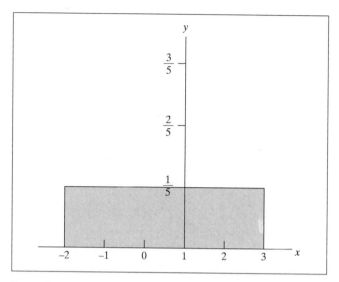

FIGURE 7.4 A special probability density.

(b) Multiplying the base of the corresponding rectangle by its height, we get

$$(3 - 1) \cdot \frac{1}{5} = 2 \cdot \frac{1}{5} = \frac{2}{5}$$

(c) The answer is the same as that to part (b), $\frac{2}{5}$. ∎

Parts (b) and (c) of the preceding example illustrate the fact that the probability that a random variable having a continuous distribution will take on a particular value is always zero. In this case, the base of the rectangle has shrunk to zero, and so has the area under the curve.

It is a consequence of measuring (rather than counting) that we must assign probability zero to any particular outcome. We claim that the probability must be zero that an individual will have a weight of *exactly* 145.27 pounds or that a horse will run a race in *exactly* 58.442 seconds. However, observe that, even though every particular outcome has probability zero, the process will still produce a value (whether or not we can measure it with extra-fine precision); apparently, then, events of probability zero not only can occur, but *must* occur when dealing with measured random quantities.

Statistical descriptions of continuous distributions, such as the mean and the standard deviation, are as important as descriptions of probability distributions or distributions of observed data. Informally, we can always picture continuous distributions as being approximated by histograms of probability distributions (see Figure 7.1) for which the mean and the standard deviation can be determined by means of the formulas of Sections 6.6 and 6.7. Then, if we choose histograms with narrower and narrower classes, the means and the standard deviations of the corresponding probability distributions will approach the mean and the standard deviation of the continuous distribution.

Since formal definitions of the mean and the standard deviation of a continuous distribution cannot be given without the use of calculus, they will be omitted in this text. Again informally, the mean μ of a continuous distribution is a measure of its *center* or *middle*, and the standard deviation σ of a continuous distribution is a measure of its *dispersion* or *spread*.

7.2
The Normal Distribution

The normal distribution is often referred to as the cornerstone of modern statistics. This is due partly to its role in the development of statistical theory and partly to the fact that distributions of observed data often have the same general pattern as normal distributions.

This pattern is shown in Figure 7.5, but it cannot be seen from such a small drawing that the bell-shaped curve extends indefinitely in both directions, coming closer and closer to the horizontal axis without ever reaching it. Fortunately, it is seldom necessary to extend the *tails* of a normal distribution very far, because the area under the curve becomes negligible once we go more than four or five standard deviations away from the mean.

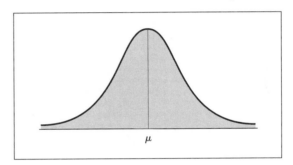

FIGURE 7.5 Normal distribution.

Since the normal distribution has a complicated mathematical equation, let us merely point out that this distribution is completely determined by the values of μ and σ. In other words, there is one and only one normal distribution with a given mean μ and a given standard deviation σ. Part (a) of Figure 7.6 shows two normal distributions with the same mean, but different standard deviations; part (b) shows two normal distributions with the same standard deviation, but different means; and part (c) shows two normal distributions with different means and different standard deviations.

In practice, we find areas under normal curves in special tables, such as Table II at the end of the book. It is not necessary to construct separate tables of normal-curve areas for all conceivable pairs of values of μ and σ. We tabulate these areas only for the normal distribution with $\mu = 0$ and $\sigma = 1$, called the **standard normal distribution**. Then we obtain areas under any normal curve by performing the change of scale (see Figure 7.7) that converts the units of measurement into **standard units** by means of the formula

Standard units

$$z = \frac{x - \mu}{\sigma}$$

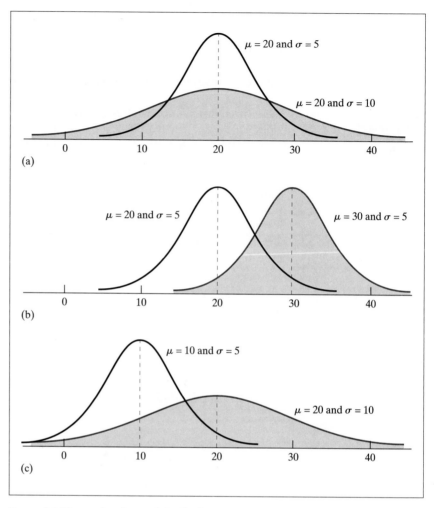

(a)

(b)

(c)

FIGURE 7.6 Three pairs of normal distributions.

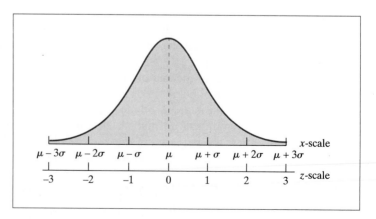

FIGURE 7.7 Change in scale to standard units.

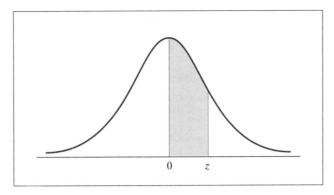

FIGURE 7.8 Tabulated areas under the graph of the standard normal distribution.

As pointed out previously, a value of z simply tells us how many standard deviations the corresponding x-value lies above or below the mean.

The entries in Table II are the areas under the standard normal curve (that is, the graph of the standard normal distribution) between the mean, $z = 0$, and $z = 0.00, 0.01, 0.02, \ldots, 3.08$, and 3.09, and also $z = 4.00$, $z = 5.00$, and $z = 6.00$. In other words, the entries in Table II are areas under the standard normal curve like that of the blue region in Figure 7.8.

Table II has no entries corresponding to negative values of z, for these are not needed by virtue of the symmetry of any normal curve about its mean.

■ **EXAMPLE**   Find the area under the standard normal curve between $z = -1.20$ and $z = 0$.

∗**Optional Solution**   This solution is provided primarily for users of graphing calculators. It can be omitted by other readers without loss of continuity. Had we wanted to use a graphing calculator in this example, we would have obtained the result shown in Figure 7.9. The "Using Technology: The Graphing Calculator" material that follows Figure 7.9 explains how to obtain this figure with a TI-83 graphing calculator. The numerical solution can be read directly from the figure and the area under the standard normal curve between $z = -1.20$ and $z = 0$ is 0.38493.   ■

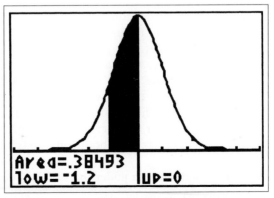

FIGURE 7.9 Area under normal curve, obtained with a TI-83 graphing calculator.

USING TECHNOLOGY : *The Graphing Calculator*

How to graph the area under the normal curve shown in Figure 7.9

(Tip: Before beginning, reset the memory to default setting—see Appendix A: TI-83 Tips for help.)

Step 1.   Set the viewing window.

> Press WINDOW
> Set   Xmin = −5
>   Xmax = 5
>   Xscl = 1
>   Ymin = −.1
>   Ymax = .4
>   Yscl = 1
>   Xres = 1

Step 2.   Select the type of display and enter the interval to be displayed.
Press 2nd, VARS to access the DISTR DRAW menu.
Highlight DRAW.
Select 1: ShadeNorm(and press ENTER.
Enter   −1.20, 0)
             (Be sure to use the (−) key for the negative sign;
             use the black , key, and use the) key to close the expression.)

Step 3.   View the graph.
Press ENTER to solve.

Solution (Without a Graphing Calculator)   Since the area under the standard normal curve between $z = -1.20$ and $z = 0$ equals the area under the standard normal curve between $z = 0$ and $z = 1.20$, we simply look up the entry corresponding to $z = 1.20$ in Table II, which is 0.3849.   ∎

Questions concerning areas under normal curves arise in various ways, and the ability to find any desired area quickly can be a big help. Although the table gives only areas between $z = 0$ and selected positive values of z, we often have to find areas to the left or to the right of given positive or negative values of z. This should not cause any difficulties provided that we remember exactly what areas are represented by the entries in Table II, and also that the standard normal curve is symmetrical about $z = 0$; so that the area under the curve to the left of $z = 0$ and the area under the curve to the right of $z = 0$ are both equal to 0.5000.

■ **EXAMPLE**   Find the area under the standard normal curve that lies
(a) to the left 0.94;
(b) to the right of $z = -0.65$;

(c) to the right of $z = 1.76$;

(d) to the left of $z = -0.85$;

(e) between $z = 0.87$ and $z = 1.28$;

(f) between $z = -0.34$ and $z = 0.62$.

Solution   (a) The area to the left of $z = 0.94$ is 0.5000 plus the entry in Table II corresponding to $z = 0.94$, that is,

$$0.5000 + 0.3264 = 0.8264 \quad \text{(see Figure 7.10)}$$

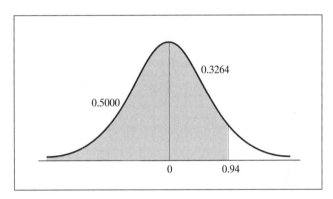

FIGURE 7.10 Area to the left of $z = 0.94$.

(b) The area to the right of $z = -0.65$ is 0.5000 plus the entry in Table II corresponding to $z = 0.65$:

$$0.5000 + 0.2422 = 0.7422 \quad \text{(see Figure 7.11)}$$

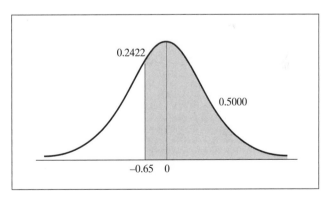

FIGURE 7.11 Area to the right of $z = -0.65$.

(c) The area to the right of $z = 1.76$ is 0.5000 minus the entry corresponding to $z = 1.76$:

$$0.5000 - 0.4608 = 0.0392 \quad \text{(see Figure 7.12)}$$

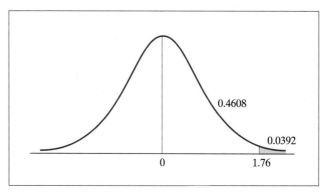

FIGURE 7.12 Area to the right of $z = 1.76$.

(d) The area to the left of $z = -0.85$ is 0.5000 minus the entry corresponding to $z = 0.85$:

$$0.5000 - 0.3023 = 0.1977 \quad \text{(see Figure 7.13)}.$$

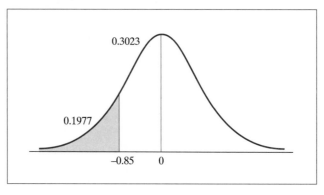

FIGURE 7.13 Area to the left of $z = -0.85$.

(e) The area between $z = 0.87$ and $z = 1.28$ is the difference between the entries corresponding to $z = 1.28$ and $z = 0.87$:

$$0.3997 - 0.3078 = 0.0919 \quad \text{(see Figure 7.14)}$$

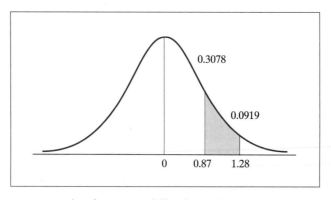

FIGURE 7.14 Area between $z = 0.87$ and $z = 1.28$.

(f) The area between $z = -0.34$ and $z = 0.62$ is the sum of the entries corresponding to $z = 0.34$ and $z = 0.62$:

$$0.1331 + 0.2324 = 0.3655 \quad \text{(see Figure 7.15)}$$ ◼

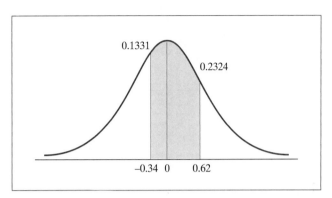

FIGURE 7.15 Area between $z = -0.34$ and $z = 0.62$.

In both of the preceding examples we dealt directly with the standard normal distribution. Now let us consider an example where μ and σ are not 0 and 1, so that we must first use the formula $z = \dfrac{x - \mu}{\sigma}$ to convert to standard units.

◼ **EXAMPLE**   Find the probabilities that a random variable will take on a value between 12 and 15 given that it has a normal distribution with

   (a) $\mu = 10$ and $\sigma = 5$;   (b) $\mu = 20$ and $\sigma = 10$.

Solution   (a) The probability is given by the area of the blue region of the upper diagram of Figure 7.16. The values of z corresponding to 12 and 15 are

$$z = \frac{12 - 10}{5} = 0.40 \quad \text{and} \quad z = \frac{15 - 10}{5} = 1.00$$

The corresponding entries in Table II are 0.1554 and 0.3413, and the probability that the random variable will take on a value between 12 and 15 is $0.3413 - 0.1554 = 0.1859$.

   (b) The probability is given by the area of the blue region of the lower diagram of Figure 7.16. The values of z corresponding to 12 and 15 are

$$z = \frac{12 - 20}{10} = -0.80 \quad \text{and} \quad z = \frac{15 - 20}{10} = -0.50$$

The entries in Table II corresponding to $z = 0.80$ and $z = 0.50$ are 0.2881 and 0.1915, and the probability that the random variable will take on a value between 12 and 15 is

$$0.2881 - 0.1915 = 0.0966$$ ◼

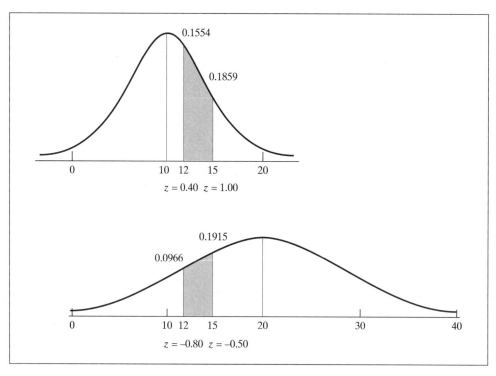

FIGURE 7.16 Areas under normal curves.

There are also problems in which we are given areas under normal curves and are asked to find the corresponding values of z. The results of the example that follows will be used extensively in subsequent chapters.

■ **EXAMPLE**   If z_α denotes the value of z for which the area under the standard normal curve to its right is equal to α (Greek lowercase *alpha*), find

(a) $z_{0.01}$;   (b) $z_{0.05}$.

Solution   (a) As can be seen from Figure 7.17, $z_{0.01}$ corresponds to an entry of $0.5000 - 0.0100 = 0.4900$ in Table II; since the nearest entry is 0.4901, corresponding to $z = 2.33$, we let $z_{0.01} = 2.33$.

(b) As can be seen from Figure 7.17, $z_{0.05}$ corresponds to an entry of $0.5000 - 0.0500 = 0.4500$ in Table II; since the two nearest entries are 0.4495 and 0.4505, corresponding to $z = 1.64$ and $z = 1.65$, we let $z_{0.05} = 1.645$. Our graphing calculator confirms that $z = 1.644853626$, or $z = 1.645$ rounded to three decimals.   ■

Table II also enables us to verify our earlier remark that for frequency distributions having the general shape of the cross section of a bell, about 68% of the values will lie within one standard deviation of the mean, about 95% will lie within two standard deviations of the mean, and about 99.7% will lie within three standard deviations of the mean. These figures apply to frequency distributions having the general shape of normal distributions, and in

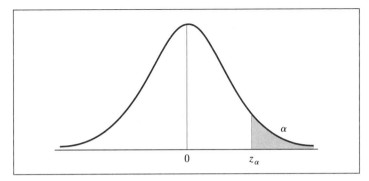

FIGURE 7.17 Diagram for determination of z_α.

Exercise 7.16 the reader will be asked to verify these percentages with the use of Table II.

Exercises

Exercises 7.1, 7.3, 7.5, 7.11, and 7.18 are practice exercises; their complete solutions are given on page 259. Each of Exercises 7.3 through 7.6 can be answered without written work and without use of the statistical tables.

7.1 Suppose that a continuous random variable takes on values on the interval from 1 to 5 and that the graph of its probability density is given by the horizontal line of Figure 7.18.

(a) What probability is represented by the blue region of the diagram and what is its value?

(b) What is the probability that the random variable will take on a value greater than 4.5? Would the answer be the same if we asked for the probability that the random variable will take on a value greater than or equal to 4.5?

(c) What is the probability that the random variable will take on a value between 1.8 and 4.2?

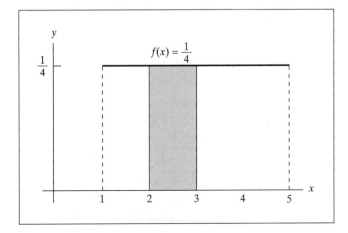

FIGURE 7.18 Diagram for Exercise 7.1.

7.2 Suppose that a continuous random variable takes on values on the interval from 0 to 4 and that the graph of its probability density is given by the blue line of Figure 7.19.
  (a) Verify that the total area under the curve is equal to 1.
  (b) What is the probability that this random variable will take on a value less than 3?
  (c) What is the probability that this random variable will take on a value between 1 and 2?

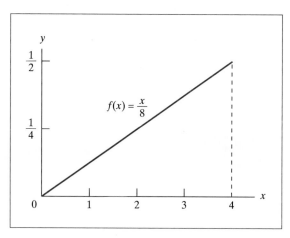

FIGURE 7.19 Diagram for Exercise 7.2.

7.3 For each of the following cases involving areas under the standard normal curve, decide whether the first area is bigger, the second area is bigger, or the two areas are equal:
  (a) the area to the right of $z = +1$ or (e) the area to the right of $z = -0.5$ or the area to the right of $z = +1.5$; the area to the right of $z = -1.5$;
  (b) the area to the left of $z = -1$ or (f) the area to the left of $z = 0$ or the area to the left of $z = -1.5$; the area to the right of $z = -0.1$;
  (c) the area to the right of $z = +1$ or (g) the area to the right of $z = 0$ or the area to the left of $z = -1.5$; the area to the left of $z = 0$;
  (d) the area to the right of $z = +1$ or (h) the area to the right of $z = -1.2$ or the area to the left of $z = -1$; the area to the left of $z = -1.2$.

7.4 For each of the following cases involving areas under the standard normal curve, decide whether the first area is bigger, the second area is bigger, or the two areas are equal:
  (a) the area between $z = 0$ and $z = 0.2$ or the area between $z = 1.1$ and $z = 1.2$;
  (b) the area between $z = -1$ and $z = -0.5$ or the area between $z = +0.5$ and $z = +1$;
  (c) the area to the left of $z = 1.5$ or the area to the right of $z = -0.5$;
  (d) the area between $z = 0$ and $z = 1$ or the area between $z = 0$ and $z = 1.4$;
  (e) the area between $z = -0.2$ and $z = +0.2$ or the area between -0.4 and $z = +0.4$;
  (f) the area between $z = -1$ and $z = +1$ or the area between $z = 0$ and $z = +2$;
  (g) the area between $z = +1$ and $z = +2$ or the area between $z = +2$ and $z = +3$;
  (h) the area between $z = -1.5$ and $z = +1$ or the area between $z = -1$ and $z = +1.5$.

7.5 For each of the following cases involving random variables with normal distributions, decide whether the first probability is bigger, the second probability is bigger, or the two probabilities are equal:

(a) for a random variable with a normal distribution with $\mu = 100$ and $\sigma = 20$, the probability of a value greater than 120 or the probability of a value less than 90;

(b) for a random variable with a normal distribution with $\mu = 100$ and $\sigma = 20$, the probability of a value greater than 120 or the probability of a value greater than 130;

(c) for a random variable with a normal distribution with $\mu = 200$ and $\sigma = 40$, the probability of a value greater than 250 or the probability of a value less than 140;

(d) for a random variable with a normal distribution with $\mu = 60$ and $\sigma = 12$, the probability of a value between 60 and 84 or the probability of a value between 48 and 72.

7.6 For each of the following cases involving random variables with normal distributions, decide whether the first probability is bigger, the second probability is bigger, or the two probabilities are equal:

(a) the probability that a random variable having the normal distribution with $\mu = 40$ and $\sigma = 5$ takes a value greater than 40, or the probability that a random variable having the normal distribution with $\mu = 50$ and $\sigma = 5$ takes a value greater than 40;

(b) the probability that a random variable having the normal distribution with $\mu = 50$ and $\sigma = 10$ takes a value less than 60, or the probability that a random variable having the normal distribution with $\mu = 500$ and $\sigma = 100$ takes a value less than 600;

(c) the probability that a random variable having the normal distribution with $\mu = 100$ and $\sigma = 5$ takes a value greater than 110, or the probability that a random variable having the normal distribution with $\mu = 108$ and $\sigma = 5$ takes a value greater than 110;

(d) the probability that a random variable having the normal distribution with $\mu = 50$ and $\sigma = 10$ takes a value less than 60, or the probability that a random variable having the normal distribution with $\mu = 50$ and $\sigma = 20$ takes a value less than 60.

7.7 In each given situation, you will receive a dollar amount generated either by process A or process B. For each process, the random variable follows a normal distribution. Decide whether you prefer process A or process B. [Only part (a) has a "right" answer.]

(a) Process A has $\mu = \$50$ and $\sigma = \$10$ and process B has $\mu = \$70$ and $\sigma = \$10$.

(b) Process A has $\mu = \$50$ and $\sigma = \$1$ and process B has $\mu = \$50$ and $\sigma = \$10$.

7.8 Find the area under the standard normal curve that lies

(a) between $z = -0.78$ and $z = 0$;   (c) to the left of $z = -1.55$;

(b) to the left of $z = 2.50$;   (d) between $z = 0.33$ and $z = 0.66$.

7.9 Find the area under the standard normal curve that lies

(a) between $z = 0$ and $z = 0.85$;

(b) to the right of $z = -0.65$;

(c) to the right of $z = 2.01$;

(d) between $z = -1.00$ and $z = -1.10$.

7.10 Find the area under the standard normal curve that lies

(a) between $z = -0.38$ and $z = 0.38$;

(b) between $z = -2.88$ and $z = 3.01$;

(c) between $z = -1.97$ and $z = -1.96$;

(d) between $z = -2.50$ and $z = 2.00$.

7.11 Find z if
   (a) the normal-curve area between 0 and z is 0.3340;
   (b) the normal-curve area to the left of z is 0.6517;
   (c) the normal-curve area to the left of z is 0.3085;
   (d) the normal-curve area between $-z$ and z is 0.9700.

7.12 Find z if
   (a) the normal-curve area between 0 and z is 0.1915;
   (b) the normal-curve area to the right of z is 0.8665;
   (c) the normal-curve area to the right of z is 0.0228;
   (d) the normal-curve area between $-z$ and z is 0.9500.

7.13 Find z if
   (a) the normal-curve area between 0 and $-z$ is 0.3508;
   (b) the normal-curve area to the left of z is 0.0838;
   (c) the normal-curve area to the left of z is 0.9812;
   (d) the normal-curve area between $-z$ and z is 0.6156.

7.14 A random variable has a normal distribution with $\mu = 69.0$ and $\sigma = 5.1$. What are the probabilities that the random variable will take on a value
   (a) less than 74.1;
   (b) greater than 63.9;
   (c) between 69.0 and 72.3;
   (d) between 66.2 and 71.8?

7.15 A random variable has a normal distribution with $\mu = 56.4$ and $\sigma = 4.8$. What are the probabilities that this random variable will take on a value
   (a) less than 65.0;
   (b) less than 46.5;
   (c) between 46.5 and 67.6;
   (d) between 50.4 and 62.4?

7.16 Find the probabilities that a random variable having a normal distribution will take on a value within
   (a) one standard deviation of the mean;
   (b) two standard deviations of the mean;
   (c) three standard deviations of the mean;
   (d) four standard deviations of the mean.

7.17 According to the definition given on page 244, $z_{\alpha/2}$ denotes the value of z for which the area under the standard normal curve to its right is equal to $\alpha/2$ and, hence, the area between $-z_{\alpha/2}$ and $z_{\alpha/2}$ is equal to $1 - \alpha$ (see Figure 7.20). Verify that
   (a) $z_{0.025} = 1.96$ corresponding to $\alpha = 0.05$;
   (b) $z_{0.005} = 2.575$ corresponding to $\alpha = 0.01$.

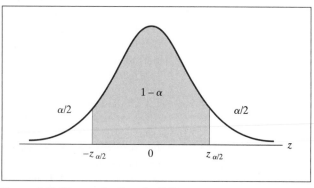

FIGURE 7.20 Diagram for Exercise 7.17.

7.18 A normal distribution has the mean $\mu = 61.6$. Find its standard deviation if 20% of the total area under the curve lies to the right of 70.0.

7.19 A normal distribution has the mean $\mu = 74.4$. Find its standard deviation if 10% of the area under the curve lies to the right of 100.0.

7.20 A random variable has a normal distribution with the standard deviation $\sigma = 10$. Find its mean if the probability is 0.8264 that it will take on a value less than 77.5.

7.21 For a certain random variable having the normal distribution, the probability is 0.33 that it will take on a value less than 245, and the probability is 0.48 that it will take on a value greater than 260. Find the mean and the standard deviation of this random variable.

7.3 Some Applications

Let us now consider some applied problems in which it will be assumed that the random variables under consideration have normal distributions.

■ **EXAMPLE** The amount of cosmic radiation to which a person is exposed while flying by jet across the United States is a random variable having a normal distribution with $\mu = 4.35$ mrem and $\sigma = 0.59$ mrem.[†] Find the probabilities that a person on such a flight will be exposed to

(a) more than 5.00 mrem of cosmic radiation;

(b) anywhere from 3.00 to 4.00 mrem of cosmic radiation.

Solution (a) This probability is given by the area of the blue region of the upper diagram of Figure 7.21, that is, the area under the curve to the right of

$$z = \frac{5.00 - 4.35}{0.59} \approx 1.10$$

Since the entry in Table II corresponding to $z = 1.10$ is 0.3643, we find that the probability is $0.5000 - 0.3643 = 0.1357$, or approximately 0.14, that a person will be exposed to more than 5.00 mrem of cosmic radiation on such a flight.

(b) This probability is given by the area of the blue region of the lower diagram of Figure 7.21, that is, the area under the curve between

$$z = \frac{3.00 - 4.35}{0.59} \approx -2.29 \quad \text{and} \quad z = \frac{4.00 - 4.35}{0.59} \approx -0.59$$

Since the entries in Table II corresponding to $z = 2.29$ and $z = 0.59$ are, respectively, 0.4890 and 0.2224, we find that the probability is $0.4890 - 0.2224 = 0.2666$, or approximately 0.27, that a person will be exposed to anywhere from 3.00 to 4.00 mrem of cosmic radiation on such a flight. ■

■ **EXAMPLE** The actual amount of instant coffee that a filling machine puts into "4-ounce" jars varies from jar to jar, and it may be looked upon as a random variable

---

[†] The unit of radiation mrem stands for *milliroentgen equivalent man*.

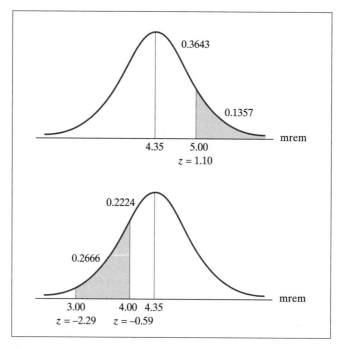

FIGURE 7.21 Diagrams for cosmic radiation example.

having a normal distribution with $\sigma = 0.04$ ounce. If only 2% of the jars are to contain less than 4 ounces of coffee, what must be the mean fill of these jars?

Solution We are given $\sigma = 0.04$, a normal-curve area (that of the blue region of Figure 7.22), and we are asked to find μ. Observe that μ must be safely above the stated weight of 4 ounces. Since the value of z for which the entry in Table II is closest to $0.5000 - 0.0200 = 0.4800$ is 2.05, we have $-2.05 = \dfrac{4.00 - \mu}{0.04}$.

Solving for μ, we get $4.00 - \mu = -2.05(0.04) = -0.082$ and then $\mu = 4.00 + 0.082 = 4.082$ ounces or 4.08 ounces to the nearest hundredth of an ounce. ∎

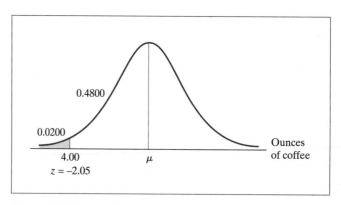

FIGURE 7.22 Diagram for instant-coffee filling machine example.

Although the normal distribution is a continuous distribution that applies to continuous random variables, it is often used to approximate distributions of random variables that can take on only a finite number of values or as many values as there are positive integers. In many situations, this will yield satisfactory results, provided that we make the **continuity correction** illustrated in the following example.

■ **EXAMPLE**   In a study of aggressive behavior, male white mice, returned to the group in which they live after four weeks of isolation, averaged 18.6 fights in the first five minutes with a standard deviation of 3.3 fights. If it can be assumed that the distribution of this random variable (the number of fights into which such a mouse gets under the stated conditions) can be approximated closely with a normal distribution, what is the probability that such a mouse will get into at least 15 fights in the first five minutes?

*Optional Solution   This solution is provided primarily for users of graphing calculators. With the exception of the explanation of the continuity correction, it can be omitted by other readers. The answer is read directly from the screen of the graphing calculator shown in Figure 7.23, and the probability is 0.89296. The probability is given by the area of the shaded region of the figure which is to the right of 14.5, not 15. The reason for this is that the number of fights in which such a mouse gets involved is a whole number. Hence, if we want to approximate the distribution of this random variable with a normal curve, we must *spread* its values over a continuous scale, and we do this by representing each whole number k by the interval from $k - \frac{1}{2}$ to $k + \frac{1}{2}$. For instance, 5 is represented by the interval from 4.5 to 5.5, 10 is represented by the interval from 9.5 to 10.5, 20 is represented by the interval from 19.5 to 20.5, and the probability of 15 or more is given by the area under the curve to the right of 14.5 (the boundary between the white and black areas of the graph). Accordingly, it follows from Figure 7.23 that the area of the black region—the probability that such a mouse will get into at least 15 fights in the first 5 minutes—is 0.89296, or approximately 0.89. ■

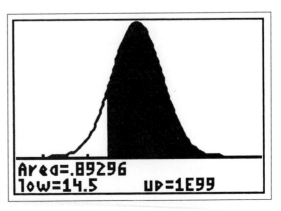

FIGURE 7.23 Normal-curve area reproduced from the display screen of a TI-83 graphing calculator.

USING TECHNOLOGY : *The Graphing Calculator*

How to graph the area under the normal curve shown in Figure 7.23

(Tip: Before beginning, reset the memory to default setting—see Appendix A: TI-83 Tips.)

Step 1.   Set the viewing window.

> Press **WINDOW**
> Set   Xmin = 2.1
>   Xmax = 35.1
>   Xscl = 6
>   Ymin = −.03
>   Ymax = .12
>   Yscl = 1
>   Xres = 1

Step 2.   Select the type of display and enter the interval to be displayed.
Press 2nd, **VARS** to access the **DISTR DRAW** menu.
Highlight **DRAW**.
Select 1: ShadeNorm(and press **ENTER**.
Enter   14.5, 1E99, 18.6, 3.3).
           (For instructions on entering "1E99", for infinity, see
           Appendix A: TI-83 Tips.)

Step 3.   View the graph.
Press **ENTER** to solve.

Solution (Without a Graphing Calculator)

To solve without a graphing calculator, we use the continuity correction and substitute $x = 14.5$ in place of 15, $\mu = 18.6$, and $\sigma = 3.3$ into the formula $z = \dfrac{x - \mu}{\sigma}$, obtaining

$$z = \frac{14.5 - 18.6}{3.3} \approx -1.24$$

Entering Table II with $z = 1.24$, we obtain 0.3925. Finally, we combine 0.3925 and 0.5000 and get the probability $0.5000 + 0.3925 = 0.8925$, which is nearly identical to the result obtained with the graphing calculator. ■

All the examples of this section dealt with random variables having normal distributions, or distributions that can be approximated closely with normal curves. Whenever we observe a value of a random variable having a normal distribution, we say that we are **sampling a normal population**.

Exercises

Exercises 7.22 and 7.29 are practice exercises; their complete solutions are given on page 260.

7.22 In an experiment to determine the amount of time required to assemble an "easy to assemble" toy, the assembly time was found to be a random variable having approximately a normal distribution with $\mu = 27.8$ minutes and $\sigma = 4.0$ minutes. What are the probabilities that this kind of toy can be assembled in
 (a) less than 25.0 minutes;
 (b) anywhere from 26.0 to 29.6 minutes?

7.23 A salesman who frequently drives from Boston to New York finds that his driving time is a random variable having roughly a normal distribution with $\mu = 4.3$ hours and $\sigma = 0.2$ hour. Find the probabilities that such a trip will take
 (a) more than 4.5 hours;
 (b) less than 4.0 hours.

7.24 With reference to Exercise 7.23, below what value (number of hours) are the fastest 25% of his trips?

7.25 Suppose that during periods of relaxation therapy the reduction of a person's oxygen consumption may be looked upon as a random variable having a normal distribution with $\mu = 38.6$ cubic centimeters (cc) per minute and $\sigma = 4.3$ cc per minute. Find the probabilities that during a period of relaxation therapy a person's oxygen consumption will be reduced by
 (a) at least 40.0 cc per minute;
 (b) anywhere from 35.0 to 45.0 cc per minute.

7.26 The lengths of the sardines received by a cannery have a mean of 4.64 inches and a standard deviation of 0.25 inch. If the distribution of these lengths can be approximated closely with a normal distribution, what percentage of all these sardines are
 (a) shorter than 4.00 inches;   (b) from 4.40 to 4.80 inches long?

7.27 With reference to Exercise 7.26, above which length lies the longest 15% of the sardines?

7.28 The management of a small ice skating rink knows that the daily numbers of skaters using the rink each morning is a random variable with a distribution that can be approximated closely by a normal distribution with the mean $\mu = 45.7$ and the standard deviation $\sigma = 4.8$. Using the continuity correction, determine the probability that the daily number of skaters using the rink during the morning session will
 (a) exceed 50 on any given day;
 (b) be less than 40 on any given day.

7.29 In a very large class in world history, the final examination grades have a mean of 66.5 and a standard deviation of 12.6. Assuming that it is reasonable to approximate the distribution of these grades with a normal distribution, what percentage of the grades should exceed 74? Assume that the grades are always whole numbers.

7.30 An airline knows from experience that the number of suitcases that get lost each week on a certain route is a random variable having approximately a normal distribution with the mean $\mu = 21.4$ and the standard deviation $\sigma = 4.5$. What are the probabilities that in any given week they will lose
 (a) exactly 18 suitcases;   (b) at most 18 suitcases?

7.31 In a municipal parking lot, the number of parking tickets issued daily for over-time parking is a random variable with $\mu = 15.3$ and $\sigma = 3.4$. What is the probability that on any day exactly 20 tickets will be issued in this parking lot?

7.32 The annual number of tornadoes in a certain state is a random variable with $\mu = 28.2$ and $\sigma = 5.1$. Approximating the distribution of this random variable with a normal distribution, find the probability that there will be at least 24 tornadoes in the state in any given year.

7.33 The number of days that patients are hospitalized at a certain hospital is a random variable having approximately a normal distribution with mean $\mu = 7.1$ and $\sigma = 3.2$. Among 1,000 patients, how many can be expected to be hospitalized anywhere from 5 to 10 days?

7.4
The Normal Approximation to the Binomial Distribution

The normal distribution, which is a continuous distribution, provides a close approximation to the binomial distribution when n, the number of trials, is large and p, the probability of success, is not too far from $\frac{1}{2}$. Figure 7.24 shows the histograms of binomial distributions with $p = \frac{1}{2}$ and $n = 2, 5, 10$, and 25, and it can be seen that with increasing n these distributions approach the symmetrical bell-shaped pattern of a normal curve. In fact, a normal distribution with $\mu = np$ and $\sigma = \sqrt{np(1 - p)}$ can often be used to approximate a binomial distribution, even when n is fairly small and p differs from $\frac{1}{2}$. A good rule of thumb is to use this approximation only when np and $n(1 - p)$ are both greater than 5;[†] symbolically, when

Conditions for the normal approximation to the binomial distribution

$$np > 5 \quad \text{and} \quad n(1 - p) > 5$$

■ **EXAMPLE** Check in each case whether the conditions for the normal approximation to the binomial distribution are satisfied.

(a) $n = 40$ and $p = \frac{1}{4}$;

(b) $n = 100$ and $p = \frac{1}{25}$;

(c) $n = 150$ and $p = 0.98$.

Solution
(a) Since $np = 40 \cdot \frac{1}{4} = 10$ and $n(1 - p) = 40 \cdot \frac{3}{4} = 30$ both exceed 5, the conditions are satisfied.

(b) Since $np = 100 \cdot \frac{1}{25} = 4$ does not exceed 5, the conditions are not satisfied.

(c) Since $n(1 - p) = 150(0.02) = 3$ does not exceed 5, the conditions are not satisfied.   ■

■ **EXAMPLE** Find the exact probability of getting 6 heads and 10 tails in 16 flips of a balanced coin, and also use a normal curve to approximate this binomial probability.

Solution Substituting $n = 16$, $p = \frac{1}{2}$, and $x = 6$ into the formula for the binomial distribution, we find that the probability of getting 6 heads and 10 tails in 16 flips of a balanced coin is

$$f(6) = \binom{16}{6}\left(\frac{1}{2}\right)^6\left(1 - \frac{1}{2}\right)^{16-6} = 8{,}008\left(\frac{1}{2}\right)^{16} = \frac{8{,}008}{65{,}536} \approx 0.1222$$

[†] Note that there are values of n and p (for instance, $n = 100$ and $p = 0.08$) for which either the normal approximation to the binomial distribution or the Poisson approximation (see page 211) can be used. This matter is discussed further on page 256.

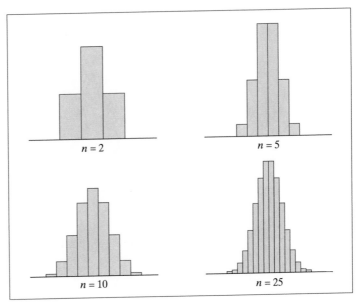

FIGURE 7.24 Binomial distributions with $p = \frac{1}{2}$ and $n = 2, 5, 10,$ and 25.

To find the normal-curve approximation to this probability, we use the continuity correction and represent 6 heads by the interval from 5.5 to 6.5 (see Figure 7.25). Since $\mu = 16 \cdot \frac{1}{2} = 8$ and $\sigma = \sqrt{16 \cdot \frac{1}{2} \cdot \frac{1}{2}} = 2$, converting 5.5 and 6.5 into standard units yields

$$z = \frac{5.5 - 8}{2} = -1.25 \quad \text{and} \quad z = \frac{6.5 - 8}{2} = -0.75$$

The entries in Table II corresponding to $z = 1.25$ and $z = 0.75$ are 0.3944 and 0.2734, and we get

$$0.3944 - 0.2734 = 0.1210$$

for the normal-curve approximation to the probability of getting 6 heads and 10 tails in 16 flips of a balanced coin. This differs by only 0.0012 from the value obtained with the formula for the binomial distribution. ∎

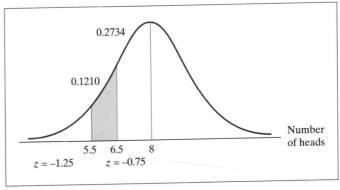

FIGURE 7.25 Normal-curve approximation to the binomial distribution.

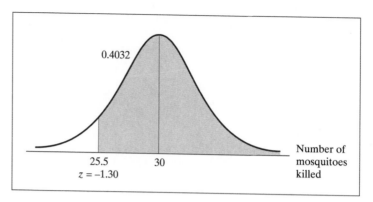

FIGURE 7.26 Normal-curve approximation to the binomial distribution.

The normal-curve approximation to the binomial distribution is particularly useful in problems where we would otherwise have to use the formula for the binomial distribution repeatedly to obtain the values of many different terms.

■ **EXAMPLE**   What is the probability that at least 26 of 50 mosquitoes will be killed by a new insect spray when the probability is 0.60 that any one of them will be killed by the spray?

Solution   If we tried to find the answer by using the formula for the binomial distribution, we would have to find the sum of the probabilities corresponding to $26, 27, 28, \ldots, 49$, and 50 successes in 50 trials. Without a computer, this would obviously involve a great deal of work; but by using the normal-curve approximation, we need only find the area of the blue region of Figure 7.26, namely, that to the right of 25.5. Observe that the mean is $0.60 \cdot 50 = 30$, so that we draw 25.5 to the left of the mean. Here, again, we are using the continuity correction since we are dealing with whole numbers. Accordingly, 26 is represented by the interval from 25.5 to 26.5, 27 is represented by the interval from 26.5 to 27.5, and so on.

Since $\mu = 50(0.60) = 30$ and $\sigma = \sqrt{50(0.60)(0.40)} \approx 3.464$, we find that in standard units 25.5 becomes

$$z = \frac{25.5 - 30}{3.464} \approx -1.30$$

The entry in Table II corresponding to $z = 1.30$ is 0.4032, and the probability of at least 26 successes in 50 trials when $p = 0.60$ is

$$0.5000 + 0.4032 = 0.9032 \qquad\qquad ■$$

To determine the error of the approximation made in this example, let us refer to the computer printout of Figure 7.27, which shows the cumulative "LESS OR =" binomial probabilities for $n = 50$ and $p = 0.60$. Since the probability of "at least 26 successes" is 1 minus the probability of "25 or less," we find that the desired probability is

$$1 - 0.0978 = 0.9022$$

```
MTB > CDF;
SUBC> BINOMIAL N=50 P=0.60

   BINOMIAL WITH N = 50 P = 0.60000

   K    P( X LESS OR = K)
   15          0.0000
   16          0.0001
   17          0.0002
   18          0.0005
   19          0.0014
   20          0.0034
   21          0.0076
   22          0.0160
   23          0.0314
   24          0.0573
   25          0.0978
   26          0.1562
```

27	0.2340
28	0.3299
29	0.4390
30	0.5535
31	0.6644
32	0.7631
33	0.8439
34	0.9045
35	0.9460
36	0.9720
37	0.9867
38	0.9943
39	0.9978
40	0.9992
41	0.9998
42	0.9999
43	1.0000

FIGURE 7.27 Modified MINITAB printout of the binomial distribution with $n = 50$ and $p = 0.60$.

and we find 0.0978 in the row corresponding to K = 25. Thus, the error of the normal approximation is only

$$0.9032 - 0.9022 = 0.0010$$

With the method of this section, we now have four ways of determining binomial probabilities:

1. We refer to a table of binomial probabilities or a computer printout of binomial probabilities.
2. We use the formula for the binomial distribution.
3. We use the Poisson approximation to the binomial distribution.
4. We use the normal approximation to the binomial distribution.

The first alternative is by far the most preferable, although it frequently requires that we go beyond the table given at the end of this book. In contrast, the formula for the binomial distribution is used only when absolutely necessary, as when it is the basis of an algebra exercise, rather than a direct computation.

As far as the approximations are concerned, they are not only subject to the rules of thumb stated previously, but their use requires a good deal of professional judgment. For instance, if either approximation can be used, we may be inclined to use the Poisson approximation to determine a probability associated with a single value of a binomial random variable, and we may be inclined to use the normal approximation to determine a probability associated with the tail (and hence many values) of a binomial distribution.

In any case, we have given the Poisson approximation to the binomial distribution mainly to *introduce* the Poisson distribution, and the normal approximation because it will be needed in Chapter 11 for large-sample inferences concerning proportions.

Exercises

Exercise 7.34 is a practice exercise; its complete solution is given on page 260.

7.34 If 20% of the loan applications received by a bank are refused, what is the probability that among 225 loan applications at least 50 will be refused?

7.35 Check in each case whether the conditions for the normal approximation to the binomial distribution are satisfied.
  (a) $n = 16$ and $p = \frac{1}{4}$;   (b) $n = 12$ and $p = 0.50$;   (c) $n = 100$ and $p = 0.96$.

7.36 Use the normal-curve approximation to find the probability of getting 5 heads and 7 tails in 12 flips of a balanced coin, and compare the result with the corresponding value given in Table I.

7.37 If 62% of all clouds seeded with silver iodide show spectacular growth, what is the probability that among 24 clouds thus seeded exactly 15 will show spectacular growth? If the value given for this probability in the table of binomial probabilities is 0.1661, what is the error of the approximation?

7.38 A dating service finds that 15% of the couples that it matches eventually get married. In the next 50 matches that the service makes, find the probabilities that
  (a) at least 6 couples marry;   (b) at most 10 couples marry.

7.39 A representative of a computer company claims that 20% of customer complaints result from the failure of customers to follow operating instructions. If this claim is correct, what is the probability that among 100 complaints more than 25 were caused by failure of customers to follow operating instructions?

7.40 A student answers each of the 48 questions on a multiple-choice test, each with four possible answers, by randomly drawing a card from an ordinary deck of 52 playing cards and checking the first, second, third, or fourth answer depending on whether the card drawn is a spade, heart, diamond, or club.
  (a) Use the normal-curve approximation to find the probability that the student will get exactly 15 correct answers.
  (b) Use the normal-curve approximation to find the probability that the student will get at least 15 correct answers.

7.41 If 80% of the students at a certain college reside in the college dormitories, what is the probability that among 200 randomly selected students more than 150 reside in the college dormitories?

7.42 To avoid accusations of sexism or worse, the author of a mathematics text decides by the flip of a balanced coin whether to use "he" or "she" whenever the occasion arises in exercises and examples. If he runs into this problem 80 times while revising one of his books, what is the probability that he will use "she" at least 48 times?

7.43 To illustrate the Law of Large Numbers, which we mentioned in connection with the frequency interpretation of probability, suppose that among all adults living in a large city we know that there are as many men as women. Using the normal-curve approximation, find the probabilities that in a random sample of adults living in this city the proportion of men will be anywhere from 0.49 to 0.51 when the number of persons in the sample is
  (a) 100;   (b) 1,000;   (c) 10,000.

7.44 Use a computer printout of the binomial distribution with $n = 50$ and $p = 0.15$ to determine the errors of the approximations of both parts of Exercise 7.38.

7.45 Use a computer printout of the binomial distribution with $n = 48$ and $p = 0.25$ to determine the errors of the approximations of both parts of Exercise 7.40.

7.46 Use a computer printout of the binomial distribution with $n = 40$ and $p = 0.30$ to determine the errors that we would make by using the normal-curve approximation to find the probabilities that a random variable having this binomial distribution will take on
 (a) a value greater than or equal to 15; (b) a value less than or equal to 13;
 (c) the value 9.

7.47 The normal-curve approximation can also be used for discrete distributions other than the binomial distribution. For instance, it is reasonable to use this approximation if the Poisson distribution has parameter λ greater than 30. This approximation uses λ as the mean and $\sqrt{\lambda}$ as the standard deviation.

Suppose that the number of false alarms received weekly by the fire department of a large city follows a Poisson distribution with $\lambda = 45$. In any given week, find the probability that
 (a) there will be 50 or more false alarms;
 (b) there will be at most 44 false alarms.

Solutions to Practice Exercises

7.1 (a) The blue region represents the probability of getting a value on the interval from 2 to 3, and its area is $\frac{1}{4} \cdot 1 = 0.25$.
 (b) The probability is $\frac{1}{4} \cdot (5 - 4.5) = 0.125$; the answer would be the same, since the probability is zero that the value of the random variable will actually equal 4.5.
 (c) The probability is $\frac{1}{4} \cdot (4.2 - 1.8) = 0.60$.

7.3 (a) First area is bigger; (e) second area is bigger;
 (b) first area is bigger; (f) second area is bigger;
 (c) first area is bigger; (g) areas are equal;
 (d) areas are equal; (h) first area is bigger.

7.5 (a) Second probability is bigger; (c) first probability is bigger;
 (b) first probability is bigger; (d) second probability is bigger.

7.11 (a) Since 0.3340 is the entry in Table II corresponding to 0.97, the answer is $z = 0.97$ or $z = -0.97$.
 (b) Since $0.6517 - 0.5000 = 0.1517$ is the entry in Table II corresponding to 0.39, the answer is $z = 0.39$.
 (c) Since $0.5000 - 0.3085 = 0.1915$ is the entry in Table II corresponding to 0.50, the answer is $z = -0.50$.
 (d) Since $\dfrac{0.9700}{2} = 0.4850$ is the entry in Table II corresponding to 2.17, the answer is $z = 2.17$ or $z = -2.17$.

7.18 Since the entry in Table II nearest to $0.5000 - 0.2000 = 0.3000$ is 0.2995, corresponding to 0.84, we get

$$z = \frac{70.0 - 61.6}{\sigma} = 0.84$$

and hence

$$0.84\sigma = 70.0 - 61.6 = 8.4 \quad \text{and} \quad \sigma = \frac{8.4}{0.84} = 10$$

7.22   (a) $z = \dfrac{25.0 - 27.8}{4.0} = -0.70$, and since the entry in Table II corresponding to 0.70 is 0.2580, the probability is

$$0.5000 - 0.2580 = 0.2420$$

(b) $z = \dfrac{29.6 - 27.8}{4.0} = 0.45$ and $z = \dfrac{26.0 - 27.8}{4.0} = -0.45$; and since the entry in Table II corresponding to 0.45 is 0.1736, the probability is

$$0.1736 + 0.1736 = 0.3472$$

7.29 Using the continuity correction, we must find the area under the curve to the right of 74.5; since $z = \dfrac{74.5 - 66.5}{12.6} \approx 0.63$ and the entry in Table II corresponding to 0.63 is 0.2357, it follows that the answer is

$$0.5000 - 0.2357 = 0.2643, \quad \text{or } 26.43\%$$

7.34 Using the continuity correction, we must find the area under the curve to the right of 49.5; since $\mu = 225(0.20) = 45$ and $\sigma = \sqrt{225(0.20)(0.80)} = 6$, so that $z = \dfrac{49.5 - 45}{6} = 0.75$, it follows that the entry in Table II corresponding to 0.75 is 0.2734 and that the required probability is

$$0.5000 - 0.2734 = 0.2266$$

8 Sampling and Sampling Distributions

The main objective of most statistical studies is to make sound generalizations, based on samples, about the parameters of populations. Note the word *sound*, because the question of when and under what conditions samples permit such generalizations is not easily answered. For instance, if we want to estimate the average amount of money that persons spend on their vacations, would we take as our sample the amounts spent by deluxe-class passengers on an around-the-world cruise? Would we attempt to predict changes in retail prices of farm products on the basis of changes in the prices of peaches alone? Obviously not, but just which vacationers and which farm products we should include in our samples, and how many of them, is neither intuitively clear nor self-evident.

     In most of the methods that we shall study in this book, it will be assumed that we are using **random sampling**, that is, a method of sampling for which every possible sample has the same probability of being selected. We pay this attention to random sampling because it permits valid, or logical, generalizations and hence is widely used in practice. In this chapter we begin with a formal definition of random sampling in Section 8.1. Then, in Section 8.2, we introduce the related concept of a sampling distribution, which tells us how quantities determined from samples may vary from sample to sample; and in Sections 8.3 and 8.4 we learn how such variations can be measured.

8.1
Random Sampling

In Section 1.3, we distinguished between populations and samples, saying that a population consists of all conceivably possible observations (instances, or occurrences) of a given phenomenon, while a sample is simply part of a population. For the work that follows, let us also distinguish between **finite populations** and **infinite populations**.

A population is finite if it consists of a finite, or fixed, number of elements (items, objects, measurements, or observations). Examples of finite populations are the net weights of 5,000 cans of baked beans in a production lot, SAT scores of all the freshmen admitted to a certain university in 1999, or the assessed values of all structures in a certain town.

In contrast, a population is infinite if there is no limit to the number of items, or values, that can be observed. This is the case, for example, when we observe repeated flips of a coin, when we sample with replacement from a population, or when we observe values of a continuous random variable, say, when we sample a normal population. The essential distinction between the two kinds of populations is that a finite population has a finite number of elements.

It should be apparent from this distinction between finite and infinite populations that the previous definition of *random sampling* applies only to finite populations—if there are infinitely many possibilities, we cannot very well speak of "every possible sample." Indeed, let us state formally that

A sample of size n from a finite population of size N is random if it is chosen in such a way that each of the $\binom{N}{n}$ possible samples has the same probability, $\dfrac{1}{\binom{N}{n}}$, of being selected.

For instance, if a finite population consists of the $N = 5$ elements $a, b, c, d,$ and e (which might be the incomes of five persons, the weights of five guinea pigs, or the prices of five commodities), there are $\binom{5}{3} = 10$ possible samples of size $n = 3$ consisting, respectively, of the elements $abc, abd, abe, acd, ace, ade, bcd, bce, bde,$ and cde. If we choose one of these samples in such a way that each sample has the probability $\frac{1}{10}$ of being selected, we call this sample a **random sample**.

This leaves the question of how random samples can be drawn in actual practice. In a simple case like the one described in the preceding paragraph, we could write each of the 10 possible samples on a slip of paper, put the slips of paper into a hat, shuffle them thoroughly, and then draw one without looking. Obviously, though, this would be impractical in a more realistically complex situation where N, and hence $\binom{N}{n}$, is quite large. In fact, we have mentioned it here only to stress the point that the selection of a random sample must depend entirely on chance.

Fortunately, we can take random samples from finite populations without having to go through the tedious process of listing all possible samples. One possibility is to write each of the N elements on a slip of paper and then take a random sample by choosing the elements to be included in the sample one at a time, making sure that in each successive drawing the remaining elements all have the same chance of being selected. It is not very difficult to show mathematically that this also leads to the same probability of $\dfrac{1}{\binom{N}{n}}$ for each possible sample. For instance, to take a random sample of size $n = 12$ from the population that consists of the sales tax collected by a city's 247 drugstores in December 1999, we could write each of the 247 figures on a slip of paper, mix them up thoroughly in a box, and then draw (without looking) 12 of the slips of paper one after the other without replacement.

Even this relatively easy procedure can be simplified further. Usually, the simplest way of taking a random sample from a finite population is to refer to a table of **random numbers**. Such a table contains the digits 0, 1, 2, 3, 4, 5, 6, 7, 8, and 9 set down as they might appear if they had been generated by a gambling device, giving each digit the same probability of $\frac{1}{10}$ of appearing at any given place in the table. Table XII at the end of the book is excerpted from such a table.

■ **EXAMPLE**   With reference to the sales tax example, we number the 247 drugstores 001, 002, 003, ..., 246, and 247. Then we arbitrarily pick a starting place in the table in Figure 8.1 and move in any direction, reading off three-digit numbers. For instance, we can read off the digits in the 26th, 27th, and 28th columns of the table, starting with the sixth row and going down the page, to obtain a random sample of 12 of the drugstores and, hence, a random sample of size 12 of the sales tax figures for December 1999.

Solution   Ignoring numbers greater than 247, it can be seen from Figure 8.1 that we get

046   230   079   022   119   150   056   064   193   232   040   146

These are the numbers of the drugstores, and the corresponding sales tax figures constitute the desired random sample. If any number had recurred when reading the values off the table, it would also have been ignored. It is necessary that we can associate the drugstores with the numbers 1, 2, ..., 247. In this case, store 046 was selected, and we have to be able to identify that particular store. ■

Some early tables of random numbers were copied from pages of census data and from tables of 20-place logarithms, but now they are prepared with the use of computers. Indeed, it is not difficult to program a computer so that a person can generate his or her own random numbers.

When lists are available and items are readily numbered, it is easy to draw samples from finite populations with the aid of published or computer-generated random numbers. Unfortunately, though, it is often impossible to proceed as in the preceding example. For instance, if we want to use a sample to estimate the mean diameter of thousands of ball bearings packed in a

48611	62866	33963	14045	79451	04934	45576
78812	03509	78673	73181	29973	18664	04555
19472	63971	37271	31445	49019	49405	46925
51266	11569	08697	91120	64156	40365	74297
55806	96275	26130	47949	14877	69594	83041
77527	81360	18180	97421	55541	90275	18213
77680	58788	33016	61173	93049	✓04694	43534
15404	96554	88265	34537	38526	67924	40474
14045	22917	60718	66487	46346	30949	03173
68376	43918	77653	04127	69930	43283	35766
93385	13421	67957	20384	58731	53396	59723
09858	52104	32014	53115	03727	98624	84616
93307	34116	49516	42148	57740	31198	70336
04794	01534	92058	03157	91758	80611	45357
86265	49096	97021	92582	61422	75890	86442
65943	79232	45702	67055	39024	57383	44424
90038	94209	04055	27393	61517	✓23002	96560
97283	95943	78363	36498	40662	94188	18202
21913	72958	75637	99936	58715	✓07943	23748
41161	37341	81838	19389	80336	46346	91895
23777	98392	31417	98547	92058	✓02277	50315
59973	08144	61070	73094	27059	69181	55623
82690	74099	77885	23813	10054	✓11900	44653
83854	24715	48866	65745	31131	47636	45137
61980	34997	41825	11623	07320	✓15003	56774
99915	45821	97702	87125	44488	77613	56823
48293	86847	43186	42951	37804	85129	28993
33225	31280	41232	34750	91097	60752	69783
06846	32828	24425	30249	78801	26977	92074
32671	45587	79620	84831	38156	74211	82752
82096	21913	75544	55228	89796	✓05694	91552
51666	10433	10945	55306	78562	89630	41230
54044	67942	24145	42294	27427	84875	37022
66738	60184	75679	38120	17640	36242	99357
55064	17427	89180	74018	44865	53197	74810
69599	60264	84549	78007	88450	✓06488	72274
64756	87759	92354	78694	63638	80939	98644
80817	74533	68407	55862	32476	✓19326	95558
39847	96884	84657	33697	39578	90197	80532
90401	41700	95510	61166	33757	✓23279	85523
78227	90110	81378	96659	37008	✓04050	04228
87240	52716	87697	79433	16336	52862	69149
08486	10951	26832	39763	02485	71688	90936
39338	32169	03713	93510	61244	73774	01245
21188	01850	69689	49426	49128	✓14660	14143
13287	82531	04388	64693	11934	35051	68576
53609	04001	19648	14053	49623	10840	31915
87900	36194	31567	53506	34304	39910	79630
81641	00496	36058	75899	46620	70024	88753
19512	50277	71508	20116	79520	06269	74173

FIGURE 8.1 Part of a table of random numbers.

crate or if we want to estimate the mean height of the trees in a forest, it would be impossible to number the ball bearings or the trees, choose random numbers, and then locate and measure the corresponding ball bearings or trees. In these and in many similar situations, we may have no choice but to proceed according to the alternative dictionary definition of the word *random*, which says that the selection should be "haphazard, without aim or purpose." With some reservations, such samples can often be treated as if they were, in fact, real random samples. We must be careful in such cases to avoid unintentional selection by size. For example, we may favor large trees when making the selection. It therefore helps to have a clear idea as to exactly which trees are to be considered for the sample. For instance, we might decide that all trees with a height of at least 6 feet and diameter of at least 2 inches are eligible to be sampled.

Thus far we have discussed random sampling only in connection with finite populations.

> *For infinite populations, we say that a sample is random if it consists of values of independent random variables having the same distribution.*

By independent, we mean that the distribution of any one of the random variables is in no way affected by the values taken on by the others. For example, this definition applies when we observe honest flips of a balanced coin. If we get, say,

$$H \quad T \quad H \quad T \quad T \quad T \quad H \quad H \quad T \quad H \quad H \quad H \quad T$$

where H and T denote heads and tails, this result is a random sample as long as the H's and T's are values of independent random variables, each having the binomial distribution with $n = 1$ and $p = \frac{1}{2}$. As we pointed out on page 262, the population that we are sampling here is infinite because there is no limit to the number of times we can flip a coin.

For another example of random sampling from an infinite population, suppose that the weight loss of persons on a certain two-week diet is a random variable having the normal distribution with $\mu = 7.4$ pounds and $\sigma = 1.3$ pounds. If we get, say, 8.3, 5.9, 7.0, 10.5, and 6.8 pounds for the weight loss of five persons who have been on the diet, these figures constitute a random sample as long as they are values of independent random variables, each having the normal distribution with $\mu = 7.4$ pounds and $\sigma = 1.3$ pounds. To judge whether this is actually the case, we would have to see how the persons were chosen, that is, how the data was collected.

Exercises

Exercises 8.1 and 8.2 are practice exercises; their complete solutions are given on page 281.

8.1 A random sample of size $n = 3$ is drawn from the finite population that consists of the elements a, b, c, d, and e. What is the probability that any specific element, say, element b, will be contained in the sample?

8.2 A college president wants to determine the attitude of the college's 7,423 students, who are numbered 0001 through 7,423, toward a small increase in tuition. If the president decides to interview 12 students concerning the proposed change, which ones would be selected if the sample were chosen by using the first four columns of the random digits table (Figure 8.1), beginning with row five?

8.3 How many different samples of size $n = 3$ can be drawn from a finite population of size
(a) 5;   (b) 10;   (c) 15?

8.4 How many different samples of size 2 can be drawn from a finite population of size
(a) 5;   (b) 10;   (c) 15?

8.5 What is the probability of each possible sample if a random sample of size 3 is to be drawn from a finite population of size
(a) 5;   (b) 10;   (c) 15?

8.6 What is the probability of each possible sample of a random sample of size 6 being drawn from a finite population of size 18?

8.7 List the $\binom{6}{2} = 15$ possible samples of size $n = 2$ that can be drawn from the finite population whose elements are denoted by $a, b, c, d, e,$ and f.

8.8 With reference to Exercise 8.7, what is the probability that a random sample of size $n = 2$ from the given finite population will include the element denoted by the letter f?

8.9 List all possible combinations of five of these six hydrocarbons: benzene, decane, hexane, octane, toluene, and xylene. If an engineer randomly selects five of these liquid hydrocarbons to study their heat content (measured in British thermal units per pound), find
(a) the probability of each possible sample;
(b) the probability that toluene will be included in the sample.

8.10 A marketing research organization wants to include 5 of the 50 states of the United States in a marketing survey. If the states are numbered 01, 02, 03, ..., 49, and 50, which ones, by number, will be included if the organization selects them by using the 14th and 15th columns of the table in Figure 8.1, going down the page and beginning with the fourth row?

8.11 A new-car dealer wants to inspect a sample of $n = 6$ of 150 new cars that are parked in parking spaces numbered 001, 002, ..., 149, and 150. Which ones will be selected if the dealer chooses the last three columns (columns 33, 34, and 35) of the table in Figure 8.1, going down the page and starting with the first row?

8.12 Two hundred sales have been recorded by salespersons on sales slips consecutively numbered from 201 to 400, inclusive. If an auditor wants to randomly select six of them for verification, which ones would be chosen by using the 11th, 12th, and 13th columns of the table in Figure 8.1, going down the page and starting with the 10th row?

8.13 A college registrar wants to verify the grade-point averages of a random sample of 15 graduating students in a graduating class of 3,852 students. If the students are numbered 0001, 0002, ..., 3,851, and 3,852, which ones (by number) will the registrar select by choosing columns 21, 22, 23, and 24 of the table in Figure 8.1, going down the page and starting with the second row?

8.14 A laboratory supply warehouse stores its inventory of 6,429 items in bins, compartments and spaces that are consecutively numbered 0001, 0002, ..., 6,428, and 6,429. To check the perpetual inventory of items that is maintained by computer, the manager decides to randomly select 20 storage spaces and manually count the items in them. Which of the bins, compartments, and spaces will be selected if the manager uses the 26th, 27th, 28th, and 29th columns of the table in Figure 8.1, going down the page and starting with the first row?

8.15 Randy and Susan are both members of a population of 60 students. A researcher is going to select 10 students from this population.
  (a) What is the probability that Randy will be in the sample?
  (b) What is the probability that Susan will be in the sample?
  (c) What is the probability that both will be in the sample?
  (d) Is your solution to (c) greater than or less than $\left(\dfrac{10}{60}\right)^2 = \dfrac{1}{36}$?

8.2
Sampling Distributions

The sample mean, the sample median, and the sample standard deviation are examples of random variables, whose values will vary from sample to sample. Their distributions, which reflect such chance variations, play an important role in statistics, and they are referred to as **sampling distributions**.

To illustrate the concept of a sampling distribution, let us construct the one for the mean of a random sample of size $n = 2$ from the finite population of size $N = 5$, whose elements are the numbers 1, 3, 5, 7, and 9. The mean and the standard deviation of this population, which we shall need later, are

$$\mu = \frac{1 + 3 + 5 + 7 + 9}{5} = 5$$

and

$$\sigma = \sqrt{\frac{(1 - 5)^2 + (3 - 5)^2 + (5 - 5)^2 + (7 - 5)^2 + (9 - 5)^2}{5}} = \sqrt{8} \approx 2.828$$

in accordance with formulas on pages 45 and 68.

This illustration is, of course, hypothetical in that we know all N values. In actual applications we never know all the values. Now, if we take a random sample of size $n = 2$ from this population, there are

$$\binom{5}{2} = 10 \text{ possibilities}$$

and they are

1 and 3	3 and 7
1 and 5	3 and 9
1 and 7	5 and 7
1 and 9	5 and 9
3 and 5	7 and 9

The means of these samples are

$$\frac{1 + 3}{2} = 2, \quad \frac{1 + 5}{2} = 3, \quad 4, \quad 5, \quad 4, \quad 5, \quad 6, \quad 6, \quad 7, \quad \text{and} \quad 8$$

If sampling is random, then each sample has the probability $\frac{1}{10}$, and we arrive at the following sampling distribution of the mean:

$\bar{x}$	Probability
2	$\frac{1}{10}$
3	$\frac{1}{10}$
4	$\frac{2}{10}$
5	$\frac{2}{10}$
6	$\frac{2}{10}$
7	$\frac{1}{10}$
8	$\frac{1}{10}$

A histogram of this probability distribution is shown in Figure 8.2.

An examination of this sampling distribution reveals some pertinent information about the chance variations of the mean of a random sample of size $n = 2$ from the given population. For instance, we find that the probability is $\frac{6}{10}$ that a sample mean will differ from the population mean $\mu = 5$ by 1 or less, and that the probability is $\frac{8}{10}$ that a sample mean will differ from the population mean $\mu = 5$ by 2 or less. (The first case corresponds to $\bar{x} = 4, 5,$ or 6, and the second case corresponds to $\bar{x} = 3, 4, 5, 6,$ or 7.) Thus, if we did not know the mean of the given population and wanted to estimate it with the mean of a random sample of two observations, this would give us some idea about the potential size of our error. Further useful information about this sampling distribution of the mean can be obtained by calculating its mean $\mu_{\bar{x}}$ and its standard deviation $\sigma_{\bar{x}}$, where the subscripts serve to distinguish between these parameters and those of the original population. Following the definitions of the mean and the variance of a probability distribution on pages 220 and 221, we get

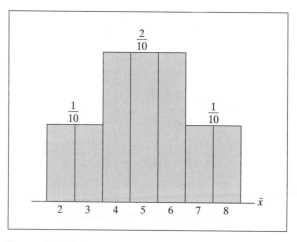

FIGURE 8.2 Sampling distribution of the mean.

$$\mu_{\bar{x}} = 2 \cdot \frac{1}{10} + 3 \cdot \frac{1}{10} + 4 \cdot \frac{2}{10} + 5 \cdot \frac{2}{10}$$

$$+ 6 \cdot \frac{2}{10} + 7 \cdot \frac{1}{10} + 8 \cdot \frac{1}{10} = 5$$

and

$$\sigma_{\bar{x}}^2 = (2-5)^2 \cdot \frac{1}{10} + (3-5)^2 \cdot \frac{1}{10}$$

$$+ (4-5)^2 \cdot \frac{2}{10} + (5-5)^2 \cdot \frac{2}{10}$$

$$+ (6-5)^2 \cdot \frac{2}{10} + (7-5)^2 \cdot \frac{1}{10}$$

$$+ (8-5)^2 \cdot \frac{1}{10} = 3$$

so that $\sigma_{\bar{x}} = \sqrt{3} \approx 1.732$.

Observe that, at least for this example,

1. $\mu_{\bar{x}}$, the mean of the sampling distribution of $\bar{x}$, equals μ, the mean of the population;

2. $\sigma_{\bar{x}}$, the standard deviation of the sampling distribution of $\bar{x}$, is smaller than σ, the standard deviation of the population.

These relationships are of fundamental importance in statistics, and we shall return to them in Section 8.3.

In the preceding example we took a very small sample from a very small population, but it would be difficult to use the same method to construct the sampling distribution of the mean of a large sample from a large population— we would have to enumerate too many possibilities. To construct the sampling distribution of the mean even for random samples of size $n = 5$ from a finite population of size $N = 100$, we would have to list

$$\binom{100}{5} = 75,287,520 \text{ possibilities}$$

So, to get an idea about the sampling distribution of the mean of a larger random sample from a larger finite population, we shall use a **computer simulation**. This means that we shall use a computer to take a number of random samples from the given population, calculate their means, and display the results in the form of a histogram. This histogram will, we hope, give us some idea about the key features of the actual sampling distribution of the mean for random samples of the given size from the given population.

■ **EXAMPLE** Use a computer to generate 100 random samples of size $n = 15$ from the finite population that consists of the integers from 1 to 1,000. Also, calculate the mean of each of the 100 samples, find their mean and their standard deviation, and group them into a distribution with the class marks 225.5, 275.5,…, and so on.[†]

Solution   Without a computer, the reader may picture this kind of simulation as follows: First, the numbers from 1 to 1,000 are written on 1,000 slips of paper (poker chips, small balls, or whatever may lend itself to drawing random samples). Then, a random sample of size $n = 15$ is drawn from this population and the values are recorded; each random sample of 15 is replaced before the next sample is drawn. This process is repeated until 100 random samples have been obtained.

Actually using an appropriate computer package, MINITAB in this case, we obtained the printout shown in Figure 8.3. Here the lines

$$MTB > SET\ C1 \quad and \quad DATA > 1:1000$$

are instructions to the computer to generate our population, the integers from 1 to 1,000, placing the values in column 1. In the next step, this population is described as having $\mu = 500.5$ and $\sigma = 289$ (to the nearest integer). Then, the 100 samples of size $n = 15$ are generated and their means are put into column C3. The printout following the instructions

$$MEAN\ C3 \quad and \quad STDEV\ C3$$

tells us that the mean of the 100 means is 495.75 and that their standard deviation is 75.748.

Finally, the 100 sample means are grouped into a distribution with the required classes, and this distribution is presented graphically in a form that might be described as a histogram lying on its side. We might say that this distribution is fairly symmetrical and bell shaped; in fact, the overall pattern seems to follow quite closely that of a normal curve. All this applies to the distribution that we obtained by means of the computer simulation involving only 100 random samples of size $n = 15$ from the population that consists of the integers from 1 to 1,000.   ∎

Note also that the results of this experiment support the two points made on page 269. Although the mean of the one hundred $\bar{x}$'s does not equal $\mu = 500.5$, its value, 495.75, is very close. Also, the standard deviation of the one hundred $\bar{x}$'s, 75.748, is smaller than the population standard deviation, $\sigma = 289$.

8.3
The Standard Error of the Mean

In most practical situations we can determine how close a sample mean might be to the mean of the population from which the sample came by referring to two theorems that express essential facts about sampling distributions. The first of these theorems expresses formally what we discovered in connection with the first sampling distribution of Section 8.2: the mean of the sampling distribution of $\bar{x}$ equals the mean of the population, and its standard deviation is smaller than the standard deviation of the population. It may be

---

[†] These class marks were chosen so that the corresponding class boundaries are "impossible" values; they are values that cannot be taken on by a mean of 15 positive integers.

```
MTB > SET C1
DATA> 1:1000
DATA> END
MTB > MEAN C1
   MEAN = 500.50
MTB > STDEV C1
   ST.DEV. = 288.82
MTB > SET C3
DATA> *
DATA> END
MTB > STORE 'E'
STOR> SAMPLE 15 C1 C2
STOR> LET K1 = AVER(C2)
STOR> STACK C3 K1 PUT INTO C3
STOR> ERASE C2
STOR> END
MTB > NOECHO
MTB > EXECUTE 'E' 100 TIMES
MTB > MEAN C3
    MEAN    =     495.75
MTB > STDEV C3
    ST.DEV. =      75.748
MTB > HIST C3 225.5 50


Histogram of C3   N = 100 N* = 1

 Midpoint Count
  225.5        0
  275.5        0
  325.5        2 **
  375.5       12 ***********
  425.5       16 ***************
  475.5       20 *******************
  525.5       26 *************************
  575.5       15 **************
  625.5        8 ********
  675.5        1 *
```

FIGURE 8.3 Computer simulation of a sampling distribution of the mean.

phrased as follows: For random samples of size n taken from a population having the mean μ and the standard deviation σ, the sampling distribution of $\bar{x}$ has the mean

Mean of the sampling distribution of $\bar{x}$

$$\mu_{\bar{x}} = \mu$$

and the standard deviation

Standard error of the mean

$$\sigma_{\bar{x}} = \frac{\sigma}{\sqrt{n}} \quad \text{or} \quad \sigma_{\bar{x}} = \frac{\sigma}{\sqrt{n}} \cdot \sqrt{\frac{N-n}{N-1}}$$

depending on whether the population is infinite or finite of size N.

It is customary to refer to $\sigma_{\bar{x}}$, the standard deviation of the sampling distribution of the mean, as the **standard error of the mean**. Its role in statistics is fundamental, since it measures the extent to which sample means can be expected to fluctuate, or vary, due to chance. What determines the size of $\sigma_{\bar{x}}$, and hence the goodness of an estimate, can be seen from the preceding formulas. Both formulas show that the standard error of the mean increases as the variability of the population increases (in fact, it is directly proportional to σ) and that it decreases as the sample size increases.

■ **EXAMPLE**   When we sample from an infinite population, what happens to the standard error of the mean (and hence to the size of the error that we are exposed to when we use $\bar{x}$ as an estimate of μ) if the sample size is increased from $n = 50$ to $n = 200$?

Solution   The ratio of the two standard errors is

$$\frac{\dfrac{\sigma}{\sqrt{200}}}{\dfrac{\sigma}{\sqrt{50}}} = \frac{\sqrt{50}}{\sqrt{200}} = \sqrt{\frac{50}{200}} = \sqrt{\frac{1}{4}} = \frac{1}{2}$$

so that the standard error of the mean is divided by 2.   ■

The factor $\sqrt{\dfrac{N-n}{N-1}}$ in the second formula for $\sigma_{\bar{x}}$ is called the **finite population correction factor**. Without it, the two formulas for $\sigma_{\bar{x}}$ (for infinite and finite populations) are the same. In practice, it is omitted unless the sample constitutes at least 5% of the population, for otherwise it is so close to 1 that it has little effect on the value of $\sigma_{\bar{x}}$.

■ **EXAMPLE**   Find the value of the finite population correction factor for $n = 100$ and
(a) $N = 10{,}000$;   (b) $N = 200$.

Solution   (a) Substituting $n = 100$ and $N = 10{,}000$, we get

$$\sqrt{\frac{N-n}{N-1}} = \sqrt{\frac{10{,}000 - 100}{10{,}000 - 1}} \approx 0.995$$

and this is so close to 1 that the correction factor can be omitted for most practical purposes.
(b) Substituting $n = 100$ and $N = 200$, we get

$$\sqrt{\frac{N-n}{N-1}} = \sqrt{\frac{200 - 100}{200 - 1}} \approx 0.709$$

so that the finite population correction factor will substantially reduce the standard error of the mean. This reflects the fact that in this case we have a great deal of information about the population; in fact, the sample constitutes half the population. ∎

Since we did not actually prove the two formulas for the standard error of the mean, let us verify the second one with reference to previous examples on pages 267 and 271.

■ **EXAMPLE** With reference to the illustration on page 267, verify that the formula for $\sigma_{\bar{x}}$ for a random sample from a finite population will also yield $\sigma_{\bar{x}} = \sqrt{3}$.

Solution Substituting $n = 2$, $N = 5$, and $\sigma = \sqrt{8}$ into the second of the two formulas for $\sigma_{\bar{x}}$, we get

$$\sigma_{\bar{x}} = \frac{\sqrt{8}}{\sqrt{2}} \cdot \sqrt{\frac{5-2}{5-1}} = \frac{\sqrt{8}}{\sqrt{2}} \cdot \sqrt{\frac{3}{4}} = \sqrt{\frac{8}{2} \cdot \frac{3}{4}} = \sqrt{3}$$ ∎

■ **EXAMPLE** With reference to the computer simulation on page 271, where we had $n = 15$, $N = 1,000$, and $\sigma = 289$, what value might we have expected for the standard deviation of the 100 sample means?

Solution Substituting $n = 15$, $N = 1,000$, and $\sigma = 289$ into the second of the two formulas for σ, we get

$$\sigma_{\bar{x}} = \frac{289}{\sqrt{15}} \cdot \sqrt{\frac{1,000-15}{1,000-1}} \approx 74.1$$

and this is fairly close to 75.748, the value that we actually obtained in the printout of Figure 8.3. ∎

Exercises

Exercises 8.16 and 8.17 are practice exercises; their complete solutions are given on pages 281 and 282.

8.16 For a sample of size $n = 2$, finite population size of $N = 5$, and elements numbered 1, 3, 5, 7, and 9, suppose that sampling is *with replacement*. List the 25 ordered samples 1 and 1, 1 and 3, 1 and 5, 1 and 7, 1 and 9, 3 and 1, 3 and 3, 3 and 5, and so on. Calculate the means of these 25 ordered samples. Then, assuming that each ordered pair has the probability of $\frac{1}{25}$, construct the sampling distribution of the mean for random samples of size $n = 2$ taken with replacement from the given population.

8.17 Calculate the standard deviation of the sampling distribution obtained in Exercise 8.16 and verify the result by substituting $n = 2$ and $\sigma = \sqrt{8}$ into the standard error formula $\sigma_{\bar{x}} = \frac{\sigma}{\sqrt{n}}$.

8.18 Random samples of size $n = 2$ are drawn from a finite population that consists of the numbers 2, 4, 6, and 8.
  (a) Calculate the mean and the standard deviation of this population.

(b) List the six possible random samples of size $n = 2$ that can be drawn from this population and calculate their means.

(c) Use the results of part (b) to construct the sampling distribution of the mean for random samples of size $n = 2$ from the given population.

(d) Calculate the standard deviation of the sampling distribution obtained in part (c) and verify the result by substituting $n = 2$, $N = 4$, and the value of σ obtained in part (a) into the second of the two standard error formulas on page 272.

8.19 Rework the preceding exercise for sampling with replacement from the given population.

8.20 When we sample from an infinite population, what happens to the standard error of the mean if the sample size is

(a) increased from 4 to 16;

(b) decreased from 16 to 4;

(c) increased from 4 to 64;

(d) decreased from 64 to 4?

8.21 When we sample from an infinite population, what happens to the standard error of the mean if the sample size is

(a) increased from 10 to 1,000;

(b) decreased from 1,000 to 10;

(c) increased from 50 to 450;

(d) decreased from 450 to 50?

8.22 What is the value of the finite population correction factor when

(a) $n = 5$ and $N = 200$;

(b) $n = 5$ and $N = 125$;

(c) $n = 60$ and $N = 1,000$?

8.23 The two formulas for $\sigma_{\bar{x}}$ are

$$\sigma_{\bar{x}} = \frac{\sigma}{\sqrt{n}} \quad \text{and} \quad \sigma_{\bar{x}} = \frac{\sigma}{\sqrt{n}} \cdot \sqrt{\frac{N-n}{N-1}}$$

For what positive value of n are these two formulas identical?

8.24 For what positive value of n is the finite population correction factor $\sqrt{\dfrac{N-n}{N-1}}$ equal to zero?

8.25 For simplicity, the finite population correction factor $\sqrt{\dfrac{N-n}{N-1}}$ can be replaced by $\sqrt{\dfrac{N-n}{N}} = \sqrt{1-f}$, where $f = \dfrac{n}{N}$ is called the **sampling fraction**. Give the values of $\sqrt{\dfrac{N-n}{N-1}}$ and $\sqrt{1-f}$ in each of the following cases:

(a) $N = 20$, $n = 5$;

(b) $N = 80$, $n = 20$;

(c) $N = 1,000$, $n = 50$;

(d) $N = 350$, $n = 60$.

8.26 Use a computer package to

(a) generate the population of integers from 1 to 800 and determine its mean and its standard deviation;

(b) take 100 random samples of size $n = 5$ from this population;

(c) determine the means of these 100 samples and calculate their mean and their standard deviation;

(d) construct a histogram of the means obtained in part (c), using the class marks 40.5, 80.5, 120.5, 160.5, ...

Also,

(e) compare the mean of the 100 sample means with that of the population;

(f) compare the standard deviation of the 100 sample means with the value that we would expect in accordance with the theory on page 272;

(g) discuss the overall shape of the histogram obtained in part (d).

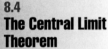 **8.27** Repeat Exercise 8.26 using sampling with replacement so that the whole procedure may be looked upon as sampling from an infinite population.

8.28 Use a computer package to

(a) generate the population of integers from 1 to 1,200 and determine its mean and its standard deviation;

(b) take 150 random samples of size $n = 25$ from this population;

(c) determine the means of these 150 samples and calculate their mean and their standard deviation;

(d) construct a histogram of the means obtained in part (c), using the class marks 30.5, 90.5, 150.5, 210.5, ...

Also,

(e) compare the mean of the 150 sample means with that of the population;

(f) compare the standard deviation of the 150 sample means with the value that we would expect in accordance with the theory on page 272;

(g) discuss the overall shape of the histogram obtained in part (d).

8.29 Repeat Exercise 8.28 using sampling with replacement so that the whole procedure may be looked upon as sampling from an infinite population.

8.4 The Central Limit Theorem

When we use a sample mean to estimate the mean of a population, we usually express our confidence in the estimate by attaching a probability statement about the size of our error. For instance, if we apply Chebyshev's theorem (as formulated in Section 6.8) to the sampling distribution of the mean, we can assert with a probability of at least $1 - \dfrac{1}{k^2}$ that the mean of a random sample will differ from the mean of the population from which it came by less than $k \cdot \sigma_{\bar{x}}$. In other words, if we use the mean of a random sample to estimate the mean of the population from which it came, we can assert with a probability of at least $1 - \dfrac{1}{k^2}$ that our error will be less than $k \cdot \sigma_{\bar{x}}$.

■ **EXAMPLE**   Based on Chebyshev's theorem with $k = 2$, what can we assert about the maximum size of our error if we use the mean of a random sample of size $n = 64$ to estimate the mean of an infinite population with $\sigma = 20$?

Solution   Substituting $\sigma = 20$ and $n = 64$ into the first of the two formulas for the standard error of the mean, we get

$$\sigma_{\bar{x}} = \frac{20}{\sqrt{64}} = 2.5$$

and it follows that we can assert with a probability of at least

$$1 - \frac{1}{2^2} = 0.75$$

that the error will be less than $k\sigma_{\bar{x}} = 2 \cdot 2.5 = 5$.   ■

Chebyshev's theorem can be used in problems like this; but if we want to be more specific about the probabilities, we have to refer to another theorem, which is called the **Central Limit theorem**. It states that for large samples the sampling distribution of the mean can be approximated closely with a normal distribution.

To formulate this more precisely, we must convert $\bar{x}$ to standard units. Since a value of a random variable is converted to standard units by subtracting from it the mean of the distribution of the random variable and then dividing by its standard deviation, we write

$$z = \frac{\bar{x} - \mu}{\sigma_{\bar{x}}}$$

where we used the same notation as in Section 8.3. Recalling that

$$\mu_{\bar{x}} = \mu \quad \text{and} \quad \sigma_{\bar{x}} = \frac{\sigma}{\sqrt{n}}$$

for random samples from infinite populations, we can now say formally that

Central Limit theorem

> If $\bar{x}$ is the mean of a random sample of size n from an infinite population, with the mean μ and the standard deviation σ, and n is large, then
>
> $$z = \frac{\bar{x} - \mu}{\sigma/\sqrt{n}}$$
>
> has approximately the standard normal distribution.

This theorem is of fundamental importance in statistics, because it justifies the use of normal-curve methods in a wide range of problems; it applies to infinite populations, and also to finite populations when n, though large, constitutes but a small portion of the population. We cannot say precisely how large n must be so that the Central Limit theorem can be applied, but unless the population distribution has a very unusual shape, $n = 30$ is usually regarded as sufficiently large. When the population that we are sampling has, itself, roughly the shape of a normal curve, the sampling distribution of the mean can be approximated closely with a normal distribution regardless of the size of n.

The Central Limit theorem can also be used for finite populations, but a precise description of the situations under which this can be done is rather complicated. The most common proper use is the case in which n is large while $\frac{n}{N}$ is small. This is the case, for example, in most political polls.

■ **EXAMPLE** Returning to our example on page 275, where we determined on the basis of Chebyshev's theorem that our error was less than 5 with a probability of at least 0.75, with what probability can we assert that our error is less than 5 on the basis of the Central Limit theorem? In that example we used the mean of a random sample of size $n = 64$ to estimate the mean of an infinite population with $\sigma = 20$.

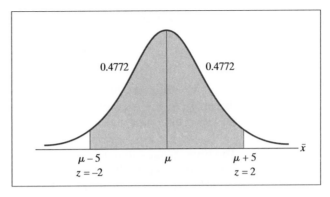

FIGURE 8.4 Sampling distribution of the mean.

Solution The desired probability is given by the area of the blue region under the curve in Figure 8.4, that is, the normal-curve area between

$$z = \frac{-5}{20/\sqrt{64}} = -2 \quad \text{and} \quad z = \frac{5}{20/\sqrt{64}} = 2$$

Since the entry in Table II corresponding to $z = 2.00$ is 0.4772, we find that the answer is

$$0.4772 + 0.4772 = 0.9544$$

Compared to "at least 0.75," we can thus make the much stronger statement that the probability is 0.9544 that the mean of a random sample of size $n = 64$ from the given population will differ from the population mean by less than 5.   ■

■ **EXAMPLE** Suppose that $\sigma = 5.5$ tons for the daily sulfur oxides emission of a certain industrial plant. What is the probability that the mean of a random sample of size $n = 40$ will differ from the mean of the population by less than 1.0 ton?

Solution The desired probability is given by the area of the blue region under the curve in Figure 8.5, namely, the normal-curve area between

$$z = \frac{-1.0}{5.5/\sqrt{40}} \approx -1.15 \quad \text{and} \quad z = \frac{1.0}{5.5/\sqrt{40}} \approx 1.15$$

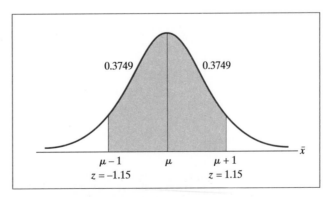

FIGURE 8.5 Sampling distribution of the mean.

Since the entry in Table II corresponding to $z = 1.15$ is 0.3749, we find that the answer is

$$0.3749 + 0.3749 = 0.7498, \text{ or approximately } 0.75$$ ■

In both of the preceding examples, we were concerned with the size of the error; in the example that follows, μ will be given and we will be asked for the probability that the mean of a sample will fall within a certain range.

■ **EXAMPLE**   The time it takes students in a cooking school to learn how to prepare a particular meal is a random variable with the mean $\mu = 3.2$ hours and the standard deviation $\sigma = 1.8$ hours. Find the probability that the average time it will take 36 students to learn how to prepare the meal is less than 3.4 hours.

Solution   Again, we are concerned with the sampling distribution of the mean; and since n is greater than 30, we can approximate it with a normal curve. Thus, the desired probability is given by the area of the blue region under the curve in Figure 8.6, namely, the normal-curve area to the left of

$$z = \frac{3.4 - 3.2}{1.8/\sqrt{36}} = \frac{0.2}{0.3} \approx 0.67$$

Since the entry in Table II corresponding to $z = 0.67$ is 0.2486, the answer is

$$0.2486 + 0.5000 = 0.7486$$ ■

The primary goal of this chapter has been to introduce the concept of a sampling distribution, and the one that we chose for this purpose was the sampling distribution of the mean. Observe, however, that instead of the mean we could have studied the median or some other statistic and investigated its chance fluctuations. So far as the corresponding theory is concerned, this would, of course, have required different formulas for the standard errors, that is, different formulas for the standard deviations of the respective sampling

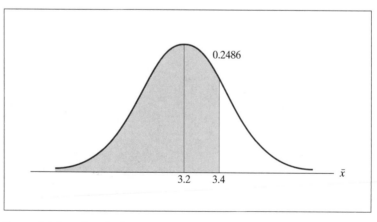

FIGURE 8.6 Sampling distribution of the mean.

distributions. For instance, for infinite populations, the **standard error of the median** is approximately

$$\sigma_{\tilde{x}} = 1.25 \frac{\sigma}{\sqrt{n}}$$

where n is the size of the sample and σ is the population standard deviation. Note that comparison of the two formulas

$$\sigma_{\bar{x}} = \frac{\sigma}{\sqrt{n}} \quad \text{and} \quad \sigma_{\tilde{x}} = 1.25 \frac{\sigma}{\sqrt{n}}$$

reflects the fact that the mean is generally more reliable than the median: it is subject to smaller chance fluctuations (see also Exercise 8.41). There are nonetheless situations, generally with medium-sized samples from nonnormal populations, in which the median is actually more reliable than the mean.

Let us also point out that there have been three occasions so far to work with standard units. In Section 3.8 we used them to illustrate an important application of the standard deviation; in Section 7.2 we used them in connection with the standard normal distribution; and in this section we used them in the formulation of the Central Limit theorem and in its applications. In general, the conversion of a value of a random variable to standard units is made with the formula

$$z = \frac{(\text{value of random variable}) - (\text{mean of random variable})}{\text{standard deviation of random variable}}$$

The mean that we subtract in the numerator may be μ or $\mu_{\bar{x}}$, and the standard deviation by which we divide may be σ or $\sigma_{\bar{x}}$, depending on whether we are dealing with one observation or with a mean. They may also be $\mu_{\tilde{x}}$ and $\sigma_{\tilde{x}}$, for example, when we are dealing with the sampling distribution of the median.

Exercises

Exercises 8.30 and 8.41 are practice exercises; their complete solutions are given on page 282.

8.30 A biologist wants to estimate the mean weight of a certain kind of animal. He knows that the standard deviation of their weights is 4.8 ounces and decides to use a random sample of size 100. Based on the Central Limit theorem, with what probability can he assert that his error will be
   (a) less than 0.60 ounce;
   (b) less than 1.20 ounces?

8.31 A university psychologist would like to estimate how many points her students can be expected to score on an objective personality test. If she uses a random sample of 49 students and assumes that the standard deviation is 8 points, what can she assert about the probability that the error will be less than 2 points if she uses
   (a) Chebyshev's theorem;
   (b) the Central Limit theorem?

8.32 A quality control inspector wants to estimate the mean width of automobile door latches that are used in the assembly of doors for new automobiles. If the inspector uses a random sample of 100 door latches and knows that their standard deviation is 2 millimeters, what can he assert about the probability that his error will be less than 0.25 millimeter if he uses
(a) Chebyshev's theorem;
(b) the Central Limit theorem?

8.33 A commuting student wants to use a sample of 36 driving times to estimate the mean time it takes to drive to the college each morning. If the standard deviation of the time required to make the trip is assumed to be 11 minutes, what can the student assert about the probability that his error will be less than 3.3 minutes if he uses
(a) Chebyshev's theorem;
(b) the Central Limit theorem?

8.34 The organizer of a large national political convention wants to use the average monetary expenditures of 100 delegates to estimate the average monetary expenditures of the population of delegates attending the convention. If the standard deviation of the expenditures is assumed to be $150, what can the organizer assert, using Chebyshev's theorem, about the probability that the error will be less than $18?

8.35 A logger wants to use the average length of a sample of 49 logs cut from a large stand of trees to estimate the length of logs produced from the population of trees in this stand. If the standard deviation of these lengths is assumed to be 1.4 feet, what can the logger assert about the probability that the error will be less than 0.1 foot using Chebyshev's theorem?

8.36 The supervisor of a waste disposal truck that collects waste in a residential neighborhood wants to estimate the mean number of cubic feet of waste collected at these residences. Suppose that the supervisor uses a sample of 30 waste pickups and knows from experience that the standard deviation is 3 cubic feet for such data. Based on the Central Limit theorem, with what probability can the supervisor assert that the error will be
(a) less than 1 cubic foot;
(b) less than 0.5 cubic foot?

8.37 A college weekly newspaper that accepts advertising wants to estimate the mean number of lines of advertising in its newspapers. Suppose that it uses a sample of 49 weekly editions and that the standard deviation is 250 lines. If the newspaper uses the Central Limit theorem, with what probability can it assert that its error will be
(a) less than 50 lines;
(b) less than 25 lines?

8.38 During the last week of the semester, students at a certain college spend on the average 4.2 hours using the school's computer terminals with a standard deviation of 1.8 hours. For a random sample of 36 students at that college, find the probabilities that the average time spent using the computer terminals during the last week of the semester is
(a) at least 4.8 hours;
(b) between 4.1 and 4.5 hours.

8.39 A particular make of car is known to show rust when it is 2.4 years old on the average, with a standard deviation of 0.8 year. If a car rental agency purchases 64 new cars of this kind, what are the probabilities that the average time it will take for these cars to show rust is
    (a) at most 2.6 years;
    (b) between 1.9 and 2.3 years?

8.40 The time that it takes a realtor to lease an apartment is 2.3 months, on the average, with a standard deviation of 1.5 months. Find the probabilities that in a sample of 36 apartments, the average time it takes to lease an apartment is at most
    (a) 1.8 months;
    (b) 2.0 months.

8.41 Show that the mean of a random sample of size $n = 64$ is as reliable an estimate of the mean of a symmetrical infinite population as the median of a random sample of size $n = 100$. (For symmetrical populations, the means of the sampling distributions of $\bar{x}$ and $\tilde{x}$ are both equal to the population mean μ.)

8.42 How large a random sample do we have to take so that its mean is as reliable an estimate of the mean of a symmetrical infinite population as the median of a random sample of size $n = 225$?

8.43 How large a random sample do we have to take so that its median is as reliable an estimate of the mean of a symmetrical infinite population as the mean of a random sample of size $n = 1{,}600$?

8.44 Show that if the mean of a random sample of size n is used to estimate the mean of an infinite population with the standard deviation σ, there is a 50–50 chance that the error will be less than the quantity $0.6745 \cdot \dfrac{\sigma}{\sqrt{n}}$, which is called the **probable error of the mean**. (*Hint:* Interpolate between the entries in Table II corresponding to $z = 0.67$ and $z = 0.68$.)

Solutions to Practice Exercises

8.1 There are ten possible samples: *abc, abd, abe, acd, ace, ade, bcd, bce, bde*, and *cde*. Six of these include *b*, so that the probability is $\frac{6}{10} = 0.6$.

8.2 Reading the values off the table, we get 5580, 1540, 1404, 6837, 0985, 0479, 6594, 2191, 4116, 2377, 5997, and 6198.

8.16 The ordered samples are 1 and 1, 1 and 3, 1 and 5, 1 and 7, 1 and 9, 3 and 1, 3 and 3, 3 and 5, 3 and 7, 3 and 9, 5 and 1, 5 and 3, 5 and 5, 5 and 7, 5 and 9, 7 and 1, 7 and 3, 7 and 5, 7 and 7, 7 and 9, 9 and 1, 9 and 3, 9 and 5, 9 and 7, and 9 and 9. The means are 1, 2, 3, 4, 5, 2, 3, 4, 5, 6, 3, 4, 5, 6, 7, 4, 5, 6, 7, 8, 5, 6, 7, 8, and 9. The sampling distribution of the mean is

$\bar{x}$	Probability
1	$\frac{1}{25}$
2	$\frac{2}{25}$
3	$\frac{3}{25}$
4	$\frac{4}{25}$
5	$\frac{5}{25}$
6	$\frac{4}{25}$
7	$\frac{3}{25}$
8	$\frac{2}{25}$
9	$\frac{1}{25}$

8.17 Since $\mu_{\bar{x}} = \mu = 5$, we get

$$\sigma_{\bar{x}}^2 = (1-5)^2 \cdot \frac{1}{25} + (2-5)^2 \cdot \frac{2}{25} + (3-5)^2 \cdot \frac{3}{25}$$

$$+ (4-5)^2 \cdot \frac{4}{25} + (5-5)^2 \cdot \frac{5}{25} + (6-5)^2 \cdot \frac{4}{25}$$

$$+ (7-5)^2 \cdot \frac{3}{25} + (8-5)^2 \cdot \frac{2}{25} + (9-5)^2 \cdot \frac{1}{25}$$

$$= 4$$

so $\sigma_{\bar{x}} = 2$. Substitution into the formula yields

$$\sigma_{\bar{x}} = \frac{\sqrt{8}}{\sqrt{2}} = \sqrt{\frac{8}{2}} = \sqrt{4} = 2$$

and the two answers agree.

8.30   (a) $\sigma_{\bar{x}} = \dfrac{4.8}{\sqrt{100}} = 0.48$, $z = \dfrac{0.60}{0.48} = 1.25$, the corresponding entry in Table

II is 0.3944, and the answer is

$$0.3944 + 0.3944 = 0.7888$$

(b) $z = \dfrac{1.20}{0.48} = 2.50$, the corresponding entry in Table II is 0.4938, and

the answer is

$$0.4938 + 0.4938 = 0.9876$$

8.41 $\sigma_{\bar{x}} = \dfrac{\sigma}{\sqrt{64}} = \dfrac{\sigma}{8}$ and $\sigma_{\tilde{x}} = 1.25 \cdot \dfrac{\sigma}{\sqrt{100}} = 1.25 \cdot \dfrac{\sigma}{10} = \dfrac{\sigma}{8}$; since the two standard errors are equal, the mean of a random sample of size $n = 64$ is as reliable as the median of a random sample of size $n = 100$.

Review: Chapters 6, 7, & 8

• •

Achievements

Having read and studied these chapters and having worked a good portion of the exercises, you should be able to:

1. Explain what is meant by *random variable* and *probability distribution*.
2. State the two rules that must be satisfied by the values of a probability distribution.
3. List the assumptions that underlie the binomial distribution.
4. Use the formula for the binomial distribution.
5. Use the table of binomial probabilities.
6. Use the formula for the hypergeometric distribution.
7. Approximate hypergeometric probabilities with binomial probabilities and know when this approximation may be used.
8. Approximate binomial probabilities with Poisson probabilities and know when this approximation may be used.
9. Use the formula for the Poisson distribution with the parameter λ.
10. Calculate multinomial probabilities.
11. Determine the mean and the standard deviation of a probability distribution.
12. Find the mean and the standard deviation of a binomial distribution.
13. Find the mean and the standard deviation of a hypergeometric distribution.
14. Find the mean and the standard deviation of a Poisson distribution.
15. Use Chebyshev's theorem as it applies to the distribution of a random variable.
16. Explain what is meant by *probability density* or *continuous distribution*.
17. Discuss the normal distribution and explain what is meant by "standard normal distribution."
18. Use the table of normal-curve areas.
19. Apply the continuity correction where needed.
20. Use the normal distribution to approximate binomial probabilities and know when this approximation may be used.
21. Explain what is meant by *random sample from a finite population*.
22. Use random numbers to select random samples from finite populations.
23. Explain what is meant by *sampling distribution* and *sampling distribution of the mean*.
24. State and apply the theorem about the mean and the standard deviation of the sampling distribution of the mean.

25. Explain what is meant by *standard error*.
26. Determine the finite population correction factor and know when it should be used.
27. State the Central Limit theorem and use it to calculate probabilities relating to the sampling distribution of the mean.

Checklist of Key Terms (with page references to their definitions)

Binomial distribution, 196, 197
Binomial probability distribution, 197
Central Limit theorem, 276
Chebyshev's theorem, 226
Computer simulation, 269
Continuity correction, 251
Continuous distribution, 234
Continuous random variable, 194
Discrete random variable, 193
Finite population, 262
Finite population correction factor, 272
Hypergeometric distribution, 205, 206
Infinite population, 262

Mean of the binomial distribution, 220
Mean of a probability distribution, 219
Multinomial distribution, 218
Normal approximation to the binomial distribution, 253
Normal curve, 234
Normal distribution, 234, 237
Poisson distribution, 211
Probability density, 234
Probability distribution, 194
Probable error of the mean, 281
Random numbers, 263
Random sample, 262
Random sampling, 261, 262

Random variable, 193
Sampling a binomial population, 201
Sampling distribution, 267
Sampling fraction, 274
Sampling a normal population, 252
Standard deviation of a probability distribution, 221
Standard error of the mean, 270, 272
Standard error of the median, 279
Standard normal distribution, 237
Standard units, 237
Variance of a probability distribution, 221

Review Exercises

R.82 A teacher hypothesizes that, in general, 3 of 10 substandard students have physical disabilities (such as poor eyesight, poor hearing, or generally poor health). Assuming that this is correct,
  (a) use the formula for the binomial distribution repeatedly to find the probabilities that among 6 substandard students, 0, 1, 2, 3, 4, 5, or 6 have physical disabilities.
  (b) Verify part (a) of this exercise by using Table I.

R.83 Using the results obtained in Exercise R.82, what is the probability that among a randomly selected group of six substandard students
  (a) none have physical disabilities;
  (b) three or more have physical disabilities;
  (c) fewer than three have physical disabilities?

R.84 With reference to Exercises R.82 and R.83, use a computer printout to work
  (a) part (a) of Exercise R.83;
  (b) part (b) of Exercise R.83;
  (c) part (c) of Exercise R.83.

R.85 The amount of coffee dispensed into plastic cups by a coin-operated vending machine is a random variable having a normal distribution with $\mu = 8.00$ fluid ounces and $\sigma = 0.23$ ounce. Find the probabilities that the vending machine will dispense
  (a) at least 7.75 fluid ounces of coffee into one cup;
  (b) from 8.25 to 8.50 fluid ounces of coffee into one cup.

R.86 A college placement officer knows from experience that the probabilities that a student will not have a job offer at the time of graduation, or have one job offer, or have more than one job offer are 0.20, 0.50, and 0.30, respectively. What is the probability that among 10 randomly selected graduates, 3 graduates will not have job offers, 5 will have one job offer, and 2 will have two or more job offers?

R.87 What is the value of the finite population correction factor if
    (a) $n = 20$ and $N = 100$;   (b) $n = 20$ and $N = 1,000$?

R.88 How many different samples of
    (a) $n = 4$ oriental rugs can be drawn from a population of $N = 10$ oriental rugs;
    (b) $n = 6$ magazines can be drawn from a population of $N = 12$ magazines;
    (c) $n = 5$ students can be drawn from a population of $N = 24$ students?

R.89 Given that the probability of obtaining a head in a flip of a balanced coin is 0.5, use Table I to find the probabilities that in 15 flips of a balanced coin we will get
    (a) up to 2 heads;   (c) exactly 6 heads;
    (b) 10 or more heads;   (d) 6 or 7 heads.

R.90 Use the normal curve approximation to find the probability of getting 6 heads in 15 flips of a balanced coin, and compare the result with the value obtained in part (c) of Exercise R.89, where the probability was obtained from Table I.

R.91 Suppose that a continuous random variable takes on values on the interval from 0 to 2 and that the graph of its probability density is as shown in Figure R.3. What are the probabilities that the random variable will take on a value
    (a) less than 0.5;   (b) between 1.2 and 1.6?

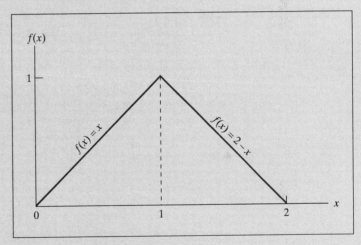

FIGURE R.3 Diagram for Exercise R.91.

R.92 According to the fire department in a large city, the average number of false alarms is a random variable with $\mu = 24.2$ and $\sigma = 2.9$. Assuming that the distribution of this random variable can be approximated closely with a normal curve, find the probability that on any given day there will be
    (a) exactly 25 false alarms;   (b) at least 25 false alarms.

R.93 An inspector wants to examine 10 washing machines from a production run of 500 washing machines that are serially numbered from 287 to 786, inclusive. Which ones, by serial number, will the inspector examine if a random sample is chosen by means of random numbers, using the first three columns of the table in Figure 8.1, going down the table and starting with the 11th row.

R.94 Suppose that the number of minutes that a swimmer spends in a swimming pool is a random variable with $\mu = 49.3$ minutes and $\sigma = 5.8$ minutes. Assuming that this random variable can be approximated closely with a normal curve, find the probabilities that a swimmer will spend
(a) at least 60 minutes in the pool;
(b) from 45.0 to 60.0 minutes in the pool.

R.95 If the number of buds produced by a certain type of lily is a random variable having the Poisson distribution with $\lambda = 3.5$, what are the probabilities that a lily of this type will produce
(a) no buds;
(b) two buds;
(c) four buds;
(d) six buds?

R.96 Find the area under the standard normal curve that lies
(a) between $z = 0$ and $z = 2.58$;
(b) to the left of $z = 1.96$;
(c) to the right of $z = -1.25$;
(d) to the right of $z = 1.00$;
(e) to the left of $z = -0.39$.

R.97 Find the mean and the variance of the binomial distribution with $n = 4$ and $p = 0.20$
(a) by looking up the probabilities in Table I and then using the formula for the mean and the variance of a probability distribution;
(b) by using the special formulas for the mean and the variance of the binomial distribution.

R.98 Check in each case whether the conditions given in Section 6.4 for the Poisson approximation to the binomial distribution are satisfied:
(a) $n = 150$ and $p = \frac{1}{10}$;
(b) $n = 450$ and $p = \frac{1}{50}$;
(c) $n = 600$ and $p = \frac{1}{100}$.

R.99 An automobile muffler that carries a lifetime guarantee lasts on average 40 months, with a standard deviation of 8.2 months. Find the probability that a random sample of 50 of these mufflers will last on the average at least 39 months.

R.100 A spotter uses chemicals, water, and steam to remove difficult stains not removed by the prespotting process in dry cleaning. If a spotter is successful 80% of the time, use the formula for the binomial distribution to determine the probability that five of six garments that are badly stained will be treated successfully by the spotter.

R.101 A zoo has a large primate collection including eight baboons and six chimpanzees. If a veterinarian randomly picks four of them for examination, what are the probabilities that these will include
(a) two baboons and two chimpanzees;
(b) one baboon and three chimpanzees?

R.102 What is the probability of each sample if a random sample of size 3 is to be drawn from a finite population of size 12?

R.103 In each case, determine whether the given values can be looked upon as the values of a probability distribution of a random variable that can take on the values 1, 2, 3, 4, and 5, and explain your answers:
(a) $f(1) = 0.18$, $f(2) = 0.20$, $f(3) = 0.22$, $f(4) = 0.20$, $f(5) = 0.18$;
(b) $f(1) = 0.05$, $f(2) = 0.05$, $f(3) = 0.05$, $f(4) = -0.05$, $f(5) = 0.90$;
(c) $f(1) = 0.07$, $f(2) = 0.23$, $f(3) = 0.50$, $f(4) = 0.19$, $f(5) = 0.01$.

R.104 A random variable has the normal distribution with $\sigma = 25$. If the probability is 0.6700 that the random variable takes a value less than 284, find the probability that it will take a value greater than 250.

R.105 If 73% of the cars traveling on a certain highway carry only one occupant, what is the probability that among 500 randomly selected cars at most 350 cars will carry only one occupant?

R.106 Determine whether the following can be probability distributions (defined in each case only for the given values of x) and explain your answers:

(a) $f(x) = \dfrac{1}{5}$ for $x = 0, 1, 2, 3, 4, 5$;

(b) $f(x) = \dfrac{x + 1}{14}$ for $x = 1, 2, 3, 4$;

(c) $f(x) = \dfrac{x(x - 2)}{2}$ for $x = 1, 2, 3$.

R.107 Check in each case whether the conditions for the normal approximation to the binomial distribution are satisfied:

(a) $n = 50$ and $p = \frac{1}{5}$;   (c) $n = 100$ and $p = 0.96$;

(b) $n = 80$ and $p = \frac{1}{40}$;   (d) $n = 200$ and $p = \frac{24}{25}$.

R.108 If the probabilities that zero, one, two, or three students are absent from a mathematics class on any given day are 0.53, 0.35, 0.09, and 0.03, find

(a) the mean of this distribution;

(b) the standard deviation of this distribution.

R.109 Find z if

(a) the normal-curve area between 0 and z is 0.4515;

(b) the normal-curve area to the right of z is 0.7967;

(c) the normal-curve area to the right of z is 0.1292.

R.110 The mean of a random sample of size $n = 81$ is used to estimate the mean annual growth of certain plants. If the standard deviation of their annual growth is $\sigma = 45$ mm, what are the probabilities that our estimate will be off either way by

(a) less than 8.5 mm;   (b) less than 2.0 mm?

R.111 Check in each case whether the condition for the binomial approximation to the hypergeometric distribution is satisfied:

(a) $a = 150$, $b = 150$, and $n = 18$;

(b) $a = 100$, $b = 300$, and $n = 15$;

(c) $a = 400$, $b = 200$, and $n = 50$.

R.112 Random samples of size $n = 2$ are drawn from the finite population that consists of the numbers 1, 2, 3, 4, 5, and 6.

(a) Verify that the mean of this finite population is $\mu = 3\frac{1}{2}$ and that its variance is $\sigma^2 = \frac{35}{12}$.

(b) List the 15 possible samples of size $n = 2$ that can be drawn without replacement from the given population, calculate their means, and, assigning each of these values the probability $\frac{1}{15}$, construct the sampling distribution of the mean for random samples of size $n = 2$ from the given population.

(c) Calculate the mean and the standard deviation of the sampling distribution obtained in part (b), and verify the result obtained for the standard deviation by substituting $n = 2$, $N = 6$, and $\sigma = \sqrt{\frac{35}{12}}$ into the second of the two standard error formulas on page 272.

R.113 President Franklin Delano Roosevelt vetoed a larger number of congressional bills than any other president in history, with 372 regular vetoes and 263 pocket vetoes. If a historian randomly selects 3 of these vetoed bills for study, find the probability that the selection consists of 2 bills that were vetoed in a regular manner and 1 that was subjected to a pocket veto by using
    (a) the formula for the hypergeometric distribution;
    (b) the binomial distribution as an approximation.

R.114 During the month of August, the daily number of persons visiting a certain tourist attraction is a random variable with $\mu = 1,200$ and $\sigma = 80$.
    (a) What does Chebyshev's theorem with $k = 7$ tell us about the number of persons who will visit the tourist attraction on an August day?
    (b) According to Chebyshev's theorem, with what probability can we assert that between 1,000 and 1,400 persons will visit the tourist attraction on an August day?

R.115 Suppose that we draw cards, one at a time and with replacement, from an ordinary deck of 52 playing cards. To determine the probability of getting 5 red kings among 156 cards thus drawn, can we use
    (a) the Poisson approximation;   (b) the normal approximation?

R.116 When we sample from an infinite population, what happens to the standard error of the mean if the sample size is
    (a) increased from 25 to 100;   (b) decreased from 400 to 40?

R.117 If a newspaper delivery service fails to deliver newspapers to 1.8% of its subscribers, use the Poisson approximation to the binomial distribution to determine the probability that among 250 subscribers, 5 will not receive their newspapers on a given day.

R.118 The average time required to perform job A is 78.5 minutes with a standard deviation of 16.2 minutes, and the average time required to perform job B is 103.2 minutes with a standard deviation of 11.3 minutes. Assuming normal distributions, what proportion of the time will job A take longer than the average job B, and what proportion of the time will job B take less time than the average job A?

9 Problems of Estimation

In statistical inference we make generalizations based on samples and, tradi-tionally, such inferences have been divided into **problems of estimation** and **tests of hypotheses**. In problems of estimation, we assign a numerical value to a population parameter, expecting or hoping that our estimate (based on a sample) will be close; or we assign an interval of values (based on a sample) expecting or hoping that it will contain the population parameter in question. In problems of hypothesis testing, we accept or reject assumptions about the parameters of populations, or perhaps their shape or form, expecting or hop-ing that what we do will not be in error.

In this chapter we shall concentrate on problems of estimation which are easy to illustrate because they arise everywhere—in science, in business, and in everyday life. In science, a biologist may want to determine the aver-age wingspan of an insect; in business, a retailer may want to determine the average income of all families living within a mile of a proposed new shop-ping center; and in everyday life, we may want to know how long it takes on the average to iron a shirt or peel a half-pound potato.

Sections 9.1, 9.2, and 9.3 are devoted to the estimation of the means of populations, Section 9.4 to the estimation of their standard deviations, and Section 9.5 to the estimation of proportions, percentages, or probabilities. Tests of hypotheses will be taken up in subsequent chapters.

To illustrate some of the problems we face in the estimation of means, let us refer to a study in which a doctor wants to determine the true average increase in the pulse rate of a person performing a certain strenuous task. The following is the data (increases in pulse rate in beats per minute) that the doctor obtained for 32 persons who performed the given task:

27	25	19	28	35	23	24	22
14	30	32	34	23	26	29	27
27	24	31	22	23	38	25	16
32	29	26	25	28	26	21	28

The mean of this sample is $\bar{x} = 26.2$ beats per minute, and in the absence of any other information, this figure may well have to serve as an estimate of the population average μ, the true average increase in the pulse rate of persons performing the given task. Observe that the estimate is the statistic $\bar{x}$, which in this example has the numerical value 26.2; it estimates the unknown parameter μ.

An estimate like this is called a **point estimate**, since it consists of a single number, or a single point on the real number scale. Although this is the most common way in which estimates are expressed, it leaves room for many questions. For instance, it does not tell us on how much information the estimate is based, and it does not tell us anything about the possible size of the error. And, of course, we must expect an error. This should be clear from our discussion of the sampling distribution of the mean in Chapter 8, where we saw that the chance fluctuations of the mean (and hence its reliability as an estimate of μ) depend on two things—the size of the sample and the size of the population standard deviation σ. Thus, we might supplement 26.2, the foregoing estimate of the true average increase in the pulse rate of persons performing the given task, with the information that the sample size is $n = 32$ and that the sample standard deviation is $s = 5.15$, as can readily be verified. Although this does not tell us the actual value of σ, the sample standard deviation can serve as an estimate of this quantity.

Scientific reports often present sample means in this way, together with the values of n and s, but this does not supply the reader of the report with a coherent picture unless he or she has had some formal training in statistics. To make the supplementary information meaningful also to the layman, let us refer to the theory of Sections 8.3 and 8.4 and the example of page 244 (see also Figure 7.20), according to which $z_{\alpha/2}$ is such that the area under the standard normal curve between $-z_{\alpha/2}$ and $z_{\alpha/2}$ is equal to $1 - \alpha$. Thus, making use of the fact that the sampling distribution of the mean of a large random sample from an infinite population is approximately a normal distribution with

$$\mu_{\bar{x}} = \mu \quad \text{and} \quad \sigma_{\bar{x}} = \frac{\sigma}{\sqrt{n}}$$

we find from Figure 9.1 that the probability is $1 - \alpha$ that a sample mean will differ from the population mean μ by at most $z_{\alpha/2} \cdot \frac{\sigma}{\sqrt{n}}$. In other words,

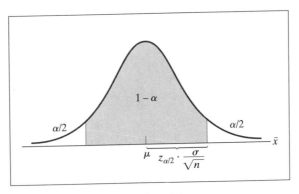

FIGURE 9.1 Sampling distribution of the mean.

Maximum error of estimate ($n \geq 30$)

> When we use $\bar{x}$ as an estimate of μ, the probability is $1 - \alpha$ that this estimate will be "off" either way by at most
>
> $$E = z_{\alpha/2} \cdot \frac{\sigma}{\sqrt{n}}$$

This result applies when n is large ($n \geq 30$) and the population is infinite, or large enough so that the finite population correction factor (see page 272) need not be used. The two values that are most commonly used for $1 - \alpha$ are 0.95 and 0.99. The reader was asked to verify in Exercise 7.17 on page 248 that $z_{0.025} = 1.96$, corresponding to $1 - \alpha = 0.95$, and $z_{0.005} = 2.575$, corresponding to $1 - \alpha = 0.99$.

■ **EXAMPLE**   An efficiency expert intends to use the mean of a random sample of size $n = 40$ to estimate the average time that it takes automobile mechanics to perform a certain task. If, based on experience, the efficiency expert can assume that $\sigma = 2.9$ minutes for such data, what can he assert with probability 0.99 about the maximum size of his error?

Solution   Substituting $n = 40$, $\sigma = 2.9$, and $z_{0.005} = 2.575$ into the formula for E, we get

$$E = 2.575 \cdot \frac{2.9}{\sqrt{40}} \approx 1.18$$

Thus, the efficiency expert can assert with probability 0.99 that his error will be at most 1.18 minutes.   ■

Suppose now that the efficiency expert of this example collects his data and gets $\bar{x} = 27.36$ minutes. Can he still assert with probability 0.99 that the error of his estimate, 27.36 minutes, is at most 1.18? After all, $\bar{x} = 27.36$ differs from the true mean by at most 1.18 minutes or it does not, and he does not know which. Actually, he can make this probability assertion, but it must be understood that the 0.99 probability applies to the *method* that he used to determine the maximum error (getting the sample data and using the formula for E) and not directly to the parameter that he is trying to estimate.

To make this distinction, it has become the custom to use the word *confidence* here instead of *probability*.

> *In general, we make probability statements about future values of random variables (say, the potential error of an estimate) and confidence statements once the data has been obtained.*

Accordingly, we would say in our example that the efficiency expert can be 99% confident that the error of his estimate, $\bar{x} = 27.36$ minutes, is at most 1.18 minutes.

In some situations, we discuss ahead of time statements that will be made after the data is collected, as in Exercises 9.12 through 9.17. In such cases it is difficult to distinguish between confidence and probability.

■ **EXAMPLE**   A quality-control specialist wants to use the mean of a random sample of size $n = 35$ to determine the average fat content of a large shipment of quarter-pound hamburgers. The value of σ is not known, but the specialist maintains that the value of σ could not possibly be larger than 0.25 ounce. What can he assert with probability 0.95 about the maximum size of his error?

Solution   Placing $n = 35$, $\sigma = 0.25$, and $z_{0.025} = 1.96$ into the formula for E, we find that

$$E = 1.96 \cdot \frac{0.25}{\sqrt{35}} = 0.083$$

The quality-control specialist will now assert, with probability 0.95, that the error in using $\bar{x}$ to estimate μ will be at most 0.083 ounce.   ■

In the preceding example the value of σ was not known, but it was possible to give a *worst-case* value. Accordingly, the maximum error of 0.083 ounce is *conservative*, and the quality-control expert can assert with a probability of at least 0.95 that the error is at most 0.083 ounce.

To be able to judge the size of the error that we might make when we use $\bar{x}$ as an estimate of μ, we must know the value of the population standard deviation σ. Since this is not the case in most practical situations, we have a serious complication. In the previous example it was possible to give a worst-case value for σ. In other situations, it may not be possible to do this, and we have no choice but to replace σ with an estimate, usually the sample standard deviation s. In general, this is considered to be reasonable provided the sample is sufficiently large, and by sufficiently large we mean again $n \geq 30$.

■ **EXAMPLE**   With reference to the pulse rates on page 290, where we had $n = 32$, $\bar{x} = 26.2$, and $s = 5.15$, what can we assert with 95% confidence about the maximum error if we use 26.2 as an estimate of the true average increase in the pulse rate of a person performing the given task?

Solution   Substituting $n = 32$, $s = 5.15$ for σ, and $z_{0.025} = 1.96$ into the formula for E, we find that we can assert with 95% confidence that the error is at most

$$E = 1.96 \cdot \frac{5.15}{\sqrt{32}} \approx 1.78 \text{ beats per minute}$$ ■

The formula for E can also be used to determine the sample size that is needed to attain a desired degree of precision. Suppose that we want to use the mean of a large random sample to estimate the mean of a population, and we want to be able to assert with probability $1 - \alpha$ that the error of this estimate will be at most some prescribed quantity E. As before, we write $E = z_{\alpha/2} \cdot \dfrac{\sigma}{\sqrt{n}}$, and upon solving this equation for n we get

Sample size for estimating μ ($n \geq 30$)

$$n = \left(\frac{z_{\alpha/2} \cdot \sigma}{E} \right)^2$$

■ **EXAMPLE**   The dean of a college wants to use the mean of a random sample to estimate the average amount of time that students take to get from one class to the next, and she wants this estimate to be in error by at most 0.25 minute with probability 0.95. If she knows from previous studies of a similar kind that it is reasonable to let $\sigma = 1.50$ minutes, how large a sample will she have to take?

Solution   Substituting $z_{0.025} = 1.96$, $\sigma = 1.50$, and $E = 0.25$ into the formula for n, we get

$$n = \left(\frac{1.96 \cdot 1.50}{0.25} \right)^2 = (11.76)^2 \approx 138.30$$

This value should be moved up to the next integer in order to satisfy the conditions. Thus, a random sample of size $n = 139$ is required for the estimate.   ■

As can be seen from the formula and also from the example, this method has the shortcoming that it cannot be used unless we know (at least approximately) the value of the standard deviation of the population whose mean we want to estimate.

Exercises

Exercises 9.1, 9.2, and 9.12 are practice exercises; their complete solutions are given on page 310.

9.1 A school district official intends to use the mean of a random sample of 100 fifth graders to estimate the mean score that all the fifth graders in the district would get if they took a certain arithmetic achievement test. If, based on experience, the official knows that $\sigma = 9.2$ for such data, what can he assert with probability 0.95 about the maximum error?

9.2 A random sample of 50 cans of peach halves has a mean weight of 16.1 ounces and a standard deviation of 0.4 ounce. If $\bar{x} = 16.1$ ounces is used as an estimate of the average weight of all the cans of peach halves in the large lot from which the sample came, with what confidence can we assert that the error of this estimate is at most 0.1 ounce?

9.3 A specialist in precision machinery wants to estimate the mean expansion of certain pistons (in inches) on the basis of a sample of 42 of these pistons. The expansion is caused by the heat generated after the engines have been started, and it is known that $\sigma = 0.020$ inch. If the specialist considers the data to constitute a random sample, what can he assert with probability 0.99 about the maximum error if he uses the mean of his sample as an estimate of the actual mean expansion of such pistons.

9.4 The commissioner of a city street department plans to estimate the mean height of a very large shipment of trees that will be planted along the streets of the city. To obtain the estimate, the commissioner will calculate the mean of a randomly selected sample of 200 trees from the shipment. The commissioner knows, from experience, that the standard deviation is $\sigma = 3.54$ inches for such data. What can be said with probability 0.95 about the maximum error of his estimate?

9.5 A random sample of 48 Danish pastries produced by a major bakery products manufacturer has a mean of 190 milligrams of sodium per pastry, with a standard deviation of 10 milligrams of sodium per pastry. If 190 milligrams of sodium is used as an estimate of the actual mean of the population of Danish pastries produced by this manufacturer, what can we assert about the maximum error

(a) with 95% confidence;
(b) with 98% confidence;
(c) with 99% confidence?

9.6 A national charitable organization would like to know how many hours, on average, it will take a volunteer worker to complete a short, self-study course in telephone solicitation procedures. One hundred randomly selected volunteer workers took an average of 7.9 hours to complete the course, with a standard deviation of 2.1 hours. What can we say with 99% confidence about the amount of time by which the charitable organization would be off if it estimated the true average time it takes a volunteer worker to complete the course as 7.9 hours?

9.7 A sample survey conducted in a metropolitan area showed that 200 families spent, on average, $233.58 per week on food eaten at home, with a standard deviation of $18.82. What can we say with 95% confidence about the maximum error if $233.58 in weekly food expenditures is used as an estimate of the average for all families in the metropolitan area?

9.8 A candy distributor wishes to determine the average water content of bottles of maple syrup from a particular New Hampshire producer. The bottles contain 12 ounces of liquid, and she decides to determine the water content of 30 of these bottles, using the sample mean as an estimate of the true population average. What can she say, with probability 0.95, about the maximum error if the largest possible standard deviation that she is willing to believe is 2.0 ounces?

9.9 The manager of a restaurant plans to introduce child-size portions for a number of menu items. As an experiment, she would like to give hamburger patties to a number of six-year-olds in an "all you can eat" situation in order to estimate the average preferred quantity. She believes that the population of children would choose, on average, somewhere between 2 ounces and 6 ounces, but she has no idea about the standard deviation. She needs to know what to say with 95% confidence about the maximum error in using the sample mean to estimate the population average if $n = 45$ children are used in the experiment.

(a) Suggest a plausible value for the standard deviation σ, and then provide a value for the maximum error using 95% confidence.

(b) Suggest a pessimistic (large) value for the standard deviation σ, and then provide a value for the maximum error using 95% confidence.

9.10 A study conducted by an airline showed that 120 of its passengers disembarking at Kennedy airport, a random sample, had to wait on the average 9.45 minutes with a standard deviation of 1.84 minutes to get their luggage. What can it say with 99% confidence about the maximum error if the airline uses $\bar{x} = 9.45$ minutes as an estimate of the true average time it takes its passengers to get their luggage when disembarking at Kennedy airport?

9.11 A random sample of 300 telephone calls made to the office of a large corporation is timed and reveals that the average call is $\bar{x} = 6.48$ minutes long, with a sample standard deviation of $s = 1.92$ minutes. What can the office manager say with 99% confidence about the maximum error if $\bar{x} = 6.48$ minutes is used as an estimate of the true length of telephone calls made to the office?

9.12 In a study of television viewing habits, it is desired to estimate the average number of hours that teenagers spend watching per week. If it is reasonable to assume that $\sigma = 3.2$ hours for data of this kind, how large a sample is needed so that one will be able to assert with 99% confidence that the sample mean is off by at most half an hour?

9.13 The marketing manager of a toy company needs to know, with reasonable accuracy, how long it takes a child of a certain age to assemble a model airplane. If preliminary studies have shown that the standard deviation is 15 minutes, how large a sample will the manager need to be able to assert with a probability of 0.95 that the sample mean will be off by at most 3 minutes?

9.14 A park ranger wants to know the average size of trout taken from a certain lake. How large a sample of trout must be taken to be able to assert with a probability of 0.98 that a sample mean will not be off by more than 0.5 inch? Assume that it is known from previous studies that $\sigma = 2.5$ inches.

9.15 Suppose that we want to estimate the average speed of cars traveling on a highway, and we want to be able to assert with probability 0.99 that the error of our estimate, the mean speed of a random sample of these cars, will be at most 3 miles per hour. How large a sample will we need if it can be assumed that $\sigma = 7.1$ miles per hour?

9.16 Before purchasing a large shipment of ground meat, a sausage manufacturer wants to be 95% confident that he is in error by no more than 2 grams in estimating the fat content (per 100 grams). If the standard deviation of the fat content (per 100 grams) is assumed to be 8 grams, on how large a sample should he base his estimate?

9.17 The ratio $E^* = \dfrac{E}{\sigma}$ is the error in standard deviation units. In some situations, lack of knowledge about σ makes it difficult to establish realistic values for E, and specifying E^* is a useful alternative. For example, requiring that E^* be 0.30 or less asks for an error of 0.30σ or less, whatever the value of σ. If a school administrator wants to estimate the average reading achievement score for her whole school by using a sample of children, how large a sample should she take if she wishes to assert with 95% confidence that E^* is at most 0.30?

**9.2
Confidence
Intervals for
Means**

Let us now introduce a different way of presenting the information provided by a sample mean together with an assessment of the error we might make when we use it to estimate the mean of the population from which the sample came. In what follows, we shall make use of the fact that for large random samples from infinite populations the sampling distribution of the mean is approximately normal, with the mean μ and the standard deviation $\dfrac{\sigma}{\sqrt{n}}$; that is,

$$z = \frac{\bar{x} - \mu}{\sigma/\sqrt{n}}$$

is a random variable having approximately the standard normal distribution. Since the probability is $1 - \alpha$ that a random variable having the standard normal distribution will take on a value between $-z_{\alpha/2}$ and $z_{\alpha/2}$ (see Figure 7.20), that is,

$$-z_{\alpha/2} < z < z_{\alpha/2}$$

we can substitute into this inequality the foregoing expression for z and get

$$-z_{\alpha/2} < \frac{\bar{x} - \mu}{\sigma/\sqrt{n}} < z_{\alpha/2}$$

If we multiply each term by $\dfrac{\sigma}{\sqrt{n}}$, subtract $\bar{x}$ from each term, and multiply each term by -1, we get

$$\bar{x} + z_{\alpha/2} \cdot \frac{\sigma}{\sqrt{n}} > \mu > \bar{x} - z_{\alpha/2} \cdot \frac{\sigma}{\sqrt{n}}$$

Note that, as a result of multiplying by -1, we had to reverse the inequality signs, which is always the case when we multiply the expressions on both sides of an inequality by a negative number. For instance, where 5 is greater than (to the right of) 2, -5 is less than (to the left of) -2.

The result we obtained can also be written as

Large-sample confidence interval for μ

$$\bar{x} - z_{\alpha/2} \cdot \frac{\sigma}{\sqrt{n}} < \mu < \bar{x} + z_{\alpha/2} \cdot \frac{\sigma}{\sqrt{n}}$$

and we can assert with probability $1 - \alpha$ that it will be satisfied for any given sample.[†] In other words, we can assert with $(1 - \alpha)$ 100% confidence that the interval from

$$\bar{x} - z_{\alpha/2} \cdot \frac{\sigma}{\sqrt{n}} \quad \text{to} \quad \bar{x} + z_{\alpha/2} \cdot \frac{\sigma}{\sqrt{n}}$$

determined on the basis of a large random sample, contains the population mean we are trying to estimate. When σ is unknown and n is at least 30, we replace σ by the sample standard deviation s. If n is at least 30, then the use of s is recommended, even if the value of σ is claimed to be known. An interval like this is called a **confidence interval**, its end points are called **confidence**

[†] Since the probability is also $1 - \alpha$ that $-z_{\alpha/2} \leq z \leq z_{\alpha/2}$, some authors substitute $\leq$ for $<$ in this confidence-interval formula. Practically speaking, this does not make any difference.

limits, and the probability $1 - \alpha$ is called the **degree of confidence**. As in Section 9.1, the values most commonly used for $1 - \alpha$ are 0.95 and 0.99, the corresponding values of $z_{\alpha/2}$ are 1.96 and 2.575, and the resulting confidence intervals are referred to as 95% and 99% confidence intervals for μ.

■ **EXAMPLE**   With reference to the previously given pulse rate data, where we had $n = 32$, $\bar{x} = 26.2$, and $s = 5.15$, construct a 95% confidence interval for the true average increase in the pulse rate of persons performing the given task.

Solution   Substituting $n = 32$, $\bar{x} = 26.2$, $s = 5.15$ for σ, and $z_{0.025} = 1.96$ into the confidence interval formula, we get

$$26.2 - 1.96 \cdot \frac{5.15}{\sqrt{32}} < \mu < 26.2 + 1.96 \cdot \frac{5.15}{\sqrt{32}}$$

or

$$24.4 < \mu < 28.0$$

for the true average increase in the pulse rate of persons performing the given task. Of course, the interval from 24.4 to 28.0 contains μ or it does not, and we really don't know which, but the 95% confidence implies that the interval was obtained by a method that works 95% of the time. To put it differently, the interval may contain μ or it may not, but if we had to bet, 95 to 5 (or 19 to 1) would be fair odds that it does.   ■

Had we wanted to calculate a 99% confidence interval in this example, we would have substituted 2.575 instead of 1.96 for $z_{\alpha/2}$, and we would have obtained $23.9 < \mu < 28.5$. Comparison of the 95% and 99% confidence intervals shows that

> *When we increase the degree of certainty, namely the degree of confidence, the confidence interval becomes wider and thus tells us less about the quantity we are trying to estimate.*

In other words, "the surer we want to be, the less we have to be sure of."
     When we use a confidence interval to estimate the mean of a population, we call this kind of estimate an **interval estimate**. In contrast to point estimates, interval estimates require no further elaboration about their reliability; this is taken care of indirectly by their width and the degree of confidence.

**9.3
Confidence
Intervals for
Means (Small
Samples)**

In the preceding section we assumed that the sample is large enough, $n \geq 30$, to treat the sampling distribution of the mean as if it were a normal distribution and, when necessary, to replace σ with s. To develop corresponding theory that applies also to small samples, it will be necessary to assume that the population we are sampling has roughly the shape of a normal distribution. We can then base our method on the **t statistic**

$$t = \frac{\bar{x} - \mu}{s/\sqrt{n}}$$

whose sampling distribution is a continuous distribution called the **t distribution**. As is shown in Figure 9.2, this distribution is symmetrical and bell shaped

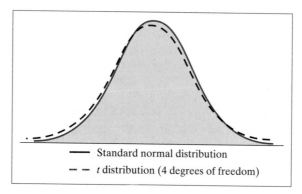

FIGURE 9.2 Standard normal distribution and t distribution.

with zero mean; in fact, its shape is very similar to that of the standard normal distribution. The exact shape of the t distribution depends on a parameter called the **number of degrees of freedom**, or simply the **degrees of freedom**, often abbreviated as df. As the distribution will be used here, the number of degrees of freedom is given by $n - 1$, the sample size less 1. For the t distribution, we define $t_{\alpha/2}$ in the same way in which we defined $z_{\alpha/2}$, so that the area under the curve to the right of $t_{\alpha/2}$ is equal to $\dfrac{\alpha}{2}$. However, $t_{\alpha/2}$ depends on $n - 1$, the number of degrees of freedom, and its value will have to be found in each case from Table III at the end of the book. Now we proceed as on page 296. Making use of the fact that the t distribution is symmetrical about $t = 0$, we find that the probability is $1 - \alpha$ that a random variable having the t distribution will take on a value between $-t_{\alpha/2}$ and $t_{\alpha/2}$ (see Figure 9.3); that is,

$$-t_{\alpha/2} < t < t_{\alpha/2}$$

Then, substituting into this inequality the foregoing expression for t, we get

$$-t_{\alpha/2} < \frac{\bar{x} - \mu}{s/\sqrt{n}} < t_{\alpha/2}$$

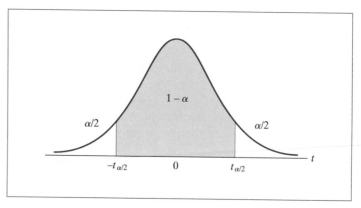

FIGURE 9.3 t distribution.

and using the same steps as on page 296, we arrive at the following **small-sample confidence interval for** μ:

Small-sample confidence interval for μ	$$\bar{x} - t_{\alpha/2} \cdot \frac{s}{\sqrt{n}} < \mu < \bar{x} + t_{\alpha/2} \cdot \frac{s}{\sqrt{n}}$$

The degree of confidence is $1 - \alpha$, and the only difference between this confidence interval formula and the large-sample formula (with s substituted for σ) is that $t_{\alpha/2}$ takes the place of $z_{\alpha/2}$.

■ **EXAMPLE**   To test the durability of a new paint for white center lines, a highway department painted test strips across heavily traveled roads in eight different locations, and electronic counters showed that they deteriorated after having been crossed by 14.26, 16.78, 13.65, 10.83, 12.64, 13.37, 16.20, and 14.94 million cars. Construct a 95% confidence interval for the average amount of traffic (car crossings) this paint can withstand before it deteriorates.

Solution   The mean and the standard deviation of these values are $\bar{x} = 14.08$ and $s = 1.92$. For $8 - 1 = 7$ degrees of freedom, we find that $t_{0.025}$ equals 2.365, and substitution into the formula yields

$$14.08 - 2.365 \cdot \frac{1.92}{\sqrt{8}} < \mu < 14.08 + 2.365 \cdot \frac{1.92}{\sqrt{8}}$$

or

$$12.47 < \mu < 15.69$$

This is the desired 95% confidence interval estimate of the average amount of traffic (millions of car crossings) that the paint can withstand before it deteriorates.   ■

A computer printout of the preceding example is shown in Figure 9.4. The difference between the confidence limits given in the printout and those given above are due to rounding.

If we had wanted the degree of confidence to be 0.97, we could not have worked this example by referring to Table III. Of course, we could have used a computer with the same software as for the printout shown in Figure 9.4, but let us illustrate here the construction of a confidence interval for the

```
MTB > SET C1
DATA> 14.26   16.78   13.65   10.83   12.64   13.37   16.20   14.94
DATA> END
MTB > TINTERVAL WITH 95 PERCENT CONFIDENCE, DATA IN C1

             N       MEAN      STDEV     SE MEAN     95.0 PERCENT C.I.
     C1      8      14.084     1.923      0.680     ( 12.476,  15.692 )
```

FIGURE 9.4 MINITAB printout for small-sample confidence interval for μ.

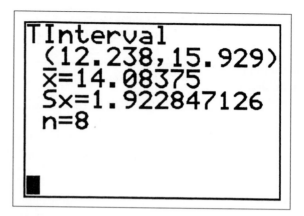

FIGURE 9.5 Small-sample 97% confidence interval for the mean reproduced from the display screen of a TI-83 graphing calculator.

mean with the use of a graphing calculator. As before, all information concerning the graphing calculator is optional and can be omitted by readers without the loss of continuity.

In Figure 9.5 the 97% confidence interval is given as (12.238, 15.929), which rounded to two decimals becomes

$$12.24 < \mu < 15.93$$

Note that this interval is wider than the corresponding 95% confidence interval.

Instructions for the use of the TI-83 graphing calculator are found in the Using Technology box on page 301.

The method we used in Section 9.1 to determine the maximum error that we risk with probability $1 - \alpha$, when we use a sample mean to estimate the mean of a population, can readily be adapted to small samples, provided that the population we are sampling has roughly the shape of a normal distribution. All we have to do is substitute s for σ and $t_{\alpha/2}$ for $z_{\alpha/2}$ in the formula for E on page 291.

■ **EXAMPLE**   In 12 test runs, an experimental engine consumed on the average 12.9 gallons of gasoline per minute with a standard deviation of 1.6 gallons. What can we assert with 99% confidence about the maximum error if we use $\bar{x} = 12.9$ gallons as an estimate of the true average gasoline consumption of the engine?

Solution   Substituting $s = 1.6$, $n = 12$, and $t_{0.005} = 3.106$ (the entry in Table III for $12 - 1 = 11$ degrees of freedom) into the modified formula for E, we get

$$E = t_{\alpha/2} \cdot \frac{s}{\sqrt{n}} = 3.106 \cdot \frac{1.6}{\sqrt{12}} \approx 1.43$$

Thus, if we use the mean $\bar{x} = 12.9$ gallons per minute as an estimate of the true average gasoline consumption of the engine, and it is reasonable to assume that the data constitutes a random sample from a normal population, we can assert with 99% confidence that the error of this estimate is at most 1.43 gallons.   ■

USING TECHNOLOGY : *The Graphing Calculator*

How to obtain the small-sample 97% confidence interval for Figure 9.5

(Tip before beginning: Be sure to clear all lists and plots—see Appendix A: TI-83 Tips.)

Step 1. Input the example data into list L1.
(Questions? See Appendix A: TI-83 Tips.)

Step 2. Select the type of statistical test you wish to use.
Press STAT.
Highlight TESTS.
Select 8: T Interval, press ENTER.
Highlight DATA.
Set   List: L1
            Freq: 1
            C-Level: .97

Step 3. Solve.
Highlight Calculate, press ENTER.

Exercises

Exercises 9.18 and 9.24 are practice exercises; their complete solutions are given on page 311.

9.18 With reference to Exercise 9.1, where we had $n = 100$ and $\sigma = 9.2$, suppose that the fifth graders in the sample averaged 62.7 on the test. Construct a 99% confidence interval for the average score that all the fifth graders in the district would get if they took the arithmetic achievement test.

9.19 The publishers of a daily newspaper would like to know how much time, on average, readers devote to the Sunday edition. A study discloses that 100 randomly selected readers devoted on average 126.5 minutes to the Sunday edition, with a standard deviation of 26.4. Construct a confidence interval for the true average time readers devote to the Sunday edition
(a) using a 0.95 degree of confidence;
(b) using a 0.99 degree of confidence.

9.20 A study made by a large gasoline service station concluded that, on average, $\bar{x} = 9.7$ gallons of gasoline were dispensed to a random sample of 120 automobiles, with a standard deviation of $s = 1.5$. Construct a 98% confidence interval for the true average number of gallons of gasoline dispensed to all automobiles that obtain their gasoline at this station.

9.21 A large hospital finds that in 50 randomly selected days it had, on average, $\bar{x} = 96.4$ patient admissions per day, with a standard deviation of $s = 12.2$.
(a) Construct a 90% confidence interval for the actual daily average number of hospital admissions.

(b) Change the number of patient admissions per day to $\bar{x} = 91.4$ and recalculate the 90% confidence interval for the true average number of patient admissions.

(c) Change the number of patient admissions per day to $\bar{x} = 101.4$ and recalculate the 90% confidence interval for the true average number of patient admissions.

9.22 Over a period of 40 weeks, the computers and related equipment operated by a company had average weekly maintenance costs of $126.45 with a standard deviation of $37.15. Construct a 95% confidence interval for the true average weekly maintenance cost of the company's computer system.

9.23 A small chain of photocopying shops wants to estimate the mean number of reams of paper used per day. In a sample of $n = 60$ randomly selected days, the average usage was 278.5 reams with a standard deviation of 46.25. Construct a 95% confidence interval for the true mean number of reams of paper used per day by the chain of photocopying shops.

9.24 In an air pollution study, an experimental station obtained a mean of 2.36 micrograms of suspended benzene-soluble organic matter per cubic meter with a standard deviation of 0.48 from a random sample of size $n = 10$.

(a) Construct a 99% confidence interval for the mean of the population sampled.

(b) What can be asserted with 95% confidence about the maximum error if $\bar{x} = 2.36$ micrograms is used as an estimate of the mean of the population sampled?

9.25 Determine the value of t from Table III, where

(a) $n = 12$ and the degree of confidence is 0.90;

(b) $n = 13$ and the degree of confidence is 0.95;

(c) $n = 14$ and the degree of confidence is 0.98;

(d) $n = 15$ and the degree of confidence is 0.99.

9.26 The owner of a vending machine has kept a record of the number of cans of soda pop sold daily by the machine. If a random sample of these records for 10 days shows average sales of 192 cans with a standard deviation of 23 cans, construct a 99% confidence interval for the true mean number of cans sold.

9.27 A sample of nine pieces of Manila rope of a certain diameter and construction, intended for nautical use, has a mean breaking strength of 41,250 pounds and a standard deviation of 1,527 pounds. What can be said with 95% confidence about the maximum error if 41,250 pounds is used as an estimate of the true average breaking strength of all such Manila rope of this diameter and construction?

9.28 With reference to Exercise 9.27, construct a 99% confidence interval for the average breaking strength of Manila rope of this diameter and construction.

9.29 In six attempts it took a locksmith 9, 14, 7, 8, 11, and 5 seconds to open a certain kind of lock. Verify that $\bar{x} = 9$ and $s = 3.16$ for these data, and construct a 95% confidence interval for the average time it takes the locksmith to open this kind of lock.

9.30 In establishing the authenticity of an ancient coin, its weight is often of critical importance. If four experts independently weighed a Phoenician tetradrachm and obtained 14.28, 14.34, 14.26, and 14.32 grams, verify that $\bar{x} = 14.30$ and $s = 0.0365$ for the given data. What can one assert with 99% confidence about the maximum error if $\bar{x} = 14.30$ grams is used as an estimate of the actual weight of the coin?

9.31 If a random sample of six students took 18, 19, 23, 19, 21, and 20 minutes to complete the registration forms for courses offered in the next semester, construct a 95% confidence interval for the mean time that it takes to complete such registration forms.

9.32 There is often some uncertainty about the level of confidence that should be used in constructing intervals. It is claimed that, in general, 99% confidence intervals are about 35% longer than the corresponding 95% intervals. Investigate this claim with the data given in Exercise 9.31, where $n = 6$.

9.33 The excess length of the 99% small-sample confidence interval over the corresponding 95% interval, expressed as a percentage of the length of the 95% interval (see Exercise 9.32), is, in fact, determined only by the sample size n. What is the excess length when
    (a) $n = 10$;   (b) $n = 20$;   (c) $n = 30$?

9.34 Use a computer package or a graphing calculator to rework Exercise 9.29.

9.35 With reference to Exercise 9.30, use a computer package or a graphing calculator to construct a 99% confidence interval for the true weight of the coin.

9.36 Use a computer package or a graphing calculator to rework Exercise 9.31.

9.4
Confidence Intervals for Standard Deviations*

On many occasions we must estimate population standard deviations, as with the substitution of s for σ to calculate a large-sample confidence interval for the mean. Let us now show how confidence intervals for σ may be determined on the basis of large samples. A technique for small samples exists, but it will not be discussed in this book.

For large samples, $n \geq 30$, we base our method on the theory that the sampling distribution of s is approximately normal with the mean σ and the standard deviation $\sigma_s = \dfrac{\sigma}{\sqrt{2n}}$, which is called the **standard error of s**. Thus, the sampling distribution of

$$z = \frac{s - \sigma}{\sigma/\sqrt{2n}}$$

is approximately the standard normal distribution, and if we substitute this expression for z into the inequality $-z_{\alpha/2} < z < z_{\alpha/2}$ and use steps similar to those on page 296, we arrive at the following large-sample confidence interval for σ:

Large-sample confidence interval for σ

$$\frac{s}{1 + \dfrac{z_{\alpha/2}}{\sqrt{2n}}} < \sigma < \frac{s}{1 - \dfrac{z_{\alpha/2}}{\sqrt{2n}}}$$

As before, the degree of confidence is $1 - \alpha$.

■ **EXAMPLE**   With reference to the example in Section 9.1 on page 290, where we obtained $\bar{x} = 26.2$ and $s = 5.15$ for the increase in the pulse rate of $n = 32$ persons performing a certain strenuous task, construct a 95% confidence interval for the standard deviation of the population sampled.

Solution   Substituting $n = 32$, $s = 5.15$, and $z_{0.025} = 1.96$ into the large-sample confidence interval formula for σ, we get

$$\frac{5.15}{1 + \dfrac{1.96}{\sqrt{64}}} < \sigma < \frac{5.15}{1 - \dfrac{1.96}{\sqrt{64}}}$$

or

$$4.14 < \sigma < 6.82$$

This is an estimate of the true variability of the increase in the pulse rate of persons performing the given task.   ■

Exercises

Exercise 9.37 is a practice exercise; its complete solution is given on page 311.

*9.37   In Exercise 9.2 we stated, "A random sample of 50 cans of peach halves has a mean weight of 16.1 ounces and a standard deviation of $s = 0.4$ ounce." Using $n = 50$ and $s = 0.4$, construct a 99% confidence interval for the standard deviation of the weights of all the cans of peach halves from which the sample came.

*9.38   If the standard deviation of the IQ scores of 60 elementary school students is $s = 10.89$, construct a 95% confidence interval for the standard deviation of the population of IQ scores from which the sample came.

*9.39   The manager of a supermarket observed that a sample of 100 randomly selected customers waited in line at the checkout counter for varying amounts of time before reaching the cashier, with a standard deviation of $s = 1.29$ minutes. Construct a 98% confidence interval for the standard deviation of the population of waiting times from which this sample came.

*9.40   In a random sample of 60 days, a newspaper dealer sold varying numbers of newspapers with a standard deviation of $s = 27$. Construct a 95% confidence interval for the actual standard deviation of the number of newspapers sold daily by this dealer.

*9.41   With reference to Exercise 9.10, where we had $n = 120$ and $s = 1.84$ minutes, construct a 99% confidence interval for the true standard deviation of the amount of time that passengers disembarking at Kennedy airport have to wait for their luggage.

9.5 The Estimation of Proportions

The information that is usually available for the estimation of a true proportion is a **sample proportion** $\dfrac{x}{n}$, where x is the number of times that an event has occurred in n trials. For instance, if 63 of 150 television viewers (interviewed in a sample survey) liked a certain new situation comedy, then

$$\frac{x}{n} = \frac{63}{150} = 0.42$$

and we can use this figure as an estimate of the true proportion of television viewers who like the new show. Since a percentage is just a proportion multiplied by 100 and a probability may be interpreted as a proportion in the long run, we could also say that we estimate that 42% of all television viewers like the new situation comedy, or that the probability is 0.42 that any one television viewer will like the new show. We have made this point to impress on the reader that the problem of estimating a true percentage or a true probability is essentially the same as that of estimating a true proportion.

Throughout this section it will be assumed that the situations satisfy (at least approximately) the conditions underlying the binomial distribution; that is, our information will consist of the number of successes observed in a given number of independent trials, and it will be assumed that for each trial the probability of a success—the parameter we want to estimate—has the constant value p. Thus, the sampling distribution of the counts on which our methods will be based is the binomial distribution with the mean $\mu = np$ and the standard deviation $\sigma = \sqrt{np(1-p)}$.

We also know that this distribution can be approximated with a normal curve when np and $n(1-p)$ are both greater than 5 (see page 253), and it follows that under these conditions

$$z = \frac{x - np}{\sqrt{np(1-p)}}$$

has approximately the standard normal distribution. If we substitute this expression for z into the inequality $-z_{\alpha/2} < z < z_{\alpha/2}$ (as on page 296) and use some relatively simple algebra, we arrive at the inequality

$$\frac{x}{n} - z_{\alpha/2}\sqrt{\frac{p(1-p)}{n}} < p < \frac{x}{n} + z_{\alpha/2}\sqrt{\frac{p(1-p)}{n}}$$

This may look like a confidence interval formula for p and, indeed, the inequality will be satisfied with probability $1 - \alpha$, but it cannot be used in practice because the unknown parameter p appears also in $\sqrt{\frac{p(1-p)}{n}}$ to the left of the first inequality sign and to the right of the second. This quantity $\sqrt{\frac{p(1-p)}{n}}$ is called the **standard error of a proportion,** as it is, in fact, the standard deviation of the sampling distribution of a sample proportion (see Exercise 9.58). To get around this difficulty, we substitute for p in $\sqrt{\frac{p(1-p)}{n}}$ the sample proportion $\frac{x}{n}$, and we have thus arrived at the following large-sample confidence interval for p:

Large-sample confidence interval for p

$$\frac{x}{n} - z_{\alpha/2}\sqrt{\frac{\frac{x}{n}\left(1 - \frac{x}{n}\right)}{n}} < p < \frac{x}{n} + z_{\alpha/2}\sqrt{\frac{\frac{x}{n}\left(1 - \frac{x}{n}\right)}{n}}$$

The degree of confidence is $1 - \alpha$. As before, the confidence interval is referred to as a $(1 - \alpha)100\%$ confidence interval. There are other methods for constructing a confidence interval for p, but this is the most common.

■ **EXAMPLE**   If 400 persons, constituting a random sample, are given a flu vaccine and 136 of them experienced some discomfort, construct a 95% large-sample confidence interval for the corresponding true proportion.

Solution   Substituting $n = 400$, $\dfrac{x}{n} = \dfrac{136}{400} = 0.34$, and $z_{0.025} = 1.96$ into the confidence interval formula, we get

$$0.34 - 1.96\sqrt{\frac{(0.34)(0.66)}{400}} < p < 0.34 + 1.96\sqrt{\frac{(0.34)(0.66)}{400}}$$

$$0.294 < p < 0.386$$

or, rounding to two decimals, $0.29 < p < 0.39$. Similar to what we said on page 291, we note that the interval from 0.29 to 0.39 contains the true proportion p or it does not, and we really don't know which, but the 95% confidence implies that the interval was obtained by a method that works 95% of the time. Note also that for $n = 400$ and p on the interval from 0.29 to 0.39, np and $n(1 - p)$ are much greater than 5, so that there can be no question about n being large enough to use the normal approximation to the binomial distribution (see page 253).   ■

Although the calculations in this example were quite easy, it may be desirable to use a computer or a graphing calculator. Thus, we shall illustrate here the construction of a confidence interval for p with the use of a graphing calculator. As before, all information concerning the graphing calculator is optional and can be omitted by readers without the loss of continuity.

In Figure 9.6 the 95% confidence interval is given as (.29358, .38642), which rounded to three decimals becomes

$$0.294 < p < 0.386$$

As should have been expected, this is identical with the result obtained before.

Directions for obtaining the large-sample confidence interval for p are given in the Using Technology box on page 307.

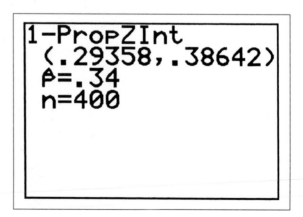

FIGURE 9.6 Large-sample confidence interval for p reproduced from the display screen of a TI-83 graphing calculator.

USING TECHNOLOGY : *The Graphing Calculator*

How to construct a confidence interval for p, Figure 9.6

Step 1.   Select the type of statistical test you wish to use.

Press STAT.
Highlight TESTS.
Scroll down to A: 1-PropZint, press ENTER.
Set   x: 136
       n: 400
       C-Level: .95

Step 2.   Solve.
Highlight Calculate, press ENTER.

The large-sample theory that we have presented here can also be used to assess the size of the error we may be making when we use a sample proportion $\frac{x}{n}$ as a point estimate of a population proportion p. In this connection, we can assert with $(1 - \alpha)100\%$ confidence, approximately, that our error is at most

Approximate maximum error of estimate in using $\frac{x}{n}$ to estimate p

$$E = z_{\alpha/2}\sqrt{\dfrac{\dfrac{x}{n}\left(1 - \dfrac{x}{n}\right)}{n}}$$

Once we have observed the value of x, we can use this formula to assess the size of our error.

■ **EXAMPLE**   With reference to the illustration on page 304 in which 63 of 150 television viewers liked a new situation comedy, what can we say with 99% confidence about the maximum error if we use $\frac{x}{n} = \frac{63}{150} = 0.42$ as an estimate of the true proportion of television viewers who like the new show?

Solution   Substituting $n = 150$, $\frac{x}{n} = 0.42$, and $z_{0.005} = 2.575$ into the formula for E, we get

$$E = 2.575\sqrt{\frac{(0.42)(0.58)}{150}} \approx 0.10$$ ■

The formula above cannot be used to select a sample size n; the following formulas will do that. As in the estimation of means, we can use the expression for the maximum error to determine how large a sample is needed to attain a desired degree of precision. If we want to assert with probability $1 - \alpha$ that a sample proportion will differ from the true proportion by not more than E, we solve the equation

$$E = z_{\alpha/2}\sqrt{\frac{p(1-p)}{n}}$$

for n, and we get

Sample size for estimating p (with some information about p)

$$n = p(1-p)\left[\frac{z_{\alpha/2}}{E}\right]^2$$

Since this formula involves p, it cannot be used unless we have some information about the possible values that p might assume. In that case, we substitute for p whichever of its values is closest to $\frac{1}{2}$. Without such information, we make use of the fact that $p(1-p)$ cannot exceed $\frac{1}{4}$ (which it equals for $p = \frac{1}{2}$) and use the formula

Sample size for estimating p (without information about p)

$$n = \frac{1}{4}\left[\frac{z_{\alpha/2}}{E}\right]^2$$

This may make the sample unnecessarily large, but, on the other hand, we can assert with a probability of *at least* $1 - \alpha$ that the error will not exceed E.

■ **EXAMPLE**   Suppose that we want to estimate what proportion of the adult population of the United States has high blood pressure, and we want to be 99% sure that the error of our estimate will not exceed 0.02. How large a sample will we need if

(a) we have no idea what the true proportion might be;

(b) we know that the true proportion lies on the interval from 0.05 to 0.20?

Solution   (a) Substituting $E = 0.02$ and $z_{0.005} = 2.575$ into the second formula, we get

$$n = \frac{1}{4}\left[\frac{2.575}{0.02}\right]^2 \approx 4{,}145$$

rounded up to the nearest integer.

(b) Substituting these same values together with $p = 0.20$ (the plausible value that is closest to $\frac{1}{2}$) into the first formula, we get

$$n = (0.20)(0.80)\left[\frac{2.575}{0.02}\right]^2 \approx 2{,}653$$

rounded up to the nearest integer. This shows how some knowledge about the values that p might take on can substantially reduce the required sample size.   ■

The methods discussed in this section are all large-sample techniques. For small samples, confidence intervals for proportions can be based on special tables, which may be found in more advanced texts.

Exercises

Exercises 9.42, 9.43, 9.52, and 9.53 are practice exercises; their complete solutions are given on pages 311, and 312.

9.42 In a sample survey, 140 of 500 persons interviewed in a large city said that they shop in the downtown area at least once a week. Construct a 99% confidence interval for the corresponding true proportion.

9.43 With reference to Exercise 9.42, what can we say with 95% confidence about the maximum error if we use the sample proportion to estimate the true proportion of persons in the given city who shop in the downtown area at least once a week?

9.44 Among 80 fish caught in a certain lake, 28 were inedible as a result of the chemical pollution of their environment. If we use the sample proportion $\frac{28}{80} = 0.35$ to estimate the corresponding true proportion, what can we assert with 99% confidence about the maximum error?

9.45 With reference to Exercise 9.44, construct a 95% confidence interval for the true proportion of fish in this lake that are inedible as a result of chemical pollution.

9.46 In a survey of educational attainment of persons 25 years old or over in the United States, 1,200 persons were interviewed and 298 were not high school graduates. Construct a 95% confidence interval of the corresponding true proportion.

9.47 With reference to Exercise 9.46, what can we say with 99% confidence about the maximum error if we use the sample proportion $\frac{298}{1,200} \approx 0.248$ to estimate the corresponding true proportion?

9.48 In a recent survey of 1,419 randomly selected households, 979 gave money in varying amounts to religious, educational, and other types of charitable organizations. Construct a 99% confidence interval for the corresponding true percentage of donors.

9.49 In a study of 2,500 workers (who had earnings), 1,030 had pension plan coverage. If $\frac{1,030}{2,500} \cdot 100 \approx 41\%$ is used as an estimate of the true percentage of workers (who had earnings) who have pension plan benefits, what can one assert with 95% confidence about the maximum error?

9.50 In a study concerning the usage of general-purpose credit cards (such as Mastercard, Visa, Optima, and Discover), 248 of 500 families with incomes of $25,000 to $49,999 annually nearly always pay off their monthly credit card balances. Construct a 95% confidence interval for the probability that a family with an annual income of $25,000 to $49,999 will nearly always pay off its monthly credit card balances.

9.51 In an analysis of labor force participation rates for wives with husbands present, but no children under the age of 18, a randomly selected sample of 1,000 such wives disclosed that 611 of them participate in the labor force. What can we say with 95% confidence about the maximum error if we estimate the probability that any such wife is a labor force participant as $\frac{611}{1,000} \approx 0.61$?

9.52 Suppose that we want to estimate what proportion of all drivers exceed the 65-mph speed limit on a stretch of road between Reno and Las Vegas. How large a sample will we need to be able to assert with probability 0.95 that the error of our estimate, the sample proportion, will be no more than 0.04?

9.53 Rework Exercise 9.52, given that the proportion of all drivers who exceed the 65-mph speed limit on the given stretch of road is at least 0.60.

9.54 A private opinion poll is engaged by a politician to estimate what proportion of her constituents favor the decriminalization of certain narcotics violations. How large a sample will the poll have to take to be able to assert with probability 0.99 that the sample proportion will be off by no more than 0.02?

9.55 Rework Exercise 9.54, given that the poll has reason to believe that the proportion will not exceed 0.30.

9.56 A large personal computer manufacturer wants to determine from a sample what proportion of households intend to purchase personal computers within the next 12 months. How large a sample will the manufacturer need to be able to assert with a probability of at least 0.95 that the sample proportion will not differ from the true proportion by more than 0.05?

9.57 With reference to Exercise 9.56, how large a sample will the manufacturer need to be able to assert with probability 0.95 that the sample proportion will be off by no more than 0.05, if he has good reason to believe that the true proportion is somewhere between 0.10 and 0.20?

9.58 The proportion of successes is simply the number of successes divided by n. Thus, the mean and the standard deviation of the sampling distribution of the sample proportion may be obtained by dividing by n the mean and the standard deviation of the sampling distribution of the number of successes. Use this argument to verify the standard error formula given on page 305.

 Solutions to Practice Exercises

9.1 The probability is 0.95 that the maximum error will be

$$E = 1.96 \cdot \frac{9.2}{\sqrt{100}} \approx 1.80$$

9.2 Substituting $n = 50$, $s = 0.4$, and $E = 0.1$ into the formula for E, we get

$$0.1 = z_{\alpha/2} \cdot \frac{0.4}{\sqrt{50}}$$

so that $z_{\alpha/2} \approx 1.77$, the corresponding entry in Table II is 0.4616, $1 - \alpha = 2(0.4616) = 0.9232$, and we can assert with 92.32% confidence that the error is at most 0.1 ounce.

9.12 Substituting $\sigma = 3.2$ and $E = 0.5$ into the formula for n, we get

$$n = \left[\frac{2.575(3.2)}{0.5} \right]^2 \approx 271.59$$

and we use the next larger integer, 272 hours.

9.18 Substituting $n = 100$, $\bar{x} = 62.7$, and $\sigma = 9.2$ into the confidence interval formula, we get

$$62.7 - 2.575 \cdot \frac{9.2}{\sqrt{100}} < \mu < 62.7 + 2.575 \cdot \frac{9.2}{\sqrt{100}}$$

$$62.7 - 2.37 < \mu < 62.7 + 2.37$$

$$60.3 < \mu < 65.1$$

9.24 (a) Substituting $\bar{x} = 2.36$, $s = 0.48$, $n = 10$, and $t_{0.005} = 3.250$ (for 9 degrees of freedom) into the small-sample confidence interval formula, we get

$$2.36 - 3.250 \cdot \frac{0.48}{\sqrt{10}} < \mu < 2.36 + 3.250 \cdot \frac{0.48}{\sqrt{10}}$$

$$1.87 \text{ micrograms} < \mu < 2.85 \text{ micrograms}$$

  (b) Substituting $s = 0.48$, $n = 10$, and $t_{0.025} = 2.262$ (for 9 degrees of freedom) into the formula for E, we get

$$E = 2.262 \cdot \frac{0.48}{\sqrt{10}} \approx 0.34 \text{ microgram}$$

*9.37 Substituting $s = 0.4$, $n = 50$, and $z_{0.005} = 2.575$ into the large-sample confidence interval formula for σ, we get

$$\frac{0.4}{1 + \dfrac{2.575}{\sqrt{100}}} < \sigma < \frac{0.4}{1 - \dfrac{2.575}{\sqrt{100}}}$$

$$0.32 \text{ ounce} < \sigma < 0.54 \text{ ounce}$$

9.42 Substituting $n = 500$, $\dfrac{x}{n} = \dfrac{140}{500} = 0.28$, and $z_{0.005} = 2.575$ into the large-sample confidence interval formula, we get

$$0.28 - 2.575\sqrt{\frac{(0.28)(0.72)}{500}} < p < 0.28 + 2.575\sqrt{\frac{(0.28)(0.72)}{500}}$$

$$0.228 < p < 0.332$$

or $0.23 < p < 0.33$ rounded to two decimals.

9.43 Substituting $n = 500$, $\dfrac{x}{n} = \dfrac{140}{500} = 0.28$, and $z_{0.025} = 1.96$ into the formula for E, we get

$$E = 1.96\sqrt{\frac{(0.28)(0.72)}{500}} \approx 0.039$$

We can assert with 95% confidence that the error is at most 0.039.

9.52 Substituting $E = 0.04$ and $z_{0.025} = 1.96$ into the second boxed formula on page 308, we get

$$n = \frac{1}{4}\left(\frac{1.96}{0.04}\right)^2 \approx 601$$

rounded up to the nearest integer.

9.53 Substituting $E = 0.04$, $z_{0.025} = 1.96$, and $p = 0.60$ into the first boxed formula on page 308, we get

$$n = (0.60)(0.40)\left(\frac{1.96}{0.04}\right)^2 \approx 577$$

rounded up to the nearest integer.

10 Tests Concerning Means

In the introduction to Chapter 9 we gave three examples of problems of estimation—one from science, one from business, and one from everyday life. They would have been tests of hypothesis, however, if the biologist had wanted to check another scientist's claim that the average wingspan of the insects is 13.4 mm, if the retailer had wanted to find out whether the average income of all families living within 1 mile of the proposed shopping center is at least $38,000, or if we wanted to check whether it really takes only 2.6 minutes to iron a shirt.

After a general introduction to tests of hypotheses in Sections 10.1 and 10.2, Sections 10.3 and 10.4 are devoted to **tests of hypotheses** concerning the mean of one population, Sections 10.5, 10.6, and 10.7 are devoted to tests of hypotheses concerning the means of two populations, and the optional Sections 10.8 and 10.9 are devoted to tests of hypotheses concerning the means of more than two populations (analysis of variance).

10.1
Tests of
Hypotheses

In Sections 9.1 through 9.3 we learned how to estimate a population mean μ by giving a confidence interval or by accompanying the point estimate $\bar{x}$ with an assessment of the possible error. Now we shall learn how to test a hypothesis about a population mean μ; that is, we shall present methods for deciding whether to accept or reject an assertion about a particular value, or particular values, of μ. As we already indicated on page 313, the biologist would be testing a hypothesis if he wanted to check (see whether to accept or reject) the claim that the average wingspan of certain insects is $\mu = 13.4$ mm. In general,

A statistical hypothesis is an assertion or conjecture about the parameter, or parameters, of a population. It may also concern the type, or nature, of a population.

Insofar as the second part of this definition is concerned, we shall test in Section 11.6 whether it is reasonable to treat a random variable as having a binomial distribution, or perhaps a Poisson distribution, and whether it is reasonable to treat a set of data as coming from a normal population. In Chapter 10 we shall be concerned only with hypotheses concerning the population mean μ.

To know exactly what to expect when a hypothesis is true, we often have to hypothesize the opposite of what we hope to prove. For instance, if we want to find out whether one method of teaching computer programming is more effective than another, we hypothesize that the two methods are equally effective. Also, if we want to determine whether one method of irrigating the soil is more expensive than another, we hypothesize that the two methods are equally expensive; and if we want to see whether a new steel alloy is stronger than ordinary steel, we hypothesize that their strength is the same. Since we hypothesize in each case that there is no difference—no difference in the effectiveness of the two teaching methods, no difference in the cost of the two methods of irrigation, and no difference in the strength of the two kinds of steel—we refer to hypotheses like these as **null hypotheses** and denote them by H_0. Actually, the term *null hypothesis* is used for any hypothesis set up primarily to see whether it can be rejected, and the idea of setting up a null hypothesis is common even in nonstatistical thinking. It is precisely what we do in criminal proceedings, where an accused is assumed to be innocent unless his or her guilt is established beyond a reasonable doubt. The assumption that the accused is not guilty is a null hypothesis. To know when to reject a null hypothesis, we must also formulate an **alternative hypothesis**. Denoted by H_A, this is the hypothesis that we accept when the null hypothesis can be rejected. For instance, in the wingspan example the alternative hypothesis might be $\mu \neq 13.4$ mm or $\mu > 13.4$ mm, and in the criminal proceedings referred to above, the alternative hypothesis would be that the accused is guilty.

■ **EXAMPLE**   A research worker wants to compare the effectiveness of advertising a department store sale with newspaper ads or with television ads. Formulate a null hypothesis and an alternative hypothesis appropriate for this situation.

Solution H_0: The two methods of advertising are equally effective.

H_A: The two methods of advertising are not equally effective.   ■

There is no question here about H_0, but depending on what the research worker hopes to show, the alternative hypothesis might also be

H_A: Newspaper ads are more effective.

or

H_A: Television ads are more effective.

It might even be that newspaper ads are twice as effective as advertising on television.

To illustrate the problems we face when testing a statistical hypothesis, let us refer again to the wingspan example, and let us suppose that the biologist wants to test the null hypothesis

$$H_0: \quad \mu = 13.4 \text{ mm}$$

against the alternative hypothesis

$$H_A: \quad \mu \neq 13.4 \text{ mm}$$

where μ is the mean wingspan of the given kind of insect. To perform this test, he decides to take a random sample of size $n = 40$ with the intention of accepting the null hypothesis if the mean of the sample falls anywhere between 13.2 and 13.6 mm; otherwise, the null hypothesis is rejected.

This provides a clear-cut criterion for accepting or rejecting the null hypothesis, but unfortunately it is not infallible. Since the decision is based on a sample, there is the possibility that the sample mean will be greater than or equal to 13.6 mm or less than or equal to 13.2 mm, even though the true mean is 13.4 mm. There is also the possibility that the sample mean will fall in the interval from 13.2 mm to 13.6 mm, even though the true mean is, say, 13.7 mm. Thus, before adopting the criterion (and, for that matter, any decision criterion) it would seem wise to investigate the chances that it will lead to a wrong decision.

Assuming that it is known from similar studies that $\sigma = 0.8$ mm for this kind of data, let us first investigate the possibility of falsely rejecting the null hypothesis. That is, assume for the sake of argument that the true mean is really 13.4 mm; then find the probability that the sample mean will be greater than or equal to 13.6 mm or less than or equal to 13.2 mm. The probability that this will happen purely due to chance is given by the combined area of the blue regions of Figure 10.1, and it can readily be determined by approximating the sampling distribution of the mean with a normal distribution. Assuming that the population sampled is large enough to be treated as if it were infinite, which seems very reasonable in this case, we have

$$\sigma_{\bar{x}} = \frac{\sigma}{\sqrt{n}} = \frac{0.8}{\sqrt{40}} \approx 0.126$$

and it follows that the dividing lines of the criterion, in standard units, are

$$z = \frac{13.2 - 13.4}{0.126} \approx -1.59 \quad \text{and} \quad z = \frac{13.6 - 13.4}{0.126} \approx 1.59$$

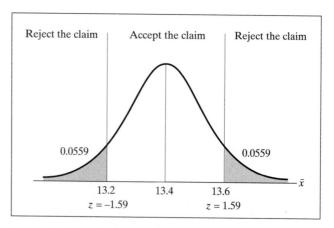

FIGURE 10.1 Test criterion where "accept the claim" and "reject the claim" refer to the corresponding intervals of $\bar{x}$.

According to Table II, the area in each tail of the sampling distribution of Figure 10.1 is $0.5000 - 0.4441 = 0.0559$, and hence the probability of getting a value in either tail of the sampling distribution is

$$0.0559 + 0.0559 = 0.1118 \approx 0.11$$

This is the probability of erroneously rejecting the null hypothesis, that is, the probability of rejecting it when it is true. Whether this is an acceptable risk is for the biologist to decide, and it will have to depend on the consequences of making this kind of error.

Let us now look at the other possibility, where the test fails to detect that the null hypothesis is false, that is, $\mu \neq 13.4$ mm. Assuming again that $\mu = 13.7$ mm, purely for the sake of argument, we find that the probability that the sample mean will fall between 13.2 and 13.6 is given by the area of the blue region of Figure 10.2, the area under the curve between 13.2 and

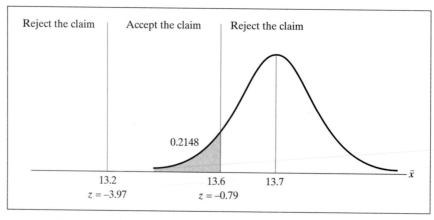

FIGURE 10.2 Test criterion.

13.6. The mean of the sampling distribution is now 13.7 and its standard deviation is again

$$\sigma_{\bar{x}} = \frac{0.8}{\sqrt{40}} \approx 0.126$$

so that the dividing lines of the criterion, in standard units, are now

$$z = \frac{13.2 - 13.7}{0.126} \approx -3.97 \quad \text{and} \quad z = \frac{13.6 - 13.7}{0.126} \approx -0.79$$

Since the area under the curve to the left of $z = -3.97$ is negligible, it follows from Table II that the area of the blue region of Figure 10.2 is $0.5000 - 0.2852 = 0.2148 \approx 0.21$. This is the probability of erroneously accepting the null hypothesis, that is, the probability of accepting it when it is false with $\mu = 13.7$ mm. Again, it is up to the biologist to decide whether this is an acceptable risk.

The situation described in this example is typical of testing a statistical hypothesis, and it may be summarized as in the following table:

	Accept H_0	Reject H_0
H_0 is true	Correct decision	Type I error
H_0 is false	Type II error	Correct decision

If the null hypothesis H_0 is true and accepted, or false and rejected, the decision is in either case correct; if it is true and rejected, or false and accepted, the decision is in either case in error.

> The error of rejecting H_0 when it is true is called a Type I error, and the probability of committing such an error is denoted by α (Greek lowercase alpha).

> The error of accepting H_0 when it is false is called a Type II error, and the probability of committing such an error is denoted by β (Greek lowercase beta).

The terms *alpha error* and *beta error* are sometimes used for Type I and Type II errors.

In our example we showed that for the given test criterion $\alpha = 0.11$. We also showed that $\beta = 0.21$ when the mean wingspan of the given kind of insect is actually 13.7 mm.

■ **EXAMPLE**   Suppose that in our example $\mu = 13.0$ mm, which the biologist does not know, and that the mean of his sample is $\bar{x} = 13.3$ mm. What type of error will he commit if he uses the criterion of Figure 10.1?

Solution   Since $\bar{x} = 13.3$ mm falls in the interval from 13.2 to 13.6 mm, he will accept the null hypothesis $\mu = 13.4$ mm even though it is false; this is a Type II error.   ■

In calculating the probability of a Type II error in our example, we chose for illustrative purposes the alternative value $\mu = 13.7$. However, in this problem, as in most others, there are infinitely many other alternatives, and for each of them there is a positive probability β of erroneously accepting H_0. So, in practice we choose some key alternative values and calculate the probabilities of committing a Type II error, as we calculated it for $\mu = 13.7$, or we sidestep the issue by using a method that will be explained in Section 10.2. If we do calculate β for various alternative values and plot these probabilities as in Figure 10.3, the resulting curve is called an **operating characteristic curve**, or simply an **OC-curve**. Such a curve provides a good overall picture of the risks to which one is exposed by a test criterion.

The operating characteristic curve shown in Figure 10.3 pertains to the wingspan example, and in Exercise 10.8 the reader will be asked to verify some of the other values of β. Since the probability of a Type II error is the probability of accepting H_0 when it is false, we "completed the picture" in Figure 10.3 by labeling the vertical scale "Probability of accepting H_0" and plotting at $\mu = 13.4$ the probability $1 - \alpha = 1 - 0.11 = 0.89$.

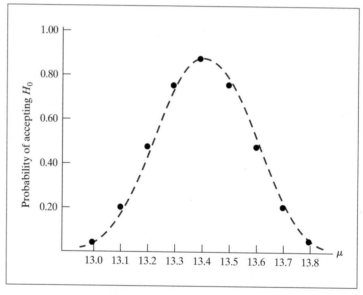

FIGURE 10.3 Operating characteristic curve.

The problems we met in this section are not limited to the particular example dealing with the wingspan of the given kind of insect; the same questions would also have arisen if we had wanted to test the hypothesis that the mean age of divorced women at the time of their divorce is 35.6 years, the hypothesis that an antibiotic is 82% effective, the hypothesis that a computer-assisted method of instruction will on the average raise a student's score on a standard achievement test by 8.2 points, and so forth.

10.2 Significance Tests

In the example of the preceding section, we had less trouble with Type I errors than with Type II errors, because we formulated H_0 as a **simple hypothesis** about μ. That is, with H_0 giving μ a single value, the probability of a Type

I error could readily be calculated. Had we formulated instead the **composite hypothesis** 13.0 mm $\le \mu \le$ 13.8 mm, the composite hypothesis $\mu \le$ 13.4 mm, or the composite hypothesis $\mu \ge$ 13.4 mm, where in each case μ can take on more than one possible value, we could not have calculated the probability of a Type I error without specifying by how much μ differs from, is less than, or is greater than 13.4 mm.

As we saw in Section 10.1, the probability of a Type I error can easily be calculated when we are given a simple hypothesis, an unambiguous criterion, and enough information to apply the sampling theory of Chapter 8. We also saw how the probabilities of Type II errors may have to be calculated for specific alternative values of μ. The following illustrates how we can sidestep Type II errors altogether. Suppose that a retailer wants to find out whether the average income of families living within 1 mile of a proposed shopping center is greater than $48,000. He has a number of these families interviewed and bases his decision on the following criterion:

> *Reject the null hypothesis $\mu =$ $48,000 and accept the alternative $\mu >$ $48,000 if the sample mean is greater than or equal to, say, $48,800; if not, reserve judgment (perhaps pending further checks).*

If we reserve judgment, there is no possibility of committing a Type II error—no matter what happens, the null hypothesis is never really accepted. This is all right in our example, because the retailer wants to know primarily whether the null hypothesis can be rejected.

The procedure we have outlined here is called a **significance test**. If the difference between what we expect under the null hypothesis and what we observe in a sample is too large to be reasonably attributed to chance, we reject the null hypothesis. If the difference between what we expect and what we observe is so small that it may well be attributed to chance, we say that the results are **not statistically significant**.

With respect to the wingspan example, the biologist could convert the criterion on page 315 into that of a significance test by writing

> *Reject the null hypothesis $\mu =$ 13.4 mm (and accept the alternative $\mu \ne$ 13.4 mm) if the mean of a random sample of size n = 40 is greater than or equal to 13.6 mm or less than or equal to 13.2 mm; otherwise, reserve judgment.*

Insofar as the rejection of the null hypothesis is concerned, the criterion has remained unchanged and the probability of a Type I error is still 0.11. However, so far as its acceptance is concerned, the biologist is now playing it safe by reserving judgment.

Reserving judgment in a significance test is similar to what happens in court proceedings when the prosecution does not have sufficient evidence to get a conviction, but it would be going too far to say that the defendant definitely did not commit the crime. In general, whether one can afford the luxury of reserving judgment in any given situation depends entirely on the nature of the problem. If a decision must be reached one way or the other, there is no way of avoiding the risk of committing a Type II error.

Since the general problem of testing hypotheses and constructing statistical decision criteria may seem confusing, at least to the beginner, it will help to proceed systematically as outlined in the following five steps:

1. We formulate a simple null hypothesis and an appropriate alternative hypothesis that is to be accepted when the null hypothesis must be rejected.

In the wingspan example, the null hypothesis was $\mu = 13.4$ mm and the alternative hypothesis was $\mu \neq 13.4$ mm (since the biologist wanted to reject 13.4 mm if this value is too high or too low). We refer to this kind of alternative as a **two-sided alternative**. In the family income example, the null hypothesis was $\mu = \$48,000$ and the alternative hypothesis was $\mu > \$48,000$ (since the retailer wants to know whether average family income in the given area exceeds $48,000). This is called a **one-sided alternative**. We can also write a one-sided alternative with the inequality going the other way. For instance, if we hope to be able to show that the mean time required to do a certain job is less than 20 minutes, we would test the null hypothesis $\mu = 20$ minutes against the alternative hypothesis $\mu < 20$ minutes.

As in the examples of the preceding paragraph, alternative hypotheses usually specify that the population mean (or whatever other parameter may be of concern) is not equal to, greater than, or less than the value assumed under the null hypothesis. For any given problem, the choice of an appropriate alternative hypothesis depends mostly on what we hope to be able to show, or better, perhaps, where we want to put the burden of proof.

■ **EXAMPLE**   A dress manufacturer whose sewing machines average 128 dresses per work shift is considering the purchase of new sewing machines. Against what alternative hypothesis would she test the null hypothesis $\mu = 128$ if

(a) she does not want to buy the new machines unless they will actually increase the output;

(b) she wants to buy the new machines (which have some other nice features) unless they will actually decrease the output.

Solution   (a) She would use the alternative hypothesis $\mu > 128$ and buy the new machines only if the null hypothesis can be rejected.

(b) She would use the alternative hypothesis $\mu < 128$ and buy the new machines unless the null hypothesis is rejected.   ■

■ **EXAMPLE**   A company uses a production process designed to make machine components that have an average thickness of 5 inches. The company suspects that the process is not maintaining its intended average.

(a) If the company wants to modify its process if the average thickness is smaller than 5 inches, what null hypothesis and alternative hypothesis should it use?

(b) If the company wants to modify its process if the average thickness is different from 5 inches, what null hypothesis and alternative hypothesis should it use?

Solution   (a) The words *smaller than* suggest that the hypothesis $\mu < 5$ inches is needed together with the hypothesis $\mu = 5$ inches. Only the second of these hypotheses is a simple hypothesis, where μ takes on a single value, so it must be the null hypothesis and we write

$$H_0: \quad \mu = 5 \text{ inches}$$

$$H_A: \quad \mu < 5 \text{ inches}$$

(b) The words *different from* suggest that the hypothesis $\mu \neq 5$ inches is needed together with the hypothesis $\mu = 5$ inches. Again, only the second of the two hypotheses is a simple hypothesis, where μ takes on a single value. Using it as the null hypothesis, we write

$$H_0: \quad \mu = 5 \text{ inches}$$

$$H_A: \quad \mu \neq 5 \text{ inches}$$ ■

In each of these examples, the hypotheses H_0 and H_A were formulated without experimental data. It is essential that H_A be expressed according to what we hope to demonstrate with subsequent data. In particular, the choice of a one-sided or two-sided version of H_A must not be suggested by the data.

It commonly happens, however, that we are presented with the data before we have had a chance to contemplate the hypotheses; in such a situation we must try to assess the motives without using the numerical data. If there is any doubt as to whether the situation calls for a one-sided or a two-sided alternative, the scrupulous action calls for the two-sided alternative. Exercise 10.17 deals with this difficulty.

2. We specify the probability of a Type I error; if possible, desired, or necessary, we may also make some specifications about the probabilities of Type II errors for specific alternatives.

The probability of a Type I error is also called the **level of significance**, and it is usually set at $\alpha = 0.05$ or $\alpha = 0.01$. Testing a null hypothesis at the level of significance $\alpha = 0.05$ simply means that we are fixing the probability of rejecting the null hypothesis even though it is true at 0.05. The decision to use $\alpha = 0.05$, $\alpha = 0.01$, or some other value depends mostly on the consequences of committing a Type I error in the given situation. Observe, however, that we cannot make the probability of a Type I error too small, for this will have the tendency to make the probability of serious Type II errors too large. In actual practice, experimenters choose α depending on the risks; in the exercises in this text, the level of significance will usually be specified.

3. Based on the sampling distribution of an appropriate statistic and the choice of the level of significance, we construct a test criterion for testing the null hypothesis against the given alternative.

4. We calculate from the data the value of the statistic on which the decision is to be based.

5. We decide whether to reject the null hypothesis, whether to accept it, or whether to reserve judgment.

Insofar as step 3 is concerned, in the wingspan example we based the criterion on the normal-curve approximation to the sampling distribution of the mean. In general, this step depends on the statistic on which we want to base the decision and on its sampling distribution. We call this statistic the **test statistic**.

Looking back at our two illustrations—the wingspan example and the family income example—we find that the construction of a test criterion depends also on the alternative hypothesis that we happen to choose. In the wingspan example, we used a **two-sided criterion** with the two-sided alternative hypothesis $\mu \neq 13.4$ mm, rejecting the null hypothesis for large or small values of the sample mean; in the family income example, we used a one-sided criterion with the one-sided alternative hypothesis $\mu > \$48,000$, rejecting the null hypothesis only for large values of the sample mean.

In general, a test is called **two-sided** or **two-tailed** if the null hypothesis is rejected for values of the test statistic falling into either tail of its sampling distribution, and it is called **one-sided** or **one-tailed** if the null hypothesis is rejected only for values of the test statistic falling into one specified tail of its sampling distribution. A two-tailed test usually goes with a two-sided criterion, and a one-tailed test usually goes with a one-sided criterion, but there are exceptions (see, for example, the tests of Section 10.9).

In connection with step 5, let us point out that we often accept null hypotheses with the tacit hope that we are not exposed to overly high risks of committing serious Type II errors. Of course, if it is necessary we can calculate enough probabilities of Type II errors to get an overall picture from the operating characteristic curve of the test criterion.

As at the end of Section 10.1, let us also point out that the concepts we have introduced here are not limited to tests concerning population means; they apply equally to tests concerning other parameters or tests concerning the nature, or form, of populations.

With the general availability of computers and statistical software, the five steps outlined above may be modified to allow for more freedom in the choice of the level of significance. How this is done will be explained in Section 10.3.

Exercises

Exercises 10.1, 10.5, and 10.9 are practice exercises; their complete solutions are given on page 360.

10.1 A pharmaceutical inspector has to examine shipments of antibiotics to check whether they possess full potency.
   (a) What type of error would he commit if he erroneously rejects the hypothesis that a given bottle of the antibiotic possesses full potency?
   (b) What type of error would he commit if he erroneously accepts the hypothesis that a given bottle of the antibiotic possesses full potency?

10.2 The manager of a sports fishing products company wants to test the null hypothesis that certain lead fishing weights have a mean weight of 4.0 ounces. Explain under what conditions the manager would commit a Type I error and under what conditions the manager would commit a Type II error.

10.3 The information printed on a package of extra-long-life electric lightbulbs states that the bulbs have an average life of 5,000 hours, and we want to test the hypothesis that the average life of such an electric lightbulb meets this specification. Explain under what conditions we would be committing
  (a) a Type I error;
  (b) a Type II error.

10.4 A real estate agent claims that the average price of a single-family residence in a large city is $195,000.
  (a) What hypothesis is she testing if she would commit a Type I error by erroneously concluding that the average price is $195,000?
  (b) What hypothesis is she testing if she would commit a Type II error by erroneously concluding that the average price is $195,000?

✓ **10.5** With reference to the wingspan illustration on page 315, suppose that the biologist increases the sample size to $n = 60$ while everything else remains unchanged.
  (a) Show that this decreases the probability of a Type I error from 0.11 to 0.05.
  (b) Show that this decreases the probability of a Type II error when $\mu = 13.7$ mm from 0.21 to 0.17.

10.6 With reference to the wingspan illustration on page 315, suppose that the biologist changes the criterion so that the hypothesis $\mu = 13.4$ mm is accepted if the sample mean falls anywhere from 13.1 to 13.7 mm, while otherwise the hypothesis will be rejected. If everything else remains the same, show that
  (a) this will decrease the probability of a Type I error from 0.11 to 0.02;
  (b) this will increase the probability of a Type II error when $\mu = 13.7$ from 0.21 to 0.50.

10.7 With reference to the wingspan illustration on page 315, suppose that the biologist increases the sample size to $n = 100$ while everything else remains unchanged.
  (a) What is the probability that the biologist will erroneously reject the null hypothesis $\mu = 13.4$ mm?
  (b) What is the probability that the biologist will erroneously accept the null hypothesis $\mu = 13.4$ mm when actually $\mu = 13.7$ mm?

10.8 With reference to the operating characteristic curve of Figure 10.3, verify that the probabilities of Type II errors are
  (a) 0.78 when $\mu = 13.5$ mm or $\mu = 13.3$ mm;
  (b) 0.50 when $\mu = 13.6$ mm or $\mu = 13.2$ mm;
  (c) 0.06 when $\mu = 13.8$ mm or $\mu = 13.0$ mm.

✓ **10.9** A department store has a salesperson whom it suspects of making more mistakes than the average of all its salespersons.
  (a) If the department store decides to let the salesperson go if this suspicion is confirmed, what null hypothesis and what alternative hypothesis should it use?
  (b) If the department store decides to let the salesperson go unless he actually makes fewer mistakes than the average of all its salespersons, what null hypothesis and what alternative hypothesis should it use?

10.10 The average drying time of a manufacturer's paint is 20 minutes. Investigating the effectiveness of a modification in the chemical composition of the paint, the manufacturer wants to test the null hypothesis $\mu = 20$ against a suitable alternative, where μ is the mean drying time of the modified paint.
  (a) What alternative hypothesis should the manufacturer use if she does not want to make the modification unless it actually decreases the drying time of the paint?

(b) What alternative hypothesis should the manufacturer use if she wants to make the modification unless it actually increases the drying time of the paint?

10.3
Tests Concerning Means

Having used tests concerning means to illustrate the basic principles of hypothesis testing, let us now see how we proceed in practice. Actually, we will depart somewhat from the procedure used in the examples given earlier in this chapter. In the wingspan example and also in the family income example, we used $\bar{x}$ as the test statistic and stated the test criterion in terms of values of $\bar{x}$. Now we will base the criterion on the statistic

Statistic for a large-sample test concerning the mean

$$z = \frac{\bar{x} - \mu_0}{\sigma/\sqrt{n}}$$

where μ_0 is the value of the mean assumed under the null hypothesis. The reason for working with this statistic, which amounts to using standard units, is that it enables us to formulate criteria that are applicable to a great variety of problems, not just one.

We refer to tests based on this statistic as **large-sample tests**, because we are using the normal-curve approximation to the sampling distribution of the mean. Thus, the test requires that $n \geq 30$, unless the population we are sampling is normal. If we are able to assume that the population is normal, the test may be used for any value of n.

■ **EXAMPLE**   A psychologist wants to determine whether the average time it takes an adult to react to a certain emergency situation is really 0.56 second, as claimed by others. From similar studies, she can assume that the variability of such measurements is given by $\sigma = 0.082$ second. Also, she decides to use a random sample of size $n = 35$ and the 0.05 level of significance. What will she conclude if her data yields $\bar{x} = 0.59$ second?

Solution   1. *Hypotheses* H_0: $\mu = 0.56$ second

H_A: $\mu \neq 0.56$ second

where the alternative hypothesis is two-sided because the psychologist will want to reject the null hypothesis if $\mu = 0.56$ second is either too high or too low.

2. *Level of significance* $\alpha = 0.05$

3. *Criterion*   Putting half of 0.05 into each tail of the sampling distribution and making use of the fact that $z_{0.025} = 1.96$, reject the null hypothesis if $z \leq -1.96$ or $z \geq 1.96$, where

$$z = \frac{\bar{x} - \mu_0}{\sigma/\sqrt{n}}$$

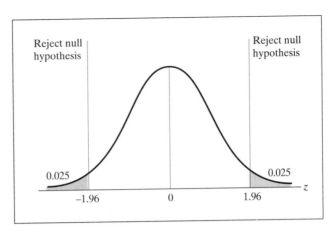

Reject null
hypothesis

Reject null
hypothesis

0.025

0.025

z

−1.96   0   1.96

FIGURE 10.4 Test criterion for reaction-time example.

Otherwise, accept the null hypothesis or reserve judgment (see also Figure 10.4). We include the dividing lines (here −1.96 and +1.96) among the values for which the null hypothesis is rejected.[†]

4. *Calculations*   Substituting $n = 35$, $\bar{x} = 0.59$, $\mu_0 = 0.56$, and $\sigma = 0.082$ into the formula for z, we get

$$z = \frac{0.59 - 0.56}{0.082/\sqrt{35}} \approx \frac{0.03}{0.0139} \approx 2.16$$

5. *Decision*   Since $z = 2.16$ exceeds 1.96, the null hypothesis must be rejected. In other words, the difference between $\bar{x} = 0.59$ and $\mu_0 = 0.56$ is too large to be attributed to chance, and this allows the psychologist to conclude that the 0.56 figure must be wrong.   ■

In this example, the value $\sigma = 0.082$ was used in the calculation of the test statistic in step 4. If the value of the sample standard deviation s had been available, then we would have used that value instead of 0.082. We prefer the data-based value of s over the prespecified assessment of σ; this is especially so when the sample size is large.

In general, for tests concerning means, the dividing lines of the criteria are as shown in Figure 10.5.

1. For the one-sided alternative, $\mu < \mu_0$, the test is one-tailed, α is placed in the left-hand tail, and the dividing line is $-z_\alpha$.

2. For the one-sided alternative, $\mu > \mu_0$, the test is one-tailed, α is placed in the right-hand tail, and the dividing line is z_α.

3. For the two-sided alternative, $\mu \neq \mu_0$, the test is two-tailed, α is divided equally between the two tails, and the dividing lines are $-z_{\alpha/2}$ and $z_{\alpha/2}$.

[†] Some authors exclude the dividing lines from the values for which the null hypothesis is rejected, but this is of no practical consequence since the standard normal distribution is a tinuous distribution.

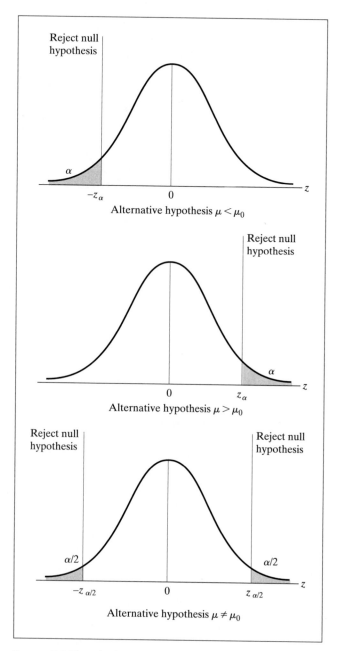

If $\alpha = 0.05$, the dividing lines, or **critical values**, of the criteria are -1.645 for the one-sided alternative in situation 1, 1.645 for the one-sided alternative in situation 2, and -1.96 and 1.96 for the two-sided alternative in situation 3. Similarly, if $\alpha = 0.01$, the dividing lines of the criteria are -2.33 for the one-sided alternative in situation 1, 2.33 for the one-sided alternative in situation 2, and -2.575 and 2.575 for the two-sided alternative in situation 3. All these values come directly from Table II (see Exercise 7.17 and the example on page 244).

In most practical situations where σ is unknown, we must substitute the sample standard deviation s. Again, this is permissible when the sample is large, that is, when $n \geq 30$.

■ **EXAMPLE**   A trucking firm suspects that the average lifetime of 28,000 miles claimed for certain tires is too high. To check the claim, the firm puts 40 of these tires on its trucks and gets a mean lifetime of 27,563 miles and a standard deviation of 1,348 miles. What can it conclude at the 0.01 level of significance if it tests the null hypothesis $\mu = 28,000$ miles against an appropriate alternative?

Solution   1. *Hypotheses* H_0: $\mu = 28,000$ miles

H_A: $\mu < 28,000$ miles

since the firm is interested in determining whether the tires may, perhaps, not last as long as claimed.

2. *Level of significance* $\alpha = 0.01$

3. *Criterion*   Reject the null hypothesis if $z \leq -2.33$, where

$$z = \frac{\bar{x} - \mu_0}{\sigma/\sqrt{n}}$$

with σ replaced by s; otherwise, accept the null hypothesis or reserve judgment.

4. *Calculations*   Substituting $n = 40$, $\bar{x} = 27,563$, $\mu_0 = 28,000$, and $s = 1,348$ for σ into the formula for z, we get

$$z = \frac{27,563 - 28,000}{1,348/\sqrt{40}} \approx -2.05$$

5. *Decision*   Since $z = -2.05$ is greater than -2.33, the null hypothesis cannot be rejected. The trucking firm's suspicion is not confirmed, and it may accept the claim about the tires or reserve judgment.   ■

With reference to the reaction-time example, let us now introduce the alternative procedure for conducting tests of significance, which we mentioned on page 321. Instead of comparing $z = 2.16$ with 1.96 (and tacitly with -1.96), we compare the total area of the blue regions of Figure 10.6, $0.0154 + 0.0154 = 0.0308$, with the specified level of significance $\alpha = 0.05$. This area is referred to as the **p-value**, prob-value, tail probability, or observed level of significance corresponding to the observed value of the test statistic. When the alternative hypothesis is two-sided, the p-value is usually the sum of two

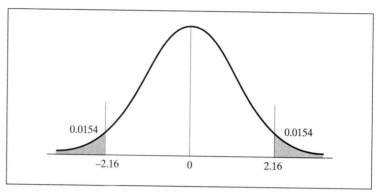

FIGURE 10.6 The *p*-value for the reaction-time example.

areas, as in Figure 10.6. When the alternative hypothesis is one-sided, the
p-value is simply the area under the curve to the right (or left) of the
observed value of the test statistic.

In general, *p*-values may be defined as follows: corresponding to an
observed value of a test statistic, the *p*-value is the lowest level of significance
at which the null hypothesis could have been rejected. In our example, the
p-value was 0.0308, and we could have rejected the null hypothesis at the
0.0308 level of significance. We certainly would have rejected the null hypothesis at the actual specified level of significance $\alpha = 0.05$.

If we wish to base tests of significance on *p*-values, steps 1 and 2 remain
the same, but steps 3, 4, and 5 must be modified as follows:

3′. We specify the test statistic.

4′. We calculate the value of the specified test statistic and the corresponding *p*-value from the sample data.

5′. We compare the *p*-value obtained in step 4′ with the level of significance
α specified in step 2. If the *p*-value is *less than or equal to* α, the null
hypothesis must be rejected; otherwise, we accept the null hypothesis or
reserve judgment.[†] Sometimes the *p*-value is determined in this step.

Let us illustrate this procedure with reference to the truck-tire mileage
example. In the solution, steps 1 and 2 remain the same, but steps 3, 4, and 5
are replaced by

3′. *Test statistic*   The test statistic is

$$z = \frac{\bar{x} - \mu_0}{\sigma/\sqrt{n}} \quad \text{with } \sigma \text{ replaced by } s$$

[†] If test criteria are modified as in the footnote on page 325, step 5′ will call for rejection of the
null hypothesis when the *p*-value is *less than* α.

4'. *Calculations* Substituting $n = 40$, $\bar{x} = 27{,}563$, $\mu_0 = 28{,}000$, and $s = 1{,}348$ for σ into the formula for z, we get

$$z = \frac{27{,}563 - 28{,}000}{1{,}348/\sqrt{40}} \approx -2.05$$

and from Table II we find that the p-value, the area under the curve to the left of z, is $0.5000 - 0.4798 = 0.0202$.

5'. *Decision* Since 0.0202 is greater than $\alpha = 0.01$, the null hypothesis cannot be rejected. As before, the trucking firm's suspicion is not confirmed.

The p-value calculation is sometimes used to avoid the formalism of hypothesis testing, especially in situations where there is no immediate consequence to an accept-or-reject decision. In such cases the p-value may be used as a measure of the relevance, or interest, of the data.

In the truck-tire example an actual decision is required; the firm must decide what brand of tires to use on its trucks. The end result must be a decision about adopting the specified brand of tires.

Consider, however, the plight of a social scientist exploring the relationships between family economics and school performance. He could be testing hundreds of hypotheses involving dozens of variables. The work is very complicated, and there are no immediate policy consequences. In this situation, the social scientist can tabulate the hypothesis tests according to their p-values. Those tests leading to the lowest p-values are the most provocative, and they will certainly be the subject of future discussion. The social scientist need not actually accept or reject the hypotheses, and the use of the p-value furnishes a convenient summary.

In the truck-tire example, the decision must be reached by specifying the level of significance, though the p-value can be used to supplement the results. On the other hand, the social scientist does not have to make any actual policy decision. He might not even specify a level of significance, finding the p-value a more convenient way to handle the data and letting it go at that.

Many experimenters relate the p-value to the common levels of significance $\alpha = 0.05$ and $\alpha = 0.01$ by means of statements such as the following: $p < 0.01$ or $0.01 < p < 0.05$. When the p-value is obtained from the normal distribution, as in Table II, this is readily done. The truck-tire examples led to $z = -2.05$ and a p-value of 0.0202. This is easily converted to the statement $0.01 < p < 0.05$, if so desired. In many areas of research, the result $0.01 < p < 0.05$ is indicated by attaching a single asterisk to the observed value of the test statistic. Correspondingly, the result $p < 0.01$ is indicated by attaching $**$ to the test statistic. For instance, in the preceding example we could have written $z = -2.05*$ as part of step 4'.

10.4
Tests Concerning Means (Small Samples)

When we do not know the value of the population standard deviation σ and the sample is small, $n < 30$, we assume, as on pages 297 and 298, that the population we are sampling has roughly the shape of a normal distribution and base our decision on the statistic

Statistic for a small-sample test concerning the mean

$$t = \frac{\bar{x} - \mu_0}{s/\sqrt{n}}$$

whose sampling distribution is the t distribution (see page 297) with $n - 1$ degrees of freedom. Of course, if it cannot be assumed that the population we are sampling has roughly the shape of a normal distribution, this small-sample procedure cannot be used.

The criteria we use for this test are those of Figure 10.5 with z replaced by t and z_α and $z_{\alpha/2}$ replaced by t_α and $t_{\alpha/2}$. As was explained on page 298, for given degrees of freedom, t_α and $t_{\alpha/2}$ are values for which the areas to their right under the corresponding t distributions are, respectively, α and $\alpha/2$. All the critical values for this **one-sample t test** may be read from Table III, with the number of degrees of freedom equal to $n - 1$.

■ **EXAMPLE**   Suppose that we want to test, on the basis of a random sample of size $n = 5$, whether the fat content of a certain kind of processed meat exceeds 30%. What can we conclude at the 0.01 level of significance if the sample values are 31.9, 30.3, 32.1, 31.7, and 30.9%?

Solution   The sample size $n = 5$ is small, and we must begin by stating that the work rests on the assumption that the values are sampled from a population that has roughly the shape of a normal distribution.

1. *Hypotheses* H_0: $\mu = 30\%$
   H_A: $\mu > 30\%$

   The alternative is one-sided because our concern is that the fat content might exceed 30%.

2. *Level of significance* $\alpha = 0.01$

3. *Criterion*   Reject the null hypothesis if $t \geq 3.747$, the value of $t_{0.01}$ for $5 - 1 = 4$ degrees of freedom, where

$$t = \frac{\bar{x} - \mu_0}{s/\sqrt{n}}$$

   (see also Figure 10.7); otherwise, accept the null hypothesis or reserve judgment.

4. *Calculations*   First calculating the mean and the standard deviation of the sample, we get $\bar{x} = 31.38$ and $s = 0.756$, and substituting these values together with $n = 5$ into the formula for t, we find that

$$t = \frac{31.38 - 30}{0.756/\sqrt{5}} \approx 4.08$$

5. *Decision*   Since $t = 4.08$ exceeds 3.747, the null hypothesis must be rejected. We believe that the mean fat content of the given kind of processed meat exceeds 30%.   ■

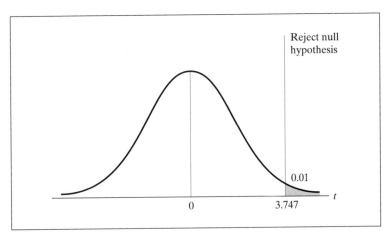

FIGURE 10.7 Test criterion for fat-content example.

Table III shows that the value obtained for t in the preceding example, 4.08, falls between $t_{0.010} = 3.747$ and $t_{0.005} = 4.604$. Thus, the corresponding p-value falls between 0.005 and 0.010, but to determine its actual value we would have to use a computer package or a special calculator. For instance, using a TI-83 graphing calculator, we obtain the result shown in Figure 10.8, which is reproduced from the display screen of the graphing calculator. The graphing calculator information shown here and elsewhere in this book is entirely optional and can be omitted by the reader without the loss of continuity. The p-value shown here is $p = .007548869$, or .0075 rounded to four decimals. Thus, if the level of significance had been, say, $\alpha = 0.009$, we would have found that the p-value, 0.0075, is less than α and, hence, that the null hypothesis could also have been rejected.

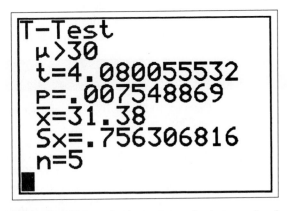

FIGURE 10.8 The p-value for a one-sample t test reproduced from the display screen of a TI-83 graphing calculator.

Directions for obtaining the *p*-value given in Figure 10.8 are in the following Using Technology box.

USING TECHNOLOGY : *The Graphing Calculator*

How to obtain the *p*-value given in Figure 10.8

(Tip before beginning: Be sure to clear all lists and plots—see Appendix A: TI-83 Tips.)

Step 1. Input the fat percentage data from the example into list L1.
(Questions? See Appendix A: TI-83 Tips.)

Step 2. Select the type of statistical test you wish to use.

Press STAT.
Highlight TESTS.
Select 2: T-TEST, press ENTER.
Highlight: DATA
Set $\mu_0 = 30$
       List: L1
       Freq: 1
Highlight $\mu: > \mu_0$

Step 3. Solve.
Highlight Calculate, press ENTER.

When the sample size is small, it is important to state the assumption that the population sampled has roughly the shape of a normal distribution. Merely stating this assumption does not make it true, of course. The assumption has minimal impact on statistical practice since the appropriateness of the normal distribution is difficult to verify with a small sample and, moreover, the *t* statistic generally makes correct decisions even when the population does not have the shape of a normal distribution. Nevertheless, sound statistical practice recommends that the assumption be stated.

Exercises

Exercises 10.11 and 10.21 are practice exercises; their complete solutions are given on page 361.

10.11 A law student, who wants to check a professor's claim that convicted embezzlers spend on the average 12.8 months in jail, takes a random sample of 60 such cases from court files, finding $\bar{x} = 11.2$ months and $s = 3.5$ months.
(a) Test the null hypothesis $\mu = 12.8$ months at the 0.01 level of significance.
(b) Find the *p*-value.
(c) If you base the hypothesis test on the *p*-value, do you reach the same conclusion with regard to significance at the 0.01 level?

10.12 The "reasonable and customary" fee for anesthesia services for surgery lasting less than 1 hour in a certain region is $425, as specified by an insurer. A particular anesthesiology office is suspected of charging, on average, more than this amount, and a random sample of 45 of their billings shows $\bar{x} = \$455$ and $s = \$44$. Can we conclude at the 0.05 level of significance that this particular anesthesiology office charges on average more than $425?

10.13 According to a production engineer, a trained worker should be able to assemble a certain type of chair in 15.0 minutes. Is this figure substantiated by the carefully timed assembly of 150 randomly selected chairs whose average assembly time is 15.4 minutes with standard deviation of 2.4 minutes? Use a two-sided test with the level of significance
   (a) $\alpha = 0.05$;
   (b) $\alpha = 0.01$.

10.14 Solve Exercise 10.13 using a one-sided test with the alternative hypothesis $H_A: \mu > 15$.

10.15 While investigating a complaint about short weighting at a cannery, an investigator finds that a random sample of 100 cans of pineapple that are labeled "net weight 907 grams" (2 pounds) actually weigh an average of only 899 grams, and it is determined that the standard deviation is $s = 29$ grams. Using an appropriate one-sided alternative and the level of significance $\alpha = 0.05$, test whether this constitutes evidence on which the investigator can base a finding that the cannery is producing cans of pineapple that contain on the average less than 907 grams.

10.16 According to the norms established for a history test, eighth graders should average 81.7 with a standard deviation of 8.5. If 100 randomly selected eighth graders from a certain school district averaged 79.6 on this test, can we conclude at the 0.05 level of significance that eighth graders from this school district can be expected to average less than the norm of 81.7 on this test? Find the p-value and use it to make the hypothesis test. Do you reach the same conclusion with regard to significance at the 0.05 level?

10.17 The average running time of a certain variety of nickel–cadmium rechargeable flashlight battery is known to be 8.5 hours. A change in the production method for this battery has been proposed, and a sample of 60 batteries produced by the new method has mean running time of 8.62 hours and a standard deviation of 0.55 hour.
   (a) Show that for the null hypothesis $H_0: \mu = 8.5$ the test statistic has the value $z = 1.69$.
   (b) If the level of significance is 0.05 and if the alternative is $H_A: \mu \neq 8.5$, what conclusion do you reach?
   (c) A chemical engineer examines the data and makes the inference that the change in production was made for the purpose of improving the battery running time. He claims that the alternative must be $H_A: \mu > 8.5$. What conclusion does he reach?
   (d) A second chemical engineer notes that the production method change might have been made for reasons such as reducing cost or increasing the number of times that the battery can be recharged. She claims that the major concern must be that the running time has not significantly worsened and that the alternative must be $H_A: \mu < 8.5$. What conclusion does she reach?
   (e) Suppose that you were given the data with no explanation as to why the change in production method was done. Suppose also that you were given no advice about the level of significance. What would you do?

10.18 A police patrol car drives a prescribed route in a neighborhood, and the dispatcher wants to know whether, if uninterrupted, the average time required by the patrol car to drive its route is 28 minutes. If, in a random sample of 36 uninterrupted rounds, the police car averaged 29.5 minutes with a sample standard deviation of 6.1 minutes, can the dispatcher reject the null hypotheses $\mu = 28$ minutes at the level of significance 0.05?

10.19 In a study of new sources of food, it is reported that a pound of a certain kind of fish yields on the average 2.45 ounces of FPC (fish-protein concentrate), which is used to enrich various food products. Is this figure supported by a study in which 30 samples of this kind of fish yielded on the average 2.48 ounces of FPC (per pound of fish) with a standard deviation of 0.07 ounce, if we use
  (a) the level of significance $\alpha = 0.05$;
  (b) the level of significance $\alpha = 0.01$?

10.20 A city manager wants to know whether the true mean number of temporary secretaries employed per day is 24. If, in a random sample of 60 days, the mean number of temporary secretaries employed is 21.6 with the standard deviation $s = 7.1$, is there, at the 0.05 level of significance, sufficient evidence to reject the null hypothesis $\mu = 24$. Use a two-sided test.

 10.21 In an experiment with a new tranquilizer, the pulse rates (per minute) of 12 patients were determined before they were given the tranquilizer and again 5 minutes later, and their pulse rates were found to be reduced on the average by 7.2 beats with a standard deviation of 1.8. At the level of significance 0.05, do we have significant evidence that the mean pulse reduction with this tranquilizer is less than 9.0 beats?

10.22 In nine test jumps, a newly designed track shoe enables a high-jumper to jump an average of 7.02 feet with a standard deviation of 0.24 foot. What does this tell us about the athlete's claim that with these new shoes he can jump on the average 7.1 feet? Test at the 0.05 level of significance whether his true average jump is less than 7.1 feet.

10.23 A new weight-reducing program that includes diet, exercise, and medication helped 10 overweight people lose 7, 9, 10, 8, 9, 12, 10, 7, 8, and 10 pounds during a three-month test period. Test at the level of significance 0.05 whether this supports the claim that participants in this program lose 10.0 pounds in a three-month period.

10.24 A randomly selected sample of 10 persons who borrowed books at a local public library showed that they borrowed 1, 5, 4, 4, 3, 6, 6, 2, 5, and 4 books. Test at the 0.05 level whether this supports the comment that the library is not lending, on average, 4.2 books per patron.

10.25 A manufacturer guarantees a certain ball bearing to have a mean outside diameter of 0.7500 inch with a standard deviation of 0.0030. If a random sample of 10 such bearings has a mean outside diameter of 0.7510, can we reject the manufacturer's guarantee with regard to the mean outside diameter at the level of significance 0.01?

10.26 The yield of alfalfa from six test plots is 1.2, 2.2, 1.9, 1.1, 1.8, and 1.4 tons per acre. Test at the level of significance $\alpha = 0.05$ whether this supports the contention that the true average yield for this kind of alfalfa is less than 2.0 tons per acre.

10.27 A random sample from a company's very extensive files shows that orders for a certain piece of machinery were filled, respectively, in 12, 10, 17, 14, 13, 18, 11, and 9 days. Use the 0.01 level of significance to test the claim that, on average, such orders are filled in 9.5 days. Choose the alternative hypothesis in such a way that rejection of the null hypothesis $\mu = 9.5$ days implies that it takes longer than that. Find the p-value and indicate whether use of the p-value leads you to the same conclusion regarding significance at the 0.01 level.

10.28 A florist is considering the effects of a new fertilizer on the fruit size of ornamental potted orange trees. The use of standard fertilizers produces, on average, fruits of diameter 4.4 cm. As an experiment, she tries the fertilizer on 20 potted orange trees and then randomly selects one orange from each tree. The diameters of these 20 selected oranges have mean 4.2 cm and standard deviation 0.5 cm.

(a) Using the 0.05 level of significance, can she conclude that the fertilizer has any effect on the diameters of the oranges?

(b) After observing that the oranges that received the new fertilizer are on average smaller, she realizes that it might be possible to advertize these as "dainty petite" oranges. Accordingly, she reformulates the alternative hypothesis as $\mu < 4.4$. At the 0.05 level of significance, does this revised alternative allow her to claim that the fertilized oranges are significantly smaller?

(c) Comment on the appropriateness of selecting H_A after seeing the data.

10.29 The standard formulation used by a certain company for making wall paint will enable a 1-gallon can to cover 250 square feet. The company's research laboratory has produced a new formulation, and the management is anxious to show that it does not cover significantly less area. Accordingly, they test the hypothesis $H_0: \mu = 250$ square feet against the alternative $H_A: \mu < 250$ square feet. The experiment involves 20 cans of the new formulation, and this sample gets mean coverage of 237 square feet and a standard deviation of 25 square feet.

(a) Give the value of the test statistic.

(b) What is the p-value associated with this statistic?

10.30 Suppose that an unscrupulous manufacturer wants "scientific proof" that a totally useless chemical additive will improve the mileage yield of gasoline.

(a) If a research group investigates this additive with one experiment, what is the probability that they will come up with a "significant result" (which they will then use to promote the additive with "scientific claims"), even though the additive is totally ineffective? (Assume that $\alpha = 0.05$ is used.)

(b) If two independent research groups investigate the additive, what is the probability that at least one of them will come up with a "significant result," even though the additive is totally ineffective? (Assume that both use $\alpha = 0.05$.)

(c) If 32 independent research groups investigate the additive, what is the probability that at least one of them will come up with a "significant result," even though the additive is totally ineffective? (Assume that all use $\alpha = 0.05$.)

10.31 Suppose that a manufacturer of pharmaceuticals would like to find a new ointment to reduce swellings. It tries 20 different medications and tests for each whether it reduces swellings at the 0.10 level of significance.

(a) What is the probability that at least one of them will "prove effective," even though all of them are totally useless?

(b) What is the probability that more than one will "prove effective," even though all of them are totally useless?

 10.32 Use a computer package or a graphing calculator to rework Exercise 10.24.

10.33 Use a computer package or a graphing calculator to rework Exercise 10.26. If the results include a *p*-value, note whether it is smaller than the level of significance α.

10.34 Use a computer package or a graphing calculator to rework Exercise 10.27. If the results include a *p*-value, note whether it is smaller than the level of significance α.

10.5 Differences Between Means

There are many problems in which we must decide whether an observed difference between two sample means can be attributed to chance, or whether it is indicative of the fact that the two samples came from populations with unequal means. For instance, we may want to know whether there really is a difference in the mean gasoline consumption of two kinds of cars, when sample data shows that one kind averaged 24.6 miles per gallon while, under the same conditions, the other kind averaged 25.7 miles per gallon. Similarly, we may want to decide on the basis of sample data whether men can perform a certain task faster than women, whether one kind of ceramic insulator is more brittle than another, whether the average diet in one country is more nutritious than that in another country, and so on.

The method we shall use to test whether an observed difference between two sample means can be attributed to chance or whether it is statistically significant is based on the following theory: If $\bar{x}_1$ and $\bar{x}_2$ are the means of two **independent random samples**, then the sampling distribution of the statistic $\bar{x}_1 - \bar{x}_2$ has the mean

$$\mu_1 - \mu_2$$

and the standard deviation

$$\sqrt{\frac{\sigma_1^2}{n_1} + \frac{\sigma_2^2}{n_2}}$$

where μ_1, μ_2, σ_1, and σ_2 are the means and the standard deviations of the two populations sampled. It is customary to refer to the standard deviation of this sampling distribution as the **standard error of the difference between two means**.

By *independent* samples, we mean that the selection of one sample is in no way affected by the selection of the other. Thus, the theory does not apply to "before and after" kinds of comparisons, nor does it apply, say, if we want to compare the caloric value of the food intake of husbands and wives. A special method for comparing the means of dependent samples is explained in Section 10.7.

To base tests of the significance between two sample means on the normal distribution, we shall have to convert to standard units, writing

$$z = \frac{\bar{x}_1 - \bar{x}_2 - (\mu_1 - \mu_2)}{\sqrt{\frac{\sigma_1^2}{n_1} + \frac{\sigma_2^2}{n_2}}}$$

where we subtracted from $\bar{x}_1 - \bar{x}_2$ the mean of its sampling distribution and then divided by the standard deviation of its sampling distribution.

Then, if we limit ourselves to large samples, $n_1 \geq 30$ and $n_2 \geq 30$, we can base the test of the null hypothesis $\mu_1 = \mu_2$ on the foregoing z statistic with $\mu_1 - \mu_2 = 0$, that is, on the statistic

Statistic for a large-sample test concerning the difference between two means

$$z = \frac{\bar{x}_1 - \bar{x}_2}{\sqrt{\dfrac{\sigma_1^2}{n_1} + \dfrac{\sigma_2^2}{n_2}}}$$

which has approximately the standard normal distribution.

Depending on whether the alternative hypothesis is $\mu_1 < \mu_2$, $\mu_1 > \mu_2$, or $\mu_1 \neq \mu_2$, the criteria we use for these significance tests of the difference between two means are again those of Figure 10.5, with $\mu_1 - \mu_2$ substituted for μ and 0 substituted for μ_0. We can refer to Figure 10.5, even though we are concerned with the sampling distribution of the difference between two means instead of the sampling distribution of the mean, because the criteria are all given in standard units.

The test we have described here is essentially a large-sample test; it is exact only when both of the populations sampled are normal. In most practical situations, where σ_1 and σ_2 are unknown, we must make the further approximation of substituting for them the sample standard deviations s_1 and s_2.

■ **EXAMPLE**   In a study designed to test whether there is a difference between the average heights of adult females born in two different countries, random samples yielded the following results:

$$n_1 = 120 \qquad \bar{x}_1 = 62.7 \qquad s_1 = 2.50$$

$$n_2 = 150 \qquad \bar{x}_2 = 61.8 \qquad s_2 = 2.62$$

where the measurements are in inches. Use the level of significance 0.05 to test the null hypothesis that the corresponding population means are equal against the alternative hypothesis that they are not equal.

Solution   1. *Hypotheses* H_0: $\mu_1 = \mu_2$
H_A: $\mu_1 \neq \mu_2$

where μ_1 and μ_2 are the unknown means of the two populations.

2. *Level of significance* $\alpha = 0.05$

3. *Criterion*   Reject the null hypothesis if $z \leq -1.96$ or $z \geq 1.96$, where

$$z = \frac{\bar{x}_1 - \bar{x}_2}{\sqrt{\dfrac{\sigma_1^2}{n_1} + \dfrac{\sigma_2^2}{n_2}}}$$

with s_1 and s_2 substituted for σ_1 and σ_2; otherwise, accept the null hypothesis or reserve judgment.

4. *Calculations*   Substituting $n_1 = 120$, $n_2 = 150$, $\bar{x}_1 = 62.7$, $\bar{x}_2 = 61.8$, $s_1 = 2.50$, and $s_2 = 2.62$ into the formula for z, we get

$$z = \frac{62.7 - 61.8}{\sqrt{\dfrac{(2.50)^2}{120} + \dfrac{(2.62)^2}{150}}} \approx 2.88$$

5. *Decision*   Since $z = 2.88$ exceeds 1.96, the null hypothesis must be rejected. In other words, the sample data shows that there is a difference between the average heights of adult females born in the two countries.

Let us now work this example using the p-value, which requires that we modify steps 3, 4, and 5.

3'. *Test statistic*

$$z = \frac{\bar{x}_1 - \bar{x}_2}{\sqrt{\dfrac{\sigma_1^2}{n_1} + \dfrac{\sigma_2^2}{n_2}}}$$

4'. *Calculations*

$$z = \frac{62.7 - 61.8}{\sqrt{\dfrac{(2.50)^2}{120} + \dfrac{(2.62)^2}{150}}} \approx 2.88$$

5'. *Decision*   The area under the normal curve to the right of $z = 2.88$ is 0.0020. This is also the area under the curve to the left of $z = -2.88$, so that the p-value is $0.0020 + 0.0020 = 0.0040$. This is smaller than $\alpha = 0.05$, so that the null hypothesis must be rejected.   ∎

It should be noted that there is a certain awkwardness about comparing means when σ_1 and σ_2 have different values (or when the sample values s_1 and s_2 are not close). Consider, for example, two normal populations with means $\mu_1 = 50$ and $\mu_2 = 52$ and standard deviations $\sigma_1 = 5$ and $\sigma_2 = 15$. Though the second population has a larger mean, it is much more likely to produce a value below 40. An investigator faced with situations like this must decide whether the comparison of μ_1 and μ_2 really addresses the relevant problem.

10.6 Differences Between Means (Small Samples)

As in Section 10.4, a small-sample test of the significance of the difference between two means may be based on an appropriate t statistic. For this test, which is used when $n_1 < 30$ or $n_2 < 30$, we must assume that we have independent random samples from populations that can be approximated closely by normal distributions with the same standard deviation. Then we can base our decision on the statistic

Statistic for a small-sample test concerning the difference between two means

$$t = \frac{\bar{x}_1 - \bar{x}_2}{s_p\sqrt{\dfrac{1}{n_1} + \dfrac{1}{n_2}}} \quad \text{where} \quad s_p = \sqrt{\frac{(n_1 - 1)s_1^2 + (n_2 - 1)s_2^2}{n_1 + n_2 - 2}}$$

whose sampling distribution is the t distribution with $(n_1 - 1) + (n_2 - 1) = n_1 + n_2 - 2$ degrees of freedom. Here s_p is the pooled **sample standard deviation**.

Observe that t an be rewritten algebraically in various ways. Of course, if we cannot assume that the two populations sampled are approximately normal and that they have the same standard deviation, this small-sample procedure cannot be used.

The criteria we use for this **two-sample t test** are again those of Figure 10.5, with t, t_α, and $t_{\alpha/2}$ substituted for z, z_α, and $z_{\alpha/2}$, with $\mu_1 - \mu_2$ substituted for μ and 0 substituted for μ_0.

■ **EXAMPLE** In five games with a 15-pound ball, a professional bowler scored 205, 220, 200, 210, and 201, and in five games with a 16-pound ball he scored 218, 204, 223, 198, and 211. At the 0.05 level of significance, can we conclude that on the average he will score higher with the 16-pound ball?

Solution Since the sample sizes are small, the work rests on the assumption that the populations sampled have roughly the shape of normal distributions with equal standard deviations.

1. *Hypotheses* H_0: $\mu_1 = \mu_2$
   H_A: $\mu_1 < \mu_2$
2. *Level of significance* $\alpha = 0.05$
3. *Criterion*   Reject the null hypothesis if $t \le -1.860$, where 1.860 is the value of $t_{0.05}$ for 8 degrees of freedom and where t is calculated by means of the formula given above; otherwise, accept the null hypothesis or reserve judgment.
4. *Calculations*   First calculating $\bar{x}_1$, $\bar{x}_2$, s_1, and s_2 for the given data, we substitute these values together with $n_1 = n_2 = 5$ into the formula for s_p, getting

$$s_p = \sqrt{\frac{4(8.17)^2 + 4(10.13)^2}{8}}$$

$$\approx 9.20$$

and

$$t = \frac{207.20 - 210.80}{9.20\sqrt{\dfrac{1}{5} + \dfrac{1}{5}}}$$

$$\approx -0.62$$

5. *Decision*   Since $t = -0.62$ is greater than -1.860, the null hypothesis cannot be rejected. There is no real evidence that the bowler's performance with the two balls is not equally good.   ■

Had the level of significance been, say, $\alpha = 0.15$ in the preceding example, we could not have obtained the value of $t_{0.15}$ from Table III. To determine the p-value and, at the same time, graph the corresponding area under the curve, we enter the data into our TI-83 graphing calculator and solve, obtaining the results shown in Figure 10.9. The graphing calculator information shown here and elsewhere in this book is entirely optional and can be omitted by the reader without the loss of continuity.

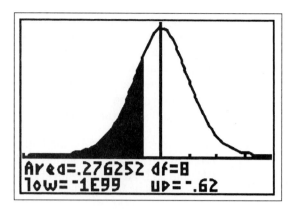

FIGURE 10.9 The *p*-value for a two-sample *t* test reproduced from the display screen of a TI-83 graphing calculator.

Since the *p*-value of .276252 in Figure 10.9 exceeds $\alpha = 0.15$, we conclude that at this level of significance, too, the null hypothesis cannot be rejected. Directions for obtaining the graph in Figure 10.9 are given in the following Using Technology box.

USING TECHNOLOGY : *The Graphing Calculator*

How to obtain the *p*-value and graph the area under the curve in Figure 10.9

(Tip before beginning: Be sure to clear all lists and plots—see Appendix A: TI-83 Tips.)

Step 1. Set the viewing window.

> Press WINDOW.
> Set   Xmin = -5
>   Xmax = 4
>   X scl = 1
>   Ymin = -.1
>   Ymax = .4
>   Yscl = 1
>   Xres = 1

Step 2. Select the type of display and enter the interval to be displayed.
Press 2nd, VARS to access the DISTR DRAW menu.
Highlight DRAW.
Select 2: Shade_t(, press ENTER.
Enter   -1E99,-0.62,8.
        (to graph and determine the area under the curve from $-\infty$ to -0.62 for the *t*-distribution with 8 degrees of freedom).

Step 3. View the graph.
Press ENTER to solve.

Finally, we note that the two-sample t test is one of the most commonly used statistical procedures and is a fundamental tool of scientific research.

10.7 Differences Between Means (Paired Data)

The methods of Sections 10.5 and 10.6 apply only when the samples are independent. For instance, they cannot be used for "before and after" kinds of comparisons or studies of differences between husbands and wives in the caloric intake of their diets. To handle data of this kind, we work with the (signed) differences of the paired data and test whether these differences may be looked on as a random sample from a population that has the mean $\mu = 0$. If the sample is small, we make the assumption that the population sampled has roughly the shape of a normal distribution and use the one-sample t test of Section 10.4; otherwise, we use the large-sample test of Section 10.3.

■ **EXAMPLE**   Use the 0.05 level of significance to test the effectiveness of an industrial safety program on the basis of the following data on the average weekly loss of labor hours due to accidents in 10 plants before and after the program was put into operation:

Plant	Before	After	Difference
1	45	36	9
2	73	60	13
3	46	44	2
4	124	119	5
5	33	35	-2
6	57	51	6
7	83	77	6
8	34	29	5
9	26	24	2
10	17	11	6

The difference column represents before minus after.

Solution   Since the sample size is small, we begin by stating that the work rests on the assumption that the differences constitute a sample from a population having roughly the shape of a normal distribution.

1. *Hypotheses* H_0: $\mu = 0$
                    H_A: $\mu > 0$

   where μ is the mean of the population of differences.

2. *Level of significance* $\alpha = 0.05$

3. *Criterion*   Reject the null hypothesis if $t \geq 1.833$, the value of $t_{0.05}$ for $10 - 1 = 9$ degrees of freedom, where

$$t = \frac{\bar{x} - \mu_0}{s/\sqrt{n}}$$

Otherwise, accept the null hypothesis or reserve judgment.

4. *Calculations*   Since $\bar{x} = 5.2$ and $s = 4.08$ for the 10 differences, substitution of these values together with $n = 10$ and $\mu_0 = 0$ into the formula for t yields

$$t = \frac{5.2 - 0}{4.08/\sqrt{10}} \approx 4.03$$

5. *Decision*   Since $t = 4.03$ exceeds 1.833, the null hypothesis must be rejected, and we conclude that the industrial safety program is effective.

We can also use the *p*-value in this example. We would replace steps 3, 4, and 5 with these:

3'. *Test statistic*   The test statistic is

$$t = \frac{\bar{x} - \mu_0}{s/\sqrt{n}} \quad \text{which has 9 degrees of freedom}$$

4'. *Calculations*   Find $t = 4.03$, as above.

5'. *Decision*   Since $4.03 > t_{0.005} = 3.250$, using 9 degrees of freedom, we conclude that $p < 0.005$. Thus, the *p*-value is less than the specified level of significance, $\alpha = 0.05$, and the null hypothesis must be rejected. ■

Exercises

Exercises 10.35, 10.41, *and* 10.44 *are practice exercises; their complete solutions are given on pages* 361 *and* 362.

10.35 A sample study was made of the number of business lunches that executives claim as deductible expenses per month. If 60 executives in the insurance industry average 12.8 such deductions with a standard deviation of 5.4 in a given month, while 50 bank executives averaged 9.2 with a standard deviation of 6.3, test at the 0.01 level of significance whether the difference between these two sample means is significant.

10.36 A consumer testing organization wants to know whether two brands of work shoes have equally long-lasting heels. Tests show that 50 randomly selected workers wore out the heels of Brand X shoes, on average, in 96 days with a standard deviation of 6.4 days, while 50 randomly selected workers wore out the heels of Brand Y shoes, on average, in 99 days with a standard deviation of 6.1 days.
   (a) Test at the level of significance $\alpha = 0.05$ whether the difference between the two average wearing lives is significant.
   (b) Compute the *p*-value and use it to decide whether the difference between the sample means is significant at the 0.05 level of significance.

10.37 Repeat Exercise 10.36 after changing the level of significance from 0.05 to 0.01. Also, compute the *p*-value and use it to decide whether the difference between the sample means is significant at the 0.01 level.

10.38 The concession manager at a minor-league baseball field wishes to compare the fans' food purchases at day games with their purchases at night games. He obtains samples of day games and of night games over a time period in which

food prices did not change. For each game selected, he computes the food purchase amount per attendee (by dividing total sales by the attendance), obtaining this information:

Type of game	Number of games	Food purchase amount per attendee	
		Mean	Standard deviation
Day	30	$4.40	$0.60
Night	35	$4.80	$0.70

Test at the 0.01 level of significance whether the difference between these two
sample means is significant. Compute the p-value and indicate whether it leads
to the same conclusion.

10.39 An office supply store claims that its rubber bands are better than those of a
competitor. A study shows that a sample of 70 of its rubber bands of a certain
type could stretch, on average, 8.6 inches before breaking, with a standard
deviation of 0.75 inch; while a sample of 50 rubber bands of the same type sold
by a competitor could stretch, on average, only 8.3 inches with a standard deviation of 0.80 inch before breaking. Use a suitable one-sided alternative and the
0.01 level of significance to test the null hypothesis $\mu_1 = \mu_2$ against the suitable
alternative hypothesis $\mu_1 > \mu_2$.

10.40 An investigation conducted in two canneries showed that in one cannery a
sample of 48 cans averaged 15.0 grams of protein per quarter-cup serving with
a standard deviation of 1.6 grams, while in the other cannery a sample of 48
cans averaged 14.5 grams of protein per quarter-cup serving with a standard
deviation of 1.4 grams. Test the null hypothesis $\mu_1 = \mu_2$ against the alternative
hypothesis $\mu_1 > \mu_2$ at the level of significance $\alpha = 0.01$.

 10.41 Measurements of the heat-producing capacity of coal from two mines yielded
the following results:

$$n_1 = 5 \qquad \bar{x}_1 = 8,160 \qquad s_1 = 252$$

$$n_2 = 5 \qquad \bar{x}_2 = 7,730 \qquad s_2 = 207$$

where the measurements are in millions of calories per ton. At the level of significance 0.05, can we conclude that the mean heat-producing capacity of coal
from the two mines is not the same? Do you reach the same conclusion if you
use the p-value?

10.42 A newspaper reporter wished to compare the workload at two motor vehicle
offices. She was able to obtain records for randomly sampled calendar days
during the previous year, obtaining the following information:

Location	Number of days	Number of motorists processed	
		Mean	Standard deviation
Downtown	12	446.5	25.8
Suburban	10	415.8	23.6

Test at the 0.05 level of significance whether the difference between these two
sample means is significant. Compute the p-value and indicate whether it leads
to the same conclusion.

10.43 Twelve measurements each of the hydrogen content (in percent number of atoms) of gases collected from the eruptions of two volcanos yielded $\bar{x}_1 = 41.5$, $\bar{x}_2 = 46.1$, $s_1 = 5.2$, and $s_2 = 6.7$. Use the level of significance $\alpha = 0.05$ to test the null hypothesis that there is no difference (with regard to hydrogen content) in the composition of the gases from the two eruptions. Provide the p-value also.

 10.44 In a study of the effectiveness of physical exercise in weight reduction, a group of 32 persons engaged in a prescribed program of physical exercise for one month showed the following results (in pounds):

Weight before	Weight after	Weight before	Weight after
209	196	170	164
178	171	153	152
169	170	183	179
212	207	165	162
180	177	201	199
192	190	179	173
158	159	243	231
180	180	144	140
211	203	179	180
193	183	202	197
245	229	169	175
188	190	187	190
201	194	213	205
222	219	174	170
190	195	196	197
199	197	201	201

Use a one-sided test and a 0.01 level of significance to determine whether the prescribed program of exercise is effective.

10.45 The following data was obtained in an experiment designed to check whether there is a systemic difference in the weights (in grams) obtained with two different scales:

Rock specimen	Scale I	Scale II
1	12.13	12.17
2	17.56	17.61
3	9.33	9.35
4	11.40	11.42
5	28.62	28.61
6	10.25	10.27
7	23.37	23.42
8	16.27	16.26
9	12.40	12.45
10	24.78	24.75

Test at the 0.01 level of significance whether the difference between the means of the weights obtained with the two scales is significant.

10.46 Twelve persons with Type II diabetes were given a new oral medication to reduce their blood sugar, and their average fasting (before breakfast) blood sugar readings before they started taking the medication and six weeks later were 128 and 117, 145 and 133, 115 and 108, 156 and 133, 129 and 111, 147 and 129, 130 and 122, 127 and 124, 120 and 103, 145 and 129, 124 and 116, and 135 and 117. Test at the 0.05 level of significance whether this medication will reduce a Type II diabetic's blood sugar by more than 10 points in six weeks. Use the 0.05 level of significance.

10.47 A random sample of 10 high-school students in Phoenix scored 81, 85, 75, 79, 83, 76, 78, 81, 84, and 78, and a random sample of 10 high-school students in Tucson scored 82, 87, 94, 88, 79, 80, 87, 83, 75, and 85 on the same social studies examination. Use a computer package or a graphing calculator to test whether the difference between the two sample means is significant at the 0.01 level of significance.

10.48 To compare two kinds of bumper guards, six of each kind were mounted on the same kind of compact car and crashed into a wall at 5 miles per hour. The repair costs for bumper guard A were 345, 290, 360, 380, 310, and 260 dollars, and those for bumper guard B were 310, 290, 315, 385, 280, and 305 dollars. Use a computer package or a graphing calculator to test at the 0.05 level of significance whether the difference between the two sample means is significant.

10.8 Differences Among *k* Means*

In the three preceding sections we developed procedures for testing the hypothesis that an observed difference between two sample means can be attributed to chance. Now we consider the problem of deciding whether observed differences among more than two sample means can be attributed to chance, or whether they indicate actual differences among the means of the populations sampled. For instance, we may want to decide on the basis of sample data whether there really is a difference in the effectiveness of three methods of teaching computer programming, or we may want to compare the yield of four varieties of wheat, or we may want to see whether there really is a difference in the average mileage obtained with five kinds of gasoline, or we may want to judge whether there really is a difference in the performance of eight different hair driers, and so on. Suppose that we actually want to compare the effectiveness of three methods of teaching the programming of a certain computer: method 1, which is straight teaching-machine instruction; method 2, which involves the personal attention of an instructor and some direct experience working with the computer; and method 3, which involves the personal attention of an instructor, but no work with the computer itself. Suppose, furthermore, that random samples of size 4 are taken from large groups of students taught by the three methods and that these students obtained the following scores in an appropriate achievement test:

Method 1: 71, 75, 65, 69
Method 2: 90, 80, 86, 84
Method 3: 72, 77, 76, 79

The means of these three samples are $\bar{x}_1 = 70$, $\bar{x}_2 = 85$, and $\bar{x}_3 = 76$, and we would like to know whether the differences among them are significant or whether they can be attributed to chance.

If μ_1, μ_2, and μ_3 are the means of the three populations sampled in this example, we shall want to test the null hypothesis $\mu_1 = \mu_2 = \mu_3$ against the alternative hypothesis that these means are not all equal. It is possible to consider specific special alternatives such as $\mu_1 < \mu_2 < \mu_3$, but we shall not do so here. This null hypothesis would be supported if the differences among the sample means, $\bar{x}_1$, $\bar{x}_2$, and $\bar{x}_3$, are small; the alternative hypothesis would be supported if at least some of the differences among the sample means are large. Thus, we need a precise measure of the discrepancies among the $\bar{x}$'s and, with it, a rule that tells us when the discrepancies are so large that the null hypothesis can be rejected.

Possible choices for such a measure are the standard deviation of the $\bar{x}$'s or their variance. To determine the latter, we first calculate the mean of the three $\bar{x}$'s, getting

$$\frac{70 + 85 + 76}{3} = 77$$

and then we find that

$$s_{\bar{x}}^2 = \frac{(70 - 77)^2 + (85 - 77)^2 + (76 - 77)^2}{3 - 1} = 57$$

where the subscript $\bar{x}$ serves to indicate that $s_{\bar{x}}^2$ measures the variation of the sample means. Let us now make two assumptions that are critical to the method by which we shall continue the analysis of our problem:

1. The populations we are sampling can be approximated closely with normal distributions.

2. These populations all have the same standard deviation σ.

With reference to our example, this means that we are assuming that (1) the test scores, for each method of teaching, are values of a random variable having (at least approximately) a normal distribution, and that (2) these random variables all have the same standard deviation σ.

With these assumptions, and if the null hypothesis $\mu_1 = \mu_2 = \mu_3$ is true, we can look upon the three samples as if they came from the same (normal) population and, hence, upon the variance of their means, $s_{\bar{x}}^2$, as an estimate of $\sigma_{\bar{x}}^2$, the square of the standard error of the mean.

Now, since $\sigma_{\bar{x}} = \dfrac{\sigma}{\sqrt{n}}$ for samples from infinite populations, we can look upon $s_{\bar{x}}^2$ as an estimate of

$$\sigma_{\bar{x}}^2 = \left(\frac{\sigma}{\sqrt{n}}\right)^2 = \frac{\sigma^2}{n}$$

where n is the size of each sample. Then, multiplying by n, we can look upon $n \cdot s_{\bar{x}}^2$ as an estimate of σ^2, and it is important to note that this estimate is based on the variation among the sample means. For our example, we thus have

$$n \cdot s_{\bar{x}}^2 = 4 \cdot 57 = 228$$

as an estimate of σ^2, the common variance of the three populations.

If σ^2 were known, we could compare $n \cdot s_{\bar{x}}^2$ with σ^2 and reject the null hypothesis that the population averages are all equal if $n \cdot s_{\bar{x}}^2$ is much larger than σ^2. However, in most practical situations σ^2 is not known, and we have no choice but to estimate it on the basis of the sample data.

Since we assumed under the null hypothesis that our three samples come from identical populations, we could use any one of the sample variances s_1^2, s_2^2, or s_3^2 as an estimate of σ^2, and we could also use their mean. Averaging, or pooling, the three sample variances in our example, we get

$$\frac{s_1^2 + s_2^2 + s_3^2}{3} = \frac{1}{3}\left(\frac{(71-70)^2 + (75-70)^2 + (65-70)^2 + (69-70)^2}{4-1} \right.$$

$$+ \frac{(90-85)^2 + (80-85)^2 + (86-85)^2 + (84-85)^2}{4-1}$$

$$\left. + \frac{(72-76)^2 + (77-76)^2 + (76-76)^2 + (79-76)^2}{4-1} \right)$$

$$= \frac{130}{9} \approx 14.44$$

We now have the following two estimates of σ^2:

$$n \cdot s_{\bar{x}}^2 = 228 \quad \text{and} \quad \frac{s_1^2 + s_2^2 + s_3^2}{3} \approx 14.44$$

It should be observed that, whereas the first estimate measures the variation among the sample means, the second estimate measures the variation within the three samples, that is, chance variation. If the $\bar{x}$'s are far apart and the first of these two estimates is much larger than the second, it thus stands to reason that the null hypothesis ought to be rejected. To put this comparison on a rigorous basis, we use the statistic

Statistic for the test concerning the differences among means

$$F = \frac{\text{variation among the samples}}{\text{variation within the samples}}$$

where the variation among the samples is measured by $n \cdot s_{\bar{x}}^2$ and the variation within the samples is measured by the mean of the three sample variances.

If the null hypothesis is true and the assumptions we made are valid, the sampling distribution of this statistic is the **F distribution**, a theoretical distribution that depends on two parameters called the **numerator and denominator degrees of freedom**. When the **F statistic** is used to compare the means of k samples of size n, the numerator and denominator degrees of freedom are, respectively, $k-1$ and $k(n-1)$.

In general, if F is close to 1 (that is, if the variation among the samples just about equals the variation within the samples and hence reflects only chance variation), the null hypothesis of equal population means cannot be rejected. However, if F is quite large (that is, if the variation among the samples is greater than we would expect it to be if it were due only to chance), the null hypothesis of equal population means will have to be rejected. To determine how large an F we need to reject the null hypothesis, we base our

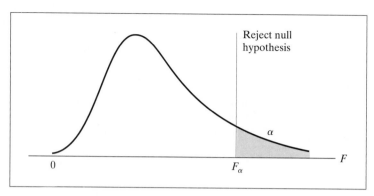

FIGURE 10.10 *F* distribution.

decision on the criterion of Figure 10.10, where F_α is such that the area under the curve to its right is equal to α. For $\alpha = 0.05$ and $\alpha = 0.01$, the values of F_α may be found in Table V at the end of the book.

Returning to our numerical example, we find that

$$F = \frac{228}{14.44} \approx 15.8$$

and since this exceeds 8.02, the value of $F_{0.01}$ for $k - 1 = 3 - 1 = 2$ and $k(n - 1) = 3(4 - 1) = 9$ degrees of freedom, the null hypothesis must be rejected at the 0.01 level of significance. In other words, we conclude that the differences among the sample means are too large to be attributed to chance.

Table V can be used to make *p*-value statements. In our example with $F = 15.8$, we observe that F exceeds $F_{0.01}$ for 2 and 9 degrees of freedom. Thus, $p < 0.01$.

Exercises

Exercise 10.49 is a practice exercise; its complete solution is given on page 363.

*10.49 An agronomist planted three test plots each with four varieties of wheat and obtained the following yields (in pounds per plot):

Variety A: 60, 61, 56
Variety B: 59, 52, 51
Variety C: 55, 55, 52
Variety D: 58, 58, 55

(a) Calculate $n \cdot s_{\bar{x}}^2$ for these data, the mean of the variances of the four samples, and the value of F.

(b) Use the 0.05 level of significance to test whether the differences among the four sample means can be attributed to chance.

*10.50 The following are the numbers of mistakes made on five occasions by three compositors setting the type for a technical report:

$$\begin{array}{lllllll}
\textit{Compositor 1:} & 10, & 13, & 9, & 11, & 12 \\
\textit{Compositor 2:} & 11, & 13, & 8, & 16, & 12 \\
\textit{Compositor 3:} & 10, & 15, & 13, & 11, & 16
\end{array}$$

(a) Calculate $n \cdot s_{\bar{x}}^2$ for these data, the mean of the variances of the three samples, and the value of F.

(b) Use the 0.05 level of significance to test whether the differences among the three sample means can be attributed to chance.

*10.51 The following are the mileages that a test driver got with four one-gallon trials for each of five brands of gasoline:

$$\begin{array}{llllll}
\textit{Brand A:} & 30, & 25, & 27, & 26 \\
\textit{Brand B:} & 29, & 26, & 29, & 28 \\
\textit{Brand C:} & 32, & 32, & 35, & 37 \\
\textit{Brand D:} & 29, & 34, & 32, & 33 \\
\textit{Brand E:} & 32, & 26, & 31, & 27
\end{array}$$

(a) Calculate $n \cdot s_{\bar{x}}^2$ for these data, the mean of the variances of the five samples, and the value of F.

(b) Use the 0.01 level of significance to test whether the differences among the five sample means can be attributed to chance.

10.9
Analysis of Variance*

Let us now look at the example of the preceding section from a different point of view, that of an **analysis of variance**. The basic idea of an analysis of variance is to express a measure of the total variation of a set of data as a sum of terms, which can be attributed to specific sources, or causes, of variation. With reference to our example, two such sources of variation would be (1) actual differences in the effectiveness of the three teaching techniques and (2) chance. In problems like this, chance variation is generally referred to as the **experimental error**, which is sometimes described also as statistical *noise*.

One's ability to perform an analysis of variance can be of great importance in scientific work, particularly when many factors produce certain results and we are interested in their individual contributions. For instance, we may want to see whether observed differences in cleansing action are due to the use of different detergents, differences in water temperature, differences in the hardness of the water, differences among the washing machines used in the experiment, and perhaps even differences among the instruments used to obtain the necessary readings. The method by which we analyze a complex situation like this is beyond the scope of this textbook, but it is a direct generalization of the work of this section.

As a measure of the total variation of kn observations consisting of k samples of size n, we shall use the **total sum of squares (SST).**[†]

$$SST = \sum_{i=1}^{k} \sum_{j=1}^{n} (x_{ij} - \bar{x}_{..})^2$$

where x_{ij} is the jth observation of the ith sample, $i = 1, 2, \ldots, k$, and $j = 1, 2, \ldots, n$. The mean of all the kn measurements or observations is $\bar{x}_{..}$, which is called the **grand mean.** Note that if we divide the total sum of squares by $kn - 1$, we get the variance of the combined data.

Letting $\bar{x}_{i.}$ denote the mean of the ith sample, $i = 1, 2, \ldots, k$, we can write the following identity, which forms the basis of a **one-way analysis of variance:**

$$SST = n \cdot \sum_{i=1}^{k} (\bar{x}_{i.} - \bar{x}_{..})^2 + \sum_{i=1}^{k} \sum_{j=1}^{n} (x_{ij} - \bar{x}_{i.})^2$$

It is customary to refer to the first term on the right-hand side of the equation, which measures the variation among the sample means, as the **treatment sum of squares, SS(Tr)**, and to the second term, which measures the variation within the samples, as the **error sum of squares, SSE.** Use of the word *treatment* is explained by the origin of many analysis-of-variance techniques in agricultural experiments in which different fertilizers, for example, were regarded as different **treatments** applied to the soil. So, we shall refer to the three teaching methods on page 345 as three different treatments, and in other problems we may refer to five nationalities as five different treatments, four different levels of education as four treatments, and so on. The word *error* in *error sum of squares* pertains to the experimental error, or chance.

The identity reads $SST = SS(Tr) + SSE$, and it can be described as a *partition* of SST. The proof of the identity requires a good deal of algebraic manipulation; however, we can easily provide a numerical illustration.

■ **EXAMPLE**   Use the scores of the students taught computer programming by three different methods to verify the identity $SST = SS(Tr) + SSE$.

Solution   Substituting the scores, the three sample means 70, 85, and 76, and the grand mean 77 into the expressions for the three sums of squares, we get

$$\begin{aligned}
SST = \quad & (71 - 77)^2 + (75 - 77)^2 + (65 - 77)^2 + (69 - 77)^2 \\
& + (90 - 77)^2 + (80 - 77)^2 + (86 - 77)^2 + (84 - 77)^2 \\
& + (72 - 77)^2 + (77 - 77)^2 + (76 - 77)^2 + (79 - 77)^2 \\
= \quad & 586
\end{aligned}$$

$$SS(Tr) = 4\left[(70 - 77)^2 + (85 - 77)^2 + (76 - 77)^2\right] = 456$$

[†] The use of double subscripts and double summations is explained briefly in Section 3.10.

and

$$SSE = \quad (71 - 70)^2 + (75 - 70)^2 + (65 - 70)^2 + (69 - 70)^2$$
$$+ (90 - 85)^2 + (80 - 85)^2 + (86 - 85)^2 + (84 - 85)^2$$
$$+ (72 - 76)^2 + (77 - 76)^2 + (76 - 76)^2 + (79 - 76)^2$$
$$= 130$$

Since $586 = 456 + 130$, we have thus shown that $SST = SS(Tr) + SSE$ for the given data. These calculations are difficult in large problems, and a simpler method will be described on page 353. ■

Examining the two terms into which the total sum of squares SST has been partitioned, we note that if we divide SS(Tr) by $k - 1$, we obtain the quantity that we denoted by $n \cdot s_{\bar{x}}^2$ on page 346. Clearly,

$$\frac{SS(Tr)}{k - 1} = \frac{n \cdot \sum_{i=1}^{k}(\bar{x}_{i.} - \bar{x}_{..})^2}{k - 1} = n \cdot \left(\frac{\sum_{i=1}^{k}(\bar{x}_{i.} - \bar{x}_{..})^2}{k - 1} \right) = n \cdot s_{\bar{x}}^2$$

This quantity, which measures the variation among the samples, is called the **treatment mean square**, and it is denoted by **MS(Tr)**. That is,

$$MS(Tr) = \frac{SS(Tr)}{k - 1}$$

Similarly, if we divide SSE by $k(n - 1)$, we obtain the mean of the k sample variances, for we can write

$$\frac{SSE}{k(n - 1)} = \frac{\sum_{i=1}^{k}\sum_{j=1}^{n}(x_{ij} - \bar{x}_{i.})^2}{k(n - 1)} = \frac{1}{k} \cdot \sum_{i=1}^{k}\left(\frac{\sum_{j=1}^{n}(x_{ij} - \bar{x}_{i.})^2}{n - 1} \right)$$

which equals

$$\frac{1}{k} \cdot \left(s_1^2 + s_2^2 + \cdots + s_k^2 \right)$$

the pooled variance from page 342. This quantity, which measures the variation within the samples, is called the **error mean square**, and it is denoted by **MSE**. Thus,

$$MSE = \frac{SSE}{k(n - 1)}$$

Since F was defined previously as the ratio of these two measures of the variation among and within the samples, we can now write

Statistic for a test concerning the differences among means

$$F = \frac{MS(Tr)}{MSE}$$

In practice, we display the work required for the determination of F in the following kind of table, called an **analysis-of-variance table**:

SOURCE OF VARIATION	DEGREES OF FREEDOM	SUM OF SQUARES	MEAN SQUARE	F
Treatments	$k - 1$	SS(Tr)	$MS(Tr) = \dfrac{SS(Tr)}{k - 1}$	$\dfrac{MS(Tr)}{MSE}$
Error	$k(n - 1)$	SSE	$MSE = \dfrac{SSE}{k(n - 1)}$	
Total	$kn - 1$	SST		

The degrees of freedom for treatments and error are the numerator and denominator degrees of freedom referred to on page 347. Note that they are also the quantities we divide into the sums of squares to obtain the corresponding mean squares. Observe that there is only one entry in the F column. By custom, we do not write a number in the Total row under Mean square.

If we make the same assumptions as in Section 10.8, the significance test is as we described it on page 347: we reject the null hypothesis $\mu_1 = \mu_2 = \cdots = \mu_k$ against the alternative that these μ's are not all equal, if the value we get for F equals or exceeds F_α for $k - 1$ and $k(n - 1)$ degrees of freedom.

■ **EXAMPLE**   Use the sums of squares from pages 350 and 351 to construct an analysis-of-variance table for our numerical example, and test the null hypothesis that the three methods of teaching computer programming are equally effective against the alternative that they are not all equally effective at the 0.01 level of significance.

Solution   Copying the values of the sums of squares from pages 350 and 351, we get

$$MS(Tr) = \frac{456}{2} = 228, \quad MSE = \frac{130}{9} \approx 14.44, \quad F = \frac{228}{14.44} \approx 15.8$$

and hence

SOURCE OF VARIATION	DEGREES OF FREEDOM	SUM OF SQUARES	MEAN SQUARE	F
Treatments	2	456	228	15.8
Error	9	130	14.44	
Total	11	586		

Since $F = 15.8$ exceeds 8.02, the value of $F_{0.01}$ for $3 - 1 = 2$ and $3(4 - 1) = 9$ degrees of freedom obtained from Table V, the null hypothesis must be rejected; in other words, we conclude that the three methods of teaching computer programming are not all equally effective. ■

The numbers that we used in our illustration were intentionally chosen so that the calculations would be easy. In actual practice, the calculation of the sums of squares can be quite tedious unless we use the following computing formulas, in which $T_{i.}$ denotes the sum of the values in the ith sample, and $T_{..}$ denotes the grand total of all the data:

Computing formulas for sums of squares (sample sizes equal)

$$SST = \sum_{i=1}^{k} \sum_{j=1}^{n} x_{ij}^2 - \frac{1}{kn} \cdot T_{..}^2$$

$$SS(Tr) = \frac{1}{n} \cdot \sum_{i=1}^{k} T_{i.}^2 - \frac{1}{kn} \cdot T_{..}^2$$

and by subtraction $\qquad SSE = SST - SS(Tr)$

■ **EXAMPLE** Use these computing formulas to verify the sums of squares obtained on pages 350 and 351.

Solution Substituting $k = 3$, $n = 4$, $T_{1.} = 280$, $T_{2.} = 340$, $T_{3.} = 304$, $T_{..} = 924$, and $\sum \sum x_{ij}^2 = 71{,}734$ into the computing formulas for the three sums of squares, we get

$$SST = 71{,}734 - \frac{1}{12}(924)^2 = 71{,}734 - 71{,}148 = 586$$

$$SS(Tr) = \frac{1}{4}(280^2 + 340^2 + 304^2) - \frac{1}{12}(924)^2$$

$$= 71{,}604 - 71{,}148 = 456$$

and
$$SSE = 586 - 456 = 130$$

These values are identical to those obtained on pages 350 and 351. ■

A computer printout generated by the SAS program for our analysis-of-variance example is shown in Figure 10.11, where the values corresponding to $SST = 586$, $SS(Tr) = 456$, and $SSE = 130$ are given in the column headed SUM OF SQUARES. Besides the degrees of freedom, the sums of squares, the mean squares, the value of F, and the p-value 0.0011, it provides other information that we shall not consider here.

Analysis-of-variance problems like the one we have discussed can also be handled quite readily by means of graphing calculators. Upon entering the numerical data that we used to obtain the analysis-of-variance table on

```
data a;
input meth $ score @@;
cards;
1 71 1 75 1 65 1 69
2 90 2 80 2 86 2 84
3 72 3 77 3 76 3 79
;
proc anova;
class meth;
model score=meth;
```

```
                        SAS
          ANALYSIS OF VARIANCE PROCEDURE

         CLASS LEVEL INFORMATION
         CLASS      LEVELS      VALUES
         METH          3         1 2 3


         NUMBER OF OBSERVATIONS IN DATA SET = 12

                             SAS
                ANALYSIS OF VARIANCE PROCEDURE
         DEPENDENT VARIABLE: SCORE
```

SOURCE	DF	SUM OF SQUARES	MEAN SQUARE	F VALUE	PR > F
MODEL	2	456.00000000	228.00000000	15.78	0.0011
ERROR	9	130.00000000	14.44444444		
CORRECTED TOTAL	11	586.00000000			

R-SQUARE	C.V.
0.778157	4.9358

ROOT MSE	SCORE MEAN
3.80058475	77.00000000

FIGURE 10.11 MINITAB printout for analysis of variance.

page 354, we get the results shown in Figure 10.12. As we can observe, the graphing calculator confirms the computer printout and also our manual calculations. Graphing calculator information shown here and elsewhere in this book is entirely optional and can be omitted by the reader without the loss of continuity.

Directions for obtaining the results shown in Figure 10.12 with a TI-83 graphing calculator are given in the Using Technology box on page 356.

The method we have discussed here applies only when the sample sizes are all equal, but minor modifications make it applicable also when the sam-

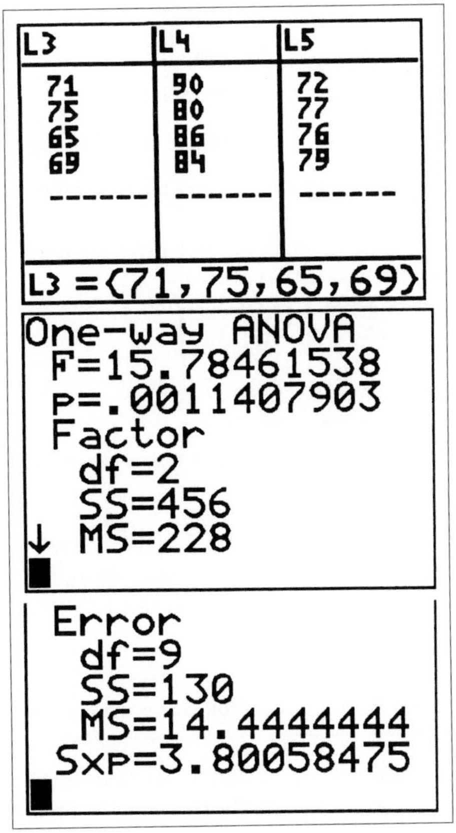

FIGURE 10.12 Analysis of variance reproduced from the display screen of a TI-83 graphing calculator.

ple sizes are not all equal. If the ith sample is of size n_i, the computing formulas for the sums of squares become

$$ \text{SST} = \sum_{i=1}^{k} \sum_{j=1}^{n_i} x_{ij}^2 - \frac{1}{N} \cdot T_{..}^2 $$

$$ \text{SS(Tr)} = \sum_{i=1}^{k} \frac{T_{i.}^2}{n_i} - \frac{1}{N} \cdot T_{..}^2 $$

$$ \text{SSE} = \text{SST} - \text{SS(Tr)} $$

where $N = n_1 + n_2 + \cdots + n_k$. The only other change is that the total number of degrees of freedom is $N - 1$, and the degrees of freedom for treatments and error are, respectively, $k - 1$ and $N - k$.

■ **EXAMPLE**   A laboratory technician wants to compare the breaking strength of three kinds of thread, and originally he planned to repeat each determination six times. Not having enough time, however, he has to base his analysis on the following results (in ounces):

Thread 1:   18.0,   16.4,   15.7,   19.6,   16.5,   18.2
Thread 2:   21.1,   17.8,   18.6,   20.8,   17.9,   19.0
Thread 3:   16.5,   17.8,   16.1

USING TECHNOLOGY : *The Graphing Calculator*

How to obtain the analysis of variance shown in Figure 10.12

(Tip before beginning: Be sure to clear all lists and plots—see Appendix A: TI-83 Tips.)

Step 1. Input the student computer programming scores from the running example into list L1, L2, and L3.

> L1 = (71, 75, 65, 69)
> L2 = (90,80,86,84)
> L3 = (72,77,76,79)

Step 2. Select the type of statistical test you wish to use.
Press STAT.
Highlight TESTS.
Scroll down to F: ANOVA(, press ENTER.
Enter the lists in which the data is stored.
Press 2nd, 1 (for L1), 2nd, 2 (for L2), 2nd, 3 (for L3).

Step 3. Solve.
Press ENTER.
Use the ▼ button to scroll down through the answer.

Perform an analysis of variance to test at the 0.05 level of significance whether the differences among the sample means are significant.

Solution   1. *Hypotheses* H_0: $\mu_1 = \mu_2 = \mu_3$
H_A: the μ's are not all equal

where μ_1, μ_2, and μ_3 are the means of the three populations.

2. *Level of significance* $\alpha = 0.05$

3. *Criterion* Reject the null hypothesis if $F \geq 3.89$, the value of $F_{0.05}$ for $k - 1 = 3 - 1 = 2$ and $N - k = 15 - 3 = 12$ degrees of freedom, where F is to be determined by an analysis of variance. Otherwise, accept H_0 or reserve judgment.

4. *Calculations*

$$T_{1.} = 104.4, \quad T_{2.} = 115.2, \quad T_{3.} = 50.4, \quad T_{..} = 270.0,$$

and

$$\Sigma \Sigma x_{ij}^2 = 4{,}897.46$$

Then, substituting these totals together with $n_1 = 6$, $n_2 = 6$, $n_3 = 3$, and $N = 15$ into the computing formulas for the sums of squares, we get

$$SST = 4{,}897.46 - \frac{1}{15}(270.0)^2 = 4{,}897.46 - 4{,}860 = 37.46$$

$$SS(Tr) = \frac{104.4^2}{6} + \frac{115.2^2}{6} + \frac{50.4^2}{3} - \frac{1}{15}(270.0)^2$$

$$= 4{,}875.12 - 4{,}860 \approx 15.12$$

and

$$\text{SSE} = 37.46 - 15.12 = 22.34$$

Then

$$\text{MS(Tr)} = \frac{15.12}{2} = 7.56, \qquad \text{MSE} = \frac{22.34}{12} \approx 1.86,$$

and

$$F = \frac{7.56}{1.86} \approx 4.06$$

All these results are shown in the following analysis-of-variance table:

SOURCE OF VARIATION	DEGREES OF FREEDOM	SUM OF SQUARES	MEAN SQUARE	F
Treatments	2	15.12	7.56	4.06
Error	12	22.34	1.86	
Total	14	37.46		

5. *Decision* Since $F = 4.06$ exceeds 3.89, the null hypothesis must be rejected; in other words, we conclude that there is a difference in the strength of the three kinds of thread.

This example could also utilize the p-value. The final three steps would be these:

3′. *Test statistic* The statistic is

$$F = \frac{\text{MS(Tr)}}{\text{MSE}}$$

4′. *Calculations* Find

$$F = \frac{7.56}{1.86} \approx 4.06$$

5′. *Decision* Since $F = 4.06$ falls between 3.89 and 6.93, the values of $F_{0.05}$ and $F_{0.01}$ for 2 and 12 degrees of freedom, we find that $0.01 < p < 0.05$. Since the level of significance was specified as $\alpha = 0.05$, we conclude, as before, that the null hypothesis must be rejected. Had the level of significance not been specified, we might simply have stated that $0.01 < p < 0.05$ or written the result as $F = 4.06*$. ∎

To get the actual size of the p-value for the preceding example, we would have to use a computer package or a special calculator. Using a TI-83 graphing calculator, we can get the p-value and, at the same time, graph the corresponding area under the curve. Entering the requisite information into the calculator, we obtain the result reproduced in Figure 10.13. As can be seen, the p-value is .045011, or .045 rounded to three decimals. This agrees with the previous comment that $0.01 < p < 0.05$. Graphical calculator information shown here and elsewhere in this book is optional and can be omitted by the reader without the loss of continuity.

Directions for obtaining the graph shown in Figure 10.13 with a TI-83 graphing calculator are given in the Using Technology box on page 358.

USING TECHNOLOGY : *The Graphing Calculator*

How to obtain the actual size of the *p*-value and graph the area under the curve shown in Figure 10.13

(Tip before beginning: Reset the memory to default settings and clear all plots—see Appendix A: TI-83 Tips.)

Step 1.   Set the viewing window.

> Press WINDOW
> Set   Xmin = 0
>   Xmax = 6
>   Xscl = 1
>   Ymin = -.25
>   Ymax = .9
>   Yscl = 1
>   Xres = 1

Step 2.   Select the type of display and enter the interval to be displayed.
Press 2nd, VARS to access the DISTR DRAW menu.
Highlight DRAW.
Select 4: ShadeF(, press ENTER.
Enter   4.06, 1E99, 2, 12)
(For instructions on entering "1E99," for infinity, see Appendix A: TI-83 Tips; 4.6 and 1E99 (for ∞) are the lower and upper limits of the interval; 2 and 12 are the degrees of freedom.)

Step 3.   View the graph.
Press ENTER to solve.

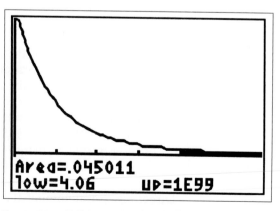

FIGURE 10.13 F distribution reproduced from the display screen of a TI-83 graphing calculator.

Exercises

Exercises 10.52 and 10.54 are practice exercises; their complete solutions are given on page 363 and 364.

 *10.52 To study the effectiveness of five different kinds of packaging, a processor of breakfast foods obtained the following data on the numbers of sales on five different days:

Packaging I:	60,	52,	56,	52,	65
Packaging II:	54,	64,	66,	54,	57
Packaging III:	55,	66,	68,	57,	55
Packaging IV:	55,	56,	70,	58,	56
Packaging V:	71,	65,	60,	59,	62

Perform an analysis of variance to test at the 0.05 level of significance whether the differences among the five sample means can be attributed to chance.

*10.53 The following are eight consecutive weeks' earnings (in dollars) of three door-to-door cosmetics salespersons employed by a firm:

Salesperson A:	309,	293,	284,	300,	306,	288,	312,	276
Salesperson B:	295,	280,	299,	310,	298,	284,	293,	287
Salesperson C:	311,	289,	296,	323,	287,	280,	303,	264

Perform an analysis of variance to test at the 0.01 level of significance whether the differences among the average weekly earnings of the three salespersons are significant. Provide also an inequality that describes the p-value.

 *10.54 The following are the weight losses, in milligrams, of certain machine parts due to friction, when they were used with three different lubricants:

Lubricant X:	12,	11,	7,	13,	9,	11,	12,	9			
Lubricant Y:	8,	10,	7,	5,	6,	10,	7,	8,	11,	7,	8
Lubricant Z:	9,	3,	7,	8,	4,	6,	6,	5			

Perform an analysis of variance to test at the 0.01 level of significance whether the differences among the three sample means can be attributed to chance.

*10.55 The following are the numbers of words per minute that a secretary typed on several occasions on four different computer keyboards:

Keyboard C:	71,	75,	69,	77,	61,	72,	71,	78
Keyboard D:	68,	71,	74,	66,	69,	67,	70,	62
Keyboard E:	75,	70,	81,	73,	78,	72		
Keyboard F:	62,	59,	71,	68,	63,	65,	72,	60, 64

Perform an analysis of variance to test at the 0.01 level of significance whether the differences among the four sample means can be attributed to chance. Give also an inequality that describes the p-value.

*10.56 To study the performance of a newly designed motorboat, it was timed over a marked course under various wind and water conditions, and the following data (in minutes) was obtained:

Calm conditions:	25,	18,	15,	21		
Moderate conditions:	24,	27,	24,	19,	17,	22
Choppy conditions:	22,	24,	27,	25,	30	

Perform an analysis of variance to test at the 0.05 level of significance whether the differences among the three sample means are significant.

*10.57 Use a computer package or a graphing calculator to rework Exercise 10.52.

*10.58 Use a computer package or a graphing calculator to rework Exercise 10.53.

*10.59 Use a computer package or a graphing calculator to rework Exercise 10.54.

*10.60 Use a computer package or a graphing calculator to rework Exercise 10.55.

✓ Solutions to Practice Exercises

10.1 We take as the null hypothesis the statement that the bottle of the antibiotic possesses full potency, since this is the usual state of affairs.
 (a) Erroneously rejecting a true null hypothesis is a Type I error.
 (b) Erroneously accepting a false null hypothesis is a Type II error.

10.5 (a) Since $\sigma_{\bar{x}} = \dfrac{0.8}{\sqrt{60}} \approx 0.103$, the dividing lines of the criterion in standard units are

$$z = \frac{13.2 - 13.4}{0.103} \approx -1.94 \quad \text{and} \quad z = \frac{13.6 - 13.4}{0.103} \approx 1.94$$

and the probability of a Type I error is

$$2(0.5000 - 0.4738) = 0.0524 \approx 0.05$$

 (b) The dividing lines of the criterion in standard units are

$$z = \frac{13.2 - 13.7}{0.103} \approx -4.85 \quad \text{and} \quad z = \frac{13.6 - 13.7}{0.103} \approx -0.97$$

and the probability of a Type II error is

$$0.5000 - 0.3340 = 0.1660 \approx 0.17$$

10.9 (a) The null hypothesis $\mu = \mu_0$, where μ is the actual number of mistakes averaged by the salesperson and μ_0 is the number of mistakes averaged by all the salespersons; the alternative hypothesis is $\mu > \mu_0$; the salesperson will be fired if the null hypothesis can be rejected.

(b) The null hypothesis $\mu = \mu_0$, and the alternative hypothesis $\mu < \mu_0$; the salesperson will be fired unless the null hypothesis can be rejected.

10.11 Here μ denotes the true average time that convicted embezzlers spend in jail.

(a) **1.** *Hypotheses* H_0: $\mu = 12.8$
H_A: $\mu \neq 12.8$

The alternative is two-sided because there is no indication of a one-sided objective.

2. *Level of significance* $\alpha = 0.01$

3. *Criterion*   Reject the null hypothesis if $z \leq -2.575$ or $z \geq 2.575$, where

$$z = \frac{\bar{x} - \mu_0}{\sigma/\sqrt{n}}$$

with s substituted for σ. Otherwise, accept the null hypothesis or reserve judgment.

4. *Calculations*

$$z = \frac{11.2 - 12.8}{3.5/\sqrt{60}} \approx -3.54$$

5. *Decision*   Since $z = -3.54$ is less than -2.575, the null hypothesis must be rejected; in other words, convicted embezzlers do not spend on the average 12.8 months in jail.

(b) Note that the value $z = -3.54$ is off the main part of Table II. We can use the figures at the bottom of the table to determine that the p-value is between $2(0.5000 - 0.4990) = 0.002$ and $2(0.5000 - 0.49997) = 0.00006$.

(c) The p-value is consistent with rejecting the null hypothesis with $\alpha = 0.01$.

10.21 The sample size $n = 12$ is small, and we must begin by stating that the work rests on the assumption that the pulse reduction values are sampled from a population that has roughly the shape of a normal distribution.

1. *Hypotheses* H_0: $\mu = 9.0$
H_A: $\mu < 9.0$

2. *Level of significance* $\alpha = 0.05$

3. *Criterion*   Reject the null hypothesis if $t \leq -1.796$, which is the value of $-t_{0.05}$ for $12 - 1 = 11$ degrees of freedom, where

$$t = \frac{\bar{x} - \mu_0}{s/\sqrt{n}}$$

4. *Calculations*

$$t = \frac{7.2 - 9.0}{1.8/\sqrt{12}} \approx -3.46$$

5. *Decision*   Since $t = -3.46$ is less than -1.796, the null hypothesis must be rejected; in other words, the tranquilizer reduces the pulse rate on the average by less than 9.0 beats.

10.35   **1.** *Hypotheses* H_0: $\mu_1 = \mu_2$
H_A: $\mu_1 \neq \mu_2$

2. *Level of significance* $\alpha = 0.01$

3. *Criterion*   Reject the null hypothesis if $z \leq -2.575$ or $z \geq 2.575$, where

$$z = \frac{\bar{x}_1 - \bar{x}_2}{\sqrt{\dfrac{\sigma_1^2}{n_1} + \dfrac{\sigma_2^2}{n_2}}}$$

with s_1 and s_2 substituted for σ_1 and σ_2. Otherwise, state that the difference between the two sample means is not significant.

4. *Calculations*

$$z = \frac{12.8 - 9.2}{\sqrt{\dfrac{5.4^2}{60} + \dfrac{6.3^2}{50}}} \approx 3.18$$

5. *Decision*   Since $z = 3.18$ exceeds 2.575, the null hypothesis must be rejected; in other words, the two kinds of executives do not average equally many business lunches.

10.41   1. *Hypotheses* H_0: $\mu_1 = \mu_2$
                                   H_A: $\mu_1 \neq \mu_2$

2. *Level of significance* $\alpha = 0.05$

3. *Criterion*   Reject the null hypothesis if $t \leq -2.306$ or $t \geq 2.306$, where 2.306 is the value of $t_{0.025}$ for $5 + 5 - 2 = 8$ degrees of freedom and t is given by the formula on page 338.

4. *Calculations*

$$s_p = \sqrt{\frac{4(252)^2 + 4(207)^2}{8}} \approx 230.6 \quad \text{and} \quad t = \frac{8{,}106 - 7{,}730}{230.6\sqrt{\dfrac{1}{5} + \dfrac{1}{5}}} \approx 2.95$$

5. *Decision*   Since $t = 2.95$ exceeds 2.306, the null hypothesis must be rejected; in other words, the average heat-producing capacity of coal from the two mines is not the same.

The p-value can also be used. The value $t = 2.95$ lies between $t_{0.010}$ and $t_{0.005}$ for 8 degrees of freedom. The alternative hypothesis is two-sided, so we note that $0.01 < p < 0.02$. This is consistent with rejecting the null hypothesis with $\alpha = 0.05$.

10.44 The differences are $13, 7, -1, 5, 3, 2, -1, 0, 8, 10, 16, -2, 7, 3, -5, 2, 6, 1, 4, 3, 2,$ $6, 12, 4, -1, 5, -6, -3, 8, 4, -1,$ and 0; their mean is $\bar{x} \approx 3.47$, their standard deviation is $s \approx 5.07$, and for these data we perform the following test:

1. *Hypotheses* H_0: $\mu = 0$
                     H_A: $\mu > 0$

where μ is the mean of the population of differences.

2. *Level of significance* $\alpha = 0.01$

3. *Criterion*   Reject the null hypothesis if $z \geq 2.33$, where

$$z = \frac{\bar{x} - \mu_0}{\sigma/\sqrt{n}}$$

using $\mu_0 = 0$ as the comparison value and with s substituted for σ; otherwise, we accept the null hypothesis or reserve judgment.

4. *Calculations*

$$z \approx \frac{3.47 - 0}{5.07/\sqrt{32}} \approx 3.87$$

5. *Decision*   Since $z \approx 3.87$ exceeds 2.33, the null hypothesis must be rejected; in other words, the prescribed program of exercise is effective.

10.49   (a) The four sample means are $\bar{x}_1 = 59$, $\bar{x}_2 = 54$, $\bar{x}_3 = 54$, and $\bar{x}_4 = 57$; their mean is

$$\frac{59 + 54 + 54 + 57}{4} = 56$$

so that

$$s_{\bar{x}}^2 = \frac{(59 - 56)^2 + (54 - 56)^2 + (54 - 56)^2 + (57 - 56)^2}{4 - 1} = 6$$

and $n \cdot s_{\bar{x}}^2 = 3 \cdot 6 = 18$; also,

$$s_1^2 = \frac{(60 - 59)^2 + (61 - 59)^2 + (56 - 59)^2}{3 - 1} = 7$$

and similarly, $s_2^2 = 19$, $s_3^2 = 3$, and $s_4^2 = 3$; so that

$$\frac{1}{4}\left(s_1^2 + s_2^2 + s_3^2 + s_4^2\right) = \frac{7 + 19 + 3 + 3}{4} = 8$$

and $F = \frac{18}{8} = 2.25$.

(b) Since $F = 2.25$ does not exceed 4.07, the value of $F_{0.05}$ for $k - 1 = 4 - 1 = 3$ and $k(n - 1) = 4(3 - 1) = 8$ degrees of freedom, the null hypothesis cannot be rejected; in other words, the differences among the four sample means are not significant.

10.52   **1.** *Hypotheses* H_0: $\mu_1 = \mu_2 = \mu_3 = \mu_4 = \mu_5$
                          H_A: the μ's are not all equal

2. *Level of significance* $\alpha = 0.05$

3. *Criterion*   Reject the null hypothesis if $F \geq 2.87$, the value of $F_{0.05}$ for $5 - 1 = 4$ and $5(5 - 1) = 20$ degrees of freedom; otherwise accept H_0 or reserve judgment.

4. *Calculations*   Since $T_1 = 285$, $T_2 = 295$, $T_3 = 301$, $T_4 = 295$, $T_5 = 317$, $T = 1{,}493$, and $\sum\sum x^2 = 89{,}933$, we get

$$\text{SST} = 89{,}933 - \frac{(1{,}493)^2}{25} = 89{,}933 - 89{,}161.96 = 771.04$$

$$\text{SS(Tr)} = \frac{285^2 + 295^2 + 301^2 + 295^2 + 317^2}{5} - 89{,}161.96$$

$$= 89{,}273 - 89{,}161.96 = 111.04$$

$$\text{SSE} = 771.04 - 111.04 = 660$$

The remainder of the arithmetic can be displayed easily in an analysis-of-variance table:

SOURCE OF VARIATION	DEGREES OF FREEDOM	SUM OF SQUARES	MEAN SQUARE	F
Treatments	4	111.04	27.76	0.84
Error	20	660.00	33	
Total	24	771.04		

5. *Decision*   Since $F = 0.84$ does not exceed 2.87, the null hypothesis cannot be rejected.

*10.54   1. *Hypotheses* H_0: $\mu_1 = \mu_2 = \mu_3$
H_A: the μ's are not all equal

2. *Level of significance* $\alpha = 0.01$

3. *Criterion*   Reject the null hypothesis if $F \geq 5.61$, the value of $F_{0.01}$ for $k - 1 = 3 - 1 = 2$ and $N - k = 27 - 3 = 24$ degrees of freedom, where F is to be determined by an analysis of variance.

4. *Calculations* $T_{1.} = 84$, $T_{2.} = 87$, $T_{3.} = 48$, $T_{..} = 219$,   and $\sum\sum x_{ij}^2 = 1{,}947$. Substituting these values together with $n_1 = 8$, $n_2 = 11$, $n_3 = 8$, and $N = 27$ into the computing formulas for the sums of squares, we get

$$\text{SST} = 1{,}947 - \frac{1}{27}(219)^2 \approx 1{,}947 - 1{,}776.33 = 170.67$$

$$\text{SS(Tr)} = \frac{84^2}{8} + \frac{87^2}{11} + \frac{48^2}{8} - \frac{1}{27}(219)^2$$

$$\approx 1{,}858.09 - 1{,}776.33 = 81.76$$

$$\text{SSE} \approx 170.67 - 81.76 = 88.91$$

The remainder of the work is shown in the following analysis-of-variance table:

SOURCE OF VARIATION	DEGREES OF FREEDOM	SUM OF SQUARES	MEAN SQUARE	F
Treatments	2	81.76	40.88	11.05
Error	24	88.91	3.70	
Total	26	170.67		

5. *Decision*   Since $F = 11.05$ exceeds 5.61, the null hypothesis must be rejected; in other words, we conclude that there is a difference in the effectiveness of the lubricants in reducing the weight loss of the machine parts. The p-value approach to this exercise requires that we compare the test statistic $F = 11.05$ to the table points $F = 3.40$ and $F = 5.61$. Since $F = 0.05$, we note that $p = 0.01$. We therefore reject the null hypothesis.

CHAPTER 11

Tests Based on Count Data

For the most part, Chapters 9 and 10 dealt with inferences based on measurements. We used measurements to estimate the means of populations and their standard deviations, and we also used them to test hypotheses about population means. The only exception was in Section 9.5, where we used sample proportions for the estimation of population proportions, percentages, or probabilities. Such data is referred to as **count data** or **enumeration data**, since it is obtained by performing counts rather than measurements.

In this chapter we shall use count data in tests of hypotheses. For instance, we may want to test the claim that 24% of all recreational skydivers counted at a certain drop zone (such as the parachutists of the cartoon) are women.

In Sections 11.1 and 11.2, we shall study tests about population proportions, which also serve to test hypotheses about percentages and probabilities. As we pointed out on page 26, a percentage is just a proportion multiplied by 100, and a probability may be interpreted as a proportion in the long run. In Sections 11.3 and 11.4, we shall study tests concerning two or more population proportions, and in Section 11.5 we shall learn how to analyze data tallied into a **two-way classification**. This kind of problem might arise, for example, when cars are rated on their fuel economy as being low, average, or high, and also on their ride being very uncomfortable, somewhat uncomfortable, fairly comfortable, or very comfortable. Finally, Section 11.6 deals with the comparison of observed and expected distributions, that is, the comparison of counts we actually got with those which we might have expected according to theory or assumptions.

11.1 Tests Concerning Proportions

In this section we shall be concerned with tests of hypotheses that enable us to decide, on the basis of sample data, whether the true value of a proportion, percentage, or probability equals a given constant. These tests make it possible, for example, to determine whether the true proportion of tenth graders who can name the two senators of their state is 0.30, whether it is true that 10% of the answers that the IRS gives to taxpayers' telephone inquiries are in error, or whether the true probability is 0.25 that a flight from Seattle to San Francisco will be late.

Questions of this kind are usually decided on the basis of the observed number of successes in n trials, or the observed proportion of successes, and it will be assumed throughout this section that these trials are independent and that the probability of a success is the same for each trial. In other words, we shall assume that we can use the binomial distribution and that we are, in fact, testing hypotheses about the parameter p of binomial populations.

When n is small, tests concerning true proportions can be based directly on tables of binomial probabilities, as is illustrated by the following examples.

▪ **EXAMPLE**   It has been claimed that at least 40% of all seniors at a large university prefer to live off campus. If 4 of 14 seniors interviewed at random prefer to live off campus, test the claim at the 0.05 level of significance. Use an appropriate one-sided criterion.

Solution   1. *Hypotheses* H_0: $p = 0.40$

H_A: $p < 0.40$

2. *Level of significance* $\alpha = 0.05$

3′. *Test statistic*   The test statistic is simply the number of successes, here four.

4′. *Calculations*   Table I shows that for $n = 14$ and $p = 0.40$ the probability of four or fewer successes is

$$0.001 + 0.007 + 0.032 + 0.085 + 0.155 = 0.280$$

and this is the sum of the areas of the blue rectangles of Figure 11.1. The p-value is thus 0.280.

5′. *Decision*   Since 0.280 is greater than $\alpha = 0.05$, the null hypothesis cannot be rejected. We accept the claim or reserve judgment.   ▪

Here we used the alternative procedure based on p-values, introduced first on page 327. We did this because, for tests based on binomial probabilities, this can greatly simplify the work, especially when a test is two-tailed, as in the example that follows.

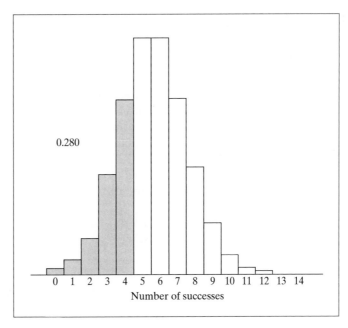

0.280

0 1 2 3 4 5 6 7 8 9 10 11 12 13 14
Number of successes

FIGURE 11.1 Binomial distribution with $p = 0.40$ and $n = 14$.

■ **EXAMPLE**   It has been claimed that 60% of all shoppers can identify a highly advertised trademark. At the 0.05 level of significance, can we reject this claim if, in a random sample of 12 shoppers, only 3 were able to identify the trademark?

Solution   1. *Hypotheses* H_0: $p = 0.60$
                                         H_A: $p \neq 0.60$

2. *Level of significance* $\alpha = 0.05$

3'. *Test statistic*   The test statistic is simply the number of successes, here three.

4'. *Calculations*   Table I shows that for $n = 12$ and $p = 0.60$ the probability of three or fewer successes is

$$0.002 + 0.012 = 0.014$$

so that the *p*-value is $2(0.014) = 0.028$.

5'. *Decision*   Since 0.028 is less than 0.05, the null hypothesis must be rejected. We reject the claim that the true percentage of shoppers who can identify the trademark is 60%.   ■

11.2
Tests Concerning Proportions (Large Samples)

For large *n*, tests concerning proportions are usually based on the normal-curve approximation to the binomial distribution. We base tests of the null hypothesis $p = p_0$ on the value we obtain for

Statistic for a large-sample test concerning proportion

$$z = \frac{x - np_0}{\sqrt{np_0(1 - p_0)}}$$

which has approximately the standard normal distribution.

To make the same continuity correction as on page 251, some statisticians substitute $x - \dfrac{1}{2}$ or $x + \dfrac{1}{2}$ for x into the formula for z, whichever makes z closer to 0. Since the effect of this correction is generally negligible when n is large, we shall not use it here. However, we shall use it later, in Chapter 13, in connection with some small samples. Note that the continuity correction does not even have to be considered, when without it the null hypothesis cannot be rejected.

The test criteria are again those of Figure 10.5 with p and p_0 substituted for μ and μ_0. For the one-sided alternative hypothesis $p < p_0$, we reject the null hypothesis if $z \leq -z_\alpha$, for the one-sided alternative hypothesis $p > p_0$, we reject the null hypothesis if $z \geq z_\alpha$, and for the two-sided alternative hypothesis $p \neq p_0$, we reject the null hypothesis if $z \leq -z_{\alpha/2}$ or $z \geq z_{\alpha/2}$.

■ **EXAMPLE**   Suppose that a nutritionist claims that at least 75% of the preschool children in a certain country have protein-deficient diets, and that a sample survey reveals that this is true for 206 preschool children in a sample of 300. Test the claim at the 0.05 level of significance.

Solution   1. *Hypotheses* H_0: $p = 0.75$
$\qquad\qquad\qquad\qquad\quad H_A$: $p < 0.75$

2. *Level of significance* $\alpha = 0.05$

3. *Criterion*   Reject the null hypothesis if $z \leq -1.645$, where

$$z = \frac{x - np_0}{\sqrt{np_0(1 - p_0)}}$$

(see also Figure 11.2); otherwise, accept it or reserve judgment.

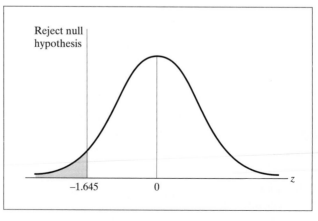

FIGURE 11.2 Test criterion for protein-deficiency example.

4. *Calculations*   Substituting $x = 206$, $n = 300$, and $p_0 = 0.75$ into the formula for z, we get

$$z = \frac{206 - 300(0.75)}{\sqrt{300(0.75)(0.25)}} \approx -2.53$$

5. *Decision*   Since $z = -2.53$ is less than -1.645, the null hypothesis must be rejected; in other words, we conclude that fewer than 75% of the preschool children in the given country have protein-deficient diets.

Had we wanted to use the *p*-value approach in this example, we would have had to determine the area under the curve to the left of $z = -2.53$. Since this area is $0.5000 - 0.4943 = 0.0057$, which is less than $\alpha = 0.05$, we would have concluded (as before) that the null hypothesis must be rejected. ■

Exercises

Exercises 11.1 and 11.7 are practice exercises; their complete solutions are given on page 391.

11.1 A physicist claims that at most 10% of all persons exposed to a certain amount of radiation will feel any effects. If, in a random sample, 4 of 13 persons feel an effect, test the claim at the 0.05 level of significance.

11.2 The program director of a television series claims that 60% of the television viewing audience recognizes its theme song and associates it with the program. To verify this claim, a random sample of 14 television viewers was interviewed, of whom 5 associated the theme song with the program. What can be concluded at the level of significance $\alpha = 0.05$ if the sample data is used to test the null hypothesis $p = 0.60$ against the alternative hypothesis $p < 0.60$?

11.3 A television critic claims that at least 80% of all viewers find the noise level of a certain commercial objectionable. If 9 out of 15 persons shown this commercial object to the noise level, what can we conclude about this claim at the 0.05 level of significance?

11.4 The dean of student affairs at a college wants to test the claim that 30% of all undergraduate students reside in the college dormitories. If 5 of 10 randomly selected undergraduate students reside in the dormitories, does this support the claim that 30% of all undergraduate students reside in the dormitories? Use the alternative hypothesis $p \neq 0.30$ and $\alpha = 0.01$.

11.5 A food processor wants to know whether the probability is really 0.60 that a customer will prefer a new kind of packaging to the old kind. If, in a random sample, 6 of 15 customers prefer the new kind of packaging to the old kind, test the null hypothesis $p = 0.60$ against the alternative hypothesis $p \neq 0.60$ at the 0.05 level of significance.

11.6 A representative of a health insurance company asserts that its claims payment procedures satisfy at least 80% of its policyholders. If in a random sample of 14 policyholders, it is determined that 8 are satisfied and 6 are dissatisfied, what can we conclude if we test this claim at the 0.05 level of significance?

11.7 In a random sample of 400 automobile accidents, it was found that 128 were due at least in part to driver fatigue. Use the 0.01 level of significance to test whether this supports the claim that 35% of all automobile accidents are due at least in part to driver fatigue.

11.8 In a wage negotiation an employer claims that 70% of the firm's employees are willing to take an increase in retirement benefits in lieu of an increase in salary. What can we conclude about the employer's hypothesis $p = 0.70$ if, in a random sample of 100 employees, there were 63 who said that they would be willing to take an increase in retirement benefits in lieu of an increase in salary? Use the two-sided alternative hypothesis $p \neq 0.70$, and the level of significance $\alpha = 0.01$.

11.9 A self-service gasoline station operator claims that at least 50% of the drivers who patronize the station are women. To check this claim, a random sample was taken, and it was found that 85 of 200 patrons were women. Test the null hypothesis $p = 0.50$ against a suitable alternative hypothesis at the 0.01 level of significance.

11.10 Repeat Exercise 11.9 with the new information that 90 (not 85) of the 200 randomly selected drivers are women. Otherwise, this exercise is the same as Exercise 11.9.

11.11 In a random sample of 500 cars making a left turn at a certain intersection, 169 pulled into the wrong lane. Test the null hypothesis that the actual proportion of drivers who make this mistake (at the given intersection) is 0.30 against the alternative hypothesis that this figure is too low. Use the 0.01 level of significance.

11.12 It has been claimed that 30% of all families moving away from California move to Arizona. If, in a random sample of the records of several large van lines, it is found that the belongings of 104 of 400 families moving away from California were shipped to Arizona, test the null hypothesis $p = 0.30$ against the alternative hypothesis $p < 0.30$ at the 0.05 level of significance.

11.13 A food chemist notes that 90% of the bricks of a certain cheddar cheese will remain mold free for three months under standard refrigeration. A variation is made in the salt content of this cheese, and a sample of 15 bricks of the modified cheese is kept under standard refrigeration for three months.
   (a) Suppose that only 11 of the bricks remain mold free. At the 5% level of significance, test the null hypothesis $p = 0.90$ against the alternative hypothesis $p \neq 0.90$.
   (b) Suppose that all 15 of the bricks remain mold free. At the 5% level of significance, test the null hypothesis $p = 0.90$ against the alternative hypothesis $p \neq 0.90$. What can you say about this experiment's ability to detect a value p larger than 0.90?

11.3 Differences Between Proportions

There are many problems in which we must decide whether an observed difference between two sample proportions can be attributed to chance, or whether it is indicative of the fact that the corresponding true proportions are unequal. For instance, we may want to decide on the basis of sample data whether there really is a difference between the proportions of persons with and without flu shots who actually catch the disease, or we may want to check on the basis of samples whether two manufacturers of electronic equipment ship equal proportions of defectives.

The method we shall use to test whether an observed difference between two sample proportions can be attributed to chance or whether it is statistically significant is based on the following theory: If x_1 and x_2 are the numbers of successes obtained in n_1 trials of one kind and n_2 of another, the trials are all independent, and the corresponding probabilities of a success are, respectively, p_1 and p_2, then the sampling distribution of $\dfrac{x_1}{n_1} - \dfrac{x_2}{n_2}$ has the mean $p_1 - p_2$ and the standard deviation

$$\sqrt{\frac{p_1(1 - p_1)}{n_1} + \frac{p_2(1 - p_2)}{n_2}}$$

It is customary to refer to this standard deviation as the **standard error of the difference between two proportions**.

When we test the null hypothesis $p_1 = p_2 (= p)$ against an appropriate alternative hypothesis, the mean of the sampling distribution of the difference between the two sample proportions is $p_1 - p_2 = 0$, and its standard deviation can be written

$$\sqrt{p(1 - p)\left(\frac{1}{n_1} + \frac{1}{n_2}\right)}$$

Since the value of p is not known, we must estimate it; this is done by **pooling** the data and substituting for p the combined sample proportion $\hat{p} = \dfrac{x_1 + x_2}{n_1 + n_2}$, which reads "$p$-hat." Then, since for large samples the sampling distribution of the difference between two proportions can be approximated closely with a normal distribution, we base the test on the statistic

Statistic for a large-sample test concerning the difference between two proportions

$$z = \frac{\dfrac{x_1}{n_1} - \dfrac{x_2}{n_2}}{\sqrt{\hat{p}(1 - \hat{p})\left(\dfrac{1}{n_1} + \dfrac{1}{n_2}\right)}} \quad \text{with} \quad \hat{p} = \frac{x_1 + x_2}{n_1 + n_2}$$

which has approximately the standard normal distribution. The test criteria are again those of Figure 10.5, with $p_1 - p_2$ substituted for μ and 0 substituted for μ_0. For the one-sided alternative hypothesis $p_1 < p_2$, we reject the null hypothesis if $z \leq -z_\alpha$, for the one-sided alternative hypothesis $p_1 > p_2$, we reject the null hypothesis if $z \geq z_\alpha$, and for the two-sided alternative hypothesis $p_1 \neq p_2$, we reject the null hypothesis if $z \leq -z_{\alpha/2}$ or $z \geq z_{\alpha/2}$.

■ **EXAMPLE**   To test the effectiveness of a new pain-relieving drug, 80 patients at a clinic were given a pill containing the drug and 80 others were given a placebo. At the 0.01 level of significance, what can we conclude about the effectiveness of the drug if in the first group 56 of the patients felt a beneficial effect, while 38 of those who received the placebo felt a beneficial effect?

Solution   1. *Hypotheses* H_0: $p_1 = p_2$

$\qquad\qquad\qquad\qquad H_A$: $p_1 > p_2$

2. *Level of significance* $\alpha = 0.01$

3. *Criterion*   Reject the null hypothesis if $z \geq 2.33$, where z is given by the formula above; otherwise, accept it or reserve judgment.

4. *Calculations*   Substituting $x_1 = 56$, $x_2 = 38$, $n_1 = 80$, $n_2 = 80$, and $\hat{p} = \dfrac{56 + 38}{80 + 80} = 0.5875$ into the formula for z, we get

$$z = \frac{\dfrac{56}{80} - \dfrac{38}{80}}{\sqrt{(0.5875)(0.4125)\left(\dfrac{1}{80} + \dfrac{1}{80}\right)}} \approx 2.89$$

5. *Decision*   Since $z = 2.89$ exceeds 2.33, the null hypothesis must be rejected; in other words, we conclude that the new pain-relieving drug is effective. ■

Exercises

Exercise 11.14 is a practice exercise; its complete solution is given on page 391.

11.14 In random samples of 200 tractors from one assembly line and 400 tractors from another, there were, respectively, 16 tractors and 20 tractors that required extensive adjustments before they could be shipped. At the 0.05 level of significance, can we conclude that there is a difference in the quality of the work of the two assembly lines?

11.15 In a random sample of 250 persons who skipped breakfast, 102 reported that they experienced midmorning fatigue, and in a random sample of 250 persons who ate breakfast, 73 reported that they experienced midmorning fatigue. Use the 0.01 level of significance to test the null hypothesis that there is no difference between the corresponding population proportions against the alternative hypothesis that midmorning fatigue is more prevalent among persons who skip breakfast.

11.16 A study of the commuting patterns of the residents of Northboro showed that, in a random sample, 141 of 300 persons commuted by privately owned automobile. At the same time, a random sample of the residents of Southboro showed that 123 of 300 persons commuted by privately owned automobile. Use the 0.05 level of significance to test the null hypothesis that there is no difference between the corresponding proportions of persons in these two large cities who commute by privately owned automobile.

11.17 In a true–false test, a test item is considered to be good if it discriminates between well-prepared students and poorly prepared students. What can we conclude about the merit of a test item that was answered correctly by 246 of 300 well-prepared students and by 165 of 300 poorly prepared students? Use the 0.05 level of significance.

11.18 In a recent study of labor force participation rates in the United States, a random sample of 1,000 men (16 years old and older) revealed that 749 participated in the labor force; and a random sample of 1,000 women (16 years old and older) revealed that 593 participated in the labor force. Test at the 0.01 level of significance whether the difference between the two sample proportions, $\dfrac{749}{1,000}$ and $\dfrac{593}{1,000}$, may be attributed to chance.

11.4
Differences Among Proportions

Following the pattern of Chapter 10, we shall now consider a test that enables us to decide whether observed differences among more than two sample proportions can be attributed to chance. For instance, if 25 of 200 brand A tires, 21 of 200 brand B tires, 32 of 200 brand C tires, and 18 of 200 brand D tires failed to last 30,000 miles, we may want to know whether the differences among the sample proportions

$$\frac{25}{200} = 0.125, \quad \frac{21}{200} = 0.105, \quad \frac{32}{200} = 0.160, \quad \text{and} \quad \frac{18}{200} = 0.090$$

can be attributed to chance, or whether they are indicative of actual differences in the quality of the tires.

To illustrate the method we use to analyze this kind of data, suppose that a survey in which independent random samples of workers in three parts of the country were asked whether they feel that unemployment or inflation is a more serious problem yielded the results shown in the following table:

	Northeast	Midwest	Southwest
Unemployment	87	73	66
Inflation	113	77	84
	200	150	150

If we let p_1, p_2, and p_3 denote the true proportions of workers in the three parts of the country who feel that unemployment is the more serious economic problem, the hypotheses we shall want to test are

$$H_0: \quad p_1 = p_2 = p_3$$

$$H_A: \quad \text{the } p\text{'s are not all equal}$$

If the null hypothesis is true, we can combine the data and estimate the common proportion of workers in the three parts of the country who feel that unemployment is the more serious economic problem as

$$\hat{p} = \frac{87 + 73 + 66}{200 + 150 + 150} = \frac{226}{500} = 0.452$$

where, again, $\hat{p}$ reads "p-hat."

With this estimate, we would expect $200(0.452) = 90.4$ of the 200 workers in the Northeast, $150(0.452) = 67.8$ of the 150 workers in the Midwest, and $150(0.452) = 67.8$ of the 150 workers in the Southwest to choose unemployment. Also, if we subtract these figures from the sizes of the respective samples, we find that $200 - 90.4 = 109.6$ of the 200 workers in the Northeast, $150 - 67.8 = 82.2$ of the 150 workers in the Midwest, and $150 - 67.8 = 82.2$ of the 150 workers in the Southwest would be expected to choose inflation.

These results are summarized in the following table, where the **expected frequencies** are shown in parentheses below the **observed frequencies**:

	Northeast	Midwest	Southwest
Unemployment	87 (90.4)	73 (67.8)	66 (67.8)
Inflation	113 (109.6)	77 (82.2)	84 (82.2)

To test the null hypothesis that the p's are all equal, we then compare the frequencies that were actually observed with the frequencies we can expect if the null hypothesis is true. Clearly, the null hypothesis should be accepted if the discrepancies between the observed and expected frequencies are small, and it should be rejected if the discrepancies between the two sets of frequencies are large.

Denoting the observed frequencies by the letter o and the expected frequencies by the letter e, we base this comparison on the following χ^2 (**chi-square**) **statistic**:

Statistic for a test concerning the differences among proportions

$$\chi^2 = \Sigma \frac{(o - e)^2}{e}$$

In words, χ^2 is the sum of the quantities obtained by dividing $(o - e)^2$ by e separately for each **cell** of the table, and for our example we get

$$\chi^2 = \frac{(87 - 90.4)^2}{90.4} + \frac{(73 - 67.8)^2}{67.8} + \frac{(66 - 67.8)^2}{67.8}$$

$$+ \frac{(113 - 109.6)^2}{109.6} + \frac{(77 - 82.2)^2}{82.2} + \frac{(84 - 82.2)^2}{82.2}$$

$$\approx 1.048$$

It remains to be seen whether this value is large enough to reject the null hypothesis $p_1 = p_2 = p_3$.

If the null hypothesis is true, the sampling distribution of the χ^2 statistic is approximately a theoretical distribution called the **chi-square distribution**. The parameter of this distribution, like that of the t distribution, is called the **number of degrees of freedom**, or simply the **degrees of freedom**, and it equals $k - 1$ when we compare k sample proportions.

Since the null hypothesis is to be rejected only when the value obtained for χ^2 is too large, we base our decision on the criterion of Figure 11.3, where χ^2_α is such that the area under the curve to its right is equal to α. Values of $\chi^2_{0.05}$ and $\chi^2_{0.01}$ for 1, 2, 3, ..., and 30 degrees of freedom are given in Table IV at the end of the book.

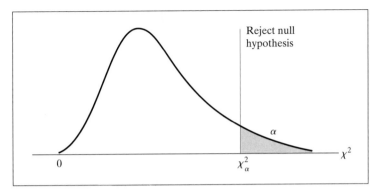

FIGURE 11.3 Chi-square distribution.

Returning to our numerical example and supposing that the level of significance is to be 0.05, we find that $\chi_{0.05}^2 = 5.991$ for $k - 1 = 2$ degrees of freedom, and that the null hypothesis cannot be rejected since $\chi^2 = 1.048$ does not exceed 5.991. In other words, the data tends to support, rather than refute, the hypothesis that the proportion of workers who feel that unemployment is a more serious economic problem than inflation is the same in all three parts of the country.

In general, if we want to test the null hypothesis $p_1 = p_2 = \cdots = p_k$ on the basis of random samples from k populations, we proceed as follows:

1. *Hypotheses* H_0: $p_1 = p_2 = \cdots = p_k$

    H_A: the p's are not all equal

2. *Level of significance* α

3. *Criterion*   Compute

$$\chi^2 = \sum \frac{(o - e)^2}{e}$$

taking the sum over all $2 \cdot k$ cells. Reject the null hypothesis if $\chi^2 \geq \chi_\alpha^2$ for $k - 1$ degrees of freedom. Otherwise, accept the null hypothesis or reserve judgment.

4. *Calculations*   We estimate $p_1 = p_2 = \cdots = p_k$, the common population proportion, as

$$\hat{p} = \frac{x_1 + x_2 + \cdots + x_k}{n_1 + n_2 + \cdots + n_k}$$

and then we multiply $\hat{p}$ by the respective sample sizes to obtain the expected frequencies for the first row of the table. Then we determine the expected frequencies for the second row of the table by subtracting those of the first row from the respective sample sizes and substitute all the observed and expected frequencies into the formula for χ^2.

5. *Decision*   Depending on the value we get for χ^2, we reject the null hypothesis, or we accept it or reserve judgment.

When we calculate the expected frequencies, there is seldom any need to carry more than two decimals, even though the entries in Table IV are given to three decimals. Also, the test we have been discussing is only an approximate test, and it should not be used when one (or more) of the expected frequencies is less than 5. If this is the case, we can sometimes combine two or more of the samples in such a way that none of the e's is less than 5.

The method we have discussed here can be used only to test the null hypothesis $p_1 = p_2 = \cdots = p_k$ against the alternative hypothesis that the p's are not all equal. However, in the special case where $k = 2$, we can use instead the method of Section 11.3 and test also against either of the alternative hypotheses $p_1 < p_2$ or $p_1 > p_2$. Indeed, for $k = 2$, the two methods are equivalent, as it can be shown that the χ^2 statistic equals the square of the z statistic obtained in accordance with the formula on page 371 (see Exercises 11.24 through 11.26).

Exercises

Exercises 11.19 *and* 11.24 *are practice exercises; their complete solutions are given on page 392.*

11.19 On page 373 we referred to a study in which 25 of 200 brand A tires, 21 of 200 brand B tires, 32 of 200 brand C tires, and 18 of 200 brand D tires fail to last 30,000 miles. Use the 0.05 level of significance to test the null hypothesis that there is no difference in the durability of the four kinds of tires.

11.20 In a survey of investor attitudes, 140 were interviewed in Mexico City, 160 in Toronto, and 250 in New York City, and 91, 104, and 170 investors, respectively, think that the prices of stocks traded on the New York Stock Exchange will rise over the next month. The remaining investors are uncertain or think that stock prices will fall over the next month. Use the level of significance $\alpha = 0.05$ to test the null hypotheses that there is no difference in the true proportion of investors in these three cities who think that stock prices will rise over the next month.

11.21 In three random samples taken from the records for motor vehicle license examinations, 32 of 200 applicants in license examination station A, 69 of 300 applicants in license examination station B, and 19 of 100 applicants in license examination station C failed to pass the license examination. Using the 0.05 level of significance, test the null hypothesis that there is no difference in the true proportion of applicants who fail to pass the license examination in the three stations.

11.22 In studying problems relating to its handling of reservations, an airline takes random samples of 60 of the complaints about reservations filed in each of four cities. If 49 of the complaints from city A, 54 of the complaints from city B, 41 of the complaints from city C, and 48 of the complaints from city D are about overbooking, test at the 0.05 level of significance whether the difference among the corresponding sample proportions can be attributed to chance.

11.23 The following table shows the results of a study in which samples of the members of five labor unions were asked whether they are for or against a certain piece of legislation:

	Union 1	Union 2	Union 3	Union 4	Union 5
For	74	81	69	75	91
Against	26	19	31	25	9

Use the 0.01 level of significance to test the null hypothesis that the corresponding true proportions are all equal.

11.24 Use the method of this section to rework Exercise 11.14 and verify that the value obtained for χ^2 equals the square of the value obtained originally for z.

11.25 Use the method of this section to rework Exercise 11.16 and verify that the value obtained for χ^2 equals the square of the value obtained originally for z.

11.26 Use the method of this section to rework Exercise 11.18 and verify that the value obtained for χ^2 equals the square of the value obtained originally for z.

11.5
Contingency Tables

The χ^2 statistic plays an important role in many other problems where information is obtained by counting rather than measuring. The method we shall describe here applies to two kinds of problems, which differ conceptually but are analyzed the same way.

In the first kind of problem we deal with trials permitting more than two possible outcomes. For instance, the weather can get better, remain the same, or get worse; an undergraduate can be a freshman, a sophomore, a junior, or a senior; a movie may be rated G, PG, PG-13, R, or NC-17. In the language of Section 6.5, we could say that we are dealing with multinomial (rather than binomial) trials.

Also, in the illustration of the preceding section, each worker might have been asked whether unemployment is a more serious economic problem than inflation, whether inflation is a more serious economic problem than unemployment, or whether he or she is undecided, and this might have resulted in the following table:

	Northeast	Midwest	Southwest
Unemployment	57	53	44
Undecided	72	40	48
Inflation	71	57	58
	200	150	150

We refer to this kind of table as a 3×3 table (where the notation 3×3 is read "3 by 3"), because it has 3 horizontal rows and 3 vertical columns; more generally, when there are r horizontal rows and c vertical columns, we refer to the table as an **$r \times c$ table**. Here, as in the table analyzed in the preceding section, the column totals, representing the sample sizes, are fixed. On the other hand, the row totals depend on the responses of the persons interviewed and hence on chance.

In the second kind of problem where the method of this section applies, the column totals as well as the row totals depend on chance. To give an example, suppose that a sociologist wants to determine whether there is a relationship between the intelligence of boys who have gone through a special job-training program and their subsequent performance in their jobs, and that a sample of 400 cases taken from very extensive files yielded the following results:

		Performance			
		Poor	Fair	Good	
	Below average	67	64	25	156
Intelligence	Average	42	76	56	174
	Above average	10	23	37	70
		119	163	118	400

This is also a 3 × 3 table, and it is mainly in connection with problems like this that $r \times c$ tables are referred to as **contingency tables**. In either kind of table, the observed frequencies are referred to as the **observed cell frequencies**.

Before we demonstrate how $r \times c$ tables are analyzed, let us examine what hypotheses we want to test. In the first example we want to test whether the probabilities that a worker will choose "unemployment," "undecided," or "inflation" are the same for the three parts of the country. Formally,

H_0: For each alternative (unemployment, undecided, and inflation), the probabilities are the same for the three parts of the country.

H_A: For at least one alternative, the probabilities are not the same for the three parts of the country.

In the other example, we want to test whether there is a dependence (relationship) between on the job performance and intelligence, and we write

H_0: The two variables under consideration are independent.

H_A: The two variables are not independent.

To show how an $r \times c$ table is analyzed, let us refer to the second of our two examples, and let us begin by illustrating the calculation of an **expected cell frequency**.

■ **EXAMPLE**   Assuming that intelligence and on-the-job performance are independent and using the data in the preceding table, how many of the boys in a random sample of 400 can be expected to have below-average intelligence and perform poorly on their jobs?

Solution   Under the assumption of independence, the probability of randomly choosing a boy whose intelligence is below average and whose on-the-job performance is poor is given by the product of the probability of choosing a boy whose intelligence is below average and the probability of choosing a boy whose on-the-job performance is poor. Using the totals of the first row and the first column to estimate these two probabilities, we get

$$\frac{67 + 64 + 25}{400} = \frac{156}{400}$$

for the probability of choosing a boy whose intelligence is below average and

$$\frac{67 + 42 + 10}{400} = \frac{119}{400}$$

for the probability of choosing a boy whose on-the-job performance is poor. Hence, we estimate the probability of choosing a boy whose intelligence is below average and whose on-the-job performance is poor as $\frac{156}{400} \cdot \frac{119}{400}$, and in a sample of size 400 we would expect to find

$$400 \cdot \frac{156}{400} \cdot \frac{119}{400} = \frac{156 \cdot 119}{400} \approx 46.4$$

boys who fit this description.   ■

In the final step of this calculation, $\dfrac{156 \cdot 119}{400}$ is just the product of the of total the first row and the total of the first column divided by the grand total for the entire table. Indeed, the argument that led to this result can be used to show that, in general, under the assumption of independence:

The expected frequency for any cell of a contingency table may be obtained by multiplying the total of the row to which it belongs by the total of the column to which it belongs and then dividing by the grand total for the entire table.

With this rule we get an expected frequency of

$$\frac{156 \cdot 163}{400} \approx 63.6$$

for the second cell of the first row, and

$$\frac{174 \cdot 119}{400} \approx 51.8 \quad \text{and} \quad \frac{174 \cdot 163}{400} \approx 70.9$$

for the first two cells of the second row.

It is not necessary to calculate all the expected cell frequencies in this way, as it can be shown that the sum of the expected frequencies for any row or column must equal the sum of the corresponding observed frequencies. Therefore, we can get some of the expected cell frequencies by subtraction from row or column totals. For instance, for our example we get

$$156 - 46.4 - 63.6 = 46.0$$

for the expected frequency of the third cell of the first row,

$$174 - 51.8 - 70.9 = 51.3$$

for the expected frequency of the third cell of the second row, and

$$119 - 46.4 - 51.8 = 20.8$$

$$163 - 63.6 - 70.9 = 28.5$$

and

$$118 - 46.0 - 51.3 = 20.7$$

for the expected frequencies of the three cells of the third row. These results are summarized in the following table, where the expected frequencies are shown in parentheses below the corresponding observed frequencies:

| | | Performance | | |
		Poor	Fair	Good
	Below average	67 (46.4)	64 (63.6)	25 (46.0)
Intelligence	Average	42 (51.8)	76 (70.9)	56 (51.3)
	Above average	10 (20.8)	23 (28.5)	37 (20.7)

From here on, the work is like that of the preceding section; we calculate the χ^2 statistic according to the formula

Statistic for an analysis of an $r \times c$ table

$$\chi^2 = \sum \frac{(o - e)^2}{e}$$

with $\frac{(o - e)^2}{e}$ calculated separately for each cell of the table. Then we reject the null hypothesis at the level of significance α if the value obtained for χ^2 exceeds χ^2_α for $(r - 1)(c - 1)$ degrees of freedom. In connection with this formula for the number of degrees of freedom, observe that after $c - 1$ of the expected frequencies have been calculated for each of $r - 1$ rows by means of the rule on page 379, all the other expected frequencies may be obtained by subtraction from row or column totals. For our example, the number of degrees of freedom is $(3 - 1)(3 - 1) = 4$, and it should be observed that after we had calculated two of the expected frequencies for each of the first two rows by means of the rule on page 379, all the others were obtained by subtraction from row or column totals.

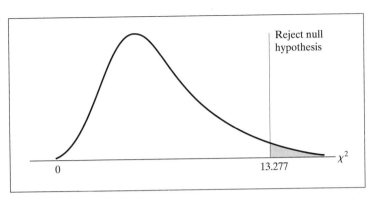

FIGURE 11.4 Test criterion for intelligence and on-the-job performance example.

■ **EXAMPLE**   With reference to our example, use the 0.01 level of significance to test the null hypothesis that intelligence and on-the-job performance are independent for boys who have gone through the special job-training program.

Solution   1. *Hypotheses* H_0:   Intelligence and on-the-job performance are independent.

H_A: Intelligence and on-the-job performance are not independent.

2. *Level of significance* $\alpha = 0.01$

3. *Criterion*   Reject the null hypothesis if $\chi^2 \geq 13.277$, the value of $\chi^2_{0.01}$ for $(3 - 1)(3 - 1) = 4$ degrees of freedom, where

$$\chi^2 = \sum \frac{(o - e)^2}{e}$$

(see also Figure 11.4); otherwise, accept it or reserve judgment.

4. *Calculations*   Copying the observed and expected cell frequencies from the table on page 380, we find that

$$\chi^2 = \frac{(67 - 46.4)^2}{46.4} + \frac{(64 - 63.6)^2}{63.6} + \frac{(25 - 46.0)^2}{46.0}$$

$$+ \frac{(42 - 51.8)^2}{51.8} + \frac{(76 - 70.9)^2}{70.9} + \frac{(56 - 51.3)^2}{51.3}$$

$$+ \frac{(10 - 20.8)^2}{20.8} + \frac{(23 - 28.5)^2}{28.5} + \frac{(37 - 20.7)^2}{20.7}$$

$$\approx 40.89$$

5. *Decision*   Since $\chi^2 = 40.89$ exceeds 13.277, the null hypothesis must be rejected; we conclude that there is a relationship between intelligence and on-the-job performance.   ■

A computer printout of the preceding chi-square analysis is shown in Figure 11.5. The difference between the values of χ^2 obtained above and in Figure 11.5 is due to rounding. Some computer programs also give the prob-

```
MTB > READ C1 C2 C3
DATA> 67   64   25
DATA> 42   76   56
DATA> 10   23   37
DATA> END
        3 ROWS READ
MTB > CHIS C1 C2 C3

Expected counts are printed below observed counts

                C1        C2        C3      Total
        1       67        64        25       156
              46.41     63.57     46.02

        2       42        76        56       174
              51.76     70.90     51.33

        3       10        23        37        70
              20.83     28.52     20.65

   Total       119       163       118       400

   ChiSq = 9.135 + 0.003 +  9.601 +
           1.842 + 0.366 +  0.425 +
           5.627 + 1.070 + 12.945 = 41.014

   df = 4
```

FIGURE 11.5 MINITAB printout for analysis of contingency table.

ability of getting a value greater than or equal to the observed value of χ^2 when the null hypothesis is true; for our example, it is about 0.00000003. Because the test is based on an approximation, this value should not be taken too literally, although the actual probability of a greater value is certainly quite small.

The method that we have used here to analyze the contingency table applies also when the column totals are fixed sample sizes (as in the illustration on page 377) and do not depend on chance. The rule according to which we multiply the row total by the column total and then divide by the grand total has to be justified in a different way, but this is of no consequence—the expected cell frequencies are determined in exactly the same way.

■ **EXAMPLE**   Analyze the 3×3 table on page 377, which pertains to a study in which workers in the Northeast, Midwest, and Southwest are asked whether unemployment or inflation is the more serious economic problem or whether they are undecided. Use the 0.01 level of significance.

Solution   1. *Hypotheses* H_0:   For each alternative (unemployment, undecided, and inflation), the probabilities are the same for the three parts of the country.

H_A: For at least one alternative, the probabilities are not the same for the three parts of the country.

2. *Level of significance* $\alpha = 0.01$

3. *Criterion*   Reject the null hypothesis if $\chi^2 \geq 13.277$, the value of $\chi^2_{0.01}$ for $(3-1)(3-1) = 4$ degrees of freedom, where

$$\chi^2 = \sum \frac{(o-e)^2}{e}$$

Otherwise, accept the null hypothesis or reserve judgment.

4. *Calculations*   The expected frequencies for the first two cells of the first two rows are

$$\frac{154 \cdot 200}{500} = 61.6, \qquad \frac{154 \cdot 150}{500} = 46.2, \qquad \frac{160 \cdot 200}{500} = 64,$$

and

$$\frac{160 \cdot 150}{500} = 48$$

by subtraction, those for the third cells of the first two rows are 46.2 and 48, and those for the third row are 74.4, 55.8, and 55.8. Then, substituting into the formula for χ^2, we get

$$\chi^2 = \frac{(57-61.6)^2}{61.6} + \frac{(53-46.2)^2}{46.2} + \frac{(44-46.2)^2}{46.2}$$

$$+ \frac{(72-64)^2}{64} + \frac{(40-48)^2}{48} + \frac{(48-48)^2}{48}$$

$$+ \frac{(71-74.4)^2}{74.4} + \frac{(57-55.8)^2}{55.8} + \frac{(58-55.8)^2}{55.8}$$

$$\approx 4.05$$

5. *Decision*   Since $\chi^2 = 4.05$ does not exceed 13.277, the null hypothesis cannot be rejected; the differences between the observed and expected frequencies may well be due to chance.   ∎

Alternatively, using a graphing calculator we could have obtained the display shown in Figure 11.6. As before, graphing calculator information shown here and elsewhere in this book is entirely optional and can be omitted by the reader without the loss of continuity. The figure shows that the *p*-value corresponding to $\chi^2 = 4.05$ and 4 degrees of freedom is .399281. Since this *p*-value exceeds $\alpha = 0.01$, we conclude (as shown earlier) that the null hypothesis cannot be rejected.

The procedure for obtaining Figure 11.6 is explained in the Using Technology: The Graphing Calculator box on page 384.

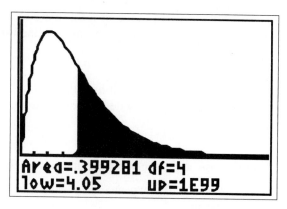

FIGURE 11.6 Chi-square distribution reproduced from the
display screen of a TI-83 graphing calculator.

USING TECHNOLOGY : *The Graphing Calculator*

How to determine the *p*-value and graph the area under the curve as shown in Figure 11.6

(Tip before beginning: Reset the memory to default settings and clear all plots—see Appendix A: TI-83 Tips.)

Step 1.   Set the viewing window.

> Press WINDOW
> Set Xmin = 0
>   Xmax = 18
>   Xscl = 1
>   Ymin = -.05
>   Ymax = .2
>   Yscl = 1
>   Xres = 1

Step 2.   Select the type of display and enter the interval to be displayed.
Press 2nd, VARS to access the DISTR DRAW menu.
Highlight DRAW.
Select 3: Shade $\chi^2($, press ENTER.
Enter   4.05, 1E99, 4)
(For instructions on entering "1E99," for infinity, see Appendix A: TI-83 Tips; the 4.05 and 1E99 (for ∞) are the lower and upper limits of the interval, and the 4 is the number of degrees of freedom.)

Step 3.   View the graph.
Press ENTER to solve.

Exercises

Exercise 11.27 is a practice exercise; its complete solution is given on page 393.

11.27 In a survey, 80 single persons, 120 married persons, and 100 widowed or divorced persons were asked whether they feel that friends and social life, job or primary activity, or health and physical condition contributes most to their general happiness. Use the results shown in the following table and the 0.05 level of significance to test whether the probabilities of the three alternatives are the same for persons who are single, married, or widowed or divorced.

	Single	*Married*	*Widowed or divorced*
Friends and social life	41	49	42
Job or primary activity	27	50	33
Health and physical condition	12	21	25
Total	80	120	100

11.28 Suppose that for 120 mental patients who did not receive psychotherapy and 120 mental patients who received psychotherapy a panel of psychiatrists determined after six months whether their condition had deteriorated, remained unchanged, or improved. Based on the results shown in the following table, test at the level of significance $\alpha = 0.05$ whether the therapy is effective.

	No therapy	*Therapy*
Deteriorated	6	11
Unchanged	65	31
Improved	49	78

11.29 The following sample data pertains to the number of men's sweaters of various sizes sold in the men's departments of a chain of department stores.

	Department Store 1	*Department Store 2*	*Department Store 3*
Small	53	49	72
Medium	108	73	170
Large	48	22	55

Use the 0.01 level of significance to test whether the three department stores sell the same proportion of small, medium, and large sweaters.

11.30 To determine the income level of borrowers who make personal loans, a large bank examines a random sample of applications for personal loans in four branches, with the following results:

	Branch 1	Branch 2	Branch 3	Branch 4
High income	28	27	22	92
Upper middle income	78	54	67	87
Lower middle income	71	99	82	11
Low income	23	20	29	10
Total	200	200	200	200

Use the 0.05 level of significance to test the null hypothesis that the four bank branches have the same proportion of high, upper middle, lower middle, and low income borrowers.

11.31 A sample survey was designed to show the weight losses, during a certain period of time, of the patrons at three weight-reducing clinics.

	Clinic 1	Clinic 2	Clinic 3
1 to 4 pounds	56	57	72
5 to 9 pounds	32	39	55
10 pounds or more	12	14	13

Use the level of significance $\alpha = 0.05$ to test the null hypothesis that there is no relationship between the number of pounds lost by patrons and the clinics that they patronized.

11.32 Tests of the fidelity and selectivity of 190 radios produced the results shown in the following table:

		Low	Fidelity Average	High
Selectivity	Low	7	12	31
	Average	35	59	18
	High	15	13	0

Use the 0.01 level of significance to test the null hypothesis that fidelity is independent of selectivity.

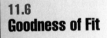

11.33 Use a computer package or a graphing calculator to rework Exercise 11.27.

11.34 Use a computer package or a graphing calculator to rework Exercise 11.28.

11.35 Use a computer package or a graphing calculator to rework Exercise 11.30.

11.36 A computer program for the analysis of a contingency table or a graphing calculator can be used to test for differences among proportions as in Section 11.4. In that case we are, in fact, analyzing a $2 \times c$ table. Use a computer package or a graphing calculator to rework the illustration of Section 11.4, which dealt with a survey in three parts of the country where workers were asked whether they feel that unemployment or inflation is a more serious problem.

11.37 Use a computer package or a graphing calculator to rework Exercise 11.22.

11.38 Use a computer package or a graphing calculator to rework Exercise 11.23.

11.6
Goodness of Fit

The χ^2 statistic can also be used to compare observed frequency distributions with distributions that we might expect according to theory or assumptions. We refer to such a comparison as a test of **goodness of fit**.

This kind of problem would arise, for example, if a quality-control engineer takes a daily sample of 10 tires coming off an assembly line, and he wants to check, on the basis of the following data, for the numbers of tires with imperfections observed on 200 days, whether it is true that 5% of all the tires have imperfections, that is, whether he is sampling a binomial population with $n = 10$ and $p = 0.05$.

Number with imperfections	Number of samples
0	138
1	53
2 or more	9

If the hypothesis is true, the probabilities of getting 0, 1, or 2 or more tires with imperfections are, respectively, 0.599, 0.315, and $0.075 + 0.010 + 0.001 = 0.086$ according to Table I, and hence the corresponding expected frequencies are $200(0.599) = 119.8$, $200(0.315) = 63.0$, and $200(0.086) = 17.2$.

To test whether the discrepancies between the observed frequencies and the expected frequencies can be attributed to chance, we use the same χ^2 statistic as in the preceding sections:

Statistic for a test of goodness of fit

$$\chi^2 = \sum \frac{(o - e)^2}{e}$$

Based on the value of this statistic, we reject the null hypothesis if $\chi^2 > \chi^2_\alpha$ for $k - m - 1$ degrees of freedom, where k is the number of terms in the summation for χ^2 and m is the number of parameters that have to be estimated on the basis of the sample data. The role of m is explained in the discussion below.

■ **EXAMPLE**   Based on the data given above, test at the 0.05 level of significance whether the quality-control engineer is sampling a binomial population with $n = 10$ and $p = 0.05$.

Solution   1. *Hypotheses* H_0: Population is binomial with $n = 10$ and $p = 0.05$.
H_A: Population is not binomial with $n = 10$ and $p = 0.05$.

2. *Level of significance* $\alpha = 0.05$

3. *Criterion*   Reject the null hypothesis if $\chi^2 \geq 5.991$, the value of $\chi^2_{0.05}$ for $k - m - 1 = 3 - 0 - 1 = 2$ degrees of freedom, where

$$\chi^2 = \sum \frac{(o - e)^2}{e}$$

Otherwise, accept the null hypothesis or reserve judgment. Observe that $m = 0$ because no parameters were estimated.

4. *Calculations*   From page 387, the observed frequencies are 138, 53, and 9, and the expected frequencies are 119.8, 63.0, and 17.2. Substituting these values into the formula for χ^2, we get

$$\chi^2 = \frac{(138 - 119.8)^2}{119.8} + \frac{(53 - 63.0)^2}{63.0} + \frac{(9 - 17.2)^2}{17.2} \approx 8.26$$

5. *Decision*   Since $\chi^2 = 8.26$ exceeds 5.991, the null hypothesis must be rejected; we conclude that the true percentage of tires with imperfections is not 5%. (In Exercise 11.43 the reader will find that the binomial distribution with $n = 10$ and $p = 0.04$ provides a much better fit to the data.)   ■

In the preceding example it was assumed under the null hypothesis that $n = 10$ and $p = 0.05$, so that none of the parameters had to be estimated and the number of degrees of freedom was

$$k - m - 1 = 3 - 0 - 1 = 2$$

However, if we want to test whether the distribution is binomial (without knowing p), we will have to estimate p from the data, and the calculation for the number of degrees of freedom is

$$k - m - 1 = 3 - 1 - 1 = 1$$

We can also test whether or not grouped data has a distribution that can be approximated closely by a normal curve (see Exercises 11.44 and 11.45); in such cases it is required that the parameters μ and σ be estimated directly from the grouped data so that $m = 2$.

Observe also that, as in the tests of the two preceding sections, the sampling distribution of the χ^2 statistic of a goodness-of-fit test is only approximately a chi-square distribution. So, if any of the expected frequencies is less than 5, we shall again have to combine some of the data, in this case adjacent classes of the distribution.

Exercises

Exercises 11.39 and 11.44 are practice exercises; their complete solutions are given on pages 393 through 395.

 11.39 To see whether a die is balanced, it is rolled 360 times and the following results are obtained: 1 showed 57 times, 2 showed 46 times, 3 showed 68 times, 4 showed 52 times, 5 showed 72 times, and 6 showed 65 times. At the 0.05 level of significance, do these results support the hypothesis that the die is balanced?

11.40 Four coins are tossed together 480 times, and 0, 1, 2, 3, and 4 heads showed 26, 104, 171, 142, and 37 times. Using the probabilities $\frac{1}{16}, \frac{4}{16}, \frac{6}{16}, \frac{4}{16}$, and $\frac{1}{16}$, test at the level of significance $\alpha = 0.05$ whether its reasonable to suppose that the coins are balanced and randomly tossed.

11.41 To determine whether the customer service department of a large store serves an equal number of customers during each hour of a six-hour afternoon, a record was kept of the number of customers who were provided with service, and the results are

Time	Customers served
1:00–1:59	30
2:00–2:59	51
3:00–3:59	47
4:00–4:59	40
5:00–5:59	51
6:00–6:59	21
Total	240

Letting the expected frequencies be $\frac{240}{6} = 40$ customers served for each of the six hours, test at the level of significance $\alpha = 0.05$ whether an equal number of customers are served each hour during the six-hour time period.

11.42 The following is the distribution of the number of calls received at the switchboard of a government building during 400 five-minute intervals:

Number of calls	Frequency
0	95
1	116
2	112
3	47
4 or more	30

Use the 0.05 level of significance to test whether the number of calls received by the switchboard in a five-minute interval is a random variable having the Poisson distribution with $\lambda = 1.5$; that is, the probabilities for 0, 1, 2, 3, or 4 or more calls are 0.22, 0.33, 0.25, 0.13, and 0.07.

11.43 If a random variable has the binomial distribution with $n = 10$ and $p = 0.04$, the probabilities of 0, 1, or 2 or more successes are 0.665, 0.277, and 0.058. Based on this information and the data on page 387, test at the 0.05 level of significance whether the quality-control engineer is sampling a binomial population with $n = 10$ and $p = 0.04$.

11.44 The following is the distribution of the grades that 100 students received on a history test:

Grade	Frequency
65–69	1
70–74	10
75–79	37
80–84	36
85–89	13
90–94	2
95–99	1

As can easily be verified, the mean and the standard deviation of this distribution are $\bar{x} = 80$ and $s = 5$. To test the null hypothesis that these data constitute a random sample from a normal population with $\mu = 80$ and $\sigma = 5$, proceed with the following steps:

(a) Find the probabilities that a random variable having the normal distribution with $\mu = 80$ and $\sigma = 5$ will take on a value less than 69.5, between 69.5 and 74.5, between 74.5 and 79.5, between 79.5 and 84.5, between 84.5 and 89.5, between 89.5 and 94.5, and greater than 94.5.

(b) Multiply the probabilities obtained in part (a) by 100, the total frequency of the observed data, to find the frequencies that we can expect under the null hypothesis for a random sample of size $n = 100$.

(c) Calculate χ^2 for the observed frequencies shown in the table and the expected frequencies obtained in part (b), and test at the 0.05 level of significance whether the null hypothesis can be rejected.

11.45 The following is the distribution obtained on page 24 for the amount of time that 80 college students engaged in leisure activities during a typical school week:

Hours	Frequency
10–14	8
15–19	28
20–24	27
25–29	12
30–34	4
35–39	1

As we showed on page 80, the mean and the standard deviation of this distribution are $\bar{x} = 20.7$ and $s = 5.4$. To test the null hypothesis that these data constitute a random sample from a normal population with $\mu = 20.7$ and $\sigma = 5.4$, proceed with the following steps:

(a) Find the probabilities that a random variable having the normal distribution with $\mu = 20.7$ and $\sigma = 5.4$ will take on a value less than 14.5, between 14.5 and 19.5, between 19.5 and 24.5, between 24.5 and 29.5, between 29.5 and 34.5, and greater than 34.5.

(b) Multiply the probabilities obtained in part (a) by 80, the total frequency of the observed data, to find the frequencies we can expect under the null hypothesis for a random sample of size $n = 80$.

(c) Calculate χ^2 for the observed frequencies shown in the table and the expected frequencies obtained in part (b), and test at the 0.05 level of significance whether the null hypothesis can be rejected.

Solutions to Practice Exercises

11.1 This solution uses the *p*-value.

1. *Hypotheses* H_0: $p = 0.10$
                    H_A: $p > 0.10$

2. *Level of significance* $\alpha = 0.05$

3′. *Test statistic*   The test statistic is simply the number feeling an effect, here four.

4′. *Calculations*   Table I shows that for $n = 13$ and $p = 0.10$ the probability of four or more successes is $0.028 + 0.006 + 0.001 = 0.035$.

5′. *Decision*   Since 0.035 is less than or equal to $\alpha = 0.05$, the null hypothesis must be rejected.

11.7

1. *Hypotheses* H_0: $p = 0.35$
                   H_A: $p \neq 0.35$

2. *Level of significance* $\alpha = 0.01$

3. *Criterion*   Reject the null hypothesis if $z \leq -2.575$ or $z \geq 2.575$, where

$$z = \frac{x - np_0}{\sqrt{np_0(1 - p_0)}}$$

Otherwise, accept the null hypothesis or reserve judgment.

4. *Calculations*   Substituting $x = 128$, $n = 400$, and $p_0 = 0.35$ into the formula for z, we get

$$z = \frac{128 - 400(0.35)}{\sqrt{400(0.35)(0.65)}} \approx -1.26$$

5. *Decision*   Since $z = -1.26$ falls between -2.575 and 2.575, the null hypothesis cannot be rejected; the data does not refute the claim.

11.14

1. *Hypotheses* H_0: $p_1 = p_2$
                   H_A: $p_1 \neq p_2$

2. *Level of significance* $\alpha = 0.05$

3. *Criterion*   Reject the null hypothesis if $z \leq -1.96$ or $z \geq 1.96$, where z is given by the formula on page 371. Otherwise, accept the null hypothesis or reserve judgment.

4. *Calculations*   Substituting $x_1 = 16$, $x_2 = 20$, $n_1 = 200$, $n_2 = 400$, and $\hat{p} = \frac{16 + 20}{200 + 400} = 0.06$ into the formula for z, we get

$$z = \frac{\frac{16}{200} - \frac{20}{400}}{\sqrt{(0.06)(0.94)\left(\frac{1}{200} + \frac{1}{400}\right)}} \approx 1.46$$

5. *Decision*   Since $z = 1.46$ falls between -1.96 and 1.96, the null hypothesis cannot be rejected; in other words, the difference between the two sample proportions is not significant.

11.19   **1.** *Hypotheses* H_0: $p_1 = p_2 = p_3 = p_4$
H_A: the p's are not all equal

2. *Level of significance* $\alpha = 0.05$

3. *Criterion*   Reject the null hypothesis if $\chi^2 \geq 7.815$, the value of $\chi^2_{0.05}$ for $4 - 1 = 3$ degrees of freedom, where

$$\chi^2 = \sum \frac{(o - e)^2}{e}$$

Otherwise, accept it or reserve judgment.

4. *Calculations*   Since

$$\hat{p} = \frac{25 + 21 + 32 + 18}{200 + 200 + 200 + 200} = \frac{96}{800} = 0.12$$

we get $200(0.12) = 24$ for each expected frequency in the first row and $200 - 24 = 176$ for each expected frequency in the second row. Then, substituting into the formula for χ^2, we get

$$\chi^2 = \frac{(25 - 24)^2}{24} + \frac{(21 - 24)^2}{24} + \frac{(32 - 24)^2}{24} + \frac{(18 - 24)^2}{24}$$

$$+ \frac{(175 - 176)^2}{176} + \frac{(179 - 176)^2}{176} + \frac{(168 - 176)^2}{176} + \frac{(182 - 176)^2}{176}$$

$$\approx 5.21$$

5. *Decision*   Since $\chi^2 = 5.21$ does not exceed 7.815, the null hypothesis cannot be rejected; we cannot conclude that there is a real difference in the durability of the tires.

11.24   **1.** *Hypotheses* H_0: $p_1 = p_2$
H_A: $p_1 \neq p_2$

2. *Level of significance* $\alpha = 0.05$

3. *Criterion*   Reject the null hypothesis if $\chi^2 \geq 3.841$, the value of $\chi^2_{0.05}$ for $2 - 1 = 1$ degrees of freedom, where

$$\chi^2 = \sum \frac{(o - e)^2}{e}$$

Otherwise, accept it or reserve judgment.

4. *Calculations*   Since $\hat{p} = \dfrac{16 + 20}{200 + 400} = 0.06$, we get $200(0.06) = 12$ and $400(0.06) = 24$ for the expected frequencies for the first row, and $200 - 12 = 188$ and $400 - 24 = 376$ for the expected frequencies for the second row. Then, substituting into the formula for χ^2, we get

$$\chi^2 = \frac{(16 - 12)^2}{12} + \frac{(20 - 24)^2}{24} + \frac{(184 - 188)^2}{188} + \frac{(380 - 376)^2}{376}$$

$$\approx 2.13$$

5. *Decision*   Since $\chi^2 = 2.13$ does not exceed 3.841, the null hypothesis cannot be rejected; in other words, the difference between the two sample proportions is not significant.

If we square 1.46, the value obtained for z in Exercise 11.14, we get $(1.46)^2 = 2.1316$, and, except for rounding, this equals the value that we obtained for χ^2.

11.27   1. *Hypotheses* H_0: For each alternative, the probabilities are the same for persons who are single, married, and widowed or divorced.

H_A: For at least one alternative, the probabilities are not the same for the three kinds of persons.

2. *Level of significance* $\alpha = 0.05$

3. *Criterion*   Reject the null hypothesis if $\chi^2 \geq 9.488$, the value of $\chi^2_{0.05}$ for $(3 - 1)(3 - 1) = 4$ degrees of freedom, where

$$\chi^2 = \sum \frac{(o - e)^2}{e}$$

Otherwise, accept it or reserve judgment.

4. *Calculations*   The expected frequencies for the first two cells of the first row are

$$\frac{132 \cdot 80}{300} = 35.2 \quad \text{and} \quad \frac{132 \cdot 120}{300} = 52.8$$

those for the first two cells of the second row are

$$\frac{110 \cdot 80}{300} \approx 29.3 \quad \text{and} \quad \frac{110 \cdot 120}{300} = 44.0$$

and, by subtraction, those for the third cells of the first two rows are 44.0 and 36.7; and those for the third row are 15.5, 23.2, and 19.3. Thus, substitution into the formula for χ^2 yields

$$\chi^2 = \frac{(41 - 35.2)^2}{35.2} + \frac{(49 - 52.8)^2}{52.8} + \frac{(42 - 44.0)^2}{44.0}$$

$$+ \frac{(27 - 29.3)^2}{29.3} + \frac{(50 - 44.0)^2}{44.0} + \frac{(33 - 36.7)^2}{36.7}$$

$$+ \frac{(12 - 15.5)^2}{15.5} + \frac{(21 - 23.2)^2}{23.2} + \frac{(25 - 19.3)^2}{19.3}$$

$$\approx 5.37$$

5. *Decision*   Since $\chi^2 = 5.37$ does not exceed 9.488, the null hypothesis cannot be rejected; there is no real evidence that the probabilities of the three alternatives are not the same for the three kinds of persons.

11.39   1. *Hypotheses* H_0: The die is balanced.

H_A: The die is not balanced.

2. *Level of significance* $\alpha = 0.05$

3. *Criterion*   Reject the null hypothesis if $x^2 \geq 11.070$, the value of $\chi^2_{0.05}$ for $6 - 1 = 5$ degrees of freedom, where

$$x^2 = \sum \frac{(o - e)^2}{e}$$

Otherwise, accept it or reserve judgment.

4. *Calculations*   The expected frequencies are all $\frac{360}{6} = 60$, so that

$$x^2 = \frac{(57 - 60)^2}{60} + \frac{(46 - 60)^2}{60} + \frac{(68 - 60)^2}{60}$$

$$+ \frac{(52 - 60)^2}{60} + \frac{(72 - 60)^2}{60} + \frac{(65 - 60)^2}{60}$$

$$\approx 8.37$$

5. *Decision*   Since $x^2 = 8.37$ does not exceed 11.070, the null hypothesis cannot be rejected; we conclude that there is no real evidence that the die is not balanced.

11.44   (a)   In standard units, the class boundaries are $z = \dfrac{69.5 - 80}{5} = -2.10,$

$z = \dfrac{74.5 - 80}{5} = -1.10, \quad z = \dfrac{79.5 - 80}{5} = -0.10, \quad z = \dfrac{84.5 - 80}{5} = 0.90,$

$z = \dfrac{89.5 - 80}{5} = 1.90,$ and $z = \dfrac{94.5 - 80}{5} = 2.90;$ the corresponding

entries in Table II are 0.4821, 0.3643, 0.0398, 0.3159, 0.4713, and 0.4891 so that the probabilities are $0.5000 - 0.4821 = 0.0179$, $0.4821 - 0.3643 = 0.1178$, $0.3643 - 0.0398 = 0.3245$, $0.0398 + 0.3159 = 0.3557$, $0.4713 - 0.3159 = 0.1554$, $0.4981 - 0.4713 = 0.0268$, and $0.5000 - 0.4981 = 0.0019$.

(b)   The expected normal curve frequencies are $100(0.0179) \approx 1.8$, $100(0.1178) \approx 11.8$, $100(0.3245) \approx 32.4$, $100(0.3557) \approx 35.6$, $100(0.1554) \approx 15.5$, $100(0.0268) \approx 2.7$, and $100(0.0019) \approx 0.2$. Thus, we have

o		*e*	
1	} 11	1.8	} 13.6
10		11.8	
37		32.4	
36		35.6	
13		15.5	
2	} 16	2.7	} 18.4
1		0.2	

where we combined some of the classes so that none of the expected frequencies is less than 5.

(c)   1.   *Hypotheses* H_0:   Data comes from a normal population.
            H_A: Data does not come from a normal population.

2.   *Level of significance* $\alpha = 0.05$

3.   *Criterion*   Reject the null hypothesis if $x^2 \geq 3.841$, the value of $\chi^2_{0.05}$ for $4 - 2 - 1 = 1$ degree of freedom, where

$$x^2 = \sum \frac{(o - e)^2}{e}$$

Otherwise, accept it or reserve judgment.

4. *Calculations*   Substituting into the formula for χ^2, we get

$$\chi^2 = \frac{(11 - 13.6)^2}{13.6} + \frac{(37 - 32.4)^2}{32.4} + \frac{(36 - 35.6)^2}{35.6} + \frac{(16 - 18.4)^2}{18.4}$$

$$\approx 1.47$$

5. *Decision*   Since $\chi^2 = 1.47$ does not exceed 3.841, the null hypothesis cannot be rejected; there is no real evidence to doubt that the data comes from a normal population.

Review: Chapters 9, 10, & 11

Having read and studied these chapters and having worked a good portion of the exercises, you should be able to:

1. Distinguish between point estimates and interval estimates.
2. Explain the difference between *probability* and *confidence*.
3. Make confidence statements about the maximum error when $\bar{x}$ is used as an estimate of μ.
4. Determine the sample size needed so that a sample mean will have a desired precision.
5. Explain what is meant by *confidence interval* and *degree of confidence*.
6. Construct large-sample confidence intervals for μ.
7. Construct small-sample confidence intervals for μ.
*8. Construct large-sample confidence intervals for σ.
9. Explain what is meant by *statistical hypothesis*, *null hypothesis*, and *alternative hypothesis*.
10. Distinguish between Type I and Type II errors.
11. Explain what is meant by the *operating characteristic curve*.
12. Distinguish between simple and composite hypotheses about parameters.
13. Explain what is meant by *significance test*, *level of significance*, and *statistically significant*.
14. List the five steps of the outline for testing a statistical hypothesis.
15. Distinguish between one-sided and two-sided alternatives and between one-tailed and two-tailed tests.
16. Explain what is meant by *p*-value.
17. Relate the value of a test statistic to its *p*-value.
18. Perform large-sample tests of the null hypothesis $\mu = \mu_0$.
19. Perform small-sample tests of the null hypothesis $\mu = \mu_0$.
20. Perform large-sample tests of the null hypothesis that two populations have equal means.
21. Perform small-sample tests of the null hypothesis that two populations have equal means.
22. Test the significance of the difference between the means of paired data.
*23. Perform a one-way analysis of variance when the sample sizes are all equal.
*24. Perform a one-way analysis of variance when the sample sizes are not all equal.

25. Construct large-sample confidence intervals for population proportions.
26. Make confidence statements about the maximum error when $\dfrac{x}{n}$ is used as an estimate of p.
27. Determine the sample size needed so that a sample proportion will have a desired precision (with and without some information about p).
28. Perform small-sample tests of the null hypothesis $p = p_0$.
29. Perform large-sample tests of the null hypothesis $p = p_0$.
30. Test for the equality of two population proportions.
31. Test for the equality of k population proportions.
32. Analyze an $r \times c$ (contingency) table.
33. Test for goodness of fit.

Checklist of Key Terms (with page references to their definitions)

Alternative hypothesis, 314
Analysis of variance, 349
Analysis of variance table, 352
Cell, 374
Chi-square distribution, 374
Chi-square statistic, 374
Composite hypothesis, 319
Confidence interval, 296
Confidence limits, 296
Contingency table, 378
Count data, 365
Critical values, 326
Degree of confidence, 297
Degrees of freedom, 347, 374, 298
Enumeration data, 365
Error mean square, 351
Error sum of squares, 350
Expected cell frequencies, 378
Expected frequencies, 374, 378
Experimental error, 349
F distribution, 347
F statistic, 347
Goodness of fit, 307
Grand mean, 350
Hypothesis, 314

Independent random samples, 336
Interval estimate, 297
Level of significance, 321
Mean square for error, 351
Mean square for treatments, 351
Null hypothesis, 314
Number of degrees of freedom, 298, 374
Numerator and denominator degrees of freedom, 347
Observed cell frequencies, 378
Observed frequencies, 374
One-sample t-test, 330
One-sided alternative, 320
One-sided test, 322
One-tailed test, 322
One-way analysis of variance, 350
Operating characteristic curve, 318
Point estimate, 290
Pooled sample standard deviation, 338
Pooling, 347, 371
Problems of estimation, 289
p-value, 327
$r \times c$ table, 377

Sample proportion, 304
Significance test, 319
Simple hypothesis, 319
Small-sample confidence interval, 299
Standard error of a proportion, 305
Standard error of s, 303
Standard error of the difference between two means, 336
Standard error of the difference between two proportions, 371
Statistically significant, 319
t statistic distribution, 297
Test statistic, 322
Tests of hypotheses, 313
Total sum of squares, 350
Treatment mean square, 351
Treatment sum of squares, 350
Treatments, 350
Two-sample t-test, 339
Two-sided alternative, 320
Two-sided criterion, 322
Two-tailed test, 322
Two-way classification, 365
Type I error, 317
Type II error, 317

Review Exercises

R.119 To compare two kinds of baseball bats, 18 baseball players were asked to swing 20 times with each kind of bat at balls pitched by a machine, and the following are the respective numbers of balls that they hit more than 300 feet: 6 and 8, 9 and 5, 4 and 4, 7 and 6, 10 and 8, 5 and 6, 9 and 7, 3 and 4, 5 and 4, 6 and 6, 12 and 9, 8 and 9, 5 and 5, 4 and 6, 9 and 6, 10 and 8, 7 and 7, and 11 and 7. Test at the 0.05 level of significance whether the difference between the means of the numbers of balls hit more than 300 feet with the two kinds of bats is significant.

R.120 The supervisor of a truck weighing station intends to use the mean of a random sample of size $n = 81$ to estimate the average weight of tractor–trailer trucks that pass a weighing station on a certain highway. If, based on experience, the supervisor can assume that $\sigma = 3,900$ pounds, what can the supervisor assert with probability 0.99 about the maximum size of the error?

R.121 A sample check of 320 persons who had filled out a warranty card for a certain brand of television set revealed that 258 were willing to purchase another electronic appliance from the same manufacturer. Construct a 95% confidence interval for the proportion of all persons in this population who would be willing to purchase another electronic appliance from this manufacturer.

R.122 A political pollster wants to determine the proportion of the population that favors a regulatory change in the automobile insurance laws in a certain state. How large a sample will she need if she wants to be able to assert with probability 0.90 that the sample proportion will differ from the population proportion by at most 0.04?

R.123 A psychologist wants to use the mean of a random sample to estimate the average time it takes mature adults to assemble a simple jigsaw puzzle, and he wants to be in error by at most 30 seconds, with a probability of 0.95. If the psychologist knows from previous studies of a similar kind that it is reasonable to let $\sigma = 100$ seconds, how large a sample will the psychologist have to take?

R.124 In a random sample of 90 persons having dinner by themselves at a French restaurant, 63 had wine with their dinner. If we use the sample proportion $\frac{63}{90} = 0.70$ to estimate the corresponding true proportion, what can we say with 95% confidence about the maximum error?

R.125 The manager of a laundry takes a random sample of size $n = 64$ of the times it takes employees to iron a shirt and obtains a mean of 2.6 minutes with $s = 0.4$ minute. Construct a 95% confidence interval for the time it takes an employee of this laundry to iron a shirt.

R.126 Referring to Exercise R.125, let us reduce the sample size from 64 to 25, but let everything else remain the same. Construct a 95% confidence interval for the true mean time it takes an employee of this laundry to iron a shirt.

R.127 Again referring to Exercise R.125, where we were given $s = 0.40$ and used this value as a substitute for σ, construct a 95% confidence interval for the standard deviation of the population sampled.

R.128 Based on the results of $n = 12$ trials, we want to test the null hypothesis $p = 0.20$ against the alternative hypothesis $p > 0.20$. If we reject the null hypothesis when the number of successes is five or more and otherwise we accept it, find
  (a) the probability of a Type I error;
  (b) the probability of a Type II error when $p = 0.30$;
  (c) the probability of a Type II error when $p = 0.50$.

R.129 Five measurements of the tar content of a certain kind of cigarette yielded 14.5, 14.2, 14.4, 14.3, and 14.6 mg/cig (milligrams per cigarette). Show that the difference between the mean of this sample, $\bar{x} = 14.4$, and the average tar content claimed by the cigarette manufacturer, $\mu = 14.0$, is significant at the 0.05 level of significance.

R.130 Six guinea pigs injected with 0.5 mg of a medication took on the average 16.8 seconds to fall asleep with a standard deviation of 2.2 seconds, while six other guinea pigs injected with 1.5 mg of the medication took on the average 14.5 seconds to fall asleep with a standard deviation of 2.6 seconds. Use the 0.05

level of significance to test whether the increase in dosage decreases the mean time it takes a guinea pig to fall asleep.

*R.131 Given that 35 one-gallon cans of a certain kind of paint covered, on the average, 452.2 square feet with a standard deviation of 22.8 square feet, construct a 95% confidence interval for σ.

R.132 Suppose that we want to estimate what proportion of women in the United States, 18 years old or older, are married, and we want to be 99% sure that the error of our estimate will not exceed 0.03. How large a sample will we need
 (a) if the Bureau of the Census data indicates that the proportion is in the interval from 0.55 to 0.60;
 (b) if we have no idea what the proportion might be?

R.133 The following table shows how many times, Monday through Friday, a bus was late arriving at a given stop in 40 weeks:

Number of times bus was late	Number of weeks
0	4
1	11
2	15
3 or more	10

Use the 0.05 level of significance to test the null hypothesis that the bus is late 30% of the time, namely, the null hypothesis that the data constitutes a random sample from a binomial population with $n = 5$ and $p = 0.30$.

R.134 In a study of the relationship between family size and school performance in junior high school, 45 children in "only child" families had an average GPA (grade point average) of 2.82 with a standard deviation of 0.34, and 60 first-born children in two-child families had an average GPA of 2.96 with a standard deviation of 0.38. At the 0.05 level of significance, is the difference between these means significant?

R.135 The following table shows how samples of the residents of three federally financed housing projects replied to the question whether they would continue to live there if they had the choice:

	Project 1	Project 2	Project 3
Yes	83	68	65
No	17	32	35

Test at the 0.05 level of significance whether the differences among the proportions of "yes" answers can be attributed to chance.

R.136 A supermarket chain is considering replacing its deposit-bottle redemption machines with a new model. If μ_0 is the average number of bottles that the old machines redeem between repairs, against what alternative should it test the null hypothesis $\mu = \mu_0$ if
 (a) it does not want to replace the old machines unless the new machines prove to be superior;
 (b) it wants to change to the new machines unless they actually turn out to be inferior?

R.137 An executive for a chain of video rental stores would like to take a sample of 50 of his stores to estimate the average number of videos rented per day. If it can be assumed that $\sigma = 40$ videos, what will he be able to assert with 99% confidence about the maximum error?

R.138 A vendor of abrasive grinding wheels suspects that the manufacturer's claimed life of $\mu = 1,000$ operating hours for these wheels is too high. To check the claim, the vendor tests 50 of these grinding wheels and gets a mean lifetime of $\bar{x} = 985$ hours and a standard deviation of $s = 53$ hours. What can the vendor conclude at the 0.05 level of significance if he tests the null hypothesis $\mu = 1,000$ hours against an appropriate alternative?

R.139 Referring to Exercise R.138, change the level of significance to $\alpha = 0.01$, and let the remaining information be unchanged. What can the vendor conclude at the 0.01 level of significance?

R.140 Referring again to Exercise R.138, change the sample size from $n = 50$ (large sample) to $n = 25$ (small sample). Let the remaining information be unchanged. What can the vendor conclude at the 0.05 level of significance?

R.141 Random samples of 12 members from each of two large congregations yielded the following results regarding their contributions in 1998, expressed as percentages of their annual incomes: $\bar{x}_1 = 4.83\%$, $\bar{x}_2 = 5.33\%$, $s_1 = 0.62\%$, and $s_2 = 0.56\%$. At the 0.01 level of significance, test whether the difference between these two sample means is significant.

R.142 A telephone sales associate for a mutual funds company claims that the mean length of telephone calls received is 5.5 minutes. If a random sample of 28 such calls had a mean of $\bar{x} = 5.1$ minutes and a standard deviation of $s = 3.2$ minutes, test the null hypothesis $\mu = 5.5$ minutes against the appropriate alternative hypothesis at the 0.05 level of significance.

R.143 Suppose that we want to test the hypothesis that solar heating unit A is more efficient than solar heating unit B. Explain under what conditions we would be committing a Type I error and under what conditions we would be committing a Type II error.

R.144 A random sample of eight daily scrap records (where scrap is expressed as a percentage of material requisitioned) yielded $\bar{x} = 5.2\%$ scrap and $s = 0.8\%$ scrap. If the mean of this sample is used to estimate the mean of the population sampled, what can we assert with 95% confidence about the maximum error?

R.145 To determine whether the students who were randomly assigned to Professor Jones's statistics class learned more slowly than students who were assigned to Professor Smith's statistics class, the department chairperson gave an examination to 10 students taught by Professor Jones and 10 taught by Professor Smith. If the students taught by Professor Jones had a mean score of 82.1 points with a standard deviation of 3.4 points, while those taught by Professor Smith had a mean score of 85.4 points with a standard deviation of 3.3 points, use a one-sided test at the level of significance 0.05 to test whether the difference in the mean scores is significant.

R.146 During the investigation of an alleged unfair trade practice, the Federal Trade Commission takes a random sample of 49 "3-ounce" candy bars from a large shipment. If the mean and the standard deviation of their weights are, respectively, 2.94 ounces and 0.12 ounce, show that, at the level of significance 0.01, the commission has grounds upon which to proceed against the manufacturer on the unfair practice of short-weight selling.

R.147 As part of an industrial training program, some trainees are instructed by method A, which is straight teaching-machine instruction, and some are instructed by method B, which also involves the personal attention of an instructor. If random samples of size 10 are taken from large groups of trainees instructed by each of these two methods, and the scores which they obtained in an appropriate achievement test are

Method A: 71, 75, 65, 69, 73, 66, 68, 71, 74, 68
Method B: 72, 77, 84, 78, 69, 70, 77, 73, 65, 75

use the 0.05 level of significance to test whether method B is more effective.

∗R.148 Samples of peanut butter produced by three different manufacturers are tested for aflatoxin content (ppb), with the following results:

Brand A: 0.5, 6.3, 1.1, 2.7, 5.5, 4.3
Brand B: 2.5, 1.8, 3.6, 5.2, 1.2, 0.7
Brand C: 3.3, 1.5, 0.4, 4.8, 2.2, 1.0

Perform an analysis of variance to test at the 0.05 level of significance whether the differences among the three sample means are significant.

R.149 To determine the effect of an advertised week-long sale on the numbers of personal computers sold, a chain of computer stores compared the number of PCs sold during the week prior to and during the week of the advertised sale, with the following results:

(1) Store number	(2) Week before	(3) Week of sale	(4) Difference
1	24	29	5
2	54	62	8
3	32	34	2
4	43	43	0
5	86	95	9
6	19	16	-3
7	38	42	4
8	41	49	8
9	28	31	3

Use the 0.05 level to test whether the advertised sale was effective in increasing sales of PCs, given that $s = 4.00$ for the nine differences of column 4.

R.150 In a study conducted at a large airport, 81 of 300 persons who had just gotten off a plane and 32 of 200 persons who were about to board a plane admitted that they were afraid of flying. At the 0.05 level of significance, test whether the difference between the corresponding sample proportions is significant.

R.151 In a French restaurant, the chef receives 26, 21, 14, 22, 18, and 20 orders for coq au vin on six different nights. Construct a 95% confidence interval for the number of orders for coq au vin that the chef can expect per night.

R.152 A general achievement test is standardized so that eighth graders should average 77.2 with a standard deviation of 4.8. If 35 eighth graders from a certain school district average 79.5 on the test, does this substantiate the claim that the eighth graders in this district are above average? Use the level of significance $\alpha = 0.05$ and assume that the data constitutes a random sample.

*R.153 Following are the numbers of pieces produced by three factory workers (who are employed to do the same job) on four consecutive days. Use the method of Section 10.8 for differences among k means and the level of significance $\alpha = 0.05$ to determine whether the difference between these means is significant.

	Anne	Bob	Cindy
	25	28	26
	24	30	21
	22	25	23
	25	29	22
Means	24	28	23

R.154 A candidate for mayor of a large city claims that he will receive 55% of the votes against the alternative that he will receive less than 55%. What decision should we make at the level of significance $\alpha = 0.01$ if a random sample of 1,000 voters disclosed that 505 favor the candidate and 495 favor his opponent.

R.155 In 10 test runs, a car ran for 18, 17, 21, 16, 19, 16, 19, 18, 19, and 17 miles with a gallon of a certain kind of gasoline. Construct a 95% confidence interval for the average number of miles that the car will get with this gasoline.

12 Regression and Correlation

The goal of many statistical investigations is to establish relationships that make it possible to predict one variable in terms of another. For example, we might want to predict the sales of a new product in terms of the amount of money spent advertising it on television, predict family expenditures on entertainment in terms of family income, or predict a college student's grade-point average based on the number of hours he or she spent studying.

It would be ideal if we could predict one quantity exactly in terms of another, but this is seldom, if ever, possible. In most instances we can predict only averages or expected values. For instance, we cannot predict exactly how much money a specific college graduate will earn 10 years after graduation, but given suitable data we can predict the average earnings of all college graduates 10 years after graduation. Similarly, we can predict the average yield of a variety of wheat in terms of the total rainfall in July, and we can predict the expected grade-point average of a student starting law school in terms of his or her IQ. This problem of predicting the average value of one variable in terms of the known value of another variable is called the problem of **regression**.

Since the relationship between the average values of one variable and the known values of another variable can be strong, weak, or even nonexistent, we must concern ourselves also with measuring the strength of such a relationship, that is, the strength of the correlation. Whereas problems of regression are taken up in Sections 12.1 through 12.3, the subject of **correlation** is treated in Sections 12.4 through 12.6.

12.1
Curve Fitting

Whenever possible, we try to express, or approximate, relationships between known quantities and quantities that are to be predicted in terms of mathematical equations. This has been very successful in the natural sciences, where it is known, for instance, that at a constant temperature the relationship between the volume, y, and the pressure, x, of a gas is given by the formula

$$y = \frac{k}{x}$$

where k is a numerical constant. Also, it has been shown that the relationship between the size of a culture of bacteria, y, and the length of time, x, it has been exposed to certain environmental conditions is given by the formula

$$y = a \cdot b^x$$

where a and b are numerical constants. More recently, equations like these have also been used to describe relationships in the behavioral sciences, the social sciences, and other fields. For instance, the first of the equations above is often used in economics to describe the relationship between price and demand, and the second has been used to describe the growth of one's vocabulary or the accumulation of wealth.

Whenever we use observed data to arrive at a mathematical equation that describes the relationship between two variables—a procedure known as **curve fitting**—we must face three kinds of problems: (1) we must decide what kind of an equation we want to use (for instance, that of a straight line or that of some other curve); (2) we must find the particular equation that is "best" in some sense; and (3) we must investigate certain questions regarding the merits of the particular equation and of the predictions made from it. The second of these problems will be discussed in some detail in Section 12.2 and the third in Section 12.3.

The first kind of problem is sometimes decided by theoretical considerations, but more often by direct inspection of the data. We plot the data on graph paper, sometimes on special graph paper with special scales, and we judge visually what kind of curve best describes their overall pattern. So far as the work in this book is concerned, we shall consider only **linear equations in two unknowns**, which are of the form

$$y = a + bx$$

where a is the **y-intercept** (the value of y for $x = 0$) and b is the **slope of the line** (the change in y that accompanies an increase of one unit in x). Ordinarily, the values of a and b are estimated from the data (by the method we shall discuss in Section 12.2), and once they have been determined we can

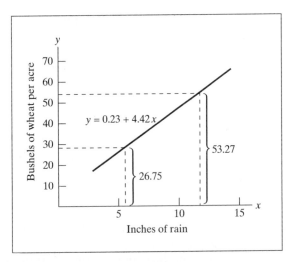

FIGURE 12.1 Graph of a linear equation.

substitute a value of x into the equation and calculate the corresponding predicted value of y.

The term *linear equation* arises from the fact that the graph of $y = a + bx$ is a straight line; that is, all pairs of values of x and y that satisfy an equation of the form $y = a + bx$ constitute points which fall on a straight line. Suppose, for instance, that

$$y = 0.23 + 4.42x$$

expresses the relationship between a certain county's September-through-August rainfall in inches, x, and its yield of wheat in bushels per acre, y. For example, for an annual rainfall of 6 inches, $x = 6$ and $y = 0.23 + 4.42(6) = 26.75$ bushels of wheat per acre, and for an annual rainfall of 12 inches, $x = 12$ and $y = 0.23 + 4.42(12) = 53.27$ bushels of wheat per acre. Observe that the points $(6, 26.75)$ and $(12, 53.27)$ lie on the straight line shown in Figure 12.1, and this is true for all other points obtained in the same way.

Since we are limiting our discussion here to linear equations, let us point out that **linear equations are useful and important not only because many relationships are actually of this form, but also because they often provide close approximations to relationships that would otherwise be difficult to describe in mathematical terms**. The problem of describing data through other kinds of curves, that is, the problem of *nonlinear curve fitting*, is discussed in the more theoretical texts listed in the bibliography at the end of the book.

12.2
The Method of
Least Squares

There are many ways in which we can fit a straight line to a set of data points, that is, to a set of paired data. Once we have decided to fit a straight line to a given set of paired data, we then face the problem of finding the equation of the particular line that in some sense provides the best possible fit. To show what is involved, let us consider the following sample data obtained in a study of the relationship between the number of years that applicants for certain foreign service jobs have studied German in high school or college and the grades that they received on a proficiency test in that language.

Number of years	Grade in test
x	*y*
3	57
4	78
4	72
2	58
5	89
3	63
4	73
5	84
3	75
2	48

If we plot the points that correspond to these 10 pairs of values, as in Figure 12.2, we observe that, even though the points do not all fall on a straight line, the overall pattern of the relationship is reasonably well described by the blue line. Thus, we feel justified in deciding that a straight line is a plausible description of the underlying relationship.

This takes care of the first kind of problem, and we must now consider the second kind, that of finding the equation of the line that in some sense provides the best fit to the data and that, it is hoped, will later yield the best possible predictions of *y* for given values of *x*. Logically, there is no limit to the number of straight lines that we can draw on a piece of paper, say, on the diagram of Figure 12.2. Some of them would fit the data so poorly that we

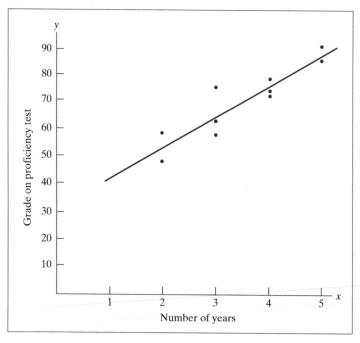

FIGURE 12.2 Data on number of years applicants studied German and their grades on a proficiency test.

cannot consider them seriously, but many others would come fairly close; and so the problem is to find the one line that fits the data *best* in some well-defined sense. If all the points actually fell on a straight line, there would be no problem, but this is an extreme case, which we rarely encounter in practice. In general, we have to be satisfied with a line having certain desirable properties, short of perfection.

The criterion that, today, is used almost exclusively for defining a best fit dates back to the early part of the nineteenth century and the work of the French mathematician Adrien Legendre. It is called the **method of least squares**, and it requires that the sum of the squares of the vertical deviations (distances) from the points to the line be as small as possible.

To explain why we do this, let us refer to Figure 12.3, where we plotted the **data points** corresponding to the data

x	y
1	6
5	5
9	7

and more or less arbitrarily *fit* a straight line; actually, we drew the line through the points $(1, 8)$ and $(9, 4)$.

Had we used this line to predict the values of y for the given values of x, that is, if we had marked $x = 1$, $x = 5$, and $x = 9$ on the horizontal scale and used the line to read off the corresponding values of y, we would have obtained 8, 6, and 4. The errors of these predictions are $6 - 8 = -2$, $5 - 6 = -1$, and $7 - 4 = 3$, and in Figure 12.3 they are the vertical distances from the points to the line.

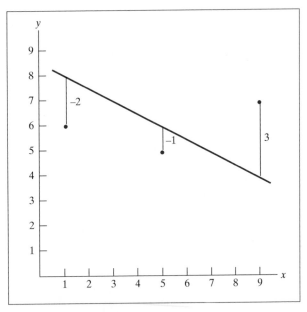

FIGURE 12.3 Errors made by using a given line to "predict" the
y-values for $x = 1$, $x = 5$, and $x = 9$.

The sum of these errors is $(-2) + (-1) + 3 = 0$, but this is not indicative of their size, and we find ourselves in a position similar to that on page 68, which led to the definition of the standard deviation. Squaring the errors as we squared the deviation from the mean on page 70, we find that the sum of the squares of the errors is $(-2)^2 + (-1)^2 + 3^2 = 14$.

Now let us fit another line to the same data points as before. The one shown in Figure 12.4 actually passes through the points $(1, 5)$ and $(9, 6)$, and judging by eye, it seems to provide a much better fit than the line of Figure 12.3. As can easily be verified, the errors of the predictions, the differences between the observed values of y and the corresponding values read off the line, are now, $6 - 5 = 1, 5 - 5.5 = -0.5$, and $7 - 6 = 1$.

The sum of these errors is $1 + (-0.5) + 1 = 1.5$, which is greater than the sum that we obtained for the errors made with the line of Figure 12.3, but this is not indicative of their size. On the other hand, the sum of the squares of the errors is now $1^2 + (-0.5)^2 + 1^2 = 2.25$, and this is much less than the 14 we obtained for the line of Figure 12.3. Indeed, this is indicative of the fact that the line of Figure 12.4 provides a better fit than the line of Figure 12.3, and we define the *best-fitting* line as the one for which the sum of the squares of the errors is as small as possible. Such a line is called a **least-squares line**.

To show how a least-squares line is actually fit to a set of paired data, let us consider n pairs of numbers $(x_1, y_1), (x_2, y_2), \ldots$, and (x_n, y_n), which might represent the heights and weights of n persons, the heights of n mothers and sons, the monthly food expenditures and mortgage payments of n families, and so on. If we write the equation of the line as $\hat{y} = a + bx$, where the symbol $\hat{y}$ is used to distinguish between the observed values of y and the corresponding values $\hat{y}$ on the line, the method of least squares requires that

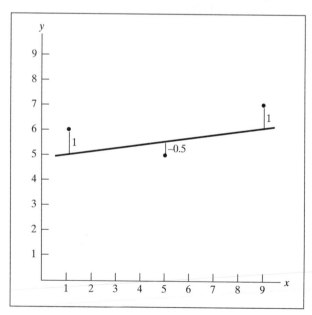

FIGURE 12.4 Errors made by using a second line to predict the
y-values for $x = 1, x = 5$, and $x = 9$.

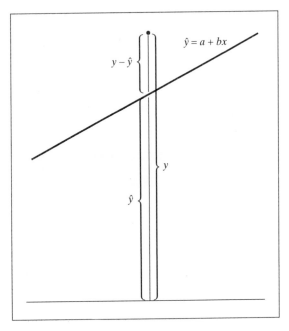

FIGURE 12.5 Least-squares line.

we minimize the sum of the squares of the differences between the y's and the $\hat{y}$'s (see Figure 12.5). This means that we must find the numerical values of the constants a and b appearing in the equation $\hat{y} = a + bx$ for which

$$\sum (y - \hat{y})^2 = \sum [y - (a + bx)]^2$$

is as small as possible. As it takes calculus or fairly tedious algebra to find the expressions for a and b that minimize this sum, let us merely state the result that they are given by the solutions for a and b of the following system of linear equations:

Normal equations

$$\sum y = na + b\left(\sum x\right)$$
$$\sum xy = a\left(\sum x\right) + b\left(\sum x^2\right)$$

In these equations, called the **normal equations**, n is the number of pairs of observations, $\sum x$ and $\sum y$ are the sums of the observed x's and y's, $\sum x^2$ is the sum of the squares of the x's, and $\sum xy$ is the sum of the products obtained by multiplying each x by the corresponding y. Since there are easier formulas for computational purposes, we generally do not fit a line numerically using these formulas; the following example illustrates nonetheless how it might be done.

■ **EXAMPLE**   Fit a least-squares line to the data pertaining to the numbers of years that certain applicants for foreign service jobs have studied German in high school or college and the grades that they received in a proficiency test in that language.

Solution   The sums needed for substitution into the normal equations are obtained by performing the calculations shown in the following table:

NUMBER OF YEARS x	TEST GRADE y	x^2	xy
3	57	9	171
4	78	16	312
4	72	16	288
2	58	4	116
5	89	25	445
3	63	9	189
4	73	16	292
5	84	25	420
3	75	9	225
2	48	4	96
Total 35	697	133	2,554

(There are many handheld calculators on which the various sums can be accumulated directly, so that there is no need to fill in all the details. Indeed, on some calculators the values of a and b can be obtained directly by recording the data and then pressing the appropriate buttons.) Substituting $\sum x = 35$, $\sum y = 697$, $\sum x^2 = 133$, $\sum xy = 2,554$, and $n = 10$ into the two normal equations, we get

$$697 = 10a + 35b$$

$$2,554 = 35a + 133b$$

and we must now solve these two simultaneous linear equations for a and b. There are several ways in which this can be done; the simplest, perhaps, is the **method of elimination**, which the reader may recall from elementary algebra. Using this method, let us eliminate a by multiplying the expressions on both sides of the first normal equation by 7, multiplying the expressions on both sides of the second normal equation by 2, and then "subtracting equals from equals." We thus get

$$4,879 = 70a + 245b$$

$$5,108 = 70a + 266b$$

and, by subtraction, $229 = 21b$. Thus, $b = \frac{229}{21} \approx 10.90$; and if we substitute this value into the first of the two original equations, we get $697 = 10a + 35(10.90)$,

$697 = 10a + 381.5$, $10a = 315.5$, and finally $a = \dfrac{315.5}{10} = 31.55$. Thus, the equation of the least-squares line is

$$\hat{y} = 31.55 + 10.90x$$ ■

As an alternative to this procedure, we will present a number of formulas that will be extremely useful in later calculations, which we will make in connection with least-squares problems.

Useful formulas for least-squares calculations

$$\bar{x} = \frac{\sum x}{n} \qquad \bar{y} = \frac{\sum y}{n}$$

$$S_{xx} = \sum x^2 - \frac{(\sum x)^2}{n} \qquad S_{yy} = \sum y^2 - \frac{(\sum y)^2}{n}$$

$$S_{xy} = \sum xy - \frac{(\sum x) \cdot (\sum y)}{n}$$

The formulas for $\bar{x}$ and $\bar{y}$ are familiar. The new formulas S_{xx}, S_{xy}, and S_{yy} will appear in many of the formulas that follow. They give us an easy numerical technique for solving the normal equations.

Solutions of normal equations

$$b = \frac{S_{xy}}{S_{xx}} \qquad a = \bar{y} - b\bar{x}$$

The computational process for finding the least-squares estimates (the solutions of the normal equations) involves several steps. First, find the values of the five sums

$$\sum x, \quad \sum y, \quad \sum x^2, \quad \sum y^2, \quad \text{and} \quad \sum xy$$

Then find the five quantities $\bar{x}$, $\bar{y}$, S_{xx}, S_{xy}, and S_{yy}. Finally, use the formulas above to find b and a. Note that you must find b first, since a is calculated in terms of b. The quantity S_{yy} has not been used yet, but it will appear in Section 12.3.

■ **EXAMPLE**   Rework the previous example, using the formulas given above.

Solution   In the table on the preceding page, we found that

$$\sum x = 35, \quad \sum y = 697, \quad \sum x^2 = 133, \quad \text{and} \quad \sum xy = 2{,}554$$

(We do not yet need S_{yy} in this problem, so we can complete the calculation of a and b without either S_{yy} or $\sum y^2$. For future reference, $\sum y^2 = 50{,}085$.)

We note that $n = 10$, so that

$$\bar{x} = \frac{35}{10} = 3.5 \quad \text{and} \quad \bar{y} = \frac{697}{10} = 69.7$$

Also, since

$$S_{xx} = 133 - \frac{(35)^2}{10} = 133 - \frac{1{,}225}{10} = 10.5$$

and

$$S_{xy} = 2{,}554 - \frac{35 \cdot 697}{10} = 2{,}554 - \frac{24{,}395}{10} = 114.5$$

it follows that

$$b = \frac{S_{xy}}{S_{xx}} = \frac{114.5}{10.5} \approx 10.90$$

and

$$a = \bar{y} - b\bar{x} = 69.7 - 10.90 \cdot 3.5 = 31.55$$ ∎

A computer printout of the preceding linear regression problem is shown in Figure 12.6. The values obtained for a and b are 31.533 and 10.905, rounded to three decimals, given in the column headed PARAMETER ESTIMATE. The differences between the results obtained above and in the printout are due to rounding.

There are relatively inexpensive handheld calculators that are programmed to solve linear regression problems. They yield the values of a and b after we enter the data and press the appropriate buttons. Since it is difficult to check for errors in entering the data and also difficult to correct errors, these calculators should be used only for very small problems.

Once we have determined the equation of a least-squares line, we can use it to make predictions.

■ **EXAMPLE**   Use the least-squares line $\hat{y} = 31.55 + 10.90x$ to predict the proficiency grade of an applicant who has studied German in high school or college for two years.

Solution   Substituting $x = 2$ into the equation, we get

$$\hat{y} = 31.55 + 10.90(2) = 53.35$$

and this is the best prediction we can make in the least-squares sense.   ∎

When we make a prediction like this, we cannot really expect that we will always hit the answer right on the nose; in fact, we cannot possibly be right when the answer has to be a whole number, as in our example, and the prediction is 53.35. With reference to this example, it would be very unreasonable to expect that every applicant who had studied German for a given

```
data A;
input x y @ @;
cards;
3 57 4 78 4 72 2 58 5 89 3 63 4 73 5 84 3 75 2 48
;
proc reg;
model y=x;
```

                                   SAS
DEP VARIABLE: Y
                          ANALYSIS OF VARIANCE

                      SUM OF   MEAN
     SOURCE   DF   SQUARES   SQUARE   F VALUE   PROB>F

     MODEL   1   1248.59524   1248.59524   39.094   0.0002
     ERROR   8   255.50476190   31.93809524
     C TOTAL   9   1504.10000

        ROOT MSE   5.65138   R-SQUARE   0.8301
        DEP MEAN   69.7   ADJ R-SQ   0.8089
        C.V.   8.108149

                           PARAMETER ESTIMATES

                     PARAMETER   STANDARD   T FOR HO:
     VARIABLE   DF   ESTIMATE   ERROR   PARAMETER=0   PROB >|T|

     INTERCEP   1   31.53333333   6.36041828   4.958   0.0011
     X   1   10.90476190   1.74405371   6.253   0.0002

FIGURE 12.6 Computer printout for linear regression.

number of years would get the same grade in the proficiency test; indeed, the data on page 406 shows that this is not the case. Thus, to make meaningful predictions based on least-squares lines, we must look at such predictions as averages, or as mathematical expectations. Interpreted in this way, we refer to least-squares lines that enable us to read off, or calculate, expected values of y for given values of x as **regression lines**. Better yet, we refer to such lines as **estimated regression lines**, since the values of a and b are determined on the basis of sample data. Questions relating to the quality of these estimates will be discussed in Section 12.3.

Exercises

Exercise 12.1 *is a practice exercise; its complete solution is given on pages* 431 *and* 432.

12.1 The following table shows the length of time that six persons have been working at an automobile inspection station and the number of cars each of them checked between noon and 1 o'clock on a given day:

Number of weeks employed	Cars checked
x	y
5	16
1	15
7	19
9	23
2	14
12	21

(a) Set up the normal equations and solve them by the method of elimination to find the equation of the least-squares line that will enable us to predict y in terms of x.

(b) Use the formulas on page 411 to check the values obtained for a and b in part (a).

(c) If a person has worked at the inspection station for 10 weeks, how many cars can we expect him or her to inspect during the given time period?

12.2 The following data pertains to the chlorine residual in a swimming pool at various times after it has been treated with chemicals:

Number of hours	Chlorine residual (parts per million)	x^2	xy
x	y		
2	1.8	4	3.8
4	1.5	16	5.5
6	1.4	36	7.4
8	1.1	64	9.1
10	0.9	100	10.4
30	6.7	220	36.7

$N = 5$

(a) Use the formulas for a and b on page 411 to find the equation of the least-squares line from which we can predict the chlorine residual in terms of the time that has elapsed since the pool was treated with chemicals.

(b) Use the equation of the least-squares line to estimate the chlorine residual in the pool five hours after it has been treated with chemicals.

12.3 The following sample data shows the demand for a product (in thousands of units) and its price (in cents) in six different market areas:

Price	Demand
19	55
23	7
21	20
15	123
16	88
18	76

(a) Fit a least-squares line from which we can predict the demand for the product in terms of its price.

(b) Estimate the demand for the product when it is priced at 20 cents.

12.4 Verify that the linear equation of the illustration on pages 404 and 405 can be obtained by fitting a least-squares line to the following data:

Rainfall (inches)	Yields of wheat (bushels per acre)
12.9	62.5
7.2	28.7
11.3	52.2
18.6	80.6
8.8	41.6
10.3	44.5
15.9	71.3
13.1	54.4

12.5 The following table shows 10 years' data on a local newspaper's annual advertising volume and its annual profit before taxes:

Annual advertising volume (1,000,000 lines)	Profit before taxes (thousands of dollars)
87.4	31,338
77.0	19,745
74.0	13,703
79.3	13,004
79.7	22,887
74.9	10,966
69.0	4,834
72.3	10,467
81.6	17,685
84.5	27,445

(a) Find the equation of the least-squares line that will enable us to predict the annual profit of the newspaper from its annual advertising volume.
(b) Predict the annual profit of the newspaper for a year in which its advertising volume is 80,500,000 lines.

12.6 The following are the numbers of hours a runner has run during each of eight weeks and the corresponding times in which she ran a mile at the end of the week:

Number of hours run	Time of mile (minutes)
13	5.2
15	5.1
18	4.9
20	4.6
19	4.7
17	4.8
21	4.6
16	4.9

(a) Find the equation of the least-squares line that will allow us to predict the runner's time for the mile from the number of hours that she ran that week.
(b) Use the equation obtained in part (a) to predict how fast she will run a mile at the end of a week in which she ran for 18 hours.

12.7 During its first five years of operation, a company's gross income from sales was 1.2, 1.9, 2.4, 3.3, and 3.5 million dollars. Fit a least-squares line and, assuming that the trend continues, predict this company's gross income from sales during its sixth year of operation.

12.8 The following is data on the IQs of 25 students, the numbers of hours they studied for a certain achievement test, and their scores on the test:

IQ	Number of hours studied	Score on test
105	8	80
98	6	62
112	10	91
102	9	77
107	14	89
95	6	65
100	18	96
110	10	85
102	15	94
96	24	91
115	8	88
105	11	85
135	9	99
92	12	76
90	18	83
109	3	70
94	5	61
114	10	86
106	15	93
121	4	92
99	16	82
103	21	95
114	15	90
107	8	79
108	12	88

Use a computer package or a graphing calculator to find the least-squares line that will enable us to predict a student's score on the test in terms of his or her IQ. These data will be used also in Exercises 12.9, 12.10, 12.19 through 12.22, 12.39, and 12.40.

12.9 With reference to the data of Exercise 12.8, use a computer package or a graphing calculator to find the least-squares line that will enable us to predict a student's score on the test in terms of the number of hours that he or she studied for the test.

12.10 With reference to the data of Exercise 12.8, use a computer package or a graphing calculator to find the least-squares line that will enable us to predict how many hours a student will study for the test, given his or her IQ.

12.3 Regression Analysis*

In the preceding section we used a least-squares line to predict that an applicant who has studied German in high school or college for two years will score 53.35 on the proficiency test; but even if we interpret the line correctly as a regression line (that is, treat predictions made from it as averages or expected values), several questions remain to be answered.

How good are the values we found for the constants a and b in the equation ŷ = a + bx? After all, a = 31.55 and b = 10.90 are only estimates based on sample data, and if we base our work on a sample of ten other applicants for the foreign service jobs, the method of least squares would almost certainly lead to different values of a and b.

How good an estimate is 53.35 of the average score of applicants who have had two years of German in high school or college?

In the first of these questions, we said that $a = 31.55$ and $b = 10.90$ are "only estimates based on sample data." This implies the existence of corresponding true values, usually denoted by α and β and referred to as **regression coefficients**. The true regression line is $y = \alpha + \beta x$. To distinguish between a and α and between b and β, we refer to a and b as the **estimated regression coefficients**, and we use the phrase **population regression coefficients** in discussing α and β.

To clarify the idea of a true regression line $y = \alpha + \beta x$, consider Figure 12.7, which shows the distributions of y for several values of x. In connection with our example, these curves represent the distributions of the proficiency scores of persons who have had, respectively, one, two, or three years of German in high school or college. To complete the picture, we can visualize similar curves corresponding to all other values of x within the range of values under consideration.

In **linear regression analysis**, we assume that the x's are known constants, and that for each x the random variable y has a certain distribution (as pictured in Figure 12.7) with the mean $\alpha + \beta x$. In **normal regression analysis**, we assume, furthermore, that these distributions are all normal distributions

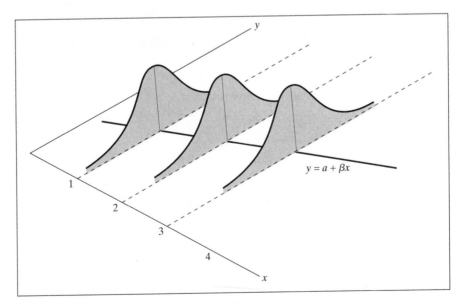

FIGURE 12.7 Distributions of *y* for given values of *x*.

with the same standard deviation σ. In other words, the distributions pictured in Figure 12.7, as well as those that we add mentally, are normal curves with means on the line $y = \alpha + \beta x$ and the same standard deviation σ.

Based on these assumptions, we can make all sorts of inferences about the regression coefficients α and β. They all require that we estimate σ, the common standard deviation of the normal distributions pictured in Figure 12.7. The estimate we shall use for this purpose is called the **standard error of estimate** and it is denoted by s_e. Its formula is

Standard error of estimate (defining formula)

$$s_e = \sqrt{\frac{\Sigma (y - \hat{y})^2}{n - 2}}$$

where, again, the y's are the observed values of y, and the $\hat{y}$'s are the corresponding values on the least-squares line. Observe that s_e^2 is the sum of the squares of the vertical deviations from the points to the line in Figure 12.5 (the quantity that we minimized by the method of least squares) divided by $n - 2$.

The formula above defines s_e, but in practice we calculate its value by means of the computing formula

Standard error of estimate (computing formula)

$$s_e = \sqrt{\frac{S_{yy} - \frac{(S_{xy})^2}{S_{xx}}}{n - 2}}$$

The symbols were defined on page 411.

■ **EXAMPLE**   Calculate s_e for the example on page 406, where we had

NUMBER OF YEARS x	TEST GRADE y
3	57
4	78
4	72
2	58
5	89
3	63
4	73
5	84
3	75
2	48
Total 35	697

Solution   In working with these numbers, we noted that $n = 10$, $\bar{x} = 3.5$, $\bar{y} = 69.7$, $S_{xx} = 10.5$, and $S_{xy} = 114.5$, but we still need the value of S_{yy}. Since we found that $\Sigma y^2 = 50{,}085$, it follows that

$$S_{yy} = 50{,}085 - \frac{(697)^2}{10} = 1{,}504.10$$

Substituting these values into the computing formula for s_e, we find that

$$s_e = \sqrt{\dfrac{1{,}504.10 - \dfrac{(114.5)^2}{10.5}}{10 - 2}} \approx 5.651 \qquad ■$$

Since α is just the y-intercept of the regression line (that is, the value of y that corresponds to $x = 0$), it is not of any special interest in most problems. We shall consider here only inferences about β. They will be based on the statistic

Statistic for inferences concerning β

$$t = \dfrac{b - \beta}{s_e} \sqrt{S_{xx}}$$

whose sampling distribution is the t distribution with $n - 2$ degrees of freedom.

■ **EXAMPLE**   Based on the data on page 418, test the hypothesis that each additional year of German in high school or college adds another 12.5 points to the expected proficiency score of an applicant. Use the alternative hypothesis $\beta \neq 12.5$ and the level of significance 0.05.

Solution   1. *Hypotheses* H_0: $\beta = 12.5$
H_A: $\beta \neq 12.5$

2. *Level of significance* $\alpha = 0.05$

3. *Criterion*   Reject the null hypothesis if $t \leq -2.306$ or $t \geq 2.306$, where 2.306 is the value of $t_{0.025}$ for $10 - 2 = 8$ degrees of freedom, and t is given by

$$t = \dfrac{b - \beta}{s_e} \sqrt{S_{xx}}$$

Otherwise, accept the null hypothesis or reserve judgment.

4. *Calculations*   Substituting $b = 10.90$, $\beta = 12.5$, $S_{xx} = 10.5$, and $s_e = 5.651$ into the formula for t, we get

$$t = \dfrac{10.90 - 12.5}{5.651} \sqrt{10.5} \approx -0.92$$

5. *Decision*   Since $t = -0.92$ falls on the interval from -2.306 to 2.306, the null hypothesis cannot be rejected; that is, the difference between $b = 10.90$ and $\beta = 12.5$ may be reasonably attributed to chance.   ■

In using the t statistic given above, the most common value tested, by far, is $\beta = 0$.

To construct confidence intervals for the regression coefficient β, we substitute for the middle term of $-t_{\alpha/2} < t < t_{\alpha/2}$ the preceding t statistic, and relatively simple algebra yields

Confidence limits for the regression coefficient β

$$b \pm t_{\alpha/2} \cdot \frac{s_e}{\sqrt{S_{xx}}}$$

The degree of confidence is $1 - \alpha$. The value of $t_{\alpha/2}$ for $n - 2$ degrees of freedom may be read from Table III. Note that the quantity

$$\frac{s_e}{\sqrt{S_{xx}}}$$

by which $t_{\alpha/2}$ is multiplied and by which $b - \beta$ is divided in the formula for the t statistic on page 419, is an estimate of the standard deviation of the sampling distribution of the least-squares estimate of β.

■ **EXAMPLE** Based on the data on page 418, construct a 99% confidence interval for β, the expected increase in the proficiency score of an applicant for each additional year of German in high school or college.

Solution Substituting $t_{0.005} = 3.355$ for $10 - 2 = 8$ degrees of freedom into the confidence-limits formula, as well as the previously determined values of b, s_e, and S_{xx}, we get

$$10.90 \pm 3.355 \cdot \frac{5.651}{\sqrt{10.5}}$$

or 10.90 ± 5.85. Thus, the 99% confidence interval for β is

$$5.05 < \beta < 16.75$$

This interval is rather wide, and this is due to two things—the magnitude of the variation as measured by s_e and the very small sample size n.   ■

Many computer packages have programs for fitting least-squares lines, and most of these programs also provide the answers for the two preceding examples, or at least they greatly facilitate the calculations. The computer printout shown in Figure 12.6 does not actually give the value of the t statistic for testing $\beta = 12.5$ or the confidence limits; but it tells us that s_e, referred to there as ROOT MSE, is 5.651, and in the column headed STANDARD ERROR in the row labeled X that the quantity

$$\frac{s_e}{\sqrt{S_{xx}}}$$

referred to above is 1.744. The computer printout agrees, within rounding, with the hypothesis test results on page 419 and the confidence interval above.

To answer the question asked on page 417 concerning the goodness of an estimate, or prediction, based on a least-squares equation, we use a method that is very similar to the one discussed on page 419. Basing our argument on another t statistic, we arrive at the following confidence limits for $\alpha + \beta x_0$, which is the mean of y when $x = x_0$:

Confidence limits for the mean of y when $x = x_0$

$$(a + bx_0) \pm t_{\alpha/2} \cdot s_e \sqrt{\frac{1}{n} + \frac{(x_0 - \bar{x})^2}{S_{xx}}}$$

The degree of confidence is $1 - \alpha$ and, as before, $t_{\alpha/2}$ for $n - 2$ degrees of freedom may be read from Table III.

■ **EXAMPLE**   Referring again to the data on page 418, suppose that the original purpose of the study was to estimate the average proficiency score of applicants who have had two years of German in high school or college. Construct a 99% confidence interval for this mean.

Solution   The value of $a + bx_0$ is $31.55 + 10.90(2) = 53.35$. Also, $s_e = 5.651$, $\bar{x} = 3.5$, $S_{xx} = 10.5$, $n = 10$, $x_0 = 2$, and 3.355 is the value of $t_{0.005}$ for $10 - 2 = 8$ degrees of freedom. We substitute these values into the confidence interval formula and get

$$53.35 \pm (3.355)(5.651)\sqrt{\frac{1}{10} + \frac{(2 - 3.5)^2}{10.5}}$$

or

$$53.35 \pm 10.63$$

Hence, the 99% confidence interval for the mean proficiency score of applicants who have had two years of German in high school or college is the interval from $53.35 - 10.63 = 42.72$ to $53.35 + 10.63 = 63.98$. ■

Exercises

Exercises 12.11 and 12.16 are practice exercises; their complete solutions are given on pages 432 and 433.

*12.11   With reference to Exercise 12.1, test the null hypothesis $\beta = 1.5$ (the hypothesis that each additional week on the job adds 1.5 to the number of cars a person can be expected to inspect in the given period of time) against the alternative hypothesis $\beta < 1.5$. Use the level of significance 0.05.

*12.12   The following table shows the assessed values and the selling prices of eight houses, constituting a random sample of all the houses sold recently in a given metropolitan area:

Assessed value (in $1,000)	Selling price (in $1,000)
40.3	63.4
72.0	118.3
32.5	55.2
44.8	74.0
27.9	48.8
51.6	81.1
80.4	123.2
58.0	92.5

Fit a least-squares line that will enable us to predict the selling prices of a house in terms of its assessed value, and test the null hypothesis $\beta = 1.30$ against the alternative hypothesis $\beta > 1.30$ at the level of significance 0.05.

*12.13 The following data shows the average numbers of hours that six students studied per week and their grade-point indexes:

Hours studied x	Grade-point index y
15	2.0
28	2.7
13	1.3
20	1.9
4	0.9
10	1.7

Test the null hypothesis $\beta = 0.10$ (the hypothesis that each additional hour of study per week will raise the expected grade-point index of the students by 0.10) against the alternative hypothesis $\beta < 0.10$ at the 0.01 level of significance.

*12.14 In a campus bar and grill, hamburger platters account for most of the food sales. The owner would like to evaluate how the price of these platters affects her weekly profit, so she experiments by varying the price during nine different weeks, with the following results:

Price	Profit
$2.00	$2,325
2.50	2,460
3.00	1,600
3.00	1,700
2.00	2,000
2.50	1,800
2.50	2,500
2.00	2,400
3.50	1,700

Fit a least-squares line to these data and construct a 95% confidence interval for the regression coefficient β.

*12.15 With reference to Exercise 12.3, construct a 99% confidence interval for the regression coefficient β.

 *12.16 With reference to Exercise 12.13, construct a 99% confidence interval for the mean grade-point index of students who study on the average 12 hours per week.

*12.17 With reference to Exercise 12.12, construct a 95% confidence interval for the average selling price of a house in the given metropolitan area that has an assessed value of $60,000.

*12.18 The following data shows the advertising expenses (expressed as a percentage of total expenses) and the net operating profit (expressed as a percentage of total sales) in a random sample of five furniture stores:

Advertising expenses x	Net operating profit y
1.1	2.5
2.8	3.7
3.1	5.2
1.6	2.9
0.7	1.4

Fit a least-squares line to these data and construct a 99% confidence interval for the mean net operating profit (expressed as a percentage of total sales) when the advertising expenses are 2.0% of total expenses.

*12.19 With reference to Exercise 12.8, use a computer package to test the null hypothesis $\beta = 0.50$ (that each unit increase in IQ adds on the average 0.50 to the score) against the alternative $\beta > 0.50$. Use the 0.05 level of significance.

*12.20 With reference to Exercise 12.8, use a computer package to construct a 95% confidence interval for β.

*12.21 With reference to Exercise 12.9, use a computer package to construct a 99% confidence interval for β (the average increase in the score for each additional hour studied).

*12.22 With reference to Exercise 12.10, use a computer package or a graphing calculator to test the null hypothesis $\beta = 0$ (that a student's IQ does not affect how many hours he or she will study for the test) against the alternative $\beta \neq 0$. Use the 0.05 level of significance.

12.4
The Coefficient of Correlation

Having learned how to fit a least-squares line to a set of paired data, let us now turn to the problem of determining how well such a line actually fits the data. Of course, we can get some idea by inspecting a diagram like that of Figure 12.2; but to show how we can be more objective, let us refer to the original data of the example dealing with the foreign service job applicants' proficiency grades in German and the number of years that they have studied the language in high school or college. Copying the data from page 406, we get

Number of years	Grade on test
x	y
3	57
4	78
4	72
2	58
5	89
3	63
4	73
5	84
3	75
2	48

and it can be seen that there are considerable differences among the y's; the smallest value is 48 and the largest value is 89. However, it can also be seen that the grade of 48 was obtained by an applicant who has studied German for two years, while the grade of 89 was obtained by an applicant who has studied German for five years. This suggests that the differences among the grades may well be due, at least in part, to the fact that the applicants have not all studied German for the same number of years. This raises the following question:

Of the total variation among the y's, how much can be attributed to chance and how much can be attributed to the relationship between the two variables x and y, that is, to the fact that the observed y's correspond to different values of x?

With reference to our example, we would thus want to know what part of the variation among the grades can be attributed to the differences in the number of years that the applicants have studied German, and what part can be attributed to all other factors (the applicants' intelligence, their health or frame of mind on the day they took the test, ..., and their luck in guessing at some of the answers), which we combine under the general heading of chance.

A convenient measure of the total variation of the observed y's is the quantity $S_{yy} = \Sigma(y - \bar{y})^2$, which is the **total sum of squares** of the y's and which is simply the variance of the y's multiplied by $n - 1$. We note that for the test grades above we had $S_{yy} = 1{,}504.10$. The calculation was done on page 418.

This takes care of the total variation of the y's, but it remains to be seen how we might measure its two parts, which are attributed, respectively, to the relationship between x and y and to chance. To this end, let us point out that, if the number of years the applicants have studied German is the only thing that affects their grades in the test and the relationship is linear, then all the data points would fall on a straight line. As is apparent from Figure 12.2, this is not the case in our example, and the extent to which the points fluctuate above and below the line provides us with an indication of the size of the chance variations. Thus, chance variation is measured by the sum of the squares of the vertical deviations from the points to the line, that is, by the quantity $\Sigma(y - \hat{y})^2$, which is called the **residual sum of squares**. This is the quantity that we minimized on page 409 in accordance with the least-squares criterion.

■ **EXAMPLE**   Calculate the residual sum of squares for the 10 test grades.

Solution   To calculate this sum of squares, we must first determine the value of $y = a + bx = 31.55 + 10.90x$ for each of the given values of x, and we get

$$31.55 + 10.90(3) = 64.25$$

$$31.55 + 10.90(4) = 75.15$$

$$\cdots\cdots\cdots\cdots\cdots\cdots\cdots$$

$$31.55 + 10.90(2) = 53.35$$

Then

$$\Sigma(y - \hat{y})^2 = (57 - 64.25)^2 + (78 - 75.15)^2 + \cdots + (48 - 53.35)^2$$

$$\approx 255.50$$ ■

The technique is painful when the sample size is large, and generally we make use of the identity

**Residual sum
of squares**

$$\sum (y - \hat{y})^2 = S_{yy} - \frac{(S_{xy})^2}{S_{xx}}$$

For the problem above, this yields

$$\sum (y - \hat{y})^2 = 1,504.10 - \frac{(114.5)^2}{10.5} \approx 255.50$$

From $S_{yy} = \sum (y - \bar{y})^2 = 1,504.10$ and $\sum (y - \hat{y})^2 = 255.50$, it follows that

$$\frac{\sum (y - \hat{y})^2}{\sum (y - \bar{y})^2} = \frac{255.50}{1,504.10} \approx 0.17$$

which is the proportion of the total variation of the grades that can be attributed to chance, while

$$1 - \frac{\sum (y - \hat{y})^2}{\sum (y - \bar{y})^2} = 1 - 0.17 = 0.83$$

is the proportion of the total variation of the grades that can be attributed to the relationship with x, that is, to the differences in the number of years that the applicants have studied German.

If we take the square root of the last proportion (the proportion of the total variation of the y's that can be attributed to the relationship with x), we obtain the statistical measure which is called the **coefficient of correlation**. It is denoted by the letter r, and its sign is chosen so that it is the same as that of the estimated regression coefficient b. Thus, for our example we get

$$r = \sqrt{0.83} \approx 0.91$$

It is of interest to note that all the quantities we have calculated so far in this section can be obtained from the computer printout of Figure 12.6. Under ANALYSIS OF VARIANCE, in the column headed SUM OF SQUARES, we find that the ERROR sum of squares is 255.50 and that the TOTAL sum of squares is 1,504.10. Also, R-SQUARE = 0.83 tells us that the square of the correlation coefficient is 0.83, and this also agrees with the result we obtained before.

It follows from the rule for the sign of r that the correlation coefficient is positive when the least-squares line has an upward slope, that is, when the relationship between x and y is such that small values of y tend to go with small values of x and large values of y tend to go with large values of x. Also, the correlation coefficient is negative when the least-squares line has a downward slope, that is, when large values of y tend to go with small values of x and small values of y tend to go with large values of x. Geometrically, examples of **positive** and **negative correlations** are shown in the first two diagrams of Figure 12.8.

Observe also that since r is defined as $\pm$ the square root of a proportion, its values must lie on the interval from -1 to $+1$. When r equals -1 or $+1$, this means that 100% of the variation among the y's can be attributed to the relationship with x; in other words, $\sum (y - \hat{y})^2 = 0$ and the points must all fall on

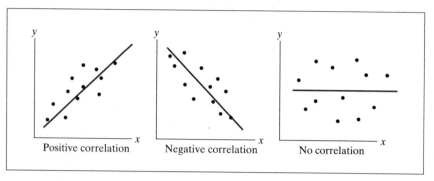

FIGURE 12.8 Types of correlation.

the least-squares line. When $r = 0$, this means that none of the variation among the y's can be attributed to the relationship with x, and we say that there is **no correlation**. This is pictured in the third diagram of Figure 12.8.

Our definition of r shows clearly the nature, or essence, of the coefficient of correlation, but in actual practice it is seldom used to determine its value. Instead, we use the computing formula

Computing formula for the coefficient of correlation

$$r = \frac{S_{xy}}{\sqrt{S_{xx} \cdot S_{yy}}}$$

which has the added advantage that it automatically gives r the correct sign. The quantities used in calculating r by this formula have been used previously and were defined on page 411.

■ **EXAMPLE**   Use the computing formula to verify the value $r = 0.91$, which we obtained for the data pertaining to the proficiency grades of 10 applicants for certain foreign service jobs and the number of years that they have studied German in high school or college.

Solution   We previously found $S_{xy} = 114.50$, $S_{xx} = 10.50$, and $S_{yy} = 1{,}504.10$. Substitution into the computing formula for r yields

$$r = \frac{114.50}{\sqrt{10.50 \cdot 1{,}504.10}} \approx 0.91$$

This agrees, as it should, with the result obtained on page 425.   ■

**12.5
The Interpretation
of r**

When r equals $+1, -1$, or 0, there is no question about the interpretation of the coefficient of correlation. As we have already indicated, it is $+1$ or -1 when all the points actually fall on a straight line, and it is 0 when none of the variation among the y's can be attributed to the relationship with x or, in other words, when knowledge of x does not help in the prediction of y. In general, the definition of r tells us that the proportion of the variation of the y's that is due to the relationship with x equals r^2, or that the percentage equals $(100\%) \cdot r^2$, and this is how we interpret the strength of the relationship implied by any value of r.

For instance, if $r = 0.80$ for one set of data and $r = 0.40$ for another, it would be very misleading to say that the correlation of 0.80 is "twice as good" or "twice as strong" as the correlation of 0.40. When $r = 0.80$, then $100\%(0.80)^2 = 64\%$ of the variation of the y's is accounted for by the relationship with x, and when $r = 0.40$, then $100\%(0.40)^2 = 16\%$ of the variation of the y's is accounted for by the relationship with x. Thus, in the sense of "percentage of variation accounted for," we can say that a correlation of 0.80 is *four times as strong* as a correlation of 0.40. By the same token, a correlation of 0.60 is *nine times as strong* as a correlation of 0.20.

There are several pitfalls in the interpretation of r. First, it is often overlooked that r measures only the strength of linear relationships; second, it should be remembered that a strong correlation (a value of r close to $+1$ or -1) does not necessarily imply a cause–effect relationship.

If r is calculated indiscriminantly, for instance, for the three sets of data in Figure 12.9, we get $r = 0.75$ in each case, but it is a meaningful measure of the strength of the relationship only in the first case. In the second case, there is a very strong relationship between the two variables, but it is not linear, and in the third case, six of the seven points actually fall on a straight line; but the seventh point is so far off that it suggests the possibility of a gross error in recording the data. Thus, before we calculate r, we should always plot the data to see whether there is reason to believe that the relationship is, in fact, linear.

The fallacy of interpreting a strong correlation as an indication of a cause–effect relationship is best explained with a few examples. A popular illustration is the strong positive correlation that has been obtained for the annual sales of chewing gum and the incidence of crime in the United States.

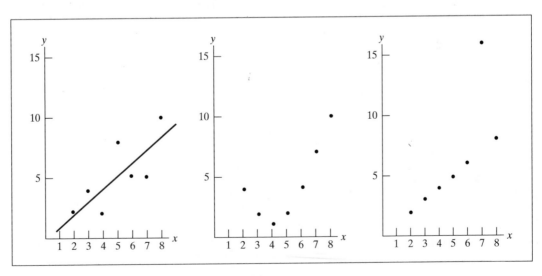

FIGURE 12.9 Three sets of paired data for which $r = 0.75$.

Obviously, one cannot conclude that the crime rate might be reduced by prohibiting the sale of chewing gum—both variables depend on the size of the population, and it is this mutual relationship with a third variable (population size) that produced the positive correlation. Another example is the strong positive correlation obtained for the number of storks seen nesting in English villages and the number of children born in the same villages. We leave it to the reader's ingenuity to explain why there might be a strong positive correlation in this case in the absence of any cause–effect relationship.

12.6 A Significance Test for r

It is sometimes overlooked that when r is calculated on the basis of sample data, we may get a fairly strong positive or negative correlation purely by chance, even though there is actually no relationship whatsoever between the two variables under consideration. Suppose, for instance, that we take a pair of dice, one red and one green, roll them a few times, and get the following results:

Red die	Green die
x	y
3	5
2	2
5	6
3	1
4	3

Calculating r for these data, we get the surprisingly high value $r = 0.66$, and this raises the question of whether anything is wrong with the assumption that there is no relationship here between x and y; after all, one die does not know what the other die is doing. To answer this question, we shall have to see whether this high value of r may be attributed to chance.

When a correlation coefficient is calculated from sample data, as in the example above, the value we obtain for r is only an estimate of a corresponding parameter, the **population correlation coefficient**, which we denote by the Greek letter ρ (lowercase *rho*). What r measures for a sample, ρ measures for a population.

To test the null hypothesis of no correlation, we shall make the same assumptions as in Section 12.3. When these assumptions are met, at least within a reasonable degree of approximation, we reject the null hypothesis of no correlation if $r \leq -r_{\alpha/2}$ or $r \geq r_{\alpha/2}$, where the value of $r_{\alpha/2}$ may be obtained from Table VI for $\alpha = 0.05$ and $\alpha = 0.01$. If the null hypothesis can be rejected, we say that there is a **significant correlation**; otherwise, we say that the value of r is not statistically significant.

■ **EXAMPLE**   Use the 0.05 level of significance to test the null hypothesis of no correlation for the example in which we rolled a pair of dice five times and obtained $r = 0.66$.

Solution   1. *Hypotheses* H_0: $\rho = 0$
H_A: $\rho \neq 0$

2. *Level of significance* $\alpha = 0.05$

3. *Criterion*   Reject the null hypothesis if $r \leq -0.878$ or $r \geq 0.878$, where 0.878 is the value of $r_{0.025}$ for $n = 5$; otherwise, state that the value of r is not significant.

4. *Calculations* $r = 0.66$

5. *Decision*   Since $r = 0.66$ falls on the interval from -0.878 to 0.878, the null hypothesis cannot be rejected; in other words, the correlation coefficient of 0.66 is not significant.   ■

■ **EXAMPLE**   Use the 0.01 level of significance to test the null hypothesis of no correlation for the example in which we obtained $r = 0.91$ for the proficiency scores of 10 applicants for certain foreign service jobs and the number of years that they have studied German in high school or college.

Solution   1. *Hypotheses* $H_0:\ \rho = 0$
$H_A:\ \rho \neq 0$

2. *Level of significance* $\alpha = 0.01$

3. *Criterion*   Reject the null hypothesis if $r \leq -0.765$ or $r \geq 0.765$, where 0.765 is the value of $r_{0.005}$ for $n = 10$; otherwise, state that the value of r is not significant.

4. *Calculations* $r = 0.91$

5. *Decision*   Since $r = 0.91$ exceeds 0.765, the null hypothesis must be rejected; in other words, there is a relationship between the two variables under consideration.   ■

Exercises

Exercises 12.23, 12.24, and 12.31 are practice exercises; their complete solutions are given on page 433.

12.23 The following data pertains to a study of the effects of environmental pollution on wildlife, in particular, the effect of DDT on the thickness of the eggshells of certain birds:

DDT residues in yolk lipids (parts per million)	Thickness of eggshell (millimeters)
117	0.49
65	0.52
393	0.37
98	0.53
122	0.49

Calculate r for these data.

12.24 With reference to Exercise 12.23, test the significance of the value obtained for r at the level of significance 0.01.

12.25 The following are the typing speeds and the reading speeds of nine secretaries:

Typing speed (words per minute)	Reading speed (words per minute)
60	370
56	551
52	528
63	348
70	645
58	454
44	503
79	618
62	500

Measure the extent of the relationship between typing speed and reading speed by calculating the coefficient of correlation.

12.26 With reference to Exercise 12.25, test the null hypothesis of no correlation at the 0.05 level of significance.

12.27 The following table shows the percentages of the vote predicted by a poll for seven candidates for the U.S. Senate in different states, x, and the corresponding percentages of the vote that they actually received, y:

Poll, x	Election, y
42	51
34	31
59	56
41	42
53	58
40	35
55	54

Calculate r for these data.

12.28 With reference to Exercise 12.27, test the null hypothesis $\rho = 0$ at the level of significance $\alpha = 0.05$.

12.29 Calculate r for the grades that 16 students received on final examinations in economics and anthropology.

Economics	Anthropology	Economics	Anthropology
51	74	45	68
68	70	73	87
72	88	93	89
97	93	66	73
55	67	20	33
73	73	91	91
95	99	74	80
74	73	80	86

12.30 With reference to Exercise 12.29, test the null hypothesis $\rho = 0$ at the level of significance $\alpha = 0.05$.

 12.31 With reference to Exercise 12.1, what percentage of the variation of the numbers of cars checked can be attributed to differences in the length of time that the persons have been working at the inspection station?

12.32 With reference to Exercise 12.4, what percentage of the variation in the yield of wheat can be attributed to differences in the amount of rain?

12.33 With reference to Exercise 12.14, what percentage of the variation of the weekly profit can be attributed to the relationship between the weekly profit and the price of the hamburger platters?

12.34 With reference to Exercise 12.25, what percentage of the variation in typing speed can be attributed to the relationship between typing speed and reading speed?

12.35 If we calculate r for each of the following sets of data, should we be surprised if we get $r = 1$ and $r = -1$, respectively? Explain.

(a)

x	y
9	2
5	1

(b)

x	y
4	8
7	2

12.36 Check in each case whether the value of r is significant at the 0.05 level of significance:
(a) $n = 12$ and $r = -0.53$;   (c) $n = 15$ and $r = -0.55$;
(b) $n = 20$ and $r = 0.58$;   (d) $n = 9$ and $r = -0.61$.

12.37 Check in each case whether the value of r is significant at the 0.01 level of significance:
(a) $n = 18$ and $r = 0.62$;   (c) $n = 16$ and $r = -0.58$;
(b) $n = 32$ and $r = -0.47$;   (d) $n = 14$ and $r = 0.63$.

12.38 State in each case whether you would expect a positive correlation, a negative correlation, or no correlation:
(a) the ages of husbands and wives;
(b) the amount of rubber on tires and the number of miles that they have been driven;
(c) income and education;
(d) shirt size and sense of humor;
(e) the number of hours that bowlers practice and their scores;
(f) hair color and one's knowledge of foreign affairs.

12.39 With reference to Exercise 12.8, use a computer package to determine the total and residual sums of squares and the value of r measuring the strength of the relationship between students' IQs and their scores on the achievement test.

12.40 With reference to Exercise 12.9, use a computer package to determine the total and residual sums of squares and the value of r measuring the strength of the relationship between the number of hours a student studies for the test and his or her score.

12.41 With reference to Exercise 12.29, use a computer package or a graphing calculator to determine r and also the percentage of the variation of the anthropology scores that can be attributed to the relationship between students' grades in the two subjects.

Solutions to Practice Exercises

12.1 (a) From the data we get $n = 6$, $\Sigma x = 36$, $\Sigma x^2 = 304$, $\Sigma y = 108$, and $\Sigma xy = 715$, so the normal equations are

$$108 = 6a + 36b$$
$$715 = 36a + 304b$$

Multiplying the expressions on both sides of the first equation by 6 and leaving the second equation as is, we get

$$648 = 36a + 216b$$
$$715 = 36a + 304b$$

and then, by subtraction, $67 = 88b$ and $b = \frac{67}{88} \approx 0.7614$. Substituting this value into the first normal equation, we get $108 = 6a + 36(0.7614)$, which yields $a = 13.4316$. Thus, the equation of the least-squares line is $\hat{y} = 13.4316 + 0.7614x$.

(b) Note first that

$$S_{xx} = \Sigma x^2 - \frac{(\Sigma x)^2}{n} = 304 - \frac{36^2}{6} = 88$$

and that

$$S_{xy} = \Sigma xy - \frac{(\Sigma x) \cdot (\Sigma y)}{n} = 715 - \frac{36 \cdot 108}{6} = 67$$

Therefore,

$$b = \frac{67}{88} \approx 0.7614$$

and since $\bar{x} = 6$ and $\bar{y} = 18$, it follows that

$$a = 18 - 0.7614 \cdot 6 = 13.4316$$

(c) Using the values of a and b obtained in part (b), we get

$$\hat{y} = 13.4316 + 0.7614(10) = 21.0456$$

or 21.05 rounded to two decimals.

12.11   **1.** *Hypotheses* H_0: $\beta = 1.5$
             H_A: $\beta < 1.5$
     2. *Level of significance* $\alpha = 0.05$
     3. *Criterion*   Reject the null hypothesis if $t \leq -2.132$, where 2.132 is the value of $t_{0.05}$ for $6 - 2 = 4$ degrees of freedom and

$$t = \frac{b - \beta}{s_e} \sqrt{S_{xx}}$$

Otherwise, accept it or reserve judgment.
     4. *Calculations*   To find the value of s_e, we use the formula on page 425, which requires S_{yy}. Since $\Sigma y^2 = 2{,}008$, we get

$$S_{yy} = 2{,}008 - \frac{108^2}{6} = 64$$

We have already found that $S_{xx} = 88$ and $S_{xy} = 67$, so that

$$s_e = \sqrt{\frac{64 - \frac{67^2}{88}}{6 - 2}} \approx 1.8020$$

Then, substituting this value together with $n = 6$, $b = 0.7614$, $\beta = 1.5$, and $S_{xx} = 88$ into the formula for t, we obtain

$$t = \frac{0.7614 - 1.5}{1.8020} \sqrt{88} \approx -3.84$$

5. *Decision*   Since $t = -3.84$ is less than -2.132, the null hypothesis must be rejected; in other words, the slope of the least-squares line is less than 1.5.

12.16   The solution to Exercise 12.13 involved the values $a = 0.7210$, $b = 0.0686$, $S_{xx} = 344$, and $s_e = 0.2720$. We substitute these together with $x_0 = 12$, $\bar{x} = 15$, and 4.604, the value of $t_{0.005}$ for $6 - 2 = 4$ degrees of freedom, into the formula for the confidence limits. This gives

$$\left[0.7210 + (0.0686)(12)\right] \pm (4.604)(0.2720)\sqrt{\frac{1}{6} + \frac{(12 - 15)^2}{344}}$$

or

$$1.544 \pm 0.550$$

rounded to three decimals. Hence, the 99% confidence interval for the mean grade-point index of students who study on the average 12 hours per week is the interval from $1.544 - 0.550 = 0.994$ to $1.544 + 0.550 = 2.094$.

12.23   The data gives $n = 5$, $\sum x = 795$, $\sum x^2 = 196{,}851$, $\sum y = 2.40$, $\sum y^2 = 1.1684$, and $\sum xy = 348.26$. From these we obtain

$$S_{xx} = 70{,}446, \quad S_{yy} = 0.0164, \quad \text{and} \quad S_{xy} = -33.34$$

We substitute these into the formula for r to get

$$r = \frac{-33.34}{\sqrt{(70{,}446)(0.0164)}} \approx -0.98$$

12.24   **1.** *Hypotheses* H_0: $\rho = 0$
$\quad\quad\quad\quad\quad\quad\quad\quad\quad H_A$: $\rho \neq 0$
2. *Level of significance* $\alpha = 0.01$
3. *Criterion*   Reject the null hypothesis if $r \leq -0.959$ or $r \geq 0.959$, where 0.959 is the value of $r_{0.005}$ for $n = 5$; otherwise, state that the value of r is not significant.
4. *Calculations* $r = -0.98$ (from the solution of Exercise 12.23).
5. *Decision*   Since $r = -0.98$ is less than -0.959, the null hypothesis must be rejected; we conclude that there is a relationship between DDT residues and eggshell thickness.

12.31   The data gives $n = 6$, $\sum x = 36$, $\sum x^2 = 304$, $\sum y = 108$, $\sum y^2 = 2{,}008$, and $\sum xy = 715$. From these we obtain

$$S_{xx} = 88, \quad S_{yy} = 64, \quad \text{and} \quad S_{xy} = 67$$

We substitute these into the formula for r to get

$$r = \frac{67}{\sqrt{(88)(64)}} \approx 0.893$$

Thus, $100\%(0.893)^2 \approx 79.7\%$ of the variation of the number of cars checked can be attributed to differences in the length of time that the persons have been working at the inspection station.

The tests that we studied in the preceding chapters required specific assumptions about the population, or populations, sampled. In many cases we assumed that the populations sampled can be approximated closely by normal distributions; sometimes we assumed that their standard deviations are known or are known to be equal; and sometimes we assumed that the samples are independent. Since there are many situations in which these assumptions cannot be met, statisticians have developed alternative techniques based on less stringent assumptions, which have become known as **nonparametric tests**.

Such tests cannot only be used under more general conditions, but they are often easier to explain and easier to understand than the standard tests that they replace. Moreover, in many nonparametric tests the computational burden is so light that they come under the heading of "quick and easy" or "shortcut" techniques. For these reasons, nonparametric tests have become quite popular, and extensive literature is devoted to their theory and application.

In Sections 13.1 through 13.3, we shall present the sign test as a nonparametric alternative to tests concerning means and differences between means of paired data. In Sections 13.4 through 13.6, we shall study methods

435

based on rank sums, which serve as nonparametric alternatives to the two-sample *t* test and the one-way analysis of variance. Indeed, ranks are often used in nonparametric tests, which makes them applicable for use with ordinal data.

In Sections 13.7 through 13.9, we shall learn how to test the randomness of a sample after the data has been selected; and in Section 13.10, we shall present a nonparametric test concerning the relationship between paired data.

13.1
The One-Sample Sign Test

The small-sample tests concerning means and differences between means, which we studied in Sections 10.4 and 10.6, are based on the assumption that the populations sampled have roughly the shape of normal distributions. When this assumption is untenable in practice, these standard tests can be replaced by any one of several nonparametric alternatives, among them the **sign test**, which we shall study in this section and in Sections 13.2 and 13.3.

Small samples usually cannot tell us whether the population sampled follows a normal distribution. A common indicator of a departure from normality is the occurrence of one or more values that straggle far away from the other sample values. If this is the case, most statisticians would be unwilling to assume that they are sampling a normal population. Without a lot of data, decisions about the appropriateness of normal distributions are quite subjective. The nonparametric procedures used in this chapter are valid under fairly general conditions. They should certainly be used when there is any doubt about the appropriateness of assuming normal distributions, and they may even be used when the assumption of normal distributions would be reasonable.

The **one-sample sign test** concerns the median $\tilde{\mu}$ of a continuous population. The probability of getting a sample value less than the median and the probability of getting a sample value greater than the median are both $\frac{1}{2}$. If the population values are symmetrically distributed about the median, then the median $\tilde{\mu}$ and the mean μ are equal. Only occasionally are we in a situation that allows us to assume the symmetry of the population. The procedures that follow will usually be described in terms of the median $\tilde{\mu}$, but we should remember that in certain situations the median and the mean are equal.

Then, to test the null hypothesis $\tilde{\mu} = \tilde{\mu}_0$ against an appropriate alternative on the basis of a random sample of size n, we replace each sample value greater than $\tilde{\mu}_0$ with a plus sign and each sample value less than $\tilde{\mu}_0$ with a minus. Then we test the null hypothesis that the total number of plus signs is the value of a random variable having the binomial distribution with $p = \frac{1}{2}$. If a sample value equals $\tilde{\mu}_0$ exactly, which is certainly possible with rounded data, it is discarded.

■ **EXAMPLE**   The following data constitutes a random sample of 15 measurements of the octane rating of a certain kind of gasoline:

$$97.5 \quad 95.2 \quad 97.3 \quad 96.0 \quad 96.8$$
$$100.3 \quad 97.4 \quad 95.3 \quad 93.2 \quad 99.1$$
$$96.1 \quad 97.6 \quad 98.2 \quad 98.5 \quad 94.9$$

Use the one-sample sign test to test the null hypothesis $\tilde{\mu} = 98.5$ against the alternative hypothesis $\tilde{\mu} < 98.5$ at the 0.05 level of significance.

Solution   Since one of the values equals 98.5 and must be discarded, the sample size for the one-sample sign test is only $n = 14$.

  1. *Hypotheses* H_0: $\tilde{\mu} = 98.5$
                        H_A: $\tilde{\mu} < 98.5$

  2. *Level of significance* $\alpha = 0.05$

  3'. *Criterion* The test statistic is x, the number of plus signs.

  4'. *Calculations*   Replacing each value greater than 98.5 with a plus sign and each value less than 98.5 with a minus sign, we get $- - - - - + - - - + - - - -$. Thus, $x = 2$.

  5'. *Decision*   Table I shows that for $n = 14$ and $p = 0.50$ the probability that $x \leq 2$ is $0.001 + 0.006 = 0.007$. The p-value is therefore 0.007. Since 0.007 is less than 0.05, the null hypothesis must be rejected.   ■

  Here we used the procedure based on the p-value. As in Section 11.1, we did this because we were able to refer directly to Table I. For the same reason, we shall do so also in the example of Section 13.2.

13.2 The Paired-Sample Sign Test

The sign test can also be used when we deal with paired data as in Section 10.7. In such paired-sample sign tests, each pair of sample values is replaced with a plus sign if the first value is greater than the second, with a minus sign if the first value is smaller than the second, and is discarded when the two values are equal. Then we proceed as in Section 13.1.

  The sign test for the paired-sample situation asks whether the population of differences between pairs has a median that is equal to zero. In most practical situations, the median is zero only when the two populations have the same distribution. The procedure that follows will therefore be described in terms of comparing the means of the two populations.

■ **EXAMPLE**   To determine the effectiveness of a new traffic control system, the number of accidents that occurred at 10 dangerous intersections during four weeks before and four weeks after the installation of the new system was observed and the following data was obtained:

3 and 1	4 and 2	2 and 3	5 and 2	3 and 3
2 and 0	3 and 2	6 and 3	1 and 2	1 and 0

Use the sign test to test the null hypothesis that the new traffic control system is only as effective as the old system at the 0.05 level of significance.

Solution   Since one of the pairs, 3 and 3, has to be discarded, the sample size for the paired-sample sign test is only $n = 9$.

  1. *Hypotheses* H_0: $\mu_1 = \mu_2$, where μ_1 and μ_2 are the mean numbers of accidents in four weeks at a dangerous intersection with the old and the new control systems

                        H_A: $\mu_1 > \mu_2$

  2. *Level of significance* $\alpha = 0.05$

3'. *Test statistic*   The statistic is x, the number of plus signs.

4'. *Calculations*   Replacing each pair of values with a plus sign if the first value is greater than the second or with a minus sign if the first value is smaller than the second, the nine unequal sample pairs yield

$$+ \quad + \quad - \quad +$$
$$+ \quad + \quad + \quad - \quad +$$

Here $x = 7$.

5'. *Decision*   Using Table I for $n = 9$ and $p = 0.50$, the probability of $x \geq 7$ is $0.070 + 0.018 + 0.002 = 0.090$. The p-value is thus 0.090. Since 0.090 exceeds 0.05, the null hypothesis cannot be rejected. In other words, we cannot conclude on the basis of this test that the new control system is more effective than the old system. ∎

The test we have described here is only one of several nonparametric ways of analyzing such paired-sample data. Another popular test used for this purpose is the Wilcoxon signed-rank test, which may be found in the books on nonparametric statistics listed in the bibliography at the end of the book.

13.3
The Sign Test
(Large Samples)

When np and $n(1 - p)$ are both greater than 5, so that we can use the normal approximation to the binomial distribution, the sign test may be based on the large-sample test of Section 11.2, namely, on the statistic

$$z = \frac{x - np_0}{\sqrt{np_0(1 - p_0)}}$$

with $p_0 = 0.50$, which has approximately the standard normal distribution. When n is small, it may be wise to use the continuity correction suggested on page 368. This is true especially if, without the continuity correction, we could *barely* reject the null hypothesis. Note that the continuity correction does not even have to be considered when *without it* the null hypothesis cannot be rejected.

■ **EXAMPLE**   The following are measurements of the ocean depths in a certain area (in fathoms): 46.4, 48.3, 51.9, 38.8, 46.5, 45.6, 52.1, 41.0, 54.2, 44.9, 52.3, 43.6, 48.7, 42.2, and 44.9. Use the sign test at the level of significance $\alpha = 0.05$ to test the null hypothesis $\tilde{\mu} = 43.0$ (the previously recorded ocean depth in that area) against the alternative hypothesis $\tilde{\mu} \neq 43.0$.

Solution   **1.** *Hypotheses* H_0: $\tilde{\mu} = 43.0$
H_A: $\tilde{\mu} \neq 43.0$

2. *Level of significance* $\alpha = 0.05$

3. *Criterion*   Reject the null hypothesis if $z \leq -1.96$ or $z \geq 1.96$, where

$$z = \frac{x - np_0}{\sqrt{np_0(1 - p_0)}}$$

with $p_0 = 0.50$; otherwise, accept it or reserve judgment.

4. *Calculations*   Replacing each value greater than 43.0 with a plus sign and each value less than 43.0 with a minus sign, we get

$$+ \ + \ + \ - \ + \ + \ + \ - \ + \ + \ + \ + \ + \ - \ +$$

Thus, $x = 12$ and substitution into the formula yields

$$z = \frac{12 - 15(0.50)}{\sqrt{15(0.50)(0.50)}} = 2.32$$

5. *Decision*   Since $z = 2.32$ exceeds 1.96, the null hypothesis must be rejected; in other words, the median ocean depth in the given area is not 43.0 fathoms, as had been previously reported. Had we used the continuity correction, we would have obtained

$$z = \frac{11.5 - 7.5}{\sqrt{15(0.50)(0.50)}} = 2.07$$

and the decision would have been the same.   ■

Exercises

Exercises 13.1 and 13.7 are practice exercises; their complete solutions are given on pages 459 and 460.

13.1 On 12 occasions, Ms. Brown had to wait 3, 6, 7, 6, 4, 8, 6, 2, 8, 6, 1, and 9 minutes for the bus that takes her to work. Use the sign test based on Table I and the 0.05 level of significance to test the null hypothesis $\tilde{\mu} = 5$ against the alternative hypothesis $\tilde{\mu} \neq 5$.

13.2 Nine women buying new eyeglasses tried on, respectively, 12, 11, 14, 15, 10, 14, 11, 8, and 12 different frames. Use the sign test at the 0.05 level of significance to test the null hypothesis $\tilde{\mu} = 10$ against the alternative hypothesis $\tilde{\mu} > 10$.

13.3 In six rounds of golf at the Paradise Valley Country Club, a professional scored 71, 69, 72, 74, 71, and 72. Use the sign test at the 0.05 level of significance to test the null hypothesis $\tilde{\mu} = 70$ against the alternative hypothesis $\tilde{\mu} > 70$.

13.4 The following is a list of the weights (in grams) of 14 packages of a certain kind of candy: 101.0, 99.8, 100.9, 103.6, 97.1, 100.0, 102.5, 100.5, 101.0, 98.2, 100.3, 102.6, 100.0, and 100.8. Use the sign test based on Table I and the level of significance $\alpha = 0.05$ to test the null hypothesis $\tilde{\mu} = 100.0$ against the alternative hypothesis $\tilde{\mu} \neq 100.0$.

13.5 The following are the numbers of passengers carried on flights 136 and 137 between Chicago and Phoenix on 12 days:

232 and 189	265 and 230	249 and 236	250 and 261
255 and 249	236 and 218	270 and 258	266 and 253
249 and 251	240 and 233	257 and 254	239 and 249

Use the sign test based on Table I and the 0.05 level of significance to test the null hypothesis $\mu_1 = \mu_2$ against the alternative hypothesis $\mu_1 > \mu_2$.

13.6 The following are the grades that 15 students received on the midterm and final examinations in a course in European history:

66 and 73   88 and 91   75 and 78   90 and 86   63 and 69
58 and 67   75 and 75   82 and 80   73 and 76   84 and 89
85 and 81   93 and 96   70 and 76   85 and 82   90 and 97

Use the sign test based on Table I and the 0.05 level of significance to test the null hypothesis $\mu_1 = \mu_2$ against the alternative hypothesis $\mu_1 < \mu_2$.

13.7 The following are the numbers of speeding tickets issued by two police officers on 20 days:

6 and 9   11 and 13   12 and 12   10 and 17   15 and 13
7 and 11   9 and 13   7 and 12   14 and 15   11 and 13
14 and 10   6 and 12   9 and 9   12 and 14   8 and 13
16 and 11 10 and 15   12 and 14   15 and 15   12 and 18

Use the sign test based on the normal approximation to the binomial distribution and the level of significance $\alpha = 0.01$ to test the null hypothesis that on the average the two police officers issue equally many speeding tickets. Work this exercise with and without the continuity correction.

13.8 The following is data for the daily sulfur oxides emission (in tons) of an industrial plant:

17   15   20   29   19   18   22   25   27   9
24   20   17   6   24   14   15   23   24   26
19   23   28   19   16   22   24   17   20   13
19   10   23   18   31   13   20   17   24   14

Use the sign test and the 0.01 level of significance to test the null hypothesis $\tilde{\mu} = 23$ against the alternative hypothesis $\tilde{\mu} < 23$. Work this exercise with and without the continuity correction.

13.9 Rework Exercise 13.4 using the sign test based on the normal approximation to the binomial distribution.

13.10 The following are the miles per gallon obtained with 40 tankfuls of a certain kind of gas:

24.1   25.0   24.8   24.3   24.2   25.3   24.2   23.6   24.5   24.4
24.5   23.2   24.0   23.8   23.8   25.3   24.5   24.6   24.0   25.2
25.2   24.4   24.7   24.1   24.6   24.9   24.1   25.8   24.2   24.2
24.8   24.1   25.6   24.5   25.1   24.6   24.3   25.2   24.7   23.3

Use the sign test at the 0.01 level of significance to test the null hypothesis $\tilde{\mu} = 24.2$ against the alternative hypothesis $\tilde{\mu} > 24.2$. Work this exercise with and without the continuity correction.

13.11 The following are the numbers of artifacts dug up by two archeologists at an ancient cliff dwelling on 30 days:

1 and 0   0 and 0   2 and 1   3 and 0   1 and 2   0 and 0
2 and 0   2 and 1   3 and 1   0 and 2   1 and 0   1 and 1
4 and 2   1 and 1   2 and 1   1 and 0   3 and 2   5 and 2
2 and 6   1 and 0   3 and 2   2 and 3   4 and 0   1 and 2
3 and 1   2 and 0   0 and 1   2 and 0   4 and 1   2 and 0

Use the sign test at the level of significance $\alpha = 0.01$ to test the null hypothesis that the two archeologists are equally good at finding artifacts against the alternative hypothesis that the first one is better.

13.12 Use the large-sample sign test to rework Exercise 10.44.

13.13 The following are the numbers of employees absent from two departments of a large firm on 25 days:

4 and 3	2 and 5	6 and 6	3 and 6	1 and 4	2 and 4	5 and 2
1 and 4	3 and 4	6 and 5	2 and 5	7 and 1	4 and 6	1 and 3
2 and 5	0 and 3	6 and 5	4 and 6	1 and 2	4 and 1	2 and 4
0 and 1	5 and 3	2 and 3	2 and 4			

Use the sign test at the 0.05 level of significance to test the null hypothesis $\mu_1 = \mu_2$ against the alternative hypothesis $\mu_1 < \mu_2$.

13.4 Rank Sums: The *U* Test

In this section we shall present a nonparametric alternative to the small-sample *t* test concerning the difference between two means. It is called the **U test**, the **Wilcoxon test**, or the **Mann–Whitney test**, named after the statisticians who contributed to its development. The different names refer to different methods of organizing the calculations, but the procedures are logically equivalent. We will be able to test the null hypothesis that the two samples come from identical populations without having to assume that the populations sampled have normal distributions; in fact, the test requires only that the populations sampled be continuous to avoid ties, and in practice it does not even matter whether this assumption is satisfied.

Suppose that you randomly select samples from each of the two populations being compared. The *U* test, as originally defined, is based on the number of times values of the sample from one population are exceeded by values from the other population. This is a somewhat complicated notion, and our work here will be described in terms of the ranks of the jointly ranked data from the two populations.

To illustrate how the *U* test is thus performed, suppose that we want to compare the grain size of sand obtained from two different locations on the moon on the basis of the following diameters (in millimeters):

Location 1:   0.37, 0.70, 0.75, 0.30, 0.45, 0.16, 0.62, 0.73, 0.33
Location 2:   0.86, 0.55, 0.80, 0.42, 0.97, 0.84, 0.24, 0.51, 0.92, 0.69

The means of these two samples are 0.49 and 0.68, and their difference is large, but it remains to be seen whether it is significant.

To perform the *U* test, we first arrange the data jointly, as if it comprises one sample, in an increasing order of magnitude. For our data, we get

0.16	0.24	0.30	0.33	0.37	0.42	0.45	0.51	0.55	0.62
1	2	1	1	1	2	1	2	2	1

0.69	0.70	0.73	0.75	0.80	0.84	0.86	0.92	0.97
2	1	1	1	2	2	2	2	2

where we indicated for each value whether it came from location 1 or location 2. Assigning the data, in this order, the ranks 1, 2, 3, ..., and 19, we find that the values of the first sample (location 1) occupy ranks 1, 3, 4, 5, 7, 10, 12, 13, and 14, while those of the second sample (location 2) occupy ranks 2, 6, 8, 9, 11, 15, 16, 17, 18, and 19. There are no ties here, but if there were, we would assign each of the tied observations the mean of the ranks that they jointly occupy. For instance, if the third and fourth values were the same, we would assign each the rank

$$\frac{3 + 4}{2} = 3.5$$

and if the ninth, tenth, and eleventh values were the same, we would assign each the rank

$$\frac{9 + 10 + 11}{3} = 10$$

Now, if there is an appreciable difference between the means of the two populations, most of the lower ranks are likely to go to the values of one sample, while most of the higher ranks are likely to go to the values of the other sample. The test of the null hypothesis that the two samples come from identical populations may thus be based on W_1, the sum of the ranks of the values of the first sample, or on W_2, the sum of the ranks of the values of the second sample. In practice, it does not matter which sample we refer to as sample 1 and which we refer to as sample 2, or whether we base the test on W_1 or W_2.[†] If the sample sizes are n_1 and n_2, the sum of W_1 and W_2 is simply the sum of the first $n_1 + n_2$ positive integers, which is known to equal

$$\frac{(n_1 + n_2)(n_1 + n_2 + 1)}{2}$$

This formula enables us to find W_2 if we know W_1, and vice versa. For our illustration, we get

$$W_1 = 1 + 3 + 4 + 5 + 7 + 10 + 12 + 13 + 14 = 69$$

and since the sum of the first 19 positive integers is

$$\frac{19 \cdot 20}{2} = 190$$

it follows that $W_2 = 190 - 69 = 121$. (This value may be checked by actually adding 2, 6, 8, 9, 11, 15, 16, 17, 18, and 19.)

When the use of **rank sums** was first proposed as a nonparametric alternative to the two-sample t test, the decision was based on W_1 or W_2. Presently, the decision is based on either of the related statistics

---

[†] When the sample sizes are unequal, it is common practice to let n_1 be the smaller of the two; this is not required, however, for the work in this book.

U_1 and U_2 statistics

$$U_1 = W_1 - \frac{n_1(n_1 + 1)}{2} \quad \text{or} \quad U_2 = W_2 - \frac{n_2(n_2 + 1)}{2}$$

or on the statistic U, which always equals the smaller of the two. The resulting tests are equivalent to those based on W_1 or W_2, but they have the advantage that they lend themselves more readily to the construction of tables of critical values. Not only do U_1 and U_2 take on values on the interval from 0 to n_1n_2—indeed, their sum is always equal to n_1n_2—but their sampling distributions are symmetrical about $\frac{n_1n_2}{2}$. The use of U, which always equals the smaller of the values of U_1 and U_2, has the added advantage that the resulting test is one-tailed and hence easier to tabulate.

Accordingly, we test the null hypothesis that the two samples come from identical populations against the alternative hypothesis that the two populations have unequal means with the following criterion:

Reject the null hypothesis if $U \leq U'_\alpha$, where U'_α is given in Table VII for $n_1 \leq 15$, $n_2 \leq 15$, and $\alpha = 0.05$ or $\alpha = 0.01$.

In the construction of Table VII, U'_α is the largest value of U for which the probability of $U \leq U'_\alpha$ is less than or equal to α, and the blank spaces indicate that the null hypothesis cannot be rejected at the given level of significance, regardless of the value that we obtain for U. More extensive tables may be found in handbooks of statistical tables; but when n_1 and n_2 are both greater than 8, it is generally considered reasonable to use the large-sample test described in Section 13.5.

Observe that the null hypothesis is rejected for small values of U. In every case, either $U = U_1$ or $U = U_2$. When U actually equals U_1, then it is sample 1 that is producing the small values. Similarly, when U is equal to U_2, then sample 2 is producing the small values.

■ **EXAMPLE**

With reference to the previous grain-size data on page 441, use the U test at the 0.05 level of significance to test the null hypothesis that the two samples come from identical populations against the alternative hypothesis that the two populations have unequal means.

Solution

1. *Hypotheses* H_0: populations are identical
   H_A: $\mu_1 \neq \mu_2$

2. *Level of significance* $\alpha = 0.05$

3. *Criterion* Reject the null hypothesis if $U \leq 20$, which is the value of U'_α for $n_1 = 9$, $n_1 = 10$, and $\alpha = 0.05$; otherwise, accept it or reserve judgment.

4. *Calculations* Having shown that $W_1 = 69$ and $W_2 = 121$, we get

$$U_1 = 69 - \frac{9 \cdot 10}{2} = 24 \qquad U_2 = 121 - \frac{10 \cdot 11}{2} = 66$$

and hence $U = 24$. Note that $U_1 + U_2 = 24 + 66 = 90$, which equals $n_1n_2 = 9 \cdot 10$.

5. *Decision*   Since $U = 24$ exceeds 20, the null hypothesis cannot be rejected; in other words, we cannot conclude that there is a real difference in the mean grain size of sand from the two locations on the moon.   ■

The test that we have described here can also be used when the alternative is $\mu_1 < \mu_2$ or $\mu_1 > \mu_2$. However, since the procedure is more complicated in that case, we shall discuss it here only for large samples in the section that follows.

13.5
Rank Sums: The *U* Test (Large Samples)*

The large-sample U test may be based on either U_1 or U_2, as given on page 443. The resulting tests will be equivalent, so that it does not matter which sample we identify as sample 1. In the description that follows, we shall use the statistic U_1.

Under the null hypothesis that the two samples come from identical populations, it can be shown that the mean and the standard deviation of the sampling distribution of U_1 are[†]

Mean and standard deviation of U_1 statistic

$$\mu_{U_1} = \frac{n_1 n_2}{2} \quad \text{and} \quad \sigma_{U_1} = \sqrt{\frac{n_1 n_2 (n_1 + n_2 + 1)}{12}}$$

Furthermore, if n_1 and n_2 are both greater than 8, the sampling distribution of U_1 can be approximated closely with a normal curve. Thus, we base the test of the null hypothesis that the two samples come from identical populations on the statistic

Statistic for the large-sample U test

$$z = \frac{U_1 - \mu_{U_1}}{\sigma_{U_1}}$$

which has approximately the standard normal distribution. If the alternative hypothesis is $\mu_1 \neq \mu_2$, we reject the null hypothesis for $z \leq -z_{\alpha/2}$ or $z \geq z_{\alpha/2}$; if the alternative hypothesis is $\mu_1 < \mu_2$, we reject the null hypothesis for $z \leq -z_\alpha$, since small values of U_1 correspond to small values of the rank sum W_1. If the alternative hypothesis is $\mu_1 > \mu_2$, we reject the null hypothesis for $z \geq z_\alpha$, since large values of U_1 correspond to large values of the rank sum W_1.

Finding ranks when n_1 and n_2 are large can be a surprisingly messy task. However, you can do this easily by first constructing a stem-and-leaf display for the values of the first sample. Then, using a different-colored pencil, continue the stem-and-leaf display for the values of the second sample.

[†] If there are ties in rank, these formulas provide only approximations, but if the number of ties is small, there is no need to make a correction.

■ **EXAMPLE**   The following are the weight gains (in pounds) of young turkeys, which are fed two different diets but are otherwise kept under identical conditions:

Diet 1:   16.3,   10.1,   10.7,   13.5,   14.9,   11.8,   14.3,   10.2,
12.0,   14.7,   23.6,   15.1,   14.5,   18.4,   13.2,   14.0

Diet 2:   21.3,   23.8,   15.4,   19.6,   12.0,   13.9,   18.8,   19.2,
15.3,   20.1,   14.8,   18.9,   20.7,   21.1,   15.8,   16.2

Use the large-sample *U* test at the 0.01 level of significance to test the null hypothesis that the two populations sampled are identical against the alternative hypothesis that, on average, the second diet produces a greater gain in weight.

Solution   1. *Hypotheses* H_0: populations are identical
H_A: $\mu_1 < \mu_2$

2. *Level of significance* $\alpha = 0.01$

3. *Criterion*   Reject the null hypothesis if $z \leq -2.33$, where

$$z = \frac{U_1 - \mu_{U_1}}{\sigma_{U_1}}$$

Otherwise, accept it or reserve judgment.

4. *Calculations*   Arranging the data according to size, we get

10.1	10.2	10.7	11.8	12.0	12.0	13.2	13.5	13.9	14.0
1	1	1	1	1	2	1	1	2	1

14.3	14.5	14.7	14.8	14.9	15.1	15.3	15.4	15.8	16.2
1	1	1	2	1	1	2	2	2	2

16.3	18.4	18.8	18.9	19.2	19.6	20.1	20.7	21.1	21.3
1	1	2	2	2	2	2	2	2	2

23.6	23.8
1	2

Below the weight gains are listed the samples from which they came. Observe that there is a tie at 12.0 pounds, involving one turkey from each diet. These are the fifth and sixth positions in the list, so each is assigned the rank

$$\frac{5 + 6}{2} = 5.5$$

The sum of the rank positions occupied by the turkeys of the first sample (diet 1) is

$$W_1 = 1 + 2 + 3 + 4 + 5.5 + 7 + 8 + 10$$

$$+ 11 + 12 + 13 + 15 + 16 + 21 + 22 + 31$$

$$= 181.5$$

and

$$U_1 = 181.5 - \frac{16 \cdot 17}{2} = 45.5$$

Since

$$\mu_{U_1} = \frac{16 \cdot 16}{2} = 128 \quad \text{and} \quad \sigma_{U_1} = \sqrt{\frac{16 \cdot 16 \cdot 33}{12}} \approx 26.53$$

it follows that

$$z = \frac{45.5 - 128}{26.53} \approx -3.11$$

5. *Decision*   Since $z = -3.11$ is less than -2.33, the null hypothesis must be rejected; in other words, we conclude that, on average, the second diet produces a greater gain in weight.   ■

Exercises

Exercises 13.14 and 13.19 are practice exercises; their complete solutions are given on pages 460 and 461.

13.14 The following are figures on the number of assaults committed in a city in six weeks in the spring and in six weeks in the fall:

> *Spring:*   46,   37,   42,   48,   38,   45
> *Fall:*   35,   30,   25,   39,   28,   32

Use the U test at the 0.01 level of significance to check the claim that, on average, there are equally many assaults per week in the given city in the spring and in the fall.

13.15 The following are the Rockwell hardness numbers obtained for six aluminum die castings randomly selected from production lot A and for eight aluminum die castings randomly selected from production lot B:

> *Production lot A:*   75,   56,   63,   70,   58,   74
> *Production lot B:*   63,   85,   77,   80,   86,   76,   72,   82

Use the U test at the 0.05 level of significance to check the claim that the average hardness of die castings from the two production lots is the same.

13.16 Tests made on two kinds of 9-volt batteries showed the following lifetimes (in hours) of continuous use:

> *Brand A:*   11.7,   12.0,   10.8,   11.1,   11.9,   12.9,   12.4
> *Brand B:*   11.5,   12.8,   13.5,   13.6,   11.1,   12.4,   13.3

Use the U test at the 0.05 level of significance to test whether the difference between the two sample means, 11.8 and 12.6, can be attributed to chance.

13.17 The following are the numbers of misprints counted on pages selected at random from two Sunday editions of a newspaper:

May 10:   12, 6, 11, 11, 15, 7
May 24:   10, 3,   6,   8,   7, 5

Use the *U* test at the level of significance $\alpha = 0.05$ to test the null hypothesis that the two samples come from identical populations against the alternative hypothesis that the two populations have unequal means.

13.18 The following are the numbers of minutes it took a sample of 15 men and 12 women to complete a short screening test given to job applicants at a large bank:

Men: 8.8, 7.8, 6.6, 10.7, 8.9, 8.4, 6.9, 6.4, 6.3, 8.0, 8.6, 8.1, 9.1, 9.7, 9.9
Women: 7.5, 8.7, 8.3, 6.2, 6.5, 7.7, 9.8, 9.6, 9.2, 10.4, 8.2, 8.5

Use the *U* test based on Table VII and the 0.01 level of significance to test the null hypothesis that the two samples come from identical populations against the alternative hypothesis that the two populations have unequal means.

*13.19 The following are the scores that samples of students from two minority groups obtained on a current events test:

Minority group 1:   70, 62, 91, 55, 72, 94, 80, 96, 73, 44, 87, 78
Minority group 2:   81, 23, 71, 30, 71, 54, 64, 93, 58, 41, 47, 56

Use the large-sample *U* test at the 0.05 level of significance to test whether students from the two minority groups can be expected to score equally well on this test.

*13.20 Use the normal approximation to the sampling distribution of *U* to rework the illustration on page 441, which dealt with the grain size of sand from two locations on the moon.

*13.21 Comparing two kinds of emergency flares, a consumer testing service obtained the following burning times (rounded to the nearest tenth of a minute):

Brand X: 17.2, 18.1, 21.2, 19.3, 14.4, 21.1, 14.6, 19.1, 18.8, 15.2, 20.3, 17.5
Brand Y: 13.6, 13.7, 11.8, 14.6, 15.2, 14.3, 22.5, 12.3, 13.5, 10.9, 14.4,   8.0

Use the large-sample *U* test at the 0.01 level of significance to see whether it is reasonable to say that, in general, brand X flares are better (last longer) than brand Y flares.

*13.22 Use the large-sample *U* test to rework Exercise 13.18.

*13.23 The following are the weekly per-capita food costs (in dollars) of families with two children chosen at random from two suburbs of a large city:

Suburb A: 78.60, 70.50, 75.38, 86.45, 67.95, 70.78, 67.89, 72.00, 64.19, 71.15
Suburb B: 62.63, 75.16, 55.35, 78.19, 71.72, 63.12, 75.91, 66.51, 63.76, 60.78

Use the large-sample *U* test at the 0.05 level of significance to check the claim that, on average, such weekly food expenditures are higher in suburb A than in suburb B.

13.6 Rank Sums: The *H* Test*

The **H test**, or **Kruskal–Wallis test**, is a rank-sum test that serves to test the null hypothesis that *k* independent random samples come from identical populations against the alternative hypothesis that the means of these populations are not all equal. Unlike the standard test that it replaces, the one-way analysis of variance of Section 10.9, it does not require the assumption that the populations sampled have, at least approximately, normal distributions.

As in the U test, the data is ranked jointly from low to high as though they constitute a single sample. Then, if R_i is the sum of the ranks assigned to the n_i values of the ith sample and $n = n_1 + n_2 + \cdots + n_k$, the H test is based on the statistic

Statistic for H test

$$H = \frac{12}{n(n+1)} \sum_{i=1}^{k} \frac{R_i^2}{n_i} - 3(n+1)$$

If the null hypothesis is true and each sample has at least five observations, the sampling distribution of H can be approximated closely with a chi-square distribution with $k - 1$ degrees of freedom. Consequently, we reject the null hypothesis that the populations sampled are identical and accept the alternative hypothesis that the means of these populations are not all equal, if the value we get for H is greater than or equal to χ_α^2 for $k - 1$ degrees of freedom.

■ **EXAMPLE**   The following are the final examination scores of samples of students who are taught German by three different methods (classroom instruction and language laboratory, only classroom instruction, and only self-study in language laboratory):

> *Method 1:*   94, 87, 91, 74, 87, 97
> *Method 2:*   85, 82, 79, 84, 61, 72, 80
> *Method 3:*   89, 67, 72, 76, 69

Use the H test at the 0.05 level of significance to test the null hypothesis that the three populations sampled are identical against the alternative hypothesis that their means are not all equal.

Solution   1. *Hypotheses* H_0:   populations are identical
H_A: population means are not all equal

2. *Level of significance* $\alpha = 0.05$

3. *Criterion*   Reject the null hypothesis if $H \geq 5.991$, the value of $\chi_{0.05}^2$ for $3 - 1 = 2$ degrees of freedom, where H is calculated in accordance with the formula above. Otherwise, accept it or reserve judgment.

4. *Calculations*   Arranging the data jointly according to size, we get 61, 67, 69, 72, 72, 74, 76, 79, 80, 82, 84, 85, 87, 87, 89, 91, 94, and 97. Assigning the data, in this order, the ranks 1, 2, 3, ..., and 18, we find that the values of the first sample occupy ranks 6, 13, 14, 16, 17, and 18, while those of the second sample occupy ranks 1, 4.5, 8, 9, 10, 11, and 12, and those of the third sample occupy ranks 2, 3, 4.5, 7, and 15. (Since the two 87's belong to the same sample, we simply assign them ranks 13 and 14.) Thus,

$$R_1 = 6 + 13 + 14 + 16 + 17 + 18 = 84$$

$$R_2 = 1 + 4.5 + 8 + 9 + 10 + 11 + 12 = 55.5$$

$$R_3 = 2 + 3 + 4.5 + 7 + 15 = 31.5$$

and it follows that

$$H = \frac{12}{18 \cdot 19} \left(\frac{84^2}{6} + \frac{55.5^2}{7} + \frac{31.5^2}{5} \right) - 3 \cdot 19 \approx 6.67$$

5. **Decision**   Since $H = 6.67$ exceeds 5.991, the null hypothesis must be rejected; in other words, we conclude that the three methods of teaching German are not all equally effective. ■

Exercises

Exercise 13.24 is a practice exercise; its complete solution is given on pages 461 and 462.

 ∗13.24 The following are the miles per gallon that a test driver got for six tankfuls each of three kinds of gasoline:

Gasoline 1:	28,	23,	26,	31,	14,	29
Gasoline 2:	21,	31,	32,	19,	27,	16
Gasoline 3:	24,	17,	21,	31,	22,	18

Use the H test at the level of significance $\alpha = 0.05$ to check the claim that there is no difference in the true average mileage yield of the three kinds of gasoline.

∗13.25 To compare four bowling balls, a professional bowler bowled five games with each ball and got the following results:

Bowling ball D:	221,	232,	207,	198,	212
Bowling ball E:	202,	225,	252,	218,	226
Bowling ball F:	210,	205,	189,	196,	216
Bowling ball G:	229,	192,	247,	220,	208

Use the H test at the 0.05 level of significance to test the null hypothesis that the bowler performs equally well with the four bowling balls.

∗13.26 Three groups of guinea pigs were injected, respectively, with 0.5 milligram (mg), 1.0 mg, and 1.5 mg of a tranquilizer, and the following are the numbers of seconds it took them to fall asleep:

0.5-mg dose:	8.2,	10.0,	10.2,	13.7,	14.0,	7.8,	12.7,	10.9	
1.0-mg dose:	9.7,	13.1,	11.0,	7.5,	13.3,	12.5,	8.8,	12.9, 7.9,	10.5
1.5-mg dose:	12.0,	7.2,	8.0,	9.4,	11.3,	9.0,	11.5,	8.5	

Use the H test at the level of significance $\alpha = 0.01$ to test the null hypothesis that the differences in dosage have no effect on the length of time it takes guinea pigs to fall asleep.

∗13.27 Use the H test to rework Exercise 10.53.

13.7
Tests of
Randomness: Runs

All the methods of inference that we have discussed are based on the assumption that the samples are random; yet there are many applications where it is difficult to decide whether this assumption is justifiable. This is true, particularly, when we have little or no control over the selection of the data, as is the case, for example, when we rely on whatever records are available to make long-range predictions of the weather, when we use whatever data is available to estimate the mortality rate of a disease, or when we use sales records for past months to make predictions of a department store's sales. None of this information constitutes a random sample in the strict sense.

There are several methods of judging the randomness of a sample on the basis of the order in which the observations are obtained; they enable us to decide, after the data has been collected, whether patterns that look suspiciously nonrandom may be attributed to chance. The technique we shall describe here and in the next two sections is based on the **theory of runs**.

A **run** is a succession of identical letters (or other kinds of symbols) that is followed and preceded by different letters or no letters at all. To illustrate, consider the following arrangement of healthy, H, and diseased, D, elm trees that were planted many years ago along a country road:

$$\underline{H\,H\,H\,H}\ \ \underline{D\,D\,D}\ \ \underline{H\,H\,H\,H\,H\,H\,H}\ \ \underline{D\,D}\ \ \underline{H\,H}\ \ \underline{D\,D\,D\,D}$$

Using underlines to combine the letters that constitute the runs, we find that there is first a run of four H's, then a run of three D's, then a run of seven H's, then a run of two D's, then a run of two H's, and finally a run of four D's.

The **total number of runs** appearing in an arrangement of this kind is often a good indication of a possible lack of randomness. If there are too few runs, we might suspect a definite grouping or clustering, or perhaps a trend; if there are too many runs, we might suspect some sort of repeated alternating, or cyclical, pattern. In the example above there seems to be a definite clustering—the diseased trees seem to come in groups—but it remains to be seen whether this is significant or whether it can be attributed to chance.

If there are n_1 letters of one kind, n_2 letters of another kind, and u runs, we base this kind of decision on the following criterion:

> *Reject the null hypothesis of randomness if*
>
> $$u \leq u'_{\alpha/2} \quad or \quad u \geq u_{\alpha/2}$$
>
> *where $u'_{\alpha/2}$ and $u_{\alpha/2}$ are given in Table* VIII *for $n_1 \leq 15$, $n_2 \leq 15$, and $\alpha = 0.05$ or $\alpha = 0.01$.*

In the construction of Table VIII, $u'_{\alpha/2}$ is the largest value of u for which the probability of $u \leq u'_{\alpha/2}$ is less than or equal to $\alpha/2$, $u_{\alpha/2}$ is the smallest value of u for which the probability of $u \geq u_{\alpha/2}$ is less than or equal to $\alpha/2$, and the blank spaces indicate that the null hypothesis of randomness cannot be rejected for values in that tail of the sampling distribution of u, regardless of the value that we obtain for u. More extensive tables for the **u test** may be found in handbooks of statistical tables, but when n_1 and n_2 are both at least 10, it is generally considered reasonable to use the large-sample test described in Section 13.8.

■ **EXAMPLE**   With reference to the arrangement of healthy and diseased elm trees given previously, use the u test at the 0.05 level of significance to test the null hypothesis of randomness against the alternative hypothesis that the arrangement is not random.

Solution

1. *Hypotheses* H_0: arrangement is random
   H_A: arrangement is not random

2. *Level of significance* $\alpha = 0.05$

3. *Criterion*   Since $n_1 = 13$, $n_2 = 9$, and $\alpha = 0.05$, we get $u'_{0.025} = 6$ and $u_{0.025} = 17$ from Table VIII; thus, the null hypothesis must be rejected if $u \leq 6$ or $u \geq 17$. Otherwise, accept it or reserve judgment.

4. *Calculations* $u = 6$, as can be seen from the data.

5. *Decision*   Since $u = 6$ is less than or equal to 6, the null hypothesis must be rejected; in other words, we conclude that the arrangement of healthy and diseased elm trees is not random. Indeed, it seems that the diseased trees come in clusters.   ■

13.8 Tests of Randomness: Runs (Large Samples)*

Under the null hypothesis that n_1 letters of one kind and n_2 letters of another kind are arranged at random, it can be shown that the mean and the standard deviation of u, the total number of runs, are

Mean and standard deviation of u

$$\mu_u = \frac{2n_1 n_2}{n_1 + n_2} + 1 \quad \text{and} \quad \sigma_u = \sqrt{\frac{2n_1 n_2 (2n_1 n_2 - n_1 - n_2)}{(n_1 + n_2)^2 (n_1 + n_2 - 1)}}$$

Furthermore, if neither n_1 nor n_2 is less than 10, the sampling distribution of u can be approximated closely with a normal curve. Thus, we base the test of the null hypothesis of randomness on the statistic

Statistic for the large-sample u test

$$z = \frac{u - \mu_u}{\sigma_u}$$

which has approximately the standard normal distribution. If the alternative hypothesis is that the arrangement is not random, we reject the null hypothesis for $z \leq -z_{\alpha/2}$ or $z \geq z_{\alpha/2}$; if the alternative hypothesis is that there is a clustering or a trend, we reject the null hypothesis for $z \leq -z_\alpha$; and if the alternative hypothesis is that there is an alternating, or cyclical, pattern, we reject the null hypothesis for $z \geq z_\alpha$.

■ **EXAMPLE**   The following is an arrangement of men, M, and women, W, lined up to purchase tickets for a rock concert:

M W M W M M M W M W M M M W W M

(cont.)   M M M W W M W M M M W M M M W W

(cont.)   W M W M M M W M W M M M M W W M

Test for randomness at the 0.05 level of significance.

Solution   1. *Hypotheses* H_0:   arrangement is random
                                   H_A: arrangement is not random

2. *Level of significance* $\alpha = 0.05$

3. *Criterion*   Reject the null hypothesis if $z \leq -1.96$ or $z \geq 1.96$, where

$$z = \frac{u - \mu_u}{\sigma_u}$$

Otherwise, accept it or reserve judgment.

4. *Calculations*   Since $n_1 = 30$, $n_2 = 18$, and $u = 27$, we get

$$\mu_u = \frac{2 \cdot 30 \cdot 18}{30 + 18} + 1 = 23.5$$

$$\sigma_u = \sqrt{\frac{2 \cdot 30 \cdot 18(2 \cdot 30 \cdot 18 - 30 - 18)}{(30 + 18)^2(30 + 18 - 1)}} \approx 3.21$$

and hence,

$$z = \frac{27 - 23.5}{3.21} \approx 1.09$$

5. *Decision*   Since $z = 1.09$ falls between -1.96 and 1.96, the null hypothesis of randomness cannot be rejected; there is no real evidence of any lack of randomness.   ■

13.9 Tests of Randomness: Runs Above and Below the Median

The u test is not limited to tests of the randomness of sequences of attributes, such as the H's and D's or M's and W's of our examples. Any sample consisting of numerical measurements or observations can be treated similarly by using the letters a and b to denote, respectively, values falling above and below the median of the sample. Numbers equal to the median are omitted. The resulting sequence of a's and b's (representing the data in its original order) can then be tested for randomness on the basis of the total number of runs of a's and b's, that is, the total number of **runs above and below the median**. Depending on the size of n_1 and n_2, we use Table VIII or the large-sample test of Section 13.8.

■ **EXAMPLE**   On 24 successive trips between two cities, a bus carried

$$
\begin{array}{cccccccccccc}
24 & 19 & 32 & 28 & 21 & 23 & 26 & 17 & 20 & 28 & 30 & 24 \\
13 & 35 & 26 & 21 & 19 & 29 & 27 & 18 & 26 & 14 & 21 & 23
\end{array}
$$

passengers. Use the total number of runs above and below the median and the level of significance $\alpha = 0.01$ to decide whether it is reasonable to treat the data as if it constitutes a random sample.

Solution   Since the median is 23.5, as can easily be verified, we get the following arrangement of values above and below the median:

$$a \ b \ a \ a \ b \ b \ a \ b \ b \ a \ a \ a \ b \ a \ a \ b \ b \ a \ a \ b \ a \ b \ b \ b$$

1. *Hypotheses* H_0: arrangement is random
   H_A: arrangement is not random

2. *Level of significance* $\alpha = 0.01$

3. *Criterion*   Since $n_1 = 12$, $n_2 = 12$, and $\alpha = 0.01$, we get $u'_{0.005} = 6$ and $u_{0.005} = 20$ from Table VIII; thus, the null hypothesis must be rejected if $u \leq 6$ or $u \geq 20$; otherwise, accept it or reserve judgment.

4. *Calculations*   As can be seen from the arrangements of a's and b's above, there are $u = 14$ runs.

5. *Decision*   Since $u = 14$ falls between 6 and 20, the null hypothesis cannot be rejected; in other words, there is no real evidence to suggest that the data cannot be treated as if it constitutes a random sample.   ■

Exercises

Exercises 13.28, 13.32, and 13.37 are practice exercises; their complete solutions are given on pages 462 and 463.

13.28 The following sequence of C's and A's shows the order in which 25 cars with California or Arizona license plates crossed the Colorado River at Blyth, California, to enter Arizona.

$$C \; A \; A \; C \; A \; C \; C \; A \; C \; C \; C \; A \; A \; C \; A \; A \; A \; A \; C \; A \; C \; C \; A \; C \; C$$

Test for randomness at the level of significance $\alpha = 0.05$.

13.29 The following is the order in which red, R, and black, B, cards were dealt to a bridge player:

$$B \; B \; B \; R \; R \; R \; R \; R \; B \; B \; R \; R \; R$$

Test for randomness at the 0.05 level of significance.

13.30 Test at the 0.01 level of significance whether the following arrangement of defective, D, and nondefective, N, pieces coming off an assembly line may be regarded as random:

$$N \; N \; N \; N \; N \; N \; N \; D \; D \; D \; D \; N \; N \; N \; D \; D \; N \; N \; N$$

13.31 A driver buys gasoline either at a Shell station, S, or at a Chevron station, C, and the following arrangement shows where he purchased gasoline (in the given order) over a certain period of time:

$$C \; C \; C \; S \; C \; S \; C \; S \; S \; C \; C \; S \; C \; S \; C \; S \; C \; S \; S \; C \; S \; C$$

Test for randomness at the 0.05 level of significance.

*∗**13.32** Representing each 0, 2, 4, 6, and 8 by the letter E and each 1, 3, 5, 7, and 9 by the letter O, check at the 0.05 level of significance whether the arrangements of the 50 digits in the first column of the random-numbers table, Table XII, may be regarded as random.

*∗**13.33** The following arrangement shows whether 50 persons interviewed consecutively in the given order are for, F, or against, A, an increase in the city sales tax:

$$A \ A \ A \ A \ A \ F \ A \ A \ F \ F \ A \ A \ A \ A \ A \ A \ A \ F \ A \ A \ A \ A \ F \ F \ F$$

(cont.) $A \ A \ A \ F \ A \ A \ F \ A \ A \ A \ A \ F \ F \ A \ A \ A \ A \ A \ A \ A \ A \ F \ A \ A \ A$

Test for randomness at the level of significance $\alpha = 0.05$.

*13.34 Use the large-sample u test to rework Exercise 13.28.

*13.35 To test whether a radio signal contains a message or constitutes random noise, an interval of time is subdivided into a number of very short intervals, and for each of these it is determined whether the signal strength exceeds, E, or does not exceed, N, a certain level of background noise. Test at the level of significance $\alpha = 0.05$ whether the following arrangement, thus obtained, may be regarded as random and hence that the signal contains no message and may be regarded as random noise:

$$E \ N \ N \ N \ E \ N \ E \ N \ N \ N \ E \ E \ N \ N \ N \ E \ E \ N \ E \ N \ E \ N \ N$$

(cont.) $E \ E \ N \ N \ N \ N \ E \ E \ N \ E \ N \ N \ E \ N \ N \ N \ E \ E \ E \ N \ N$

(cont.) $E \ N \ E \ N \ N \ N \ N \ E \ N$

*13.36 Mentally simulate 50 flips of a coin, and test at the 0.05 level of significance whether the resulting sequence of H's and T's (heads and tails) may be regarded as random.

13.37 The following are the numbers of students absent from a school on 24 consecutive days:

38	31	32	27	28	30	26	33	36	30	28	35
33	31	29	35	31	33	31	28	30	28	25	29

Test for randomness at the 0.05 level of significance.

13.38 The following are the numbers of business lunches that an insurance agent had in 30 consecutive months:

6	7	5	6	8	6	8	6	6	4	3	2	4	4	3
4	7	5	6	8	6	6	3	4	2	5	4	4	3	7

Discarding the three values that equal the median, 5, test for randomness at the level of significance $\alpha = 0.01$.

*13.39 The following are the examination grades of 40 students in the order in which they finished an examination:

75	95	77	93	89	83	69	77	92	88	62	64	91	72
76	83	50	65	84	67	63	54	58	76	70	62	65	41
63	55	32	58	61	68	54	28	35	49	82	60		

Test for randomness at the level of significance $\alpha = 0.05$.

*13.40 The total number of retail stores opening for business and also quitting business within the calendar years 1967–1999 in a large city were

108	103	109	107	125	142	147	122	116	153	144
162	143	126	145	129	134	137	143	150	148	152
125	106	112	139	132	122	138	148	155	146	158

Making use of the fact that the median is 138, test at the 0.05 level of significance whether there is a significant trend. Work this exercise with and without the continuity correction.

13.41 The following are the weights in pounds of tomatoes harvested from 50 consecutive tomato plants:

5.1	5.0	5.2	5.8	6.6	4.9	5.2	6.1	7.1	5.8
6.4	6.1	5.2	7.0	5.4	5.6	6.9	5.0	5.9	5.3
5.9	6.1	7.2	5.5	6.2	6.4	6.0	7.4	6.7	6.0
5.3	6.5	6.7	5.5	6.9	6.5	6.8	5.8	6.4	7.4
7.2	7.6	7.8	6.6	7.8	5.9	6.7	7.0	6.8	7.2

Making use of the fact that the median is 6.3 pounds, test for randomness at the level of significance $\alpha = 0.05$ by using runs above and below the median.

13.42 A habitual coffee drinker consumed these numbers of cups on each of the 30 days of April:

5	3	4	3	5	4	2	2	4	4
3	4	3	5	6	6	6	7	5	4
3	4	3	3	3	2	2	4	4	5

Test for randomness at the level of significance $\alpha = 0.05$ by using runs above and below the median.

13.10
Rank Correlation

Since the significance test for r of Section 12.6 is based on very stringent assumptions, we sometimes use a nonparametric alternative that can be applied under much more general conditions. This test of the null hypothesis of no correlation is based on the **rank-correlation coefficient**, often called **Spearman's rank-correlation coefficient** and denoted by r_S.

The calculation of the rank-correlation coefficient for a given set of n pairs of x's and y's requires several steps. We first rank the x's among themselves from low to high (or high to low). Then we rank the y's in the same way. Then we find the d's, the differences between the ranks, and substitute into the formula

Rank-correlation coefficient

$$r_S = 1 - \frac{6\left(\sum d^2 \right)}{n(n^2 - 1)}$$

When there are ties in rank, we proceed as before and assign to each of the tied observations the mean of the ranks that they jointly occupy.

■ **EXAMPLE**   The following are the numbers of hours that 10 students studied for an examination and the grades that they received:

Number of hours studied	Grade in examination
x	y
9	56
5	44
11	79
13	72
10	70
5	54
18	94
15	85
2	33
8	65

Calculate r_S.

Solution   Ranking the x's among themselves from low to high and also the y's, we get the ranks shown in the first two columns of the following table:

RANK OF x	RANK OF y	d	d²
5	4	1.0	1.00
2.5	2	0.5	0.25
7	8	−1.0	1.00
8	7	1.0	1.00
6	6	0.0	0.00
2.5	3	−0.5	0.25
10	10	0.0	0.00
9	9	0.0	0.00
1	1	0.0	0.00
4	5	−1.0	1.00
		Total	4.50

Then, determining the d's and their squares and substituting $n = 10$ and $\Sigma d^2 = 4.50$ into the formula for r_S, we get

$$r_S = 1 - \frac{6(4.50)}{10(10^2 - 1)} \approx 0.97$$

    As can be seen from this example, r_S is easy to compute manually, and this is why it is sometimes used instead of r when no calculator is available. When there are no ties, r_S actually equals the correlation coefficient r calculated for the two sets of ranks; when ties exist, there may be a small (but usually negligible) difference. Of course, by using ranks instead of the original data, we lose some information, but this is usually offset by the rank-correlation coefficient's computational ease.

When we use r_S to test the null hypothesis of no correlation between two variables x and y, we do not have to make any assumptions about the nature of the populations sampled. Under the null hypothesis of no correlation—indeed, the null hypothesis that the x's and y's are randomly matched—the sampling distribution of r_S has the mean 0 and the standard deviation

$$\sigma_{r_S} = \frac{1}{\sqrt{n-1}}$$

Since this sampling distribution can be approximated with a normal distribution even for relatively small values of n, we base the test of the null hypothesis of no correlation on the statistic

Statistic for testing the significance of r_S

$$z = \frac{r_S - 0}{1/\sqrt{n-1}} = r_S \sqrt{n-1}$$

which has approximately the standard normal distribution.

■ **EXAMPLE**   With reference to the preceding example, where we had $n = 10$ and $r_S = 0.97$, test the significance of this value of r_S at the 0.01 level of significance.

Solution   1. *Hypotheses* H_0: $\rho = 0$ (no correlation)
H_A: $\rho \neq 0$

2. *Level of significance* $\alpha = 0.01$

3. *Criterion*   Reject the null hypothesis if $z \leq -2.575$ or $z \geq 2.575$, where

$$z = r_S \sqrt{n-1}$$

Otherwise, accept it or reserve judgment.

4. *Calculations*   For $n = 10$ and $r_S = 0.97$, we get

$$z = 0.97 \sqrt{10 - 1} \approx 2.91$$

5. *Decision*   Since $z = 2.91$ exceeds 2.575, the null hypothesis must be rejected; in other words, we conclude that there is a relationship between study time and examination grades in the populations sampled.   ■

Exercises

Exercises 13.43 and 13.44 are practice exercises; their complete solutions are given on page 463.

13.43 Calculate r_S for the following data representing the statistics grades, x, and psychology grades, y, of a sample of 15 students:

x	y	x	y
75	70	70	83
69	64	93	91
73	70	77	79
96	89	85	73
73	84	93	84
52	70	87	85
57	66		
61	63		
71	68		

13.44 Test at the level of significance $\alpha = 0.05$ whether the value obtained for r_S in Exercise 13.43 is significant.

13.45 Calculate r_S for the following sample data representing the number of minutes it took 12 mechanics to assemble a piece of machinery in the morning, x, and in the late afternoon, y:

x	y
10.8	15.1
16.6	16.8
11.1	10.9
10.3	14.2
12.0	13.8
15.1	21.5
13.7	13.2
18.5	21.1
17.3	16.4
14.2	19.3
14.8	17.4
15.3	19.0

13.46 Use the level of significance $\alpha = 0.05$ to test whether the value obtained for r_S in Exercise 13.45 is significant.

13.47 Ten weeks' sales of a downtown department store, x, and its suburban branch, y, are

x	y
71	49
64	31
58	24
80	68
63	30
69	40
76	62
60	22
66	35
55	16

where the units are $10,000. Calculate r_S.

13.48 Assuming that the data of Exercise 13.47 may be looked upon as random samples from the two stores' sales, use the level of significance $\alpha = 0.01$ to test whether the value obtained for r_S is significant.

13.49 In Exercise 12.27 we gave the percentages of the vote predicted by a poll for seven candidates for the U.S. Senate in different states, x, and the corresponding percentages of the vote that they actually received, y, as

x	y
42	51
34	31
59	56
41	42
53	58
40	35
55	54

Calculate r_S and compare it with the value of r obtained for these data in Exercise 12.27.

13.50 If a sample of $n = 20$ pairs of data yielded $r_S = 0.41$, is this rank-correlation coefficient significant at the 0.05 level of significance?

13.51 The following table shows how a panel of nutrition experts and a panel of heads of household ranked 15 breakfast foods on their palatability:

Breakfast food	Nutrition experts	Heads of household
I	7	5
II	3	4
III	11	8
IV	9	14
V	1	2
VI	4	6
VII	10	12
VIII	8	7
IX	5	1
X	13	9
XI	12	15
XII	2	3
XIII	15	10
XIV	6	11
XV	14	13

Calculate r_S as a measure of the consistency of the two rankings.

13.52 The following are the ranks that three judges gave to the work of 10 artists:

Judge A:	5,	8,	4,	2,	3,	1,	10,	7,	9,	6
Judge B:	3,	10,	1,	4,	2,	5,	6,	7,	8,	9
Judge C:	8,	5,	6,	4,	10,	2,	3,	1,	7,	9

Calculate r_S for each pair of rankings and decide
(a) which two judges are most alike in their opinions about these artists;
(b) which two judges differ the most in their opinions about these artists.

Solutions to Practice Exercises

13.1   **1.** *Hypotheses* H_0: $\tilde{\mu} = 5$
H_A: $\tilde{\mu} \neq 5$

2. *Level of significance* $\alpha = 0.05$
3'. *Test statistic*   The statistic is x, the number of plus signs.
4'. *Calculations*   We find that x, the number of plus signs, is equal to 8.
5'. *Decision*   Using Table I for $n = 12$ and $p = 0.50$, the probability of $x \geq 8$ is 0.194. This corresponds to half the p-value, so that $p = 0.388$. Since 0.388 exceeds 0.05, the null hypothesis cannot be rejected.

13.7 Since three of the pairs (12 and 12, 9 and 9, and 15 and 15) have to be discarded, the sample size for the sign test is only $n = 17$.

1. *Hypotheses* $H_0:\ \mu_1 = \mu_2$
    $H_A:\ \mu_1 \neq \mu_2$
2. *Level of significance* $\alpha = 0.01$
3. *Criterion*   Reject the null hypothesis if $z \leq -2.575$ or $z \geq 2.575$, where

$$z = \frac{x - np_0}{\sqrt{np_0(1 - p_0)}}$$

with $p_0 = 0.50$. Otherwise, accept it or reserve judgment.

4. *Calculations*   Replacing each positive difference with a plus sign and each negative difference with a minus sign, and discarding the three pairs of equal values, we get

$$- - - + - - - - - + - - - + - - -$$

where the number of plus signs is 3. Thus, we get

$$z = \frac{3 - 17(0.50)}{\sqrt{17(0.50)(0.50)}} \approx -2.67$$

5. *Decision*   Since -2.67 is less than -2.575, the null hypothesis must be rejected; in other words, we conclude that, on average, the two policemen do not issue equally many speeding tickets per day. Had we used the continuity correction, we would have obtained

$$z = \frac{3.5 - 17(0.50)}{\sqrt{17(0.50)(0.50)}} \approx -2.43$$

Since this value falls between -2.575 and 2.575, we now find that the null hypothesis should not have been rejected.

13.14 1. *Hypotheses* $H_0:$ populations are identical
    $H_A:\ \mu_1 \neq \mu_2$
2. *Level of significance* $\alpha = 0.01$
3. *Criterion*   Reject the null hypothesis if $U \leq 2$, which is the value of U'_α for $n_1 = 6$, $n_2 = 6$, and $\alpha = 0.01$; otherwise, accept it or reserve judgment.
4. *Calculations*   Arranging the data jointly according to size, we get 25, 28, 30, 32, 35, 37, 38, 39, 42, 45, 46, and 48. Assigning the data in this order the ranks 1, 2, 3, ..., and 12, we find that the values of the first sample (spring) occupy ranks 6, 7, 9, 10, 11, and 12, while those of the second sample (fall) occupy ranks 1, 2, 3, 4, 5, and 8. Thus,

$$W_1 = 6 + 7 + 9 + 10 + 11 + 12 = 55$$
$$W_2 = 1 + 2 + 3 + 4 + 5 + 8 = 23$$

so that

$$U_1 = 55 - \frac{6 \cdot 7}{2} = 34$$

$$U_2 = 23 - \frac{6 \cdot 7}{2} = 2$$

and $U = 2$.

5. *Decision*   Since $U = 2$ is less than or equal to 2, the null hypothesis must be rejected; in other words, we conclude that, on average, there are not equally many assaults per week in the spring and in the fall.

*13.19   **1.** *Hypotheses* H_0: populations are identical
                                    H_A: $\mu_1 \neq \mu_2$

2. *Level of significance* $\alpha = 0.05$

3. *Criterion*   Reject the null hypothesis if $z \leq -1.96$ or $z \geq 1.96$, where

$$z = \frac{U_1 - \mu_{U_1}}{\sigma_{U_1}}$$

Otherwise, accept it or reserve judgment.

4. *Calculations*   Arranging the data jointly according to size, we get 23, 30, 41, 44, 47, 54, 55, 56, 58, 62, 64, 70, 71, 71, 72, 73, 78, 80, 81, 87, 91, 93, 94, and 96. Assigning the data in this order the ranks 1, 2, 3, ..., and 24, we find that the values of the first sample (minority group 1) occupy ranks 4, 7, 10, 12, 15, 16, 17, 18, 20, 21, 23, and 24, so that

$$W_1 = 4 + 7 + 10 + 12 + 15 + 16 + 17 + 18 + 20 + 21 + 23 + 24$$

$$= 187$$

and

$$U_1 = 187 - \frac{12 \cdot 13}{2} = 109$$

Since

$$\mu_{U_1} = \frac{12 \cdot 12}{2} = 72 \quad \text{and} \quad \sigma_{U_1} = \sqrt{\frac{12 \cdot 12 \cdot 25}{12}} \approx 17.32$$

it follows that

$$z = \frac{109 - 72}{17.32} \approx 2.14$$

5. *Decision*   Since $z = 2.14$ exceeds 1.96, the null hypothesis must be rejected; in other words, students from the two minority groups cannot be expected to score equally well on the test.

*13.24   **1.** *Hypotheses* H_0: populations are identical
                                    H_A: population means are not all equal

2. *Level of significance* $\alpha = 0.05$

3. *Criterion*   Reject the null hypothesis if $H \geq 5.991$, which is the value of $\chi^2_{0.05}$ for $3 - 1 = 2$ degrees of freedom; otherwise, accept it or reserve judgment.

4. *Calculations*   Arranging the data jointly according to size, we get 14, 16, 17, 18, 19, 21, 21, 22, 23, 24, 26, 27, 28, 29, 31, 31, 31, and 32. Assigning the data in this order the ranks 1, 2, 3, ..., and 18, we find that the values of the first sample occupy ranks 1, 9, 11, 13, 14, and 16, while those of the second sample occupy ranks 2, 5, 6.5, 12, 16, and 18, and those of the third sample occupy ranks 3, 4, 6.5, 8, 10, and 16. Thus,

$$R_1 = 1 + 9 + 11 + 13 + 14 + 16 = 64$$

$$R_2 = 2 + 5 + 6.5 + 12 + 16 + 18 = 59.5$$

$$R_3 = 3 + 4 + 6.5 + 8 + 10 + 16 = 47.5$$

and it follows that

$$H = \frac{12}{18 \cdot 19} \left(\frac{64^2}{6} + \frac{59.5^2}{6} + \frac{47.5^2}{6} \right) - 3 \cdot 19 \approx 0.85$$

5. *Decision* Since $H = 0.85$ is less than 5.991, the null hypothesis cannot be rejected; in other words, the data tend to support the claim that there is no difference in the true average mileage yield of the three kinds of gasoline.

13.28 1. *Hypotheses* H_0: arrangement is random
                          H_A: arrangement is not random
2. *Level of significance* $\alpha = 0.05$
3. *Criterion* Since $n_1 = 13$, $n_2 = 12$, and $\alpha = 0.05$, we get $u'_{0.025} = 8$ and $u_{0.025} = 19$ from Table VIII; thus, the null hypothesis must be rejected if $u \leq 8$ or $u \geq 19$.
4. *Calculations* $u = 15$
5. *Decision* Since $u = 15$ falls between 8 and 19, the null hypothesis cannot be rejected; in other words, there is no evidence to suspect that the arrangement is not random.

∗13.32 1. *Hypotheses* H_0: arrangement is random
                            H_A: arrangement is not random
2. *Level of significance* $\alpha = 0.05$
3. *Criterion* Reject the null hypothesis if $z \leq -1.96$ or $z \geq 1.96$, where

$$z = \frac{u - \mu_u}{\sigma_u}$$

Otherwise, accept it or reserve judgment.
4. *Calculations* Since there are $n_1 = 28$ even digits and $n_2 = 22$ odd digits, and $u = 23$, we get

$$\mu_u = \frac{2 \cdot 28 \cdot 22}{28 + 22} + 1 \approx 25.64$$

$$\sigma_u = \sqrt{\frac{2 \cdot 28 \cdot 22(2 \cdot 28 \cdot 22 - 28 - 22)}{(28 + 22)^2(28 + 22 - 1)}} \approx 3.45$$

and hence $z = \dfrac{23 - 25.64}{3.45} \approx -0.77$.

5. *Decision* Since $z = -0.77$ falls between -1.96 and 1.96, the null hypothesis cannot be rejected; there is no reason to suspect any lack of randomness.

13.37 1. *Hypotheses* H_0: arrangement is random
                          H_A: arrangement is not random

2. *Level of significance* $\alpha = 0.05$
3. *Criterion*   Since the median is 30.5, $n_1 = 12$, $n_2 = 12$, and $u'_{0.025} = 7$ and $u_{0.025} = 19$, according to Table VIII; thus, the null hypothesis must be rejected if $u \leq 7$ or $u \geq 19$. Otherwise, accept the null hypothesis or reserve judgment.
4. *Calculations*   The arrangement of values above and below the median is

$$a\ a\ a\ b\ b\ b\ b\ a\ a\ b\ b\ a\ a\ a\ b\ a\ a\ a\ a\ b\ b\ b\ b\ b$$

so that $u = 8$.
5. *Decision*   Since $u = 8$ falls between 7 and 19, the null hypothesis cannot be rejected; in other words, there is no significant deviation from randomness.

13.43 Ranking the x's among themselves and also the y's, we get the ranks shown in the first two columns of the following table:

RANK OF x	RANK OF y	d	d^2
9	6	3	9
4	2	2	4
7.5	6	1.5	2.25
15	14	1	1
7.5	11.5	−4	16
1	6	−5	25
2	3	−1	1
3	1	2	4
6	4	2	4
5	10	−5	25
13.5	15	−1.5	2.25
10	9	1	1
11	8	3	9
13.5	11.5	2	4
12	13	−1	1
		Total	108.5

Then, determining the d's and their squares and substituting $n = 15$ and $\Sigma d^2 = 108.5$ into the formula for r_S, we get

$$r_S = 1 - \frac{6(108.5)}{15(15^2 - 1)} \approx 0.81$$

13.44 1. *Hypotheses* H_0: $\rho = 0$ (no correlation)
 H_A: $\rho \neq 0$
2. *Level of significance* $\alpha = 0.05$
3. *Criterion*   Reject the null hypothesis if $z \leq -1.96$ or $z \geq 1.96$, where

$$z = r_S \sqrt{n - 1}$$

Otherwise, accept it or reserve judgment.
4. *Calculations*   For $n = 15$ and $r_S = 0.81$, we get

$$z = 0.81 \sqrt{15 - 1} \approx 3.03$$

5. *Decision*   Since $z = 3.03$ exceeds 1.96, the null hypothesis must be rejected; in other words, there is a relationship between students' grades in the two subjects.

Review: Chapters 12 & 13

. . . .

Achievements

Having read and studied these chapters and having worked a good proportion of the exercises, you should be able to

1. Explain the method of least squares.
2. Fit a least-squares line by solving the two normal equations.
3. Fit a least-squares line by using the special formulas for a and b.
4. Use the equation of a least-squares line to predict a value of y.
5. Explain what is meant by *regression line*.
*6. State the assumptions underlying normal regression analysis.
*7. Calculate the standard error of estimate.
*8. Test hypotheses about the regression coefficient β.
*9. Construct confidence intervals for the regression coefficient β.
*10. Construct confidence limits for the mean of y when $x = x_0$.
11. Define the correlation coefficient in terms of the residual and total sums of squares.
12. Explain the difference between positive and negative correlation.
13. Use the computing formula to calculate r.
14. Use r to judge the strength of a linear relationship.
15. Use values of r to compare the strength of two relationships.
16. Test the significance of a value of r.
17. Explain what is meant by *nonparametric tests*.
18. Perform a one-sample sign test.
19. Apply the sign test to paired data.
20. Perform a large-sample sign test.
21. Use the U test as a nonparametric alternative to the two-sample t test (concerning the difference between two means).
*22. Perform a large-sample U test.
*23. Use the H test as a nonparametric alternative to a one-way analysis of variance.
24. Test for randomness on the basis of the total number of runs.
*25. Perform a large-sample test of randomness based on the total number of runs.
26. Test for randomness on the basis of runs above and below the median.
27. Calculate the rank-correlation coefficient.
28. Test the significance of a rank-correlation coefficient.

Checklist of Key Terms (with page references to their definitions)

Coefficient of correlation, 424–425
Correlation, 404, 425
Curve fitting, 404
Data points, 407
Estimated regression
    coefficients, 417
Estimated regression line, 413
H test, 447
Kruskal–Wallis test, 447
Least-squares line, 408
Linear equation in two
    unknowns, 404
Linear regression analysis, 417
Mann–Whitney test, 441
Method of elimination, 410
Method of least squares, 407
Negative correlation, 425

No correlation, 426
Nonparametric tests, 435
Normal equations, 409
Normal regression analysis, 417
One-sample sign test, 436
Paired-sample sign test, 437
Population correlation
    coefficient, 428
Population regression
    coefficients, 417
Positive correlation, 425
Rank-correlation coefficient, 455
Rank sums, 442
Regression, 403
Regression coefficients, 417
Regression line, 413
Residual sum of squares, 424

Run, 450
Runs above and below the
    median, 452
Significant correlation, 428
Sign test, 436
Slope of the line, 404
Spearman's rank-correlation
    coefficient, 455
Standard error of estimate, 418
Theory of runs, 450
Total number of runs, 450
Total sum of squares, 424
u test, 450
U test, 441
Wilcoxon test, 441

Review Exercises

R.156 If $r = 0.56$ for one set of paired data and $r = 0.97$ for another set of paired data, compare the strengths of the two relationships.

R.157 The following are the high school averages, x, and first-year-college grade-point averages, y, of 10 students:

x	y
3.0	2.6
2.7	2.4
3.8	3.9
2.6	2.1
3.2	2.6
3.4	3.3
2.8	2.2
3.1	3.2
3.5	2.8
3.3	2.5

Fit a least-squares line that will enable us to predict first-year-college grade-point averages in terms of high school averages, and use it to predict the first-year-college grade-point average of a student with a high school average of 3.5.

∗R.158 With reference to Exercise R.157, construct a 95% confidence interval for the regression coefficient β.

R.159 If a set of $n = 42$ paired data yields $r = 0.33$, test the null hypothesis of no correlation
 (a) at the 0.05 level of significance;
 (b) at the 0.01 level of significance.

R.160 A sample of 30 suitcases carried by an airline on transoceanic flights weighed the following (in pounds):

$$
\begin{array}{cccccccccc}
32 & 46 & 48 & 27 & 35 & 52 & 66 & 41 & 49 & 36 \\
50 & 44 & 48 & 36 & 40 & 35 & 63 & 42 & 52 & 40 \\
38 & 36 & 43 & 41 & 49 & 52 & 44 & 60 & 31 & 35 \\
\end{array}
$$

Use the sign test at the 0.01 level of significance to test the null hypothesis $\widetilde{\mu} = 37$ pounds against the alternative hypothesis $\widetilde{\mu} > 37$ pounds.

R.161 The following are measurements of the strength of samples of two kinds of fishing lines (in pounds):

Fishing line 1: 12.0, 11.5, 11.8, 11.3, 12.2, 11.7, 11.9,
Fishing line 2: 11.1, 11.6, 11.4, 12.1, 10.5, 10.0, 10.8

Use the U test at the 0.05 level of significance to test the null hypothesis that the two samples come from identical populations against the alternative hypothesis that the two populations have unequal means.

R.162 The following sequence shows whether a certain member was present, *P*, or absent, *A*, at 20 consecutive meetings of a fraternal organization:

$$P\ P\ P\ P\ P\ P\ P\ A\ P\ P\ P\ P\ P\ P\ P\ A\ A\ A\ A$$

Test for randomness at the 0.05 level of significance.

R.163 The following are the scores of 12 golfers on two consecutive Sundays:

68 and 71   73 and 76   70 and 73   74 and 71   69 and 72   72 and 74
67 and 70   72 and 68   71 and 72   73 and 74   68 and 69   70 and 72

Use the sign test at the 0.05 level of significance to test the null hypothesis $\mu_1 = \mu_2$ against the alternative hypothesis $\mu_2 > \mu_1$ (perhaps due to heavy winds on the second Sunday).

R.164 The following are the numbers of hours that 10 persons (interviewed as part of a sample survey) spent reading books or magazines, *x*, and watching television, *y*, during the preceding week:

x	*y*
4	18
8	12
9	9
3	15
2	27
10	12
5	12
1	19
5	25
7	18

Calculate the coefficient of correlation.

R.165 With reference to Exercise R.164, test the significance of r at the 0.05 level of significance.

R.166 The following are the closing prices of a commodity on 20 consecutive trading days (in dollars):

$$
\begin{array}{cccccccccc}
378 & 379 & 379 & 378 & 377 & 376 & 374 & 374 & 373 & 373 \\
374 & 375 & 376 & 376 & 376 & 375 & 374 & 374 & 373 & 374 \\
\end{array}
$$

Test for randomness at the 0.01 level of significance.

R.167 State in each case whether you would expect a positive correlation, a negative correlation, or no correlation (a zero correlation):
  (a) the price of gasoline and the occupancy rate of motels;
  (b) exposure to the sun and the incidence of skin cancer;
  (c) shoe size and years of education;
  (d) mintage figures and the values of old coins;
  (e) blood pressure and eye color;
  (f) baseball players' batting averages and their salaries.

R.168 The following are the numbers of minutes it took two ambulance services to reach the scenes of accidents:

Ambulance service 1:   9.3,   5.5,   13.1,   10.0,   7.6,   9.2,   11.2,   6.4,   14.0,   10.3,
Ambulance service 2: 12.7,   6.6,   9.1,   4.5,   7.2,   6.4,   7.5

Use the *U* test at the 0.05 level of significance to test the null hypothesis that the two samples come from identical populations against the alternative hypothesis that the two populations have unequal means.

R.169 The following are the numbers of persons who attended a "singles only" dance on 12 consecutive Saturdays:

152 188 149 212 103 146 177 158 201 175 188 162

Use the sign test based on Table I and the 0.05 level of significance to test the null hypothesis $\tilde{\mu} = 149$ against the alternative hypothesis $\tilde{\mu} > 149$.

*****R.170** The following sequence shows whether a television news program had at least 12% of the viewing audience, *A*, or less than 12%, *L*, on 36 consecutive weekday evenings:

L L L L A A L L L A L L L A A A A L

(cont.)   *A L L L A A L L L L L A L L L L L A*

Test for randomness at the 0.05 level of significance.

R.171 The following data was obtained in a study of the relationship between the resistance (ohms), *x*, and the failure time (minutes), *y*, of certain overloaded resistors:

x	*y*
33	38
41	45
48	40
46	44
35	33
30	32
50	48

Fit a least-squares line and use it to predict the failure time of such a resistor when the resistance is 40 ohms.

*****R.172** With reference to Exercise R.171, construct a 95% confidence interval for the mean failure time of such a resistor when the resistance is 40 ohms.

*****R.173** The following are the numbers of minutes that patients had to wait in the offices of four doctors:

Doctor 1:	20,	25,	38,	31,	23
Doctor 2:	19,	12,	21,	24,	9
Doctor 3:	8,	10,	27,	25,	14
Doctor 4:	17,	25,	28,	21,	15

Use the H test at the 0.05 level of significance to test the null hypothesis that the four samples come from identical populations against the alternative hypothesis that the means of the populations are not all equal.

R.174 Calculate r for the data of Exercise R.171 and test whether it is significant at the 0.01 level of significance.

R.175 The following are the numbers of burglaries committed in two suburbs of a large city on 24 days:

20 and 17	12 and 9	25 and 21	18 and 14	12 and 15	27 and 23
22 and 18	20 and 15	18 and 18	19 and 21	24 and 22	17 and 23
20 and 11	16 and 22	13 and 9	18 and 15	20 and 17	13 and 18
22 and 16	25 and 21	20 and 13	12 and 6	18 and 9	17 and 25

Use the large-sample sign test at the 0.05 level of significance to test the null hypothesis that on the average there are equally many burglaries in the two suburbs against the alternative hypothesis that there are not equally many.

R.176 The following are the batting averages, x, and home runs hit, y, by 15 baseball players during the first half of a baseball season:

x	y
0.252	12
0.305	6
0.299	4
0.303	15
0.285	2
0.191	2
0.283	16
0.272	6
0.310	8
0.266	10
0.215	0
0.211	3
0.272	14
0.244	6
0.320	7

Calculate the rank-correlation coefficient and check whether it is significant at the 0.05 level of significance.

R.177 Test at the 0.05 level of significance whether the following sequence of cars observing, O, and exceeding, E, the 65-mph speed limit may be regarded as random:

$$O\ O\ O\ E\ E\ O\ E\ O\ O\ O\ O\ E\ E\ E\ O\ E\ O\ O\ O\ O\ E\ O$$

∗R.178 The following is data on the percentage kill of two kinds of insecticides used against mosquitos:

Insecticide A:	41.9, 46.9, 44.6, 43.9, 42.0, 44.0, 41.0, 43.1, 39.0, 45.2, 44.6, 42.0
Insecticide B:	45.7, 39.8, 42.8, 41.2, 45.0, 40.2, 40.2, 41.7, 37.4, 38.8, 41.7, 38.7

Use the large-sample U test at the 0.05 level of significance to test the null hypothesis that the two samples come from identical populations against the alternative hypothesis that on the average the first insecticide is more effective than the second.

R.179 Rework Exercise R.178 using Table VII and the alternative hypothesis that on the average the two insecticides are not equally effective.

R.180 The following are the processing times (minutes), x, and hardness readings, y, of certain machine parts:

x	y
20	282
34	275
19	171
10	142
24	145
31	340
25	282
13	105
29	233

Fit a least-squares line and use it to predict the hardness reading of a part that has been processed for 25 minutes.

∗R.181 With reference to Exercise R.180, test the null hypothesis $\beta = 9.5$ against the alternative hypothesis $\beta < 9.5$ at the 0.01 level of significance.

R.182 With reference to Exercise R.180, what percentage of the variation among the hardness readings can be attributed to differences in processing times?

R.183 If $r_S = 0.34$ for a set of $n = 50$ paired data, is this rank-correlation coefficient significant at the 0.01 level of significance?

Appendix A: TI-83 Tips

■ KEYS   Each button (key) on the TI-83 can perform up to three functions.

The primary key function is written in white on the key. To use the primary function, simply press the key.

2nd key   The secondary function is written in yellow above the key. To use a secondary function, you must first press the yellow 2nd key.

■ EXAMPLE   To turn the calculator off, press 2nd ON for OFF.

alpha key   The alpha function is written in green above the key. To use an alpha function, you must first press the green ALPHA key. Most alpha functions simply print a letter or variable on the screen.

■ SCREENS
home screen   When you first turn on the calculator you are in the home screen, where you can perform computations and other commands. You can use the calculator keys to bring up other screens and menus that allow you to do more than is directly available on the keyboard. You can return to the home screen at any time by pressing 2nd MODE for QUIT.

■ EXAMPLE
graph screen   To get to the graphing screen, press GRAPH.
(The GRAPH key is in the upper right-hand corner of the keyboard.)

■ EXAMPLE   To return to the home screen, press 2nd MODE for QUIT.

■ MENUS   One type of screen you will use often is a menu. Menus allow you to do more than is directly available on the keyboard. You can move around these menus using the arrow keys located in the upper right-hand corner of the keyboard.

Some of the most common statistics menus are:

STAT EDIT	(press STAT)
STAT CALC	(press STAT ▶)
STAT TESTS	(press STAT ▶ ▶)
DISTR	(press 2nd VARS)
DISTR DRAW	(press 2nd VARS ▶)
STAT PLOT	(press 2nd Y=)
VARS Statistics	(press VARS 5)
MATH PRB	(press MATH ◀)

■ **CLEARING LISTS** Statistics are based on data, which you can enter into lists on the TI-83 (see Data entry, below). Before entering new data, you may want to clear out old data.

To clear all lists, press 2nd + for **MEM**.
Then press 4 to select 4: ClrAllLists; then press **ENTER**.

You should see: ClrAllLists
                                                                      Done

Note that you can also select 4:ClrAllLists using the down arrow key to highlight 4: and pressing **ENTER**. Then you would need to press **ENTER** one more time to execute the ClrAllLists command.

■ **CLEARING PLOTS** Graphical representation of data is important in statistics. Before plotting a new set of data, you should clear old plots.

If you want to turn off statistics plots only, without affecting anything else, press 2nd Y= for **STAT PLOT**.
Press 4 for 4: PlotsOff; then press **ENTER**.

You can also turn off plots by resetting the calculator defaults (see Resetting defaults, below).

■ **RESETTING DEFAULTS** You can restore the initial or default settings on the TI-83 without losing your data or programs.

To reset defaults, press 2nd + for **MEM**.
Press 5 for 5: Reset....
Press 2 for 2: Defaults...; then press 2 for 2: Reset.

CAUTION: Be careful not to reset all memory or you will lose all data and programs!

Resetting defaults will turn off all plots and reset the graphing screen to its standard scale (see Setting graph dimensions, below). Other defaults that you probably never changed will also be reset.

trouble shooting Any time you find that your calculator seems to be acting strange, try resetting defaults. You may have previously changed some of the defaults accidentally. Resetting defaults may just fix the problem.

■ **DATA ENTRY** To enter data on the TI-83, you need to use the List Editor. You can access the List Editor by pressing **STAT ENTER** for **EDIT**.

■ **EXAMPLE** To enter the following data set

91   59   75   66   60   69   61   72   90   89

1. Press STAT ENTER for STAT EDIT.

   You should see

   This screen is the List Editor. The lists should be empty. If not, clear the lists (see Clearing lists, above).

2. Press 91.

   You will see L1(1) = 91 at the bottom of the screen.

3. Press ENTER.

   91 is entered into L1.

4. Press 59 ENTER.

   59 is entered into L1.

5. Enter the rest of the data by typing in each number and pressing ENTER after each number. When you are done, you should see

   Notice that you may not see all of the data. When you reach the bottom of the screen, the data scrolls off the screen at the top. You can use the up arrow key to move back up the data list and see the numbers that scrolled off the screen. You can use the down arrow key to move back down through the list. You can use the left and right arrow keys to move between lists.

correcting mistakes If you entered any incorrect numbers, use the arrow keys to highlight the incorrect list entry, retype the entry, and press ENTER. Note that the highlighted entry is always shown at the bottom of the screen.

negative sign If your data includes a negative number, you need to use the gray negative sign (−) key to enter the number rather than the blue subtraction key.

■ **EXAMPLE** Change the first data entry in the above data set to −91.

   1. Use the arrow keys to highlight the 91.

   2. Press the (−) key, then press 91 ENTER.

      The first entry in the list should change to −91.

If you got an error message, you probably used the subtraction key. Return to the List Editor and try again.

■ **SETTING GRAPH DIMENSIONS**   The standard graphing screen shows the rectangular coordinate system with *x*-values going from –10 to 10 and *y*-values going from –10 to 10.

window settings   To see the graphing screen settings, press WINDOW.
You will see the current scale settings for the graphing screen.

> Xmin = smallest *x*-value shown in the graphing screen.
>
> Xmax = largest *x*-value shown in the graphing screen.
>
> Xscl = distance between successive scale marks on the *x*-axis.
>
> Ymin = smallest *y*-value shown in the graphing screen.
>
> Ymax = largest *y*-value shown in the graphing screen.
>
> Yscl = distance between successive scale marks on the *y*-axis.
>
> Xres = resolution of graph with Xres = 1 being the most improved.

You can change the settings using the down arrow key to access each line and making a new entry on that line.

graphs of data   Choosing appropriate dimensions for a graph depends on the data and type of graph you are working with. If you make a graph using one list of data, Xmin should be smaller than the smallest number in the list and Xmax should be larger than the largest number in the list. If you are using two data lists, then Ymin should be smaller than the smallest number in the second list and Ymax should be larger than the largest number in the second list.

standard normal distribution   The TI-83 allows you to draw common probability distributions. These graphs and their dimensions are based on information about the distribution rather than actual data. The most common of these distributions is the normal distribution. For a normal distribution, virtually the entire distribution falls within 5 standard deviations of the mean. You can use the following window settings for graphs of the standard normal distribution:

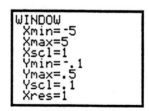

To see the new graph dimensions, press GRAPH.

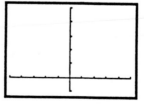

■ **INFINITY**   In statistics, you can indicate that there is no highest number in a function by using infinity. Since infinity is not a number, there is no key for infinity on the calculator. However, you can use instead the largest number the calculator can hold, which is 1×10^{99}, meaning 1 followed by 99 zeros.

To enter 1×10^{99}, press 1.
Press 2nd, then press the comma key for EE, and then press 99.
You should see 1E99

1E99 is engineering notation for 1×10^{99}.

minus infinity   You can use -11×10^{99} for negative infinity. Press the negative sign (–) key, then press 1; press 2nd followed by the comma key for EE, and then press 99.

■ **EXPRESSIONS WITH**   Many of the functions on the calculator automatically give an opening paren-
PARENTHESES   thesis. Be sure to include a closing parenthesis at the end of the expression.

■ **EXAMPLE**   To sketch the standard normal distribution:

1. Set the graph dimensions (see standard normal distribution, above).
2. Press 2nd VARS ▶ for DISTR DRAW.
3. Press ENTER for 1:ShadeNorm(.
4. To get just the distribution with no shading, press 0,0)
   (You are telling the calculator to shade from 0 to 0.)
   You should see 1:ShadeNorm(0,0)
5. Press ENTER.
   You should see

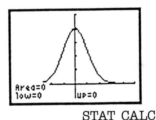

■ **MOST COMMON**   STAT EDIT   STAT CALC
STATISTICS MENUS   (press STAT)   (press STAT)

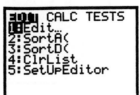

STAT TESTS
(press STAT ▶ ▶)

```
EDIT CALC TESTS
1▣Z-Test…
2:T-Test…
3:2-SampZTest…
4:2-SampTTest…
5:1-PropZTest…
6:2-PropZTest…
7↓ZInterval…
```

```
EDIT CALC TESTS
7↑ZInterval…
8:TInterval…
9:2-SampZInt…
0:2-SampTInt…
A:1-PropZInt…
B:2-PropZInt…
C↓X²-Test…
```

```
EDIT CALC TESTS
C↑X²-Test…
D:2-SampFTest…
E:LinRegTTest…
F:ANOVA(
```

DISTR
(press 2nd VARS)

```
DISTR DRAW
1▣normalpdf(
2:normalcdf(
3:invNorm(
4:tpdf(
5:tcdf(
6:X²pdf(
7↓X²cdf(
```

```
DISTR DRAW
7↑X²cdf(
8:Fpdf(
9:Fcdf(
0:binompdf(
A:binomcdf(
B:poissonpdf(
C↓poissoncdf(
```

```
DISTR DRAW
C↑poissoncdf(
D:geometpdf(
E:geometcdf(
```

DISTR DRAW
(press 2nd VARS ▶)

```
DISTR DRAW
1▣ShadeNorm(
2:Shade_t(
3:ShadeX²(
4:ShadeF(
```

STAT PLOT
(press 2nd Y=)

```
STAT PLOTS
1▣Plot1…Off
   ⊾ L1   L2   ▫
2:Plot2…Off
   ⊾ L1   L2   ▫
3:Plot3…Off
   ⊾ L1   L2   ▫
4↓PlotsOff
```

```
STAT PLOTS
2↑Plot2…Off
   ⊾ L1   L2   ▫
3:Plot3…Off
   ⊾ L1   L2   ▫
4:PlotsOff
5▣PlotsOn
```

VARS Statistics
(press VARS 5)

```
XY Σ EQ TEST PTS
1▣n
2:x̄
3:Sx
4:σx
5:ȳ
6:Sy
7↓σy
```

```
XY Σ EQ TEST PTS
7↑σy
8:minX
9:maxX
0:minY
A:maxY
```

MATH PRB
(press MATH ◀)

```
MATH NUM CPX PRB
1▣rand
2:nPr
3:nCr
4:!
5:randInt(
6:randNorm(
7:randBin(
```

Bibliography

A. PROBABILITY AND STATISTICS FOR THE LAYMAN

BROOK, R. J., ARNOLD, G. C., HASSARD, T. H., and PRINGLE, R. M., eds., *The Fascination of Statistics*. New York: Marcel Dekker, Inc., 1986.

CAMPBELL, S. K., *Flaws and Fallacies in Statistical Thinking*. Englewood Cliffs, N.J.: Prentice Hall, 1974.

FEDERER, W. T., *Statistics and Society*. New York: Marcel Dekker, Inc., 1991.

GARVIN, A. D., *Probability in Your Life*. Portland, Me: J. Weston Walch Publisher, 1978.

GONICK, L., and SMITH, W., *The Cartoon Guide to Statistics*. New York: Harper-Collins, 1993.

HOLLANDER, M., and PROSCHAN, F., *The Statistical Exorcist*: *Dispelling Statistics Anxiety*. New York: Marcel Dekker, Inc., 1984.

HOOKE, R., *How to Tell the Liars from the Statisticians*. New York: Marcel Dekker, Inc., 1983.

HUFF, D., *How to Lie with Statistics*. New York: W. W. Norton & Company, Inc., 1993.

HUFF, D., and GEIS, I., *How to Take a Chance*. New York: W. W. Norton & Company, Inc., 1959.

JOFFE, A., and SPIRER, H. F., *Misused Statistics*: *Straight Talk for Twisted Numbers*. New York: Marcel Dekker, Inc., 1986.

KIMBLE, G. A., *How to Use (and Misuse) Statistics*. Englewood Cliffs, N.J.: Prentice Hall, 1978.

LARSEN, R. J., and STROUP, D. F., *Statistics in the Real World*. New York: Macmillan Publishing Co., Inc., 1976.

LEVINSON, H. C., *Chance, Luck, and Statistics*. New York: Dover Publications, Inc., 1963.

MORONEY, M. J., *Facts from Figures*. New York: Penguin Books, 1956.

MOSTELLER, F., PIETERS, R. S., KRUSKAL, W. H., RISING, G. R., LINK, R. F., CARLSON, R., and ZELINKA, M., *Statistics by Example*. Reading, Mass.: Addison-Wesley Publishing Company, Inc., 1973.

PAULOS, J. A., *Innumeracy*. New York: Random House, 1990.

REICHMAN, W. J., *Use and Abuse of Statistics*. New York: Penguin Books, 1971.

RUNYON, R. P., *Winning with Statistics*. Reading, Mass.: Addison-Wesley Publishing Company, Inc., 1977.

TANUR, J. M. (ed.), *Statistics*: *A Guide to the Unknown*. Belmont, Calif.: Wadsworth Publishing Company, 1989.

WANG, C., *Sense and Nonsense of Statistical Inference*. New York: Marcel Dekker, Inc., 1992.

WEAVER, W., *Lady Luck*: *The Theory of Probability*. New York: Dover Publications, Inc., 1982.

B. SOME BOOKS ON THE THEORY OF PROBABILITY AND STATISTICS

FREUND, J. E., *Introduction to Probability*. New York: Dover Publications, Inc., 1993 reprint.

GOLDBERG, S., *Probability—An Introduction*. New York: Dover Publications, Inc., 1987.

HODGES, J. L., and LEHMANN, E. L., *Basic Concepts of Probability and Statistics*, 2nd rev. ed. Boca Raton, Fla.: Holden Day, 1970.

HOEL, P., PORT, S. C., and STONE, C. J., *Introduction to Probability*. Boston: Houghton Mifflin Company, 1971.

MENDENHALL, W., and SCHAEFFER, R. L., *Mathematical Statistics with Applications*. North Scituate, Mass.: Duxbury Press, 1973.

MILLER, I., and MILLER, M., *John E. Freund's Mathematical Statistics*, 6th ed. Upper Saddle River, N. J.: Prentice Hall, 1998.

MOSTELLER, F., ROURKE, R. E. K., and THOMAS, G. B., *Probability with Statistical Applications*, 2nd ed. Reading, Mass.: Addison-Wesley Publishing Company, Inc., 1970.

RICE, J. A., *Mathematical Statistics and Data Analysis*, 2nd ed. Belmont, Calif.: Wadsworth Publishing Co., 1995.

ROSS, S., *A First Course in Probability*, 5th ed. Upper Saddle River, N.J.: Prentice Hall, 1998.

SCHAEFFER, R. L., and MENDENHALL, W., *Introduction to Probability*: *Theory and Applications*. North Scituate, Mass.: Duxbury Press, 1975.

C. SOME BOOKS DEALING WITH SPECIAL TOPICS

BOX, G. E. P., HUNTER, W. G., and HUNTER, J. S., *Statistics for Experimenters*. New York: John Wiley & Sons, Inc., 1978.

CHATTERJEE, S., PRICE, B., *Regression Analysis by Example*, 2nd ed. New York: John Wiley and Sons, Inc., 1991.

CLEVELAND, W. S., *The Elements of Graphing Data*, rev. ed. Summit, N.J.: Hobart Press, 1994.

EVERITT, B. S., *The Analysis of Contingency Tables*, 2nd ed. New York: Chapman & Hall, 1992.

GIBBONS, J. D., *Nonparametric Statistical Inference*. New York: Marcel Dekker, 1985.

HARTWIG, F., and DEARING, B. E., *Exploratory Data Analysis*. Beverly Hills, Calif.: Sage Publications, Inc., 1979.

LEHMANN, E. L., *Nonparametrics*: *Statistical Methods Based on Ranks*. San Francisco: Holden-Day, Inc., 1975.

MORGAN, L., *Statistics Handbook for the TI-83*. Austin, Tex.: Texas Instruments, 1997.

MOSTELLER F., *Fifty Challenging Problems in Probability*. New York: Dover Publications, Inc., 1987.

MOSTELLER, F., and ROURKE, R. E. K., *Sturdy Statistics*. Reading, Mass.: Addison-Wesley Publishing Company, Inc., 1973.

NOETHER, G. E., *Introduction to Statistics*: *The Nonparametric Way*. New York: Springer-Verlag, 1990.

WILLIAMS, W. H., *A Sampler on Sampling*. New York: John Wiley & Sons, Inc., 1978.

D. SOME GENERAL REFERENCE WORKS AND TABLES

FREUND, J. E., and WILLIAMS, F. J., *Dictionary/Outline of Basic Statistics*. New York: Dover Publications, Inc., 1991 reprint.

HAUSER, P. M., and LEONARD, W. R., *Government Statistics for Business Use*, 2nd ed. New York: John Wiley & Sons, Inc., 1956.

KENDALL, M. G., and BUCKLAND, W. R., *A Dictionary of Statistical Terms*, 4th ed. London: Longman Group Ltd, 1982.

NATIONAL BUREAU OF STANDARDS, *Tables of the Binomial Probability Distribution*. Washington, D.C.: Government Printing Office, 1950.

OWEN, D. B., *Handbook of Statistical Tables*. Reading, Mass.: Addison-Wesley Publishing Company, Inc., 1962.

RAND CORPORATION, *A Million Random Digits with 100,000 Normal Deviates*. New York: The Free Press, 1955.

ROMIG, H. G., *50–100 Binomial Tables*. New York: John Wiley & Sons, Inc., 1953.

Statistical Tables

TABLE I   BINOMIAL PROBABILITIES[†]

n	x	0.05	0.1	0.2	0.3	0.4	0.5	0.6	0.7	0.8	0.9	0.95
2	0	0.902	0.810	0.640	0.490	0.360	0.250	0.160	0.090	0.040	0.010	0.002
	1	0.095	0.180	0.320	0.420	0.480	0.500	0.480	0.420	0.320	0.180	0.095
	2	0.002	0.010	0.040	0.090	0.160	0.250	0.360	0.490	0.640	0.810	0.902
3	0	0.857	0.729	0.512	0.343	0.216	0.125	0.064	0.027	0.008	0.001	
	1	0.135	0.243	0.384	0.441	0.432	0.375	0.288	0.189	0.096	0.027	0.007
	2	0.007	0.027	0.096	0.189	0.288	0.375	0.432	0.441	0.384	0.243	0.135
	3		0.001	0.008	0.027	0.064	0.125	0.216	0.343	0.512	0.729	0.857
4	0	0.815	0.656	0.410	0.240	0.130	0.062	0.026	0.008	0.002		
	1	0.171	0.292	0.410	0.412	0.346	0.250	0.154	0.076	0.026	0.004	
	2	0.014	0.049	0.154	0.265	0.346	0.375	0.346	0.265	0.154	0.049	0.014
	3		0.004	0.026	0.076	0.154	0.250	0.346	0.412	0.410	0.292	0.171
	4			0.002	0.008	0.026	0.062	0.130	0.240	0.410	0.656	0.815
5	0	0.774	0.590	0.328	0.168	0.078	0.031	0.010	0.002			
	1	0.204	0.328	0.410	0.360	0.259	0.156	0.077	0.028	0.006		
	2	0.021	0.073	0.205	0.309	0.346	0.312	0.230	0.132	0.051	0.008	0.001
	3	0.001	0.008	0.051	0.132	0.230	0.312	0.346	0.309	0.205	0.073	0.021
	4			0.006	0.028	0.077	0.156	0.259	0.360	0.410	0.328	0.204
	5				0.002	0.010	0.031	0.078	0.168	0.328	0.590	0.774
6	0	0.735	0.531	0.262	0.118	0.047	0.016	0.004	0.001			
	1	0.232	0.354	0.393	0.303	0.187	0.094	0.037	0.010	0.002		
	2	0.031	0.098	0.246	0.324	0.311	0.234	0.138	0.060	0.015	0.001	
	3	0.002	0.015	0.082	0.185	0.276	0.312	0.276	0.185	0.082	0.015	0.002
	4		0.001	0.015	0.060	0.138	0.234	0.311	0.324	0.246	0.098	0.031
	5			0.002	0.010	0.037	0.094	0.187	0.303	0.393	0.354	0.232
	6				0.001	0.004	0.016	0.047	0.118	0.262	0.531	0.735
7	0	0.698	0.478	0.210	0.082	0.028	0.008	0.002				
	1	0.257	0.372	0.367	0.247	0.131	0.055	0.017	0.004			
	2	0.041	0.124	0.275	0.318	0.261	0.164	0.077	0.025	0.004		
	3	0.004	0.023	0.115	0.227	0.290	0.273	0.194	0.097	0.029	0.003	
	4		0.003	0.029	0.097	0.194	0.273	0.290	0.227	0.115	0.023	0.004
	5			0.004	0.025	0.077	0.164	0.261	0.318	0.275	0.124	0.041
	6				0.004	0.017	0.055	0.131	0.247	0.367	0.372	0.257
	7					0.002	0.008	0.028	0.082	0.210	0.478	0.698
8	0	0.663	0.430	0.168	0.058	0.017	0.004	0.001				
	1	0.279	0.383	0.336	0.198	0.090	0.031	0.008	0.001			
	2	0.051	0.149	0.294	0.296	0.209	0.109	0.041	0.010	0.001		
	3	0.005	0.033	0.147	0.254	0.279	0.219	0.124	0.047	0.009		
	4		0.005	0.046	0.136	0.232	0.273	0.232	0.136	0.046	0.005	
	5			0.009	0.047	0.124	0.219	0.279	0.254	0.147	0.033	0.005
	6			0.001	0.010	0.041	0.109	0.209	0.296	0.294	0.149	0.051
	7				0.001	0.008	0.031	0.090	0.198	0.336	0.383	0.279
	8					0.001	0.004	0.017	0.058	0.168	0.430	0.663

[†] All values omitted in Table I are 0.0005 or less.

TABLE I   BINOMIAL PROBABILITIES (continued)

n	x	\(p\) 0.05	0.1	0.2	0.3	0.4	0.5	0.6	0.7	0.8	0.9	0.95
9	0	0.630	0.387	0.134	0.040	0.010	0.002					
	1	0.299	0.387	0.302	0.156	0.060	0.018	0.004				
	2	0.063	0.172	0.302	0.267	0.161	0.070	0.021	0.004			
	3	0.008	0.045	0.176	0.267	0.251	0.164	0.074	0.021	0.003		
	4	0.001	0.007	0.066	0.172	0.251	0.246	0.167	0.074	0.017	0.001	
	5		0.001	0.017	0.074	0.167	0.246	0.251	0.172	0.066	0.007	0.001
	6			0.003	0.021	0.074	0.164	0.251	0.267	0.176	0.045	0.008
	7				0.004	0.021	0.070	0.161	0.267	0.302	0.172	0.063
	8					0.004	0.018	0.060	0.156	0.302	0.387	0.299
	9						0.002	0.010	0.040	0.134	0.387	0.630
10	0	0.599	0.349	0.107	0.028	0.006	0.001					
	1	0.315	0.387	0.268	0.121	0.040	0.010	0.002				
	2	0.075	0.194	0.302	0.233	0.121	0.044	0.011	0.001			
	3	0.010	0.057	0.201	0.267	0.215	0.117	0.042	0.009	0.001		
	4	0.001	0.011	0.088	0.200	0.251	0.205	0.111	0.037	0.006		
	5		0.001	0.026	0.103	0.201	0.246	0.201	0.103	0.026	0.001	
	6			0.006	0.037	0.111	0.205	0.251	0.200	0.088	0.011	0.001
	7			0.001	0.009	0.042	0.117	0.215	0.267	0.201	0.057	0.010
	8				0.001	0.011	0.044	0.121	0.233	0.302	0.194	0.075
	9					0.002	0.010	0.040	0.121	0.268	0.387	0.315
	10						0.001	0.006	0.028	0.107	0.349	0.599
11	0	0.569	0.314	0.086	0.020	0.004						
	1	0.329	0.384	0.236	0.093	0.027	0.005	0.001				
	2	0.087	0.213	0.295	0.200	0.089	0.027	0.005	0.001			
	3	0.014	0.071	0.221	0.257	0.177	0.081	0.023	0.004			
	4	0.001	0.016	0.111	0.220	0.236	0.161	0.070	0.017	0.002		
	5		0.002	0.039	0.132	0.221	0.226	0.147	0.057	0.010		
	6			0.010	0.057	0.147	0.226	0.221	0.132	0.039	0.002	
	7			0.002	0.017	0.070	0.161	0.236	0.220	0.111	0.016	0.001
	8				0.004	0.023	0.081	0.177	0.257	0.221	0.071	0.014
	9				0.001	0.005	0.027	0.089	0.200	0.295	0.213	0.087
	10					0.001	0.005	0.027	0.093	0.236	0.384	0.329
	11							0.004	0.020	0.086	0.314	0.569
12	0	0.540	0.282	0.069	0.014	0.002						
	1	0.341	0.377	0.206	0.071	0.017	0.003					
	2	0.099	0.230	0.283	0.168	0.064	0.016	0.002				
	3	0.017	0.085	0.236	0.240	0.142	0.054	0.012	0.001			
	4	0.002	0.021	0.133	0.231	0.213	0.121	0.042	0.008	0.001		
	5		0.004	0.053	0.158	0.227	0.193	0.101	0.029	0.003		
	6			0.016	0.079	0.177	0.226	0.177	0.079	0.016		
	7			0.003	0.029	0.101	0.193	0.227	0.158	0.053	0.004	
	8			0.001	0.008	0.042	0.121	0.213	0.231	0.133	0.021	0.002
	9				0.001	0.012	0.054	0.142	0.240	0.236	0.085	0.017
	10					0.002	0.016	0.064	0.168	0.283	0.230	0.099
	11						0.003	0.017	0.071	0.206	0.377	0.341
	12							0.002	0.014	0.069	0.282	0.540

TABLE I   BINOMIAL PROBABILITIES (continued)

n	x	0.05	0.1	0.2	0.3	0.4	p 0.5	0.6	0.7	0.8	0.9	0.95
13	0	0.513	0.254	0.055	0.010	0.001						
	1	0.351	0.367	0.179	0.054	0.011	0.002					
	2	0.111	0.245	0.268	0.139	0.045	0.010	0.001				
	3	0.021	0.100	0.246	0.218	0.111	0.035	0.006	0.001			
	4	0.003	0.028	0.154	0.234	0.184	0.087	0.024	0.003			
	5		0.006	0.069	0.180	0.221	0.157	0.066	0.014	0.001		
	6		0.001	0.023	0.103	0.197	0.209	0.131	0.044	0.006		
	7			0.006	0.044	0.131	0.209	0.197	0.103	0.023	0.001	
	8			0.001	0.014	0.066	0.157	0.221	0.180	0.069	0.006	
	9				0.003	0.024	0.087	0.184	0.234	0.154	0.028	0.003
	10				0.001	0.006	0.035	0.111	0.218	0.246	0.100	0.021
	11					0.001	0.010	0.045	0.139	0.268	0.245	0.111
	12						0.002	0.011	0.054	0.179	0.367	0.351
	13							0.001	0.010	0.055	0.254	0.513
14	0	0.488	0.229	0.044	0.007	0.001						
	1	0.359	0.356	0.154	0.041	0.007	0.001					
	2	0.123	0.257	0.250	0.113	0.032	0.006	0.001				
	3	0.026	0.114	0.250	0.194	0.085	0.022	0.003				
	4	0.004	0.035	0.172	0.229	0.155	0.061	0.014	0.001			
	5		0.008	0.086	0.196	0.207	0.122	0.041	0.007			
	6		0.001	0.032	0.126	0.207	0.183	0.092	0.023	0.002		
	7			0.009	0.062	0.157	0.209	0.157	0.062	0.009		
	8			0.002	0.023	0.092	0.183	0.207	0.126	0.032	0.001	
	9				0.007	0.041	0.122	0.207	0.196	0.086	0.008	
	10				0.001	0.014	0.061	0.155	0.229	0.172	0.035	0.004
	11					0.003	0.022	0.085	0.194	0.250	0.114	0.026
	12					0.001	0.006	0.032	0.113	0.250	0.257	0.123
	13						0.001	0.007	0.041	0.154	0.356	0.359
	14							0.001	0.007	0.044	0.229	0.488
15	0	0.463	0.206	0.035	0.005							
	1	0.366	0.343	0.132	0.031	0.005						
	2	0.135	0.267	0.231	0.092	0.022	0.003					
	3	0.031	0.129	0.250	0.170	0.063	0.014	0.002				
	4	0.005	0.043	0.188	0.219	0.127	0.042	0.007	0.001			
	5	0.001	0.010	0.103	0.206	0.186	0.092	0.024	0.003			
	6		0.002	0.043	0.147	0.207	0.153	0.061	0.012	0.001		
	7			0.014	0.081	0.177	0.196	0.118	0.035	0.003		
	8			0.003	0.035	0.118	0.196	0.177	0.081	0.014		
	9			0.001	0.012	0.061	0.153	0.207	0.147	0.043	0.002	
	10				0.003	0.024	0.092	0.186	0.206	0.103	0.010	0.001
	11				0.001	0.007	0.042	0.127	0.219	0.188	0.043	0.005
	12					0.002	0.014	0.063	0.170	0.250	0.129	0.031
	13						0.003	0.022	0.092	0.231	0.267	0.135
	14							0.005	0.031	0.132	0.343	0.366
	15								0.005	0.035	0.206	0.463

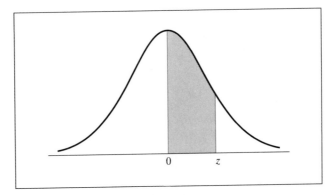

The entries in Table II are the probabilities that a random variable having the standard normal distribution will take on a value between 0 and z. They are given by the area of the gray region under the curve in the figure.

TABLE II   NORMAL-CURVE AREAS

z	.00	.01	.02	.03	.04	.05	.06	.07	.08	.09
0.0	.0000	.0040	.0080	.0120	.0160	.0199	.0239	.0279	.0319	.0359
0.1	.0398	.0438	.0478	.0517	.0557	.0596	.0636	.0675	.0714	.0753
0.2	.0793	.0832	.0871	.0910	.0948	.0987	.1026	.1064	.1103	.1141
0.3	.1179	.1217	.1255	.1293	.1331	.1368	.1406	.1443	.1480	.1517
0.4	.1554	.1591	.1628	.1664	.1700	.1736	.1772	.1808	.1844	.1879
0.5	.1915	.1950	.1985	.2019	.2054	.2088	.2123	.2157	.2190	.2224
0.6	.2257	.2291	.2324	.2357	.2389	.2422	.2454	.2486	.2517	.2549
0.7	.2580	.2611	.2642	.2673	.2704	2734	.2764	.2794	.2823	.2852
0.8	.2881	.2910	.2939	.2967	.2995	.3023	.3051	.3078	.3106	.3133
0.9	.3159	.3186	.3212	.3238	.3264	.3289	.3315	.3340	.3365	.3389
1.0	.3413	.3438	.3461	.3485	.3508	.3531	.3554	.3577	.3599	.3621
1.1	.3643	.3665	.3686	.3708	.3729	.3749	.3770	.3790	.3810	.3830
1.2	.3849	.3869	.3888	.3907	.3925	.3944	.3962	.3980	.3997	.4015
1.3	.4032	.4049	.4066	.4082	.4099	.4115	.4131	.4147	.4162	.4177
1.4	.4192	.4207	.4222	.4236	.4251	.4265	.4279	.4292	.4306	.4319
1.5	.4332	.4345	.4357	.4370	.4382	.4394	.4406	.4418	.4429	.4441
1.6	.4452	.4463	.4474	.4484	.4495	.4505	.4515	.4525	.4535	.4545
1.7	.4554	.4564	.4573	.4582	.4591	.4599	.4608	.4616	.4625	.4633
1.8	.4641	.4649	.4656	.4664	.4671	.4678	.4686	.4693	.4699	.4706
1.9	.4713	.4719	.4726	.4732	.4738	.4744	.4750	.4756	.4761	.4767
2.0	.4772	.4778	.4783	.4788	.4793	.4798	.4803	.4808	.4812	.4817
2.1	.4821	.4826	.4830	.4834	.4838	.4842	.4846	.4850	.4854	.4857
2.2	.4861	.4864	.4868	.4871	.4875	.4878	.4881	.4884	.4887	.4890
2.3	.4893	.4896	.4898	.4901	.4904	.4906	.4909	.4911	.4913	.4916
2.4	.4918	.4920	.4922	.4925	.4927	.4929	.4931	.4932	.4934	.4936
2.5	.4938	.4940	.4941	.4943	.4945	.4946	.4948	.4949	.4951	.4952
2.6	.4953	.4955	.4956	.4957	.4959	.4960	.4961	.4962	.4963	.4964
2.7	.4965	.4966	.4967	.4968	.4969	.4970	.4971	.4972	.4973	.4974
2.8	.4974	.4975	.4976	.4977	.4977	.4978	.4979	.4979	.4980	.4981
2.9	.4981	.4982	.4982	.4983	.4984	.4984	.4985	.4985	.4986	.4986
3.0	.4987	.4987	.4987	.4988	.4988	.4989	.4989	.4989	.4990	.4990

Also, for $z = 4.0$, 5.0, and 6.0, the areas are 0.49997, 0.4999997, and 0.499999999.

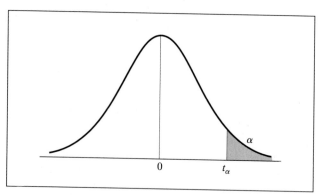

The entries in Table III are values for which the area to their right under the t distribution with given degrees of freedom (the gray area in the figure) is equal to α.

TABLE III   VALUES OF t [†]

d.f.	$t_{0.050}$	$t_{0.025}$	$t_{0.010}$	$t_{0.005}$	d.f.
1	6.314	12.706	31.821	63.657	1
2	2.920	4.303	6.965	9.925	2
3	2.353	3.182	4.541	5.841	3
4	2.132	2.776	3.747	4.604	4
5	2.015	2.571	3.365	4.032	5
6	1.943	2.447	3.143	3.707	6
7	1.895	2.365	2.998	3.499	7
8	1.860	2.306	2.896	3.355	8
9	1.833	2.262	2.821	3.250	9
10	1.812	2.228	2.764	3.169	10
11	1.796	2.201	2.718	3.106	11
12	1.782	2.179	2.681	3.055	12
13	1.771	2.160	2.650	3.012	13
14	1.761	2.145	2.624	2.977	14
15	1.753	2.131	2.602	2.947	15
16	1.746	2.120	2.583	2.921	16
17	1.740	2.110	2.567	2.898	17
18	1.734	2.101	2.552	2.878	18
19	1.729	2.093	2.539	2.861	19
20	1.725	2.086	2.528	2.845	20
21	1.721	2.080	2.518	2.831	21
22	1.717	2.074	2.508	2.819	22
23	1.714	2.069	2.500	2.807	23
24	1.711	2.064	2.492	2.797	24
25	1.708	2.060	2.485	2.787	25
26	1.706	2.056	2.479	2.779	26
27	1.703	2.052	2.473	2.771	27
28	1.701	2.048	2.467	2.763	28
29	1.699	2.045	2.462	2.756	29
inf.	1.645	1.960	2.326	2.576	inf.

[†] Richard A. Johnson and Dean W. Wichern, *Applied Multivariate Statistical Analysis*, © 1982, p. 582. Adapted by permission of Prentice Hall, Upper Saddle River, N.J.

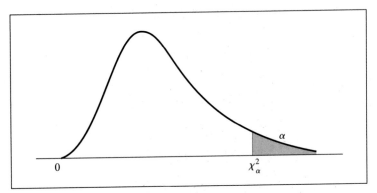

The entries in Table IV are values for which the area to their right under the chi-square distribution with given degrees of freedom (the gray area in the figure) is equal to α.

TABLE IV   VALUES OF $\chi^{2\dagger}$

d.f.	$\chi^2_{0.05}$	$\chi^2_{0.01}$	d.f.
1	3.841	6.635	1
2	5.991	9.210	2
3	7.815	11.345	3
4	9.488	13.277	4
5	11.070	15.086	5
6	12.592	16.812	6
7	14.067	18.475	7
8	15.507	20.090	8
9	16.919	21.666	9
10	18.307	23.209	10
11	19.675	24.725	11
12	21.026	26.217	12
13	22.362	27.688	13
14	23.685	29.141	14
15	24.996	30.578	15
16	26.296	32.000	16
17	27.587	33.409	17
18	28.869	34.805	18
19	30.144	36.191	19
20	31.410	37.566	20
21	32.671	38.932	21
22	33.924	40.289	22
23	35.172	41.638	23
24	36.415	42.980	24
25	37.652	44.314	25
26	38.885	45.642	26
27	40.113	46.963	27
28	41.337	48.278	28
29	42.557	49.588	29
30	43.773	50.892	30

$\dagger$ Based on Table 8 of *Biometrika Tables for Statisticians*, Vol. I, Cambridge, Cambridge University Press, 1954, by permission of the *Biometrika* trustees.

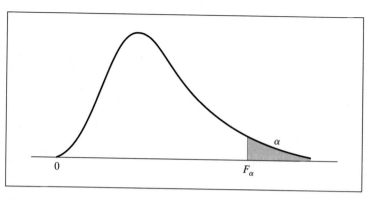

The entries in Table V are values for which the area to their right under the F distribution with given degrees of freedom (the gray area in the figure) is equal to α.

TABLE V VALUES OF $F_{0.05}$ [†]

Degrees of freedom for numerator

	1	2	3	4	5	6	7	8	9	10	12	15	20	24	30	40	60	120	∞
1	161	200	216	225	230	234	237	239	241	242	244	246	248	249	250	251	252	253	254
2	18.5	19.0	19.2	19.2	19.3	19.3	19.4	19.4	19.4	19.4	19.4	19.4	19.4	19.5	19.5	19.5	19.5	19.5	19.5
3	10.1	9.55	9.28	9.12	9.01	8.94	8.89	8.85	8.81	8.79	8.74	8.70	8.66	8.64	8.62	8.59	8.57	8.55	8.53
4	7.71	6.94	6.59	6.39	6.26	6.16	6.09	6.04	6.00	5.96	5.91	5.86	5.80	5.77	5.75	5.72	5.69	5.66	5.63
5	6.61	5.79	5.41	5.19	5.05	4.95	4.88	4.82	4.77	4.74	4.68	4.62	4.56	4.53	4.50	4.46	4.43	4.40	4.37
6	5.99	5.14	4.76	4.53	4.39	4.28	4.21	4.15	4.10	4.06	4.00	3.94	3.87	3.84	3.81	3.77	3.74	3.70	3.67
7	5.59	4.74	4.35	4.12	3.97	3.87	3.79	3.73	3.68	3.64	3.57	3.51	3.44	3.41	3.38	3.34	3.30	3.27	3.23
8	5.32	4.46	4.07	3.84	3.69	3.58	3.50	3.44	3.39	3.35	3.28	3.22	3.15	3.12	3.08	3.04	3.01	2.97	2.93
9	5.12	4.26	3.86	3.63	3.48	3.37	3.29	3.23	3.18	3.14	3.07	3.01	2.94	2.90	2.86	2.83	2.79	2.75	2.71
10	4.96	4.10	3.71	3.48	3.33	3.22	3.14	3.07	3.02	2.98	2.91	2.85	2.77	2.74	2.70	2.66	2.62	2.58	2.54
11	4.84	3.98	3.59	3.36	3.20	3.09	3.01	2.95	2.90	2.85	2.79	2.72	2.65	2.61	2.57	2.53	2.49	2.45	2.40
12	4.75	3.89	3.49	3.26	3.11	3.00	2.91	2.85	2.80	2.75	2.69	2.62	2.54	2.51	2.47	2.43	2.38	2.34	2.30
13	4.67	3.81	3.41	3.18	3.03	2.92	2.83	2.77	2.71	2.67	2.60	2.53	2.46	2.42	2.38	2.34	2.30	2.25	2.21
14	4.60	3.74	3.34	3.11	2.96	2.85	2.76	2.70	2.65	2.60	2.53	2.46	2.39	2.35	2.31	2.27	2.22	2.18	2.13
15	4.54	3.68	3.29	3.06	2.90	2.79	2.71	2.64	2.59	2.54	2.48	2.40	2.33	2.29	2.25	2.20	2.16	2.11	2.07
16	4.49	3.63	3.24	3.01	2.85	2.74	2.66	2.59	2.54	2.49	2.42	2.35	2.28	2.24	2.19	2.15	2.11	2.06	2.01
17	4.45	3.59	3.20	2.96	2.81	2.70	2.61	2.55	2.49	2.45	2.38	2.31	2.23	2.19	2.15	2.10	2.06	2.01	1.96
18	4.41	3.55	3.16	2.93	2.77	2.66	2.58	2.51	2.46	2.41	2.34	2.27	2.19	2.15	2.11	2.06	2.02	1.97	1.92
19	4.38	3.52	3.13	2.90	2.74	2.63	2.54	2.48	2.42	2.38	2.31	2.23	2.16	2.11	2.07	2.03	1.98	1.93	1.88
20	4.35	3.49	3.10	2.87	2.71	2.60	2.51	2.45	2.39	2.35	2.28	2.20	2.12	2.08	2.04	1.99	1.95	1.90	1.84
21	4.32	3.47	3.07	2.84	2.68	2.57	2.49	2.42	2.37	2.32	2.25	2.18	2.10	2.05	2.01	1.96	1.92	1.87	1.81
22	4.30	3.44	3.05	2.82	2.66	2.55	2.46	2.40	2.34	2.30	2.23	2.15	2.07	2.03	1.98	1.94	1.89	1.84	1.78
23	4.28	3.42	3.03	2.80	2.64	2.53	2.44	2.37	2.32	2.27	2.20	2.13	2.05	2.01	1.96	1.91	1.86	1.81	1.76
24	4.26	3.40	3.01	2.78	2.62	2.51	2.42	2.36	2.30	2.25	2.18	2.11	2.03	1.98	1.94	1.89	1.84	1.79	1.73
25	4.24	3.39	2.99	2.76	2.60	2.49	2.40	2.34	2.28	2.24	2.16	2.09	2.01	1.96	1.92	1.87	1.82	1.77	1.71
30	4.17	3.32	2.92	2.69	2.53	2.42	2.33	2.27	2.21	2.16	2.09	2.01	1.93	1.89	1.84	1.79	1.74	1.68	1.62
40	4.08	3.23	2.84	2.61	2.45	2.34	2.25	2.18	2.12	2.08	2.00	1.92	1.84	1.79	1.74	1.69	1.64	1.58	1.51
60	4.00	3.15	2.76	2.53	2.37	2.25	2.17	2.10	2.04	1.99	1.92	1.84	1.75	1.70	1.65	1.59	1.53	1.47	1.39
120	3.92	3.07	2.68	2.45	2.29	2.18	2.09	2.02	1.96	1.91	1.83	1.75	1.66	1.61	1.55	1.50	1.43	1.35	1.25
∞	3.84	3.00	2.60	2.37	2.21	2.10	2.01	1.94	1.88	1.83	1.75	1.67	1.57	1.52	1.46	1.39	1.32	1.22	1.00

Degrees of freedom for denominator

[†] Reproduced from M. Merrington and C. M. Thompson, "Tables of percentage points of the inverted beta (F) distribution," *Biometrika*, Vol. 33 (1943), by permission of the *Biometrika* trustees.

TABLE V   VALUES OF $F_{0.01}$[†] (continued)

Degrees of freedom for numerator

	1	2	3	4	5	6	7	8	9	10	12	15	20	24	30	40	60	120	∞
1	4,052	5,000	5,403	5,625	5,764	5,859	5,928	5,982	6,023	6,056	6,106	6,157	6,209	6,235	6,261	6,287	6,313	6,339	6,366
2	98.5	99.0	99.2	99.2	99.3	99.3	99.4	99.4	99.4	99.4	99.4	99.4	99.4	99.5	99.5	99.5	99.5	99.5	99.5
3	34.1	30.8	29.5	28.7	28.2	27.9	27.7	27.5	27.3	27.2	27.1	26.9	26.7	26.6	26.5	26.4	26.3	26.2	26.1
4	21.2	18.0	16.7	16.0	15.5	15.2	15.0	14.8	14.7	14.5	14.4	14.2	14.0	13.9	13.8	13.7	13.7	13.6	13.5
5	16.3	13.3	12.1	11.4	11.0	10.7	10.5	10.3	10.2	10.1	9.89	9.72	9.55	9.47	9.38	9.29	9.20	9.11	9.02
6	13.7	10.9	9.78	9.15	8.75	8.47	8.26	8.10	7.98	7.87	7.72	7.56	7.40	7.31	7.23	7.14	7.06	6.97	6.88
7	12.2	9.55	8.45	7.85	7.46	7.19	6.99	6.84	6.72	6.62	6.47	6.31	6.16	6.07	5.99	5.91	5.82	5.74	5.65
8	11.3	8.65	7.59	7.01	6.63	6.37	6.18	6.03	5.91	5.81	5.67	5.52	5.36	5.28	5.20	5.12	5.03	4.95	4.86
9	10.6	8.02	6.99	6.42	6.06	5.80	5.61	5.47	5.35	5.26	5.11	4.96	4.81	4.73	4.65	4.57	4.48	4.40	4.31
10	10.0	7.56	6.55	5.99	5.64	5.39	5.20	5.06	4.94	4.85	4.71	4.56	4.41	4.33	4.25	4.17	4.08	4.00	3.91
11	9.65	7.21	6.22	5.67	5.32	5.07	4.89	4.74	4.63	4.54	4.40	4.25	4.10	4.02	3.94	3.86	3.78	3.69	3.60
12	9.33	6.93	5.95	5.41	5.06	4.82	4.64	4.50	4.39	4.30	4.16	4.01	3.86	3.78	3.70	3.62	3.54	3.45	3.36
13	9.07	6.70	5.74	5.21	4.86	4.62	4.44	4.30	4.19	4.10	3.96	3.82	3.66	3.59	3.51	3.43	3.34	3.25	3.17
14	8.86	6.51	5.56	5.04	4.70	4.46	4.28	4.14	4.03	3.94	3.80	3.66	3.51	3.43	3.35	3.27	3.18	3.09	3.00
15	8.68	6.36	5.42	4.89	4.56	4.32	4.14	4.00	3.89	3.80	3.67	3.52	3.37	3.29	3.21	3.13	3.05	2.96	2.87
16	8.53	6.23	5.29	4.77	4.44	4.20	4.03	3.89	3.78	3.69	3.55	3.41	3.26	3.18	3.10	3.02	2.93	2.84	2.75
17	8.40	6.11	5.19	4.67	4.34	4.10	3.93	3.79	3.68	3.59	3.46	3.31	3.16	3.08	3.00	2.92	2.83	2.75	2.65
18	8.29	6.01	5.09	4.58	4.25	4.01	3.84	3.71	3.60	3.51	3.37	3.23	3.08	3.00	2.92	2.84	2.75	2.66	2.57
19	8.19	5.93	5.01	4.50	4.17	3.94	3.77	3.63	3.52	3.43	3.30	3.15	3.00	2.92	2.84	2.76	2.67	2.58	2.49
20	8.10	5.85	4.94	4.43	4.10	3.87	3.70	3.56	3.46	3.37	3.23	3.09	2.94	2.86	2.78	2.69	2.61	2.52	2.42
21	8.02	5.78	4.87	4.37	4.04	3.81	3.64	3.51	3.40	3.31	3.17	3.03	2.88	2.80	2.72	2.64	2.55	2.46	2.36
22	7.95	5.72	4.82	4.31	3.99	3.76	3.59	3.45	3.35	3.26	3.12	2.98	2.83	2.75	2.67	2.58	2.50	2.40	2.31
23	7.88	5.66	4.76	4.26	3.94	3.71	3.54	3.41	3.30	3.21	3.07	2.93	2.78	2.70	2.62	2.54	2.45	2.35	2.26
24	7.82	5.61	4.72	4.22	3.90	3.67	3.50	3.36	3.26	3.17	3.03	2.89	2.74	2.66	2.58	2.49	2.40	2.31	2.21
25	7.77	5.57	4.68	4.18	3.86	3.63	3.46	3.32	3.22	3.13	2.99	2.85	2.70	2.62	2.53	2.45	2.36	2.27	2.17
30	7.56	5.39	4.51	4.02	3.70	3.47	3.30	3.17	3.07	2.98	2.84	2.70	2.55	2.47	2.39	2.30	2.21	2.11	2.01
40	7.31	5.18	4.31	3.83	3.51	3.29	3.12	2.99	2.89	2.80	2.66	2.52	2.37	2.29	2.20	2.11	2.02	1.92	1.80
60	7.08	4.98	4.13	3.65	3.34	3.12	2.95	2.82	2.72	2.63	2.50	2.35	2.20	2.12	2.03	1.94	1.84	1.73	1.60
120	6.85	4.79	3.95	3.48	3.17	2.96	2.79	2.66	2.56	2.47	2.34	2.19	2.03	1.95	1.86	1.76	1.66	1.53	1.38
∞	6.63	4.61	3.78	3.32	3.02	2.80	2.64	2.51	2.41	2.32	2.18	2.04	1.88	1.79	1.70	1.59	1.47	1.32	1.00

Degrees of freedom for denominator

TABLE VI   CRITICAL VALUES OF r

n	$r_{0.025}$	$r_{0.005}$	n	$r_{0.025}$	$r_{0.005}$
3	0.997		18	0.468	0.590
4	0.950	0.999	19	0.456	0.575
5	0.878	0.959	20	0.444	0.561
6	0.811	0.917	21	0.433	0.549
7	0.754	0.875	22	0.423	0.537
8	0.707	0.834	27	0.381	0.487
9	0.666	0.798	32	0.349	0.449
10	0.632	0.765	37	0.325	0.418
11	0.602	0.735	42	0.304	0.393
12	0.576	0.708	47	0.288	0.372
13	0.553	0.684	52	0.273	0.354
14	0.532	0.661	62	0.250	0.325
15	0.514	0.641	72	0.232	0.302
16	0.497	0.623	82	0.217	0.283
17	0.482	0.606	92	0.205	0.267

TABLE VII   CRITICAL VALUES OF $U^†$

Values of $U'_{0.05}$

n_1 \ n_2	2	3	4	5	6	7	8	9	10	11	12	13	14	15
2							0	0	0	0	1	1	1	1
3				0	1	1	2	2	3	3	4	4	5	5
4			0	1	2	3	4	4	5	6	7	8	9	10
5		0	1	2	3	5	6	7	8	9	11	12	13	14
6		1	2	3	5	6	8	10	11	13	14	16	17	19
7		1	3	5	6	8	10	12	14	16	18	20	22	24
8	0	2	4	6	8	10	13	15	17	19	22	24	26	29
9	0	2	4	7	10	12	15	17	20	23	26	28	31	34
10	0	3	5	8	11	14	17	20	23	26	29	33	36	39
11	0	3	6	9	13	16	19	23	26	30	33	37	40	44
12	1	4	7	11	14	18	22	26	29	33	37	41	45	49
13	1	4	8	12	16	20	24	28	33	37	41	45	50	54
14	1	5	9	13	17	22	26	31	36	40	45	50	55	59
15	1	5	10	14	19	24	29	34	39	44	49	54	59	64

Values of $U'_{0.01}$

n_1 \ n_2	3	4	5	6	7	8	9	10	11	12	13	14	15
3							0	0	0	1	1	1	2
4				0	0	1	1	2	2	3	3	4	5
5			0	1	1	2	3	4	5	6	7	7	8
6		0	1	2	3	4	5	6	7	9	10	11	12
7		0	1	3	4	6	7	9	10	12	13	15	16
8		1	2	4	6	7	9	11	13	15	17	18	20
9	0	1	3	5	7	9	11	13	16	18	20	22	24
10	0	2	4	6	9	11	13	16	18	21	24	26	29
11	0	2	5	7	10	13	16	18	21	24	27	30	33
12	1	3	6	9	12	15	18	21	24	27	31	34	37
13	1	3	7	10	13	17	20	24	27	31	34	38	42
14	1	4	7	11	15	18	22	26	30	34	38	42	46
15	2	5	8	12	16	20	24	29	33	37	42	46	51

† This table in based on Table 11.4 of Donald B. Owen, *Handbook of Statistical Tables*, © 1962, Addison-Wesley, Reading, Massachusetts. Reprinted with permission.

TABLE VIII   CRITICAL VALUES OF $u^{\dagger}$

Values of $u_{0.025}$

n_1 \ n_2	4	5	6	7	8	9	10	11	12	13	14	15
4		9	9									
5	9	10	10	11	11							
6	9	10	11	12	12	13	13	13	13			
7		11	12	13	13	14	14	14	14	15	15	15
8		11	12	13	14	14	15	15	16	16	16	16
9			13	14	14	15	16	16	16	17	17	18
10			13	14	15	16	16	17	17	18	18	18
11			13	14	15	16	17	17	18	19	19	19
12			13	14	16	16	17	18	19	19	20	20
13				15	16	17	18	19	19	20	20	21
14				15	16	17	18	19	20	20	21	22
15				15	16	18	18	19	20	21	22	22

Values of $u'_{0.025}$

n_1 \ n_2	2	3	4	5	6	7	8	9	10	11	12	13	14	15
2											2	2	2	2
3			2	2	2	2	2	2	2	2	2	2	2	3
4			2	2	2	3	3	3	3	3	3	3	3	3
5			2	2	3	3	3	3	3	4	4	4	4	4
6		2	2	3	3	3	3	4	4	4	4	5	5	5
7		2	2	3	3	3	4	4	5	5	5	5	5	6
8		2	3	3	3	4	4	5	5	5	6	6	6	6
9		2	3	3	4	4	5	5	5	6	6	6	7	7
10		2	3	3	4	5	5	5	6	6	7	7	7	7
11		2	3	4	4	5	5	6	6	7	7	7	8	8
12	2	2	3	4	4	5	6	6	7	7	7	8	8	8
13	2	2	3	4	5	5	6	6	7	7	8	8	9	9
14	2	2	3	4	5	5	6	7	7	8	8	9	9	9
15	2	3	3	4	5	6	6	7	7	8	8	9	9	10

† This table is adapted, by permission, from F. S. Swed and C. Eisenhart, "Tables for testing randomness of grouping in a sequence of alternatives," *Annals of Mathematical Statistics*. Vol. 14.

TABLE VIII   CRITICAL VALUES OF u (continued)

n_1 \ n_2	5	6	7	8	9	10	11	12	13	14	15
					Values of $u_{0.005}$						
5		11									
6	11	12	13	13							
7		13	13	14	15	15	15				
8		13	14	15	15	16	16	17	17	17	
9			15	15	16	17	17	18	18	18	19
10			15	16	17	17	18	19	19	19	20
11			15	16	17	18	19	19	20	20	21
12				17	18	19	19	20	21	21	22
13				17	18	19	20	21	21	22	22
14				17	18	19	20	21	22	23	23
15					19	20	21	22	22	23	24

n_1 \ n_2	3	4	5	6	7	8	9	10	11	12	13	14	15
						Values of $u'_{0.005}$							
3										2	2	2	2
4						2	2	2	2	2	2	2	3
5				2	2	2	2	3	3	3	3	3	3
6			2	2	2	3	3	3	3	3	3	4	4
7			2	2	3	3	3	3	4	4	4	4	4
8		2	2	3	3	3	3	4	4	4	5	5	5
9		2	2	3	3	3	4	4	5	5	5	5	6
10		2	3	3	3	4	4	5	5	5	5	6	6
11		2	3	3	4	4	5	5	5	6	6	6	7
12	2	2	3	3	4	4	5	5	6	6	6	7	7
13	2	2	3	3	4	5	5	5	6	6	7	7	7
14	2	2	3	4	4	5	5	6	6	7	7	7	8
15	2	3	3	4	4	5	6	6	7	7	7	8	8

TABLE IX   FACTORIALS

n	$n!$
0	1
1	1
2	2
3	6
4	24
5	120
6	720
7	5,040
8	40,320
9	362,880
10	3,628,800
11	39,916,800
12	479,001,600
13	6,227,020,800
14	87,178,291,200
15	1,307,674,368,000

TABLE X   BINOMIAL COEFFICIENTS

n	$\binom{n}{0}$	$\binom{n}{1}$	$\binom{n}{2}$	$\binom{n}{3}$	$\binom{n}{4}$	$\binom{n}{5}$	$\binom{n}{6}$	$\binom{n}{7}$	$\binom{n}{8}$	$\binom{n}{9}$	$\binom{n}{10}$
0	1										
1	1	1									
2	1	2	1								
3	1	3	3	1							
4	1	4	6	4	1						
5	1	5	10	10	5	1					
6	1	6	15	20	15	6	1				
7	1	7	21	35	35	21	7	1			
8	1	8	28	56	70	56	28	8	1		
9	1	9	36	84	126	126	84	36	9	1	
10	1	10	45	120	210	252	210	120	45	10	1
11	1	11	55	165	330	462	462	330	165	55	11
12	1	12	66	220	495	792	924	792	495	220	66
13	1	13	78	286	715	1287	1716	1716	1287	715	286
14	1	14	91	364	1001	2002	3003	3432	3003	2002	1001
15	1	15	105	455	1365	3003	5005	6435	6435	5005	3003
16	1	16	120	560	1820	4368	8008	11440	12870	11440	8008
17	1	17	136	680	2380	6188	12376	19448	24310	24310	19448
18	1	18	153	816	3060	8568	18564	31824	43758	48620	43758
19	1	19	171	969	3876	11628	27132	50388	75582	92378	92378
20	1	20	190	1140	4845	15504	38760	77520	125970	167960	184756

If necessary, use the identity $\binom{n}{r} = \binom{n}{n-r}$.

TABLE XI   VALUES OF e^{-x}

x	e^{-x}	x	e^{-x}	x	e^{-x}	x	e^{-x}
0.0	1.000	2.5	0.082	5.0	0.0067	7.5	0.00055
0.1	0.905	2.6	0.074	5.1	0.0061	7.6	0.00050
0.2	0.819	2.7	0.067	5.2	0.0055	7.7	0.00045
0.3	0.741	2.8	0.061	5.3	0.0050	7.8	0.00041
0.4	0.670	2.9	0.055	5.4	0.0045	7.9	0.00037
0.5	0.607	3.0	0.050	5.5	0.0041	8.0	0.00034
0.6	0.549	3.1	0.045	5.6	0.0037	8.1	0.00030
0.7	0.497	3.2	0.041	5.7	0.0033	8.2	0.00028
0.8	0.449	3.3	0.037	5.8	0.0030	8.3	0.00025
0.9	0.407	3.4	0.033	5.9	0.0027	8.4	0.00023
1.0	0.368	3.5	0.030	6.0	0.0025	8.5	0.00020
1.1	0.333	3.6	0.027	6.1	0.0022	8.6	0.00018
1.2	0.301	3.7	0.025	6.2	0.0020	8.7	0.00017
1.3	0.273	3.8	0.022	6.3	0.0018	8.8	0.00015
1.4	0.247	3.9	0.020	6.4	0.0017	8.9	0.00014
1.5	0.223	4.0	0.018	6.5	0.0015	9.0	0.00012
1.6	0.202	4.1	0.017	6.6	0.0014	9.1	0.00011
1.7	0.183	4.2	0.015	6.7	0.0012	9.2	0.00010
1.8	0.165	4.3	0.014	6.8	0.0011	9.3	0.00009
1.9	0.150	4.4	0.012	6.9	0.0010	9.4	0.00008
2.0	0.135	4.5	0.011	7.0	0.0009	9.5	0.00008
2.1	0.122	4.6	0.010	7.1	0.0008	9.6	0.00007
2.2	0.111	4.7	0.009	7.2	0.0007	9.7	0.00006
2.3	0.100	4.8	0.008	7.3	0.0007	9.8	0.00006
2.4	0.091	4.9	0.007	7.4	0.0006	9.9	0.00005

TABLE XII   RANDOM NUMBERS[†]

04433	80674	24520	18222	10610	05794	37515
60298	47829	72648	37414	75755	04717	29899
67884	59651	67533	68123	17730	95862	08034
89512	32155	51906	61662	64130	16688	37275
32653	01895	12506	88535	36553	23757	34209
95913	15405	13772	76638	48423	25018	99041
55864	21694	13122	44115	01601	50541	00147
35334	49810	91601	40617	72876	33967	73830
57729	32196	76487	11622	96297	24160	09903
86648	13697	63677	70119	94739	25875	38829
30574	47609	07967	32422	76791	39725	53711
81307	43694	83580	79974	45929	85113	72268
02410	54905	79007	54939	21410	86980	91772
18969	75274	52233	62319	08598	09066	95288
87863	82384	66860	62297	80198	19347	73234
68397	71708	15438	62311	72844	60203	46412
28529	54447	58729	10854	99058	18260	38765
44285	06372	15867	70418	57012	72122	36634
86299	83430	33571	23309	57040	29285	67870
84842	68668	90894	61658	15001	94055	36308
56970	83609	52098	04184	54967	72938	56834
83125	71257	60490	44369	66130	72936	69848
55503	52423	02464	26141	68779	66388	75242
47019	76273	33203	29608	54553	25971	69573
84828	32592	79526	29554	84580	37859	28504
68921	08141	79227	05748	51276	57143	31926
36458	96045	30424	98420	72925	40729	22337
95752	59445	36847	87729	81679	59126	59437
26768	47323	58454	56958	20575	76746	49878
42613	37056	43636	58085	06766	60227	96414
95457	30566	65482	25596	02678	54592	63607
95276	17894	63564	95958	39750	64379	46059
66954	52324	64776	92345	95110	59448	77249
17457	18481	14113	62462	02798	54977	48349
03704	36872	83214	59337	01695	60666	97410
21538	86497	33210	60337	27976	70661	08250
57178	67619	98310	70348	11317	71623	55510
31048	97558	94953	55866	96283	46620	52087
69799	55380	16498	80733	96422	58078	99643
90595	61867	59231	17772	67831	33317	00520
33570	04981	98939	78784	09977	29398	93896
15340	93460	57477	13898	48431	72936	78160
64079	42483	36512	56186	99098	48850	72527
63491	05546	67118	62063	74958	20946	28147
92003	63868	41034	28260	79708	00770	88643
52360	46658	66511	04172	73085	11795	52594
74622	12142	68355	65635	21828	39539	18988
04157	50079	61343	64315	70836	82857	35335
86003	60070	66241	32836	27573	11479	94114
41268	80187	20351	09636	84668	42486	71303

[†] Based on parts of *Tables of 105,000 Random Decimal Digits*. Interstate Commerce Commission, Bureau of Transport Economics and Statistics, Washington, D.C.

TABLE XII   RANDOM NUMBERS (continued)

48611	62866	33963	14045	79451	04934	45576
78812	03509	78673	73181	29973	18664	04555
19472	63971	37271	31445	49019	49405	46925
51266	11569	08697	91120	64156	40365	74297
55806	96275	26130	47949	14877	69594	83041
77527	81360	18180	97421	55541	90275	18213
77680	58788	33016	61173	93049	04694	43534
15404	96554	88265	34537	38526	67924	40474
14045	22917	60718	66487	46346	30949	03173
68376	43918	77653	04127	69930	43283	35766
93385	13421	67957	20384	58731	53396	59723
09858	52104	32014	53115	03727	98624	84616
93307	34116	49516	42148	57740	31198	70336
04794	01534	92058	03157	91758	80611	45357
86265	49096	97021	92582	61422	75890	86442
65943	79232	45702	67055	39024	57383	44424
90038	94209	04055	27393	61517	23002	96560
97283	95943	78363	36498	40662	94188	18202
21913	72958	75637	99936	58715	07943	23748
41161	37341	81838	19389	80336	46346	91895
23777	98392	31417	98547	92058	02277	50315
59973	08144	61070	73094	27059	69181	55623
82690	74099	77885	23813	10054	11900	44653
83854	24715	48866	65745	31131	47636	45137
61980	34997	41825	11623	07320	15003	56774
99915	45821	97702	87125	44488	77613	56823
48293	86847	43186	42951	37804	85129	28993
33225	31280	41232	34750	91097	60752	69783
06846	32828	24425	30249	78801	26977	92074
32671	45587	79620	84831	38156	74211	82752
82096	21913	75544	55228	89796	05694	91552
51666	10433	10945	55306	78562	89630	41230
54044	67942	24145	42294	27427	84875	37022
66738	60184	75679	38120	17640	36242	99357
55064	17427	89180	74018	44865	53197	74810
69599	60264	84549	78007	88450	06488	72274
64756	87759	92354	78694	63638	80939	98644
80817	74533	68407	55862	32476	19326	95558
39847	96884	84657	33697	39578	90197	80532
90401	41700	95510	61166	33757	23279	85523
78227	90110	81378	96659	37008	04050	04228
87240	52716	87697	79433	16336	52862	69149
08486	10951	26832	39763	02485	71688	90936
39338	32169	03713	93510	61244	73774	01245
21188	01850	69689	49426	49128	14660	14143
13287	82531	04388	64693	11934	35051	68576
53609	04001	19648	14053	49623	10840	31915
87900	36194	31567	53506	34304	39910	79630
81641	00496	36058	75899	46620	70024	88753
19512	50277	71508	20116	79520	06269	74173

TABLE XII RANDOM NUMBERS (continued)

24418	23508	91507	76455	54941	72711	39406
57404	73678	08272	62941	02349	71389	45605
77644	98489	86268	73652	98210	44546	27174
68366	65614	01443	07607	11826	91326	29664
64472	72294	95432	53555	96810	17100	35066
88205	37913	98633	81009	81060	33449	68055
98455	78685	71250	10329	56135	80647	51404
48977	36794	56054	59243	57361	65304	93258
93077	72941	92779	23581	24548	56415	61927
84533	26564	91583	83411	66504	02036	02922
11338	12903	14514	27585	45068	05520	56321
23853	68500	92274	87026	99717	01542	72990
94096	74920	25822	98026	05394	61840	83089
83160	82362	09350	98536	38155	42661	02363
97425	47335	69709	01386	74319	04318	99387
83951	11954	24317	20345	18134	90062	10761
93085	35203	05740	03206	92012	42710	34650
33762	83193	58045	89880	78101	44392	53767
49665	85397	85137	30496	23469	42846	94810
37541	82627	80051	72521	35342	56119	97190
22145	85304	35348	82854	55846	18076	12415
27153	08662	61078	52433	22184	33998	87436
00301	49425	66682	25442	83668	66236	79655
43815	43272	73778	63469	50083	70696	13558
14689	86482	74157	46012	97765	27552	49617
16680	55936	82453	19532	49988	13176	94219
86938	60429	01137	86168	78257	86249	46134
33944	29219	73161	46061	30946	22210	79302
16045	67736	18608	18198	19468	76358	69203
37044	52523	25627	63107	30806	80857	84383
61471	45322	35340	35132	42163	69332	98851
47422	21296	16785	66393	39249	51463	95963
24133	39719	14484	58613	88717	29289	77360
67253	67064	10748	16006	16767	57345	42285
62382	76941	01635	35829	77516	98468	51686
98011	16503	09201	03523	87192	66483	55649
37366	24386	20654	85117	74078	64120	04643
73587	83993	54176	05221	94119	20108	78101
33583	68291	50547	96085	62180	27453	18567
02878	33223	39199	49536	56199	05993	71201
91498	41673	17195	33175	04994	09879	70337
91127	19815	30219	55591	21725	43827	78862
12997	55013	18662	81724	24305	37661	18956
96098	13651	15393	69995	14762	69734	89150
97627	17837	10472	18983	28387	99781	52977
40064	47981	31484	76603	54088	91095	00010
16239	68743	71374	55863	22672	91609	51514
58354	24913	20435	30965	17453	65623	93058
52567	65085	60220	84641	18273	49604	47418
06236	29052	91392	07551	83532	68130	56970

TABLE XII   RANDOM NUMBERS (continued)

94620	27963	96478	21559	19246	88097	44926
60947	60775	73181	43264	56895	04232	59604
27499	53523	63110	57106	20865	91683	80688
01603	23156	89223	43429	95353	44662	59433
00815	01552	06392	31437	70385	45863	75971
83844	90942	74857	52419	68723	47830	63010
06626	10042	93629	37609	57215	08409	81906
56760	63348	24949	11859	29793	37457	59377
64416	29934	00755	09418	14230	62887	92683
63569	17906	38076	32135	19096	96970	75917
22693	35089	72994	04252	23791	60249	83010
43413	59744	01275	71326	91382	45114	20245
09224	78530	50566	49965	04851	18280	14039
67625	34683	03142	74733	63558	09665	22610
86874	12549	98699	54952	91579	26023	81076
54548	49505	62515	63903	13193	33905	66936
73236	66167	49728	03581	40699	10396	81827
15220	66319	13543	14071	59148	95154	72852
16151	08029	36954	03891	38313	34016	18671
43635	84249	88984	80993	55431	90793	62603
30193	42776	85611	57635	51362	79907	77364
37430	45246	11400	20986	43996	73122	88474
88312	93047	12088	86937	70794	01041	74867
98995	58159	04700	90443	13168	31553	67891
51734	20849	70198	67906	00880	82899	66065
88698	41755	56216	66852	17748	04963	54859
51865	09836	73966	65711	41699	11732	17173
40300	08852	27528	84648	79589	95295	72895
02760	28625	70476	76410	32988	10194	94917
78450	26245	91763	73117	33047	03577	62599
50252	56911	62693	73817	98693	18728	94741
07929	66728	47761	81472	44806	15592	71357
09030	39605	87507	85446	51257	89555	75520
56670	88445	85799	76200	21795	38894	58070
48140	13583	94911	13318	64741	64336	95103
36764	86132	12463	28385	94242	32063	45233
14351	71381	28133	68269	65145	28152	39087
81276	00835	63835	87174	42446	08882	27067
55524	86088	00069	59254	24654	77371	26409
78852	65889	32719	13758	23937	90740	16866
11861	69032	51915	23510	32050	52052	24004
67699	01009	07050	73324	06732	27510	33761
50064	39500	17450	18030	63124	48061	59412
93126	17700	94400	76075	08317	27324	72723
01657	92602	41043	05686	15650	29970	95877
13800	76690	75133	60456	28491	03845	11507
98135	42870	48578	29036	69876	86563	61729
08313	99293	00990	13595	77457	79969	11339
90974	83965	62732	85161	54330	22406	86253
33273	61993	88407	69399	17301	70975	99129

Answers to Odd-Numbered Exercises

Note: Answers that are given in the Solutions to Practice Exercises at the end of each chapter of the text are not repeated here.

CHAPTER 1

1.1 Practice exercise.

1.3 Single, 9; married, 6; widow or widower, 4; divorced, 5.

1.5 (a) If we perform a study of all freshmen entering the university that year.

(b) If we use the data to apply to other university freshmen or to other years.

1.7 Practice exercise.

1.9 (a) There may be a systematic defect, say, if due to some quirk in the production procedure every tenth can is improperly sealed; then examining every fiftieth can will produce either all good cans or all bad cans.

(b) Since graduates with lower incomes are less likely to return the questionnaires, the alumni's office estimate is apt to be too high.

(c) There may well be some confusion, since the trade name XEROX is used as a generic term.

1.11 (a) Descriptive; (b) generalization; (c) descriptive; (d) generalization.

1.13 Practice exercise.

1.15 (a) Descriptive; (b) generalization; (c) descriptive; (d) descriptive; (e) generalization; (f) generalization.

CHAPTER 2

2.1 Practice exercise.

2.3

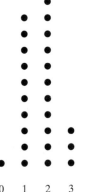

501

2.5   (a)   Paint defects   · · · · · · · · · · · · · · · · · · ·

Loose or missing parts   · · · ·

All other defects   · · ·

Electrical connections   · ·

Broken   ·

(b) There are more paint defects than all the other defects combined.

2.7

16	9
17	0 5
18	1 3 6 7 7
19	0 2 4 4 6 8 9
20	3 4 5 7
21	2 6 8
22	3 6

2.9   (a) 10, 11, 11, 12, 15, 17, 18.

(b) 120, 122, 123, 123, 125.

(c) 0.60, 0.62, 0.63, 0.66.

(d) 1.50, 1.51, 1.53, 1.54, 1.56.

2.11

5 ·	7 9
6 *	0 1 1 2 4
6 ·	5 5 5 6 6 7 7 8 8 9
7 *	0 1 2 2 3 3 4
7 ·	6 6 8
8 *	1 3
8 ·	5

2.13

1	6 6 6 7 8 9 9
2	1 2 2 3
3	1

2.15   Practice exercise.

2.17   One possibility is 0–249,999; 250,000–499,999; 500,000–749,999; 750,000–999,999; 1,000,000–1,249,999; 1,250,000–1,499,999; 1,500,000–1,749,999; 1,750,000–1,999,999.

2.19   (a) No; (b) yes; (c) no; (d) yes.

2.21   Practice exercise.

2.23   (a) 0, 20, 40, 60, 80, 100, 120, and 140.

(b) 19, 39, 59, 79, 99, 119, 139, and 159.

(c) 9.5, 29.5, 49.5, 69.5, 89.5, 109.5, 129.5, and 149.5.

(d) 20.

2.25   Practice exercise.

2.27   (a) $\dfrac{36 + 45}{2} = 40.5$, $\dfrac{45 + 54}{2} = 49.5$, 58.5, 67.5, 76.5, 85.5, and the lower boundary of the first class is $40.5 - 9 = 31.5$; similarly, the upper boundary of the last class is 94.5.

(b) 32–40, 41–49, 50–58, 59–67, 68–76, 77–85, and 86–94.

2.29 There is no provision for 11 and 31. Also, there is ambiguity because 23 can be put into the fourth or fifth class.

2.31 (a)

Grams	Percent
80–89	4
90–99	6
100–109	18
110–119	26
120–129	26
130–139	14
140–149	6
	100

(b)

Grams	Cumulative percentage
Less than or equal to 79	0
Less than or equal to 89	4
Less than or equal to 99	10
Less than or equal to 109	28
Less than or equal to 119	54
Less than or equal to 129	80
Less than or equal to 139	94
Less than or equal to 149	100

2.33 (a)

Age	Cumulative frequency
Less than or equal to 19	0
Less than or equal to 24	129
Less than or equal to 29	350
Less than or equal to 34	660
Less than or equal to 39	823
Less than or equal to 44	928
Less than or equal to 49	990
Less than or equal to 54	1,000

(b)

Age	Cumulative frequency
20 or more	1,000
25 or more	871
30 or more	650
35 or more	340
40 or more	177
45 or more	72
50 or more	10
55 or more	0

2.35 0, 1, 2, 3, 4, 5, or 6 students were absent from 6, 14, 9, 6, 3, 1, and 1 lectures, respectively.

2.37 The frequencies corresponding to excellent, very good, good, fair, poor, and very poor are 3, 9, 20, 6, 1, and 1, respectively.

2.39

Dollars	Cumulative frequency
Less than 1	0
Less than 21	18
Less than 41	80
Less than 61	143
Less than 81	186
Less than 101	200

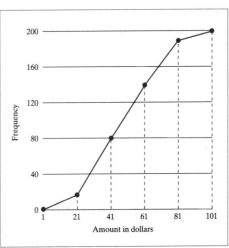

	Weight (lb)	**Cumulative frequency**
2.41	Less than 90	0
	Less than 100	4
	Less than 110	27
	Less than 120	76
	Less than 130	114
	Less than 140	131
	Less than 150	137
	Less than 160	140

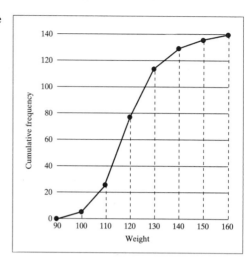

2.43 The central angles are 16.7°, 230.5°, 53.2°, 37.4°, 19.1°, 3.2°.

2.45 The central angles are 264°, 76°, 12°, 8°.

CHAPTER 3

3.1 Practice exercise.

3.3 (a) It would be a parameter if, say, the information were provided to the treasurer for the purpose of payment.

(b) It would be a statistic if the information were provided to the treasurer as a sample to assist in the preparation of budgets for the future.

3.5 74.0 years.

3.7 1.25%.

3.9 Two grandparents. As used here, the term "average criminal" is too vague.

3.11 10,752 pounds, which is less that the maximum load of 12,000 pounds.

3.13 $\bar{x} = 7.3$ hours. It is useful, but it conceals the variation of the numbers around their average.

3.15 $\bar{x} = 85°$. It is useful, but it conceals the variation of the numbers around their average.

3.17 Practice exercise.

3.19 15,700 pounds.

3.21 0.295.

3.23 $6,108.

3.25 Practice exercise.

3.27 (a) The median is the value of the 20th item.

(b) The median is the mean of the values of the 75th and 76th items.

3.29 The median is 5 meals.

3.31 The median is 59.5 power failures.

3.33 The median is 2.

3.35 The median is 68.5.

3.37 Practice exercise.

3.41   The midranges are 29.8, 30.0, and 30.3, so that the manufacturer of car C can use the midrange to substantiate the claim that its car performed best in the test.

3.43   The Q_1, median, and Q_3 positions are 5, 9.5, and 14 counting from left to right.

3.45   Practice exercise.

3.47   (a) $Q_1 = 129$ minutes, and $Q_3 = 150.5$ minutes; (b) $Q_1 = 129$ minutes and $Q_3 = 150.5$ minutes.

(c)

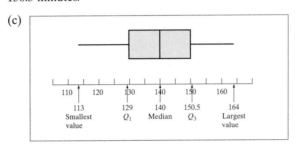

3.49   (a) $4\frac{1}{3}$ eggs; (b) 4 eggs; (c) 5 eggs; (d) 4 eggs; (e) 5 eggs.

3.51   The mode is 3 divorces per thousand of population.

3.53   The mode is 5 deliveries.

3.55   Practice exercise.

3.57   The modal choice is Independent.

3.59   Practice exercise.

3.61   For stock A, the range is $\frac{7}{8}$ and, for stock B, the range is $\frac{5}{8}$. Thus, stock B is less variable.

3.63   The range is 2.84 kilograms per cubic meter.

3.65   (a) 0.76%; (b) 0.76%. s is the same by both methods.

3.67 $s = 2.26$ seconds.

3.69 $s = 21.4$ vetos.

3.71 $\sigma = 10.91$ billions of kWh.

3.73 $s = 2.06$ and the range is 4.

3.75 $\bar{x} = 14.127$ and $s = 8.106$.

3.77   Practice exercise.

3.79   (a) At least 96%; (b) at least 98.44%; (c) at least 99%; (d) at least 99.75%.

3.81   (a) About 68%; (b) about 95%; (c) about 99.7%.

3.83   At least 88.9%. For a bell-shaped distribution, it is about 99.7%.

3.85   It would be wiser to sell stock D, which, at that time, is selling at a higher level within its range.

3.87   (a) Since $z = 1.48$ and $z = 1.39$ for the two universities, the student is in a relatively better position with respect to the first university.

(b) Since $z = 3.10$ and $z = 3.18$ for the two universities, the student is in a relatively better position with respect to the second university.

3.89   The first student is relatively more consistent.

3.91   Practice exercise.

3.93 $\bar{x} = \$7,500$; $S_{xx} = 522.0$, and $s = \$4,400$. Since the class marks were raised by 0.05, you might want to correct $\bar{x}$ to \$7,450. (The value of s should not be corrected.)

3.95   Practice exercise.

3.97   \$19,045.

3.99   Practice exercise.

3.101 $Q_1 = 16.05$ cents and $Q_3 = 22.42$ cents.

3.103   (a) Symmetrical; (b) positive skewness; (c) negative skewness.

3.105 $P_{20} = 64.12\%$; $P_{80} = 86.17\%$.

3.107   Practice exercise.

3.109 $s = 0.48$, and $SK = 1.81$.

3.111   The distribution is U shaped because the number of H's is apt to stay ahead of the number of T's once it gets ahead, and vice versa.

3.113   (a) $x_1 + x_2 + x_3 + x_4 + x_5 + x_6$; (b) $y_1 + y_2 + y_3 + y_4 + y_5$; (c) $x_1 y_1 + x_2 y_2 + x_3 y_3$; (d) $x_1 f_1 + x_2 f_2 + x_3 f_3 + x_4 f_4 + x_5 f_5 + x_6 f_6 + x_7 f_7 + x_8 f_8$; (e) $x_3^2 + x_4^2 + x_5^2 + x_6^2 + x_7^2$; (f) $(x_1 + y_1) + (x_2 + y_2) + (x_3 + y_3) + (x_4 + y_4)$.

3.115   Practice exercise.

3.117   (a) 27; (b) 35; (c) 137; (d) 587.

3.119   (a) 8, −1, and 0; (b) 4, 5, 2, and −4; (c) 7.

3.121   No.

REVIEW EXERCISES FOR CHAPTERS 1, 2, AND 3

R.1   (a) 5,767 air miles; (b) 5,564 air miles.

R.3   Statements (a) and (d) are descriptive, while statements (b) and (c) are generalizations.

R.5   (a) The data would constitute a population if the dean wants to determine the average number of failing grades by faculty members in the academic year 1999–2000.

(b) The data would constitute a sample if the dean wants to predict how many failing grades the faculty members will give in future years.

R.7   The mean is not a good average in this case since its value is greatly enlarged by one large value from California.

R.9 $\bar{x} = 23.82$ mistakes; (b) $s = 3.41$ mistakes.

R.11

12	4					
13	0	5				
14	1	2	6	9		
15	1	3	5	6	8	9
16	2	2	5			
17	3	7				
18	2					
19						
20	4					

R.13   (a) $s = 0.0044$ ppm; (b) the median is 0.0625 ppm and the range is 0.012 ppm.

R.15   (a) 25; (b) 151.

R.17 $SK = 1.98$.

R.19   (a) 18 and 17, so bus A averaged more passengers per run.

(b) 4.64 and 2.65; (c) 25.8% and 15.6%, so the number of passengers is relatively more variable for bus A.

R.21   The smallest value is 9, $Q_1 = 22$, $\tilde{x} = 33$, $Q_3 = 39$, and the largest value is 45.

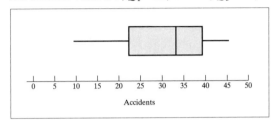

R.23   (a) 50, 100, 150, 200, 250, 300, and 350; (b) 99, 149, 199, 249, 299, 349, and 399; (c) 49.5, 99.5, 149.5, 199,5, 249.5, 299.5, 349.5, and 399.5; (d) 74.5, 124.5, 174.5, 224.5, 274.5, 324.5, and 374.5; (e) 50.

R.25 $s \approx 0.654$ second.

R.27   Since $z = \dfrac{40{,}000 - 30{,}000}{4{,}000} = 2.5$ for Mr. Ames and

$z = \dfrac{48{,}000 - 36{,}000}{6{,}000} = 2.0$ for Mr. Brown, the Ames's are relatively better off with respect to the families in their neighborhoods.

R.29   (a) This is the histogram:

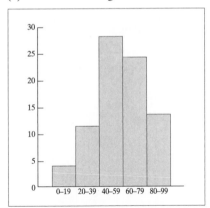

(b) This is the cumulative "less than" distribution:

NUMBER OF VEHICLES	CUMULATIVE FREQUENCY
Less than 20	4
Less than 40	15
Less than 60	43
Less than 80	67
Less than 100	80

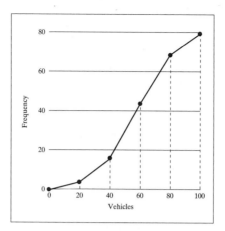

R.31 Here is a possible drawing for the box-and-whisker plot.

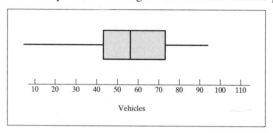

R.33 (a) $\bar{x} = 945.8$ and the median is 352.

(b) $\bar{x} = 345.8$ and the median is 352.

(c) The printing error strongly affected the mean, but not the median.

R.35 (a)

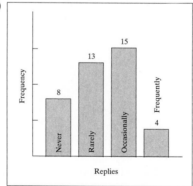

(b) The modal reply is "occasionally."

R.37

2	6	8							
3	1	4	6	8	9				
4	0	1	3	3	5	6	8	8	
5	1	2	2	3	3	4	5	7	7	7	8
6	0	1	1	2	3	4	4	7	9	9
7	0	1	2	5					

R.39   Median is 53.5°F, $Q_1 = 43$°F, and $Q_3 = 62.5$°F. The highest and lowest values, by inspection, are 75°F and 26°F.

CHAPTER 4

4.1   Practice exercise.

4.3   In two cases he will be exactly $1.00 ahead.

4.5   In five of the nine outcomes, the two union officials are not of the same sex.

4.7   Practice exercise.

4.9   In 32 ways.

4.11   In 8 different ways.

4.13   In 1,000 different ways.

4.15   In 1,024 ways.

4.17   In 180 different ways.

4.19   There are 6 different ways.

4.21   Practice exercise.

4.23   (a) False; (b) true; (c) false.

4.25   Practice exercise.

4.27   (a) 840; (b) 2,401.

4.29   In 9! = 362,880 different ways.

4.31   (a) To find the number of ways in which n objects can be arranged in a circle, we arbitrarily fix the position of one of the objects and calculate the number of ways in which the remaining positions can be filled. Thus, the number of circular permutations is $(n - 1)!$

   (b) 5! = 120.

   (c) 3! = 6.

4.33   (a) 840; (b) 3,360; (c) 90,720; (d) 34,650.

4.35   Practice exercise.

4.37   In 330 different ways.

4.39   (a) 210 ways; (b) 5,040 ways.

4.41   (a) 56; (b) 112; (c) 48; (d) 4.

4.43   In 2,450 different ways.

4.45   In 1,500,625 ways.

4.47   1, 6, 15, 20, 15, 6, and 1; 1, 7, 21, 35, 35, 21, 7, and 1; 1, 8, 28, 56, 70, 56, 28, 8, and 1.

4.49   Practice exercise.

4.51   (a) $\frac{1}{26}$; (b) $\frac{1}{2}$; (c) $\frac{4}{13}$; (d) $\frac{1}{4}$; (e) $\frac{3}{4}$; (f) $\frac{12}{13}$.

4.53 $\frac{1}{8}, \frac{3}{8}, \frac{3}{8}$, and $\frac{1}{8}$.

4.55   (a) $\frac{1}{3}$; (b) $\frac{1}{2}$.

4.57   (a) $\frac{44}{91}$; (b) $\frac{198}{455}$.

4.59   (a) $\frac{506}{1,785}$; (b) $\frac{276}{595}$.

4.61   Practice exercise.

4.63   0.421.

4.65   0.241.

4.69   Practice exercise.

4.71 $0.20.

4.73   (a) −$0.0526; (b) −$0.0526; (c) −$0.0526.

4.75 $1.00.

4.77   1.92 times.

4.79   Practice exercise.

4.81 $p = 0.16$.

4.83   The optimal strategy calls for the purchase of three boats.

4.85   Practice exercise.

4.87   Continuing the operation will maximize the company's expected profit.

4.89   Not building the factory will maximize the expected profit.

4.91   The profit table (in thousands) is the following:

		Number of Tenants				
		0	1	2	3	4
Number of stores fitted with indoor furnishings	0	0	0	0	0	0
	1	−12	18	18	18	18
	2	−24	6	36	36	36
	3	−36	−6	24	54	54
	4	−48	−18	12	42	72

The expected profits using the expert's probabilities for furnishing 0, 1, 2, 3, and 4 stores are 0, 15, 18, 12, and 3 thousand dollars, respectively. The recommendation is to furnish 2 stores, with expected profit $18,000.

CHAPTER 5

5.1   Practice exercise.

5.3 $D' = \{1, 2, 6\}$, which is the event that we roll a 1, 2, or 6. $D \cup E = \{2, 3, 4, 5, 6\}$, which is the event that we do not roll a 1. $D \cap E = \{4\}$, which is the event that we roll a 4.

5.5   (a) $M \cup N = \{Q\clubsuit, K\clubsuit, Q\diamondsuit, K\diamondsuit, Q\heartsuit, K\heartsuit, 10\spadesuit, J\spadesuit, Q\spadesuit, K\spadesuit\}$ is the event that we draw a queen, a king, or the 10 or jack of spades.

(b) $M \cap N = \{Q\spadesuit, K\spadesuit\}$ is the event that we draw the queen or king of spades.

(c) $M' = \{A\clubsuit, 2\clubsuit, \ldots, J\clubsuit, A\diamondsuit, 2\diamondsuit, \ldots, J\diamondsuit, A\heartsuit, 2\heartsuit, \ldots, J\heartsuit, A\spadesuit, 2\spadesuit, \ldots, J\spadesuit\}$ is the event that we do not draw a queen or a king.

(d) $N' = \{A\clubsuit, \ldots, K\clubsuit, A\diamondsuit, \ldots, K\diamondsuit, A\heartsuit, \ldots, K\heartsuit, A\spadesuit, \ldots, 9\spadesuit\}$ is the event that we do not draw the 10, jack, queen, or king of spades.

(e) $M' \cup N' = \{A\clubsuit, \ldots, K\clubsuit, A\diamondsuit, \ldots, K\diamondsuit, A\heartsuit, \ldots, K\heartsuit, A\spadesuit, \ldots, J\spadesuit\}$ is the event that we do not draw the queen or king of spades.

(f) $M' \cap N' = \{A\clubsuit, \ldots, J\clubsuit, A\diamondsuit, \ldots, J\diamondsuit, A\heartsuit, \ldots, J\heartsuit, A\spadesuit, \ldots, 9\spadesuit\}$ is the event that we do not draw a queen, a king, the 10 of spades, or the jack of spades.

5.7 A and B are not mutually exclusive; A and C are mutually exclusive; B and C are not mutually exclusive.

5.9 (a) Not mutually exclusive; (b) not mutually exclusive; (c) mutually exclusive; (d) mutually exclusive; (e) mutually exclusive; (f) not mutually exclusive.

5.11 (a) Mutually exclusive. A grade cannot be both passing and failing.

(b) Not mutually exclusive. Many medical doctors are also psychiatrists.

(c) Not mutually exclusive. In many commercial aircraft, telephones are available to passengers.

(d) Mutually exclusive. An elected president is a Democrat or Republican (or some other political party), but not both.

(e) Mutually exclusive. A person may be born in one city or the other, but not both.

5.13 (a) Regions 1 and 2 together represent the event that a person vacationing in Southern California visits Disneyland.

(b) Regions 2 and 3 together represent the event that a person vacationing in Southern California visits Disneyland or Universal Studios, but not both.

(c) Regions 2 and 4 together represent the event that a person vacationing in Southern California does not visit Universal Studios.

5.15 (a) Regions 1 and 3 together represent the event that the murder suspect is guilty.

(b) Regions 1 and 4 together represent the event that either the murder suspect is guilty and allowed out on bail or is not guilty and not allowed out on bail.

(c) Regions 3 and 4 together represent the event that the murder suspect is not allowed out on bail.

5.17 (a) The number of books is adequate, the number of periodicals is adequate, and the chairs are comfortable.

(b) The number of periodicals is adequate and the chairs are comfortable, but the number of books is not adequate.

(c) The chairs are comfortable, but the numbers of books and periodicals are not adequate.

(d) The number of books is not adequate, the number of periodicals is not adequate, and the chairs are not comfortable.

(e) The number of books is adequate and the chairs are comfortable.

(f) The number of periodicals is adequate, but the number of books is not adequate.

(g) The chairs are comfortable.

(h) The chairs are not comfortable.

5.19 Practice exercise.

5.21 (a) $P(J')$; (b) $P(J \cap M)$; (c) $P(J' \cap M')$; (d) $P(J \cup M)$.

5.23 (a) The probability cannot be greater than 1, and it is given as 1.09.

(b) The sum of the two probabilities cannot be greater than 1, and it is 1.10.

(c) The sum of the two probabilities is $0.30 + 0.40 = 0.70$, and not 0.90.

(d) The sum of the two probabilities should be 1, but it is given as $0.55 + 0.35 = 0.90$.

5.25 (a) $1 - 0.45 = 0.55$; (b) $1 - 0.30 = 0.70$; (c) $0.45 + 0.30 = 0.75$; (d) 0; (e) $P(Q' \cap R) = P(R) = 0.30$; (f) $P(Q \cap R') = P(Q) = 0.45$.

5.27 (a) 0.25; (b) 0.55; (c) 0.45.

5.29 1. Since $0 \leq s \leq n$, division by n yields $0 \leq \dfrac{s}{n} \leq 1$.

2. If an event is certain to occur, $s = n$, so its probability is 1; if an event is certain not to occur, then $s = 0$ and the probability is 0.

3. The sum of the first two is equal to the third.

4. $s + n - s = n$, so the probabilities sum to 1.

5.31 Practice exercise.

5.33 The odds cannot all be right, since they correspond to probabilities of $\frac{2}{3}, \frac{1}{5}$, and $\frac{1}{10}$, whose sum does not equal 1.

5.35 The probabilities are not consistent.

5.37 The demonstration proceeds as follows:

$$a(1 - p) = bp$$
$$a - ap = bp$$
$$a = ap + bp$$
$$a = p(a + b)$$
$$p = \frac{a}{a + b}$$

5.39 (a) The even money bet is attractive.

(b) Attractive.

(c) Unattractive.

5.41 0.76.

5.43 (a) 0.60; (b) 0.25; (c) 0.34.

5.45 Practice exercise.

5.47 (a) 0.34; (b) 0.44; (c) 0.22.

5.49 Practice exercise.

5.51 0.09.

5.53 $\frac{1}{16}$.

5.55 Practice exercise.

5.57 (a) $P(Q|W)$; (b) $P(W'|Q)$; (c) $P(Q'|W')$.

5.59 (a) $P(W|E)$; (b) $P(E|H')$; (c) $P(W|H \cap E)$.

5.61 Practice exercise.

5.63 (a) 0.706; (b) 0.338; (c) 0.494; (d) 0.168; (e) 0.747; (f) 0.570.

5.65 (a) $\frac{5}{12}$; (b) $\frac{3}{23}$; (c) $\frac{3}{20}$; (d) $\frac{4}{11}$; (e) $\frac{19}{42}$; (f) $\frac{8}{23}$.

5.67 (a) The probability that a young child who received a low grade in a manual dexterity test will not be able to dress without assistance.

(b) The probability that a young child who cannot dress without assistance will receive a low grade in the manual dexterity test.

(c) The probability that a young child who did not receive a low grade in the manual dexterity test will not be able to dress without assistance.

(d) The probability that a young child who can dress without assistance will receive a low grade in the manual dexterity test.

(e) The probability that a young child who can dress without assistance will not receive a low grade in the manual dexterity test.

5.69 If F and R denote the events that such a résumé has all relevant facts and that it will be reviewed by the entire placement staff, then $P(R|F) = \frac{0.58}{0.82} \approx 0.707$.

5.71 (a) Baker completed 0.60, and Abel completed 0.55.

(b) Baker completed 0.375, and Abel completed 0.20.

(c) Baker completed 0.43 (to two figures) overall, while Abel completed 0.48.

5.73 $P(L|A) = \frac{0.63}{0.75} = 0.84 = P(L)$.

5.75 (a) $\frac{1}{216}$; (b) $\frac{1}{216}$; (c) $\frac{125}{216}$.

5.77 (a) $\frac{64}{225}$; (b) $\frac{8}{29}$.

5.79 (a) $\frac{1}{64}$; (b) $\frac{125}{216}$.

5.81 $\frac{1}{22}$.

5.83 (a) 0.196; (b) 0.084.

5.85 Practice exercise.

5.87 0.590.

5.89 0.810.

5.91 0.136.

REVIEW EXERCISES FOR CHAPTERS 4 AND 5

R.41 400 choices.

R.43 The 15 points of the sample space are shown in the following diagram:

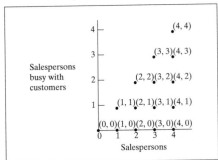

R.45 2.28 calls.

R.47 (a) $\{A, D\}$; (b) $\{C, E\}$; (c) $\{B\}$.

R.49 It is worthwhile to pay 50 cents.

R.51 3,600 ways.

R.53 (a) 24; (b) 9; (c) 8; (d) 6; (e) 0; (f) 1.

R.55 (a) 48; (b) 12; (c) 24.

R.57 (a) 0.57; (b) 0.64; (c) 0.36.

R.59   120 ways.

R.61   0.747.

R.63   Use $P(A|R') = \frac{160}{310}$, $P(A \cap R') = \frac{160}{380}$, and $P(R') = \frac{310}{380}$, and that $\frac{\frac{160}{380}}{\frac{310}{380}} = \frac{160}{310}$.

R.65   The odds are 3 to 2 against it.

R.67   Region 1 represents the event that the school's football team is rated among the top 20 by both AP and UPI. Region 2 represents the event that the school's football team is rated among the top 20 by AP, but not by UPI. Region 3 represents the event that the school's football team is rated among the top 20 by UPI, but not by AP. Region 4 represents the event that the school's football team is rated among the top 20 by neither AP nor UPI.

R.69 $\frac{7}{19} \approx 0.368$.

R.71   (a) 210; (b) 5; (c) 100.

R.73   (a) Region 5; (b) regions 1 and 2 together; (c) regions 3, 5, and 6 together; (d) regions 1, 3, 4, and 6 together.

R.75   (a) 0.80; (b) 0.60; (c) 0.50.

R.77   (a) 0.35; (b) 0.43; (c) 0.42; (d) 0.45.

R.79   0.49.

R.81   (a) $P(A|B) = \frac{0.13}{0.20} = 0.65 = P(A)$.

(b) $P(A|B') = \frac{0.65 - 0.13}{1 - 0.20} = \frac{0.52}{0.80} = 0.65 = P(A)$.

(c) $P(B|A) = \frac{0.13}{0.65} = 0.20 = P(B)$.

(d) $P(B|A') = \frac{0.20 - 0.13}{1 - 0.65} = \frac{0.07}{0.35} = 0.20 = P(B)$.

CHAPTER 6

6.1   Practice exercise.

6.3   (a) No; (b) no; (c) yes; (d) no.

6.5   Practice exercise.

6.7   Practice exercise.

6.9   0.096.

6.11   0.229.

6.13   (a) 0.302; (b) 0.302.

6.15   (a) 0.943 or 0.941; (b) 0.004; (c) 0.497.

6.17   0.196; (b) 0.304; (c) 0.153; (d) 0.151.

6.19   (a) 0.099; (b) 0.545; (c) 0.179.

6.21   (a) 0.9596; (b) 0.3414.

6.23   (a) 0.2968 (no); (b) 0.8866 (no); (c) 0.9940 (yes); (d) $n = 21$.

6.25   Practice exercise.

6.27   0.524.

6.29   0.484.

6.31   (a) 0.119; (b) 0.476; (c) 0.357; (d) 0.048.

6.33   Practice exercise.

6.35   0.005.

6.37   0.206.

6.39   (a) No; (b) no; (c) yes.

6.41   0.171.

6.43   0.043.

6.45   (a) $n \le 0.05(a + b)$, $n \ge 100$, and $n \frac{a}{a + b} < 10$.

(b) Use $n \frac{a}{a + b}$ for np in the Poisson approximation.

6.47   0.154.

6.49   (a) 0.449; (b) 0.359; (c) 0.144; (d) 0.048.

6.51   0.010.

6.53   (a) 0.433470; (b) 0.371163; (c) 0.797748.

6.55   Practice exercise.

6.57   0.117.

6.59   (a) 0.207; (b) 0.207; (c) the multinomial distribution with $k = 2$ is the binomial distribution.

6.61   1.78.

6.63 $\mu = 3.11$ and $\sigma = 1.406$.

6.65   (a) 1.7078; (b) 1.7078.

6.67   Practice exercise.

6.69   (a) 1.25; (b) 1.25.

6.71   (a) 0.9413; (b) 0.9487.

6.73   (a) $\mu = 450$, $\sigma = 15$; (b) $\mu = 67.5$, $\sigma = 7.5$; (c) $\mu = 226.8$, $\sigma = 12.6$.

6.75 $\mu = 0.9375$.

6.77   Practice exercise.

6.79   (a) At least 0.94; (b) at least 0.84; (c) at least 0.87.

6.81   (a) The probability is at least $\frac{15}{16}$ that there will be between 90 and 162 rainy days.

(b) Since $k = 3$, the probability is at least $\frac{8}{9} \approx 0.889$.

CHAPTER 7

7.1   Practice exercise.

7.3   Practice exercise.

7.5   Practice exercise.

7.7   (a) Process B is preferable; (b) no "right" answer.

7.9   (a) 0.3023; (b) 0.7422; (c) 0.0222; (d) 0.0230.

7.11   Practice exercise.

7.13 $z = 1.04$ or $z = -1.04$; (b) $z = -1.38$; (c) $z = 2.08$; (d) $z = 0.87$ or $z = -0.87$.

7.15   (a) 0.9633; (b) 0.0197; (c) 0.9704; (d) 0.7888.

7.17   (a) Since $0.5000 - 0.0250 = 0.4750$ corresponds to $z = 1.96$ in Table II, we get $z_{0.025} = 1.96$.

(b) Since $0.5000 - 0.0050 = 0.4950$, and 0.4949 and 0.4951 are the entries in Table II corresponding to $z = 2.57$ and $z = 2.58$, we get $z_{0.005} = \frac{2.57 + 2.58}{2} = 2.575$.

7.19 20.

7.21 $\mu = 258.47$; $\sigma = 30.61$.

7.23 (a) 0.1587; (b) 0.0668.

7.25 (a) 0.3707; 0.7314.

7.27 4.90 inches.

7.29 Practice exercise.

7.31 0.0445.

7.33 646 patients.

7.35 (a) Not satisfied; (b) satisfied; (c) not satisfied.

7.37 0.1662; the error of approximation is 0.0001.

7.39 0.0838.

7.41 0.9535.

7.43 (a) 0.2358; (b) 0.4908; (c) 0.9556.

7.45 (a) −0.0056; (b) −0.0034.

7.47 (a) 0.2514; (b) 0.4701.

CHAPTER 8

8.1 Practice exercise.

8.3 (a) 10; (b) 120; (c) 455.

8.5 (a) 0.1000; (b) 0.0083; (c) 0.0022.

8.7 The 15 possible samples are *ab, ac, ad, ae, af, bc, bd, be, bf, cd, ce, cf, de, df,* and *ef.*

8.9 Benzene, decane, hexane, octane, and toluene; decane, hexane, octane, toluene, and xylene; hexane, octane, toluene, xylene, and benzene; octane, toluene, xylene, benzene, and decane; toluene, xylene, benzene, decane, and hexane; xylene, benzene, decane, hexane, and octane.

(a) $\frac{1}{6}$; (b) $\frac{5}{6}$.

8.11 Car 041, 137, 074, 022, 149, and 143.

8.13 Student 2,997, 1,487, 3,852, 0372, 2,705, 1,005, 3,113, 0732, 3,780, 3,815, 2,742, 1,764, 3,247, 3,375, and 3,700.

8.15 (a) $\frac{1}{6}$; (b) $\frac{1}{6}$; (c) $\frac{3}{118}$; (d) less than $\frac{1}{36}$.

8.17 Practice exercise

8.19 (a) $\mu = 5$ and $\sigma = \sqrt{5} \approx 2.236$.

(b)

Sample		Mean	Sample		Mean
2	2	2	6	2	4
2	4	3	6	4	5
2	6	4	6	6	6
2	8	5	6	8	7
4	2	3	8	2	5
4	4	4	8	4	6
4	6	5	8	6	7
4	8	6	8	8	8

(c)

Sample mean	Probability
2	$\frac{1}{16}$
3	$\frac{2}{16}$
4	$\frac{3}{16}$
5	$\frac{4}{16}$
6	$\frac{3}{16}$
7	$\frac{2}{16}$
8	$\frac{1}{16}$

      (d) Both calculation give $\sqrt{2.5} \approx 1.581$.

8.21   (a) The standard error is divided by 10; (b) the standard error is multiplied by 10; (c) the standard error is divided by 3; and (d) the standard error is multiplied by 3.

8.23   The formulas are identical for $n = 1$.

8.25   (a) 0.889 or 0.866: (b) 0.871 or 0.866; (c) 0.975 or 0.975; (d) 0.912 or 0.910.

8.31   (a) At least $1 - \frac{1}{(1.75)^2} \approx 0.673$; (b) 0.9198.

8.33   (a) At least $1 - \frac{1}{(1.80)^2} \approx 0.691$; (b) 0.9282.

8.35   Chebyshev's theorem provides no useful information.

8.37   (a) 0.8384; (b) 0.5160.

8.39   (a) 0.9772; (b) 0.1587.

8.41   Practice exercise.

8.43 $n = 2,500$.

REVIEW EXERCISES FOR CHAPTERS 6, 7, AND 8

R.83   (a) 0.1176; (b) 0.2556; (c) 0.7442.

R.85   (a) 0.8621; (b) 0.1229.

R.87   (a) 0.899; (b) 0.990.

R.89   (a) 0.003; (b) 0.151; (c) 0.153; (d) 0.349.

R.91   (a) 0.125; (b) 0.24.

R.93   659, 411, 599, 619, 482, 332, 326, 516, 540, and 667.

R.95   (a) 0.030; (b) 0.184; (c) 0.188; (d) 0.077.

R.97   (a) $\mu = 0.8$, $\sigma^2 = 0.65$; (b) $\mu = 0.8$, $\sigma^2 = 0.64$.

R.99   0.8051.

R.101   (a) 0.420; (b) 0.160.

R.103   (a) No, since the sum of the probabilities is less than 1.

            (b) No, since $f(4)$ is negative.

            (c) Yes, since the values are all on the interval from 0 to 1 and their sum is 1.

R.105   0.0721.

R.107   (a) Yes; (b) no; (c) no; (d) yes.

R.109   (a) $z = 1.66$ or -1.66; (b) -0.83; (c) $z = 1.13$.

R.111   (a) No; (b) yes; (c) no.

R.113   (a) 0.4273; (b) 0.4265.

R.115   (a) Yes; (b) yes.

R.117   0.1707.

CHAPTER 9

9.1   Practice exercise.

9.3   0.008 inch.

9.5   (a) 2.829 milligrams; (b) 3.363 milligrams; (c) 3.717 milligrams.

9.7 $2.608.

9.9   (a) 0.44 ounce; (b) 0.58 ounce.

9.11   0.29 minute.

9.13   97 children.

9.15   38 cars.

9.17   We should use $n = 43$.

9.19   (a) 121.33 minutes $< \mu <$ 131.67 minutes.
(b) 119.70 minutes $< \mu <$ 133.30 minutes.

9.21   (a) 93.56 admissions $< \mu <$ 99.24 admissions.
(b) 88.56 admissions $< \mu <$ 94.24 admission.
(c) 98.56 admissions $< \mu <$ 104.24 admissions.

9.23   266.80 reams $< \mu <$ 290.20 reams.

9.25   (a) 1.796; (b) 2.179; (c) 2.650; (d) 2.977.

9.27   1,173.75 pounds.

9.29 $5.68 < \mu < 12.32$ seconds.

9.31 $18.1 < \mu < 21.9$ minutes.

9.33   (a) 1.44; (b) 1.37; (c) 1.35.

9.35   Computer or graphing calculator exercise.

9.37   Practice exercise.

9.39   1.11 minutes $< \sigma <$ 1.54 minutes.

9.41   1.58 minutes $< \sigma <$ 2.21 minutes.

9.43   Practice exercise.

9.45 $0.245 < p < 0.455$.

9.47   0.032.

9.49   1.93%.

9.51   0.03.

9.53   Practice exercise.

9.55   3,482.

9.57   246.

CHAPTER 10

10.1   Practice exercise.

10.3   (a) We would commit a Type I error if we erroneously reject the hypothesis that the average life of the bulb meets specifications.

   (b) We would commit a Type II error if we erroneously accept the null hypothesis that the light bulb meets specifications.

10.5   Practice exercise.

10.7   (a) 0.01; (b) 0.11.

10.9   Practice exercise.

10.11   Practice exercise.

10.13   (a) $z = 2.04$. Since $z = 2.04$ exceeds 1.96, the null hypothesis must be rejected.

   (b) Since $z = 2.04$ falls between -2.575 and 2.575, we cannot reject the null hypothesis.

10.15 $z = -2.76$. Since $z = -2.76$ is less than -1.645, the null hypothesis must be rejected.

10.17   (a) 1.69.

   (b) Since $z = 1.69$ falls between $z \leq -1.96$ or $z \geq 1.96$, we cannot reject the null hypothesis, concluding that the running time has not significantly changed.

   (c) Since $z = 1.69 > 1.645$, the engineer will reject the null hypothesis, concluding that the new process has significantly improved the running time.

   (d) Since $z = 1.69 > -1.645$, the second engineer cannot reject the null hypothesis and conclude that there is no evidence that the running time has worsened.

   (e) If you have no explanation about the purpose of the new process, you must use the two-sided alternative $H_A: \mu \neq 8.5$. Without a level of significance you can use the p value. Since the value in Table II corresponding to $z = 1.69$ is 0.4545, the p value is $2(0.5 - 0.4545) = 0.0910$.

10.19   (a) $z = 2.35$. Since $z = 2.35$ exceeds 1.96, the null hypothesis must be rejected; we conclude that the true average yield of FPC per pound is not 2.45 ounces.

   (b) Since $z = 2.35$ falls between -2.575 and 2.575, the null hypothesis cannot be rejected.

10.21   Practice exercise.

10.23 $t = -2.03$. Since -2.03 is less than -1.833, the null hypothesis must be rejected.

10.25 $t = 1.05$. since $t = 1.05$ falls between -3.250 and 3.250, the null hypothesis cannot be rejected.

10.27 $t = 3.09$. Since $t = 3.09$ is greater than 2.998, the null hypothesis must be rejected; we conclude that the average time to fill orders exceeds 9.5 days. Since $t_{0.01} = 2.998$ and $t_{0.005} = 3.449$, and $t = 3.09$ falls between these two values, we can write $0.005 < p < 0.01$.

10.29   (a) $t = -2.33$; (b) $0.01 < p < 0.025$.

10.31   (a) 0.8784; (b) 0.6082.

10.35   Practice exercise.

10.37 $z = 2.40$. Since $z = -2.40$ falls between -2.575 and 2.575, the null hypothesis cannot be rejected.

10.39 $z = 2.08$. Since 2.08 is less than 2.33, the null hypothesis cannot be rejected. We accept the null hypothesis or reserve judgment.

10.41   Practice exercise.

10.43 $t = -1.88$. Since $t = -1.88$ falls between -2.074 and 2.074, the null hypothesis cannot be rejected.

10.45 $t = -2.20$. since $t = -2.20$ falls between -3.205 and 3.205, the null hypothesis cannot be rejected.

10.47 $t = -1.99$. Since $t = -1.99$ falls between -2.878 and 2.878, the null hypothesis cannot be rejected.

10.49   Practice exercise.

10.51   (a) 34, 5.2, and $F = 6.54$.

   (b) The differences among the sample means are significant.

10.53

SOURCE OF VARIATION	DEGREES OF FREEDOM	SUM OF SQUARES	MEAN SQUARE	F
Treatments	2	31.5833	15.7916	0.08
Error	21	4,182.3750	199.1607	
Total	23	4,213.9583		

Since $F = 0.08$ does not exceed 5.78, the null hypothesis cannot be rejected; the differences among the average weekly earnings may be attributed to chance.

10.55

SOURCE OF VARIATION	DEGREES OF FREEDOM	SUM OF SQUARES	MEAN SQUARE	F
Treatments	3	412.6446	137.5482	6.84
Error	27	543.0973	20.1147	
Total	30	955.7419		

Since $F = 6.84$ exceeds 4.68, the null hypothesis must be rejected; we conclude that the secretary does not type with equal speed on the four typewriters.

CHAPTER 11

11.1   Practice exercise.

11.3   The probability of nine or fewer successes is 0.061. since 0.061 exceeds 0.05, the null hypothesis cannot be rejected. The data does not refute the claim.

11.5   The probability of six or fewer successes is 0.094, and the probability of six or more successes is 0.967. Since neither probability is less than or equal to 0.025, the null hypothesis cannot be rejected.

11.7   Practice exercise.

11.9 $z = -2.12$. Since -2.12 is greater than -2.33, the null hypothesis cannot be rejected. The data do not refute the claim.

11.11 $z = 1.85$. Since $z = 1.85$ does not exceed 2.33, the null hypothesis cannot be rejected.

11.13   (a) The probability of 11 or fewer successes is 0.055, and the probability of 11 or more successes is 0.988. Since neither probability is less than or equal to 0.025, the null hypothesis cannot be rejected.

   (b) This is similar to (a), but with $x = 15$, the probability of 15 or fewer successes is 1.000 and the probability of 15 or more successes is 0.206, and the null hypothesis cannot be rejected. Thus, even getting a 100% success rate in the experiment does not lead us to conclude that $p \neq 0.90$.

11.15 $z = 2.72$. Since $z = 2.72$ exceeds 2.33, the null hypothesis must be rejected.

11.17 $z = 7.118$. Since $z = 7.118$ exceeds 1.645, the null hypothesis must be rejected.

11.19 Practice exercise.

11.21 $\chi^2 = 3.75$. Since $\chi^2 = 3.75$ is less than 5.991, the null hypothesis cannot be rejected.

11.23 $\chi^2 = 16.55$. Since $\chi^2 = 16.55$ exceeds 13.277, the null hypothesis must be rejected.

11.25 $\chi^2 = 2.19$. Since $\chi^2 = 2.19$ does not exceed 3.841, the null hypothesis cannot be rejected.

11.27 Practice exercise.

11.29 $\chi^2 = 7.58$. Since $\chi^2 = 7.58$ does not exceed 13.277, the null hypothesis cannot be rejected.

11.31 $\chi^2 = 1.87$. Since $\chi^2 = 1.87$ does not exceed 9.488, the null hypothesis cannot be rejected.

11.39 Practice exercise.

11.41 $\chi^2 = 18.80$. Since $\chi^2 = 18.80$ exceeds 11.070, we reject the null hypothesis.

11.43 $\chi^2 = 0.875$. Since $\chi^2 = 0.875$ does not exceed 5.991, the null hypothesis cannot be rejected.

11.45 $\chi^2 = 1.77$. Since $\chi^2 = 1.77$ does not exceed 3.841, the null hypothesis cannot be rejected.

REVIEW EXERCISES FOR CHAPTERS 9, 10, AND 11

R.119 $t = 1.84$. Since $t = 1.84$ falls between -2.110 and 2.110, the null hypothesis cannot be rejected.

R.121 (0.763, 0.850).

R.123 43

R.125 $2.50 < \mu < 2.70$.

R.127 $0.34 < \sigma < 0.48$.

R.129 $t = 5.66$. Since $t = 5.66$ exceeds 2.776, the null hypothesis must be rejected.

R.131 18.5 square feet $< \sigma < 29.8$ square feet.

R.133 $\chi^2 = 4.32$. Since $\chi^2 = 4.32$ is less than 7.815, the null hypothesis cannot be rejected.

R.135 $\chi^2 = 9.23$. Since $\chi^2 = 9.23$ exceeds 5.991, the null hypothesis must be rejected.

R.137 $E = 14.6$.

R.139 $z = -2.00$. Since $z = -2.00$ is greater than $z = -2.33$, we cannot reject the null hypothesis.

R.141 $t = -2.08$. Since $t = -2.08$ falls between -2.819 and 2.819, we cannot reject the null hypothesis.

R.143 We would be committing a Type I error if we erroneously reject the null hypothesis that unit A is more efficient than unit B. We would be committing a Type II error if we erroneously accept the null hypothesis that unit A is more efficient than unit B.

R.145 $t = -2.20$. Since $t = -2.20$ is less than -1.734, the null hypothesis must be rejected.

R.147 $t = -1.99$. Since $t = -1.99$ is less than -1.734, the null hypothesis must be rejected.

R.149 $t = 3.00$. Since $t = 3.00$ exceeds 1.860, the null hypothesis must be rejected.

R.151   15.95 orders $< \mu <$ 24.39 orders.

R.153 $F = 7.41$. $F = 7.41$ exceeds 4.26, the null hypothesis must be rejected.

R.155   16.88 miles $< \mu <$ 19.12 miles.

CHAPTER 12

12.1   Practice exercise.

12.3   (a) $\hat{y} = 325.49 - 14.14x$; (b) $\hat{y} = 42.69$ or approximately 43,000 units.

12.5   (a) $\hat{y} = -877{,}177.277 + 1{,}338.781x$; (b) \$20,594.79, or about \$20,600.

12.7 $\hat{y} = 0.66 + 0.60x$, where the years are numbered $x = 1$, $x = 2$, $x = 3$, $x = 4$, and $x = 5$; $\hat{y} = 4.26$ or \$4,260,000.

12.9 $\hat{y} = 70.950 + 1.126x$.

12.11   Practice exercise.

12.13 $t = -2.14$. Since $t = -2.14$ is not less than -3.747, the null hypothesis cannot be rejected.

12.15 $-20.67 < \beta < -7.61$.

12.17   \$92,897 to \$98,283.

12.19 $t = 0.250$; the null hypothesis cannot be rejected.

12.21 $0.177 < \beta < 2.076$.

12.23   Practice exercise.

12.25 $r = 0.352$.

12.27 $r = 0.885$.

12.29 $r = 0.916$.

12.31   Practice exercise.

12.33   45%.

12.35   Neither answer should come as a surprise, since we can always draw a straight line passing through two given points and, hence, get a perfect fit. (a) $r = 1$ since the larger value of x corresponds to the larger value of y. (b) $r = -1$ since the larger value of x corresponds to the smaller value of y.

12.37   (a) $r = 0.62$, significant; (b) $r = -0.47$, significant; (c) $r = -0.58$, not significant; (d) $r = 0.63$, not significant.

12.39 $r = 0.514$.

12.41 $r = 0.916$, so $(0.916)^2(100\%) \approx 83.9\%$.

CHAPTER 13

13.1   Practice exercise.

13.3 $x = 5$; the null hypothesis cannot be rejected.

13.5 $x = 9$; the null hypothesis cannot be rejected.

13.7   Practice exercise.

13.9 $z = 1.73$. Since $z = 1.73$ falls between -1.96 and 1.96, the null hypothesis cannot be rejected.

13.11 $z = 2.75$. Since $z = 2.75$ exceeds 2.33, the null hypothesis must be rejected.

13.13 $z = -2.04$. Since $z = -2.04$ is less than -1.645, the null hypothesis must be rejected.

13.15 $U = 5.5$. Since $U = 5.5$ is less than 8, the null hypothesis must be rejected.

13.17 $U = 6$. Since $U = 6$ is greater than 5, the null hypothesis cannot be rejected.

13.19   Practice exercise.

13.21 $z = 3.20$. Since $z = 3.20$ exceeds 2.33, the null hypothesis must be rejected.

13.23 $z = 1.51$. Since $z = 1.51$ is less than 1.645, the null hypothesis cannot be rejected.

13.25 $H = 4.51$. Since $H = 4.51$ is less than 7.815, the null hypothesis cannot be rejected.

13.27 $H = 0.245$. Since $H = 0.245$ does not exceed 9.210, the null hypothesis cannot be rejected.

13.29   Since $\mu = 4$ falls between 3 and 11, the null hypothesis of randomness cannot be rejected.

13.31 $\mu = 17$. Since $\mu = 17$ equals 17, the null hypothesis of randomness must be rejected.

13.33 $z = -0.88$. Since $z = -0.88$ falls between -1.96 and 1.96, the null hypothesis of randomness cannot be rejected.

13.35 $z = 0.38$. Since $z = 0.38$ falls between -1.96 and 1.96, the null hypothesis of randomness cannot be rejected.

13.37   Practice exercise.

13.39 $z = -2.88$. Since $z = -2.88 < -1.96$ the null hypothesis of randomness must be rejected. There seems to be a trend, with grades decreasing as the students take longer to finish the examination.

13.41 $z = -0.57$. Since $z = -0.57$ falls between -1.96 and 1.96, the null hypothesis of randomness cannot be rejected.

13.43   Practice exercise.

13.45 $r_s = 0.65$.

13.47 $r_s = 0.99$.

13.49 $r_s = 0.893$ compared to $r = 0.885$. The two values are close.

13.51 $r_s = 0.75$.

REVIEW EXERCISES FOR CHAPTERS 12 AND 13

R.157 $\hat{y} = -1.07 + 1.22x$; 3.2.

R.159   (a) $r = 0.33$ is significant; (b) $r = 0.33$ is not significant.

R.161 $U = 9$. Since $U = 9$ is greater than 8, the null hypothesis cannot be rejected.

R.163 $x = 2$; since 0.019 is less than 0.05, the null hypothesis must be rejected.

R.165 $r = -0.65$. Since $r = -0.65$ is less than -0.632, the null hypothesis must be rejected.

R.167   (a) Negative correlation; (b) positive correlation; (c) zero correlation; (d) negative correlation; (e) zero correlation; (f) positive correlation.

R.169   From Table I, the probability of 9 or more successes is 0.032. Since 0.032 is less than 0.05, the null hypothesis must be rejected.

R.171 $\hat{y} = 14.247 + 0.637(40) = 39.727$ minutes.

R.173 $H = 5.03$. Since $H = 5.03$ does not exceed 7.815, the null hypothesis cannot be rejected.

R.175 $z = 2.29$. Since $z = 2.29$ exceeds 1.96, the null hypothesis must be rejected.

R.177 $u = 11$. Since $u = 11$ falls between 6 and 16, the null hypothesis cannot be rejected.

R.179 $U = 35$. Since $U = 35$ is less than 37, the null hypothesis must be rejected.

R.181 $t = -0.805$. Since $t = 0.805$ is greater than -2.998, the null hypothesis cannot be rejected.

R.183 $z = 2.38$. Since $z = 2.38$ falls between -2.575 and 2.575, the null hypothesis cannot be rejected.

Index

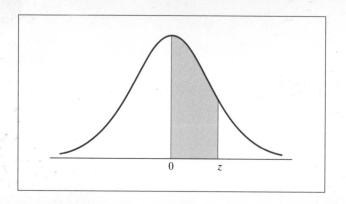

TABLE II NORMAL-CURVE AREAS

z	.00	.01	.02	.03	.04	.05	.06	.07	.08	.09
0.0	.0000	.0040	.0080	.0120	.0160	.0199	.0239	.0279	.0319	.0359
0.1	.0398	.0438	.0478	.0517	.0557	.0596	.0636	.0675	.0714	.0753
0.2	.0793	.0832	.0871	.0910	.0948	.0987	.1026	.1064	.1103	.1141
0.3	.1179	.1217	.1255	.1293	.1331	.1368	.1406	.1443	.1480	.1517
0.4	.1554	.1591	.1628	.1664	.1700	.1736	.1772	.1808	.1844	.1879
0.5	.1915	.1950	.1985	.2019	.2054	.2088	.2123	.2157	.2190	.2224
0.6	.2257	.2291	.2324	.2357	.2389	.2422	.2454	.2486	.2517	.2549
0.7	.2580	.2611	.2642	.2673	.2704	2734	.2764	.2794	.2823	.2852
0.8	.2881	.2910	.2939	.2967	.2995	.3023	.3051	.3078	.3106	.3133
0.9	.3159	.3186	.3212	.3238	.3264	.3289	.3315	.3340	.3365	.3389
1.0	.3413	.3438	.3461	.3485	.3508	.3531	.3554	.3577	.3599	.3621
1.1	.3643	.3665	.3686	.3708	.3729	.3749	.3770	.3790	.3810	.3830
1.2	.3849	.3869	.3888	.3907	.3925	.3944	.3962	.3980	.3997	.4015
1.3	.4032	.4049	.4066	.4082	.4099	.4115	.4131	.4147	.4162	.4177
1.4	.4192	.4207	.4222	.4236	.4251	.4265	.4279	.4292	.4306	.4319
1.5	.4332	.4345	.4357	.4370	.4382	.4394	.4406	.4418	.4429	.4441
1.6	.4452	.4463	.4474	.4484	.4495	.4505	.4515	.4525	.4535	.4545
1.7	.4554	.4564	.4573	.4582	.4591	.4599	.4608	.4616	.4625	.4633
1.8	.4641	.4649	.4656	.4664	.4671	.4678	.4686	.4693	.4699	.4706
1.9	.4713	.4719	.4726	.4732	.4738	.4744	.4750	.4756	.4761	.4767
2.0	.4772	.4778	.4783	.4788	.4793	.4798	.4803	.4808	.4812	.4817
2.1	.4821	.4826	.4830	.4834	.4838	.4842	.4846	.4850	.4854	.4857
2.2	.4861	.4864	.4868	.4871	.4875	.4878	.4881	.4884	.4887	.4890
2.3	.4893	.4896	.4898	.4901	.4904	.4906	.4909	.4911	.4913	.4916
2.4	.4918	.4920	.4922	.4925	.4927	.4929	.4931	.4932	.4934	.4936
2.5	.4938	.4940	.4941	.4943	.4945	.4946	.4948	.4949	.4951	.4952
2.6	.4953	.4955	.4956	.4957	.4959	.4960	.4961	.4962	.4963	.4964
2.7	.4965	.4966	.4967	.4968	.4969	.4970	.4971	.4972	.4973	.4974
2.8	.4974	.4975	.4976	.4977	.4977	.4978	.4979	.4979	.4980	.4981
2.9	.4981	.4982	.4982	.4983	.4984	.4984	.4985	.4985	.4986	.4986
3.0	.4987	.4987	.4987	.4988	.4988	.4989	.4989	.4989	.4990	.4990

Also, for z = 4.0, 5.0, and 6.0, the areas are 0.49997, 0.4999997, and 0.499999999.